Geological History of Greenland

GEUS

Geological History of Greenland

Four billion years of Earth evolution

Niels Henriksen

GEOLOGICAL HISTORY OF GREENLAND
Four billion years of Earth evolution

The present book is a translation of the Danish original 'Grønlands geologiske udvikling fra urtid til nutid' published in 2005 and reprinted in 2006 (ISBN 87-7871-163-0).

Text and selection of illustrations: Niels Henriksen
Layout and graphics: Carsten E. Thuesen
Drawings and 3-D illustrations: Carsten E. Thuesen, Eva Melskens, Annabeth Andersen, Margareta Christoffersen, Lis Duegaard, Jette Halskov, Henrik Klinge Pedersen, Stefan Sølberg and Helle Zetterwall
Digital and photographic work: Jakob Lautrup, Benny M. Schark and Peter K. Warna-Moors
Production and distribution advice: Henrik Højmark Thomsen and Bent B. Katz
Cover design: Carsten E. Thuesen

Translation from Danish into English: James A. Chalmers and Clark R.L. Friend
Geological quality control: A.K. Higgins, Jon R. Ineson and W. Stuart Watt
Editorial secretaries: Jane Holst and Esben W. Glendal

Publisher: Geological Survey of Denmark and Greenland (GEUS)
Printers: Schultz Grafisk, Albertslund, Denmark
Printed: June 2008, 1st English edition, number printed 4000

Cover photograph: Mountain wall on the south side of Ymer Ø, above the fjord Antarctic Sund in North-East Greenland. The photograph was taken from a position close to that illustrated on pages 66–67, and shows a succession of 600–700 million years old sediments comprising brightly coloured sandstones and mudstones. The rocks were folded during the Caledonian mountain building episode about 420 million years ago. Photo: Jakob Lautrup, GEUS.

Frontispiece: Fjord wall about 500 m high in South Greenland, showing strongly folded and metamorphosed dark schists cut by a network of light-coloured, flat-lying granitic veins and sheets. These basement rock complexes were formed during the Ketilidian mountain building episode about 1800 million years ago. Part of this photograph is shown on page 51 where the Ketilidian orogenic deformation is described in more detail. Photo: Adam A. Garde, GEUS.

Keywords: Greenland geology, Earth evolution, geological history, mineral potential, hydrocarbon potential, Archaean crust, Proterozoic crust, Palaeozoic fold belts, Palaeogene volcanism, sedimentary basins, offshore geology, Ice Age.

ISBN: 978-87-7871-211-0

© De Nationale Geologiske Undersøgelser for Danmark og Grønland (GEUS), 2008
Geological Survey of Denmark and Greenland (GEUS), 2008
Ministry of Climate and Energy
Øster Voldgade 10, DK-1350 Copenhagen K, Denmark

Copyright: No part of this publication may be reproduced or utilised in any form or by any means, electronic or mechanical, including photocopying, recording or by any information storage and retrieval system, without prior written permission from the publisher, GEUS. Disputes on copyright and other intellectual property rights shall be governed by Danish law and be subject to Danish jurisdiction.

This book can be ordered through any **bookshop**. Further information can be obtained from **GEUS** at the address above, Tel: +45 38142000, Web: www.geus.dk, Email: bogsalg@geus.dk

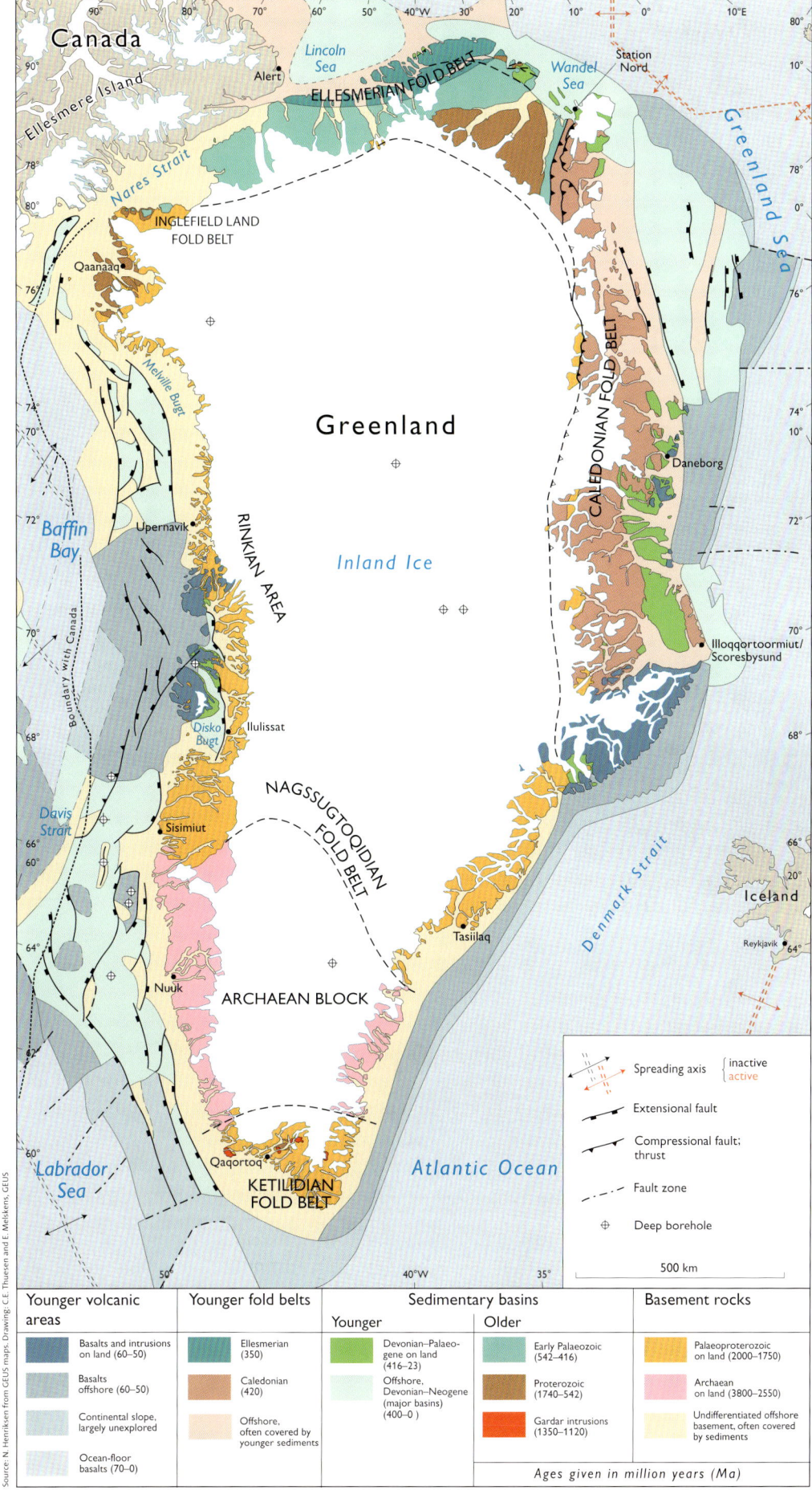

Summary of the geology of Greenland

The island of Greenland, together with its offshore continental shelf, records a history of geological development that spans more than 3800 million years (Ma). The oldest areas constitute a basement shield composed of strongly folded gneissic rocks representing the root zones of Archaean (3800–3600 and 3000–2550 Ma) and Proterozoic (2000–1750 Ma) fold belts (orogenic belts). These belts are now welded together to form a stable coherent block.

The extensive basement shield is surrounded by sedimentary basins that developed in three discrete periods – the Proterozoic (1740–542 Ma), the Cambrian–Silurian (542–416 Ma) and the Devonian–Neogene (416–0 Ma). Two coast-parallel mountain chains formed in the Early Palaeozoic, one in North-East Greenland (the Caledonian fold belt about 420 Ma) and the other in North Greenland (the Ellesmerian fold belt about 350 Ma). Volcanic units occur locally in the Proterozoic sedimentary formations, but most of the exposed volcanic rocks were erupted much later as lavas formed in connection with continental break-up when the North Atlantic began to open about 60–55 Ma ago.

The offshore continental shelves can be perceived as a continuation of the land area. They comprise crystalline basement covered by younger sediments (400–0 Ma) and Palaeogene basalts. Farther away from the coast, continental rocks pass into ocean-floor rocks, comprising volcanic and associated sedimentary material formed in connection with sea-floor spreading.

Mineral resources within the land areas include gold, lead-zinc, diamonds and rare earth elements, as well as building and construction materials. Possibilities for oil and gas reserves are found in the younger sedimentary basins offshore North-East and West Greenland.

The ice-free land area comprises approximately 410 000 km², while the economic zone of the continental shelf is about 825 000 km². About 81% of Greenland is covered by ice (about 1 755 000 km²).

A simplified geological map of Greenland with a classification of rock types in the ice-free land area and the offshore areas. The Inland Ice covers the whole of the central part of Greenland and there is very limited information on the geology beneath it. However, it is probable that most of the area comprises Archaean and Palaeoproterozoic basement rocks, similar to those in the ice-free area.

CONTENTS

Place name map and Preface	8
1. Greenland's geological evolution	11
2. Geology in Greenland	17
3. Landscapes	23
4. The crystalline basement	29
The Archaean basement	36
The younger part of the basement	40
The Nagssugtoqidian in southern West Greenland	42
Fold belts in central and northern West Greenland	46
The Ketilidian fold belt in South Greenland	50
5. The Gardar Province	55
6. Basin deposition	61
Older continental basins in North and North-West Greenland	70
Older marine basins in North-East Greenland	74
Older marine basins in North Greenland	82
7. Fold belts in North and North-East Greenland	91
The Caledonian fold belt in North-East Greenland	96
The Ellesmerian fold belt in North Greenland	104

Younger sedimentary basins	109
The Devonian basin in North-East Greenland	114
The Wandel Sea Basin in North Greenland	118
Rift basins in East Greenland	122
The Nuussuaq Basin in West Greenland	128
The Kangerlussuaq Basin in South-East Greenland	132
Palaeogene volcanism	137
Plateau basalts in the Nuussuaq Basin	142
The East Greenland volcanic province	146
Geology offshore	155
Geology offshore North and East Greenland	166
Geology offshore West Greenland	174
The Ice Age	181
Mineral resources	193
Early mining in Greenland	204
Mineralisation in Greenland	210
Oil and gas	219
Hydrocarbon potential of North and East Greenland	226
Hydrocarbon potential of West Greenland	232
Background and acknowledgements	242
Further reading	244
Glossary	246
Acronyms – for organisations and projects	265
Subject index	266

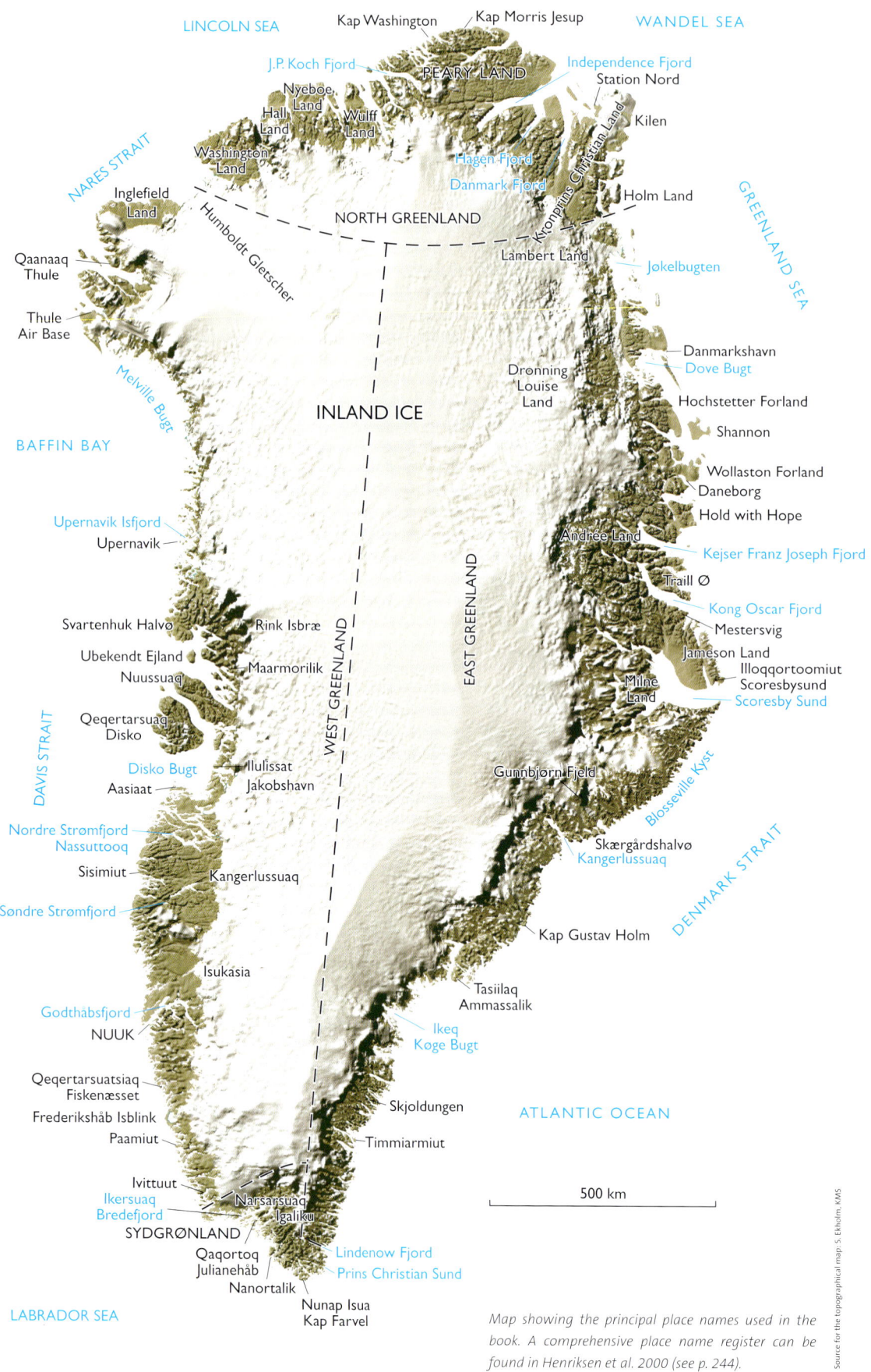

Map showing the principal place names used in the book. A comprehensive place name register can be found in Henriksen et al. 2000 (see p. 244).

PREFACE

The Earth Sciences study the structure, composition and development of the Earth, together with studies of related processes and materials. Early studies were confined to the surface of the Earth, but technological advances in geophysics have meant that today much of the Earth's crust and upper mantle are accessible for study.

Investigation and exploration of Greenland's geology started during the early discovery voyages in the first half of the nineteenth century, but systematic studies first began about a hundred years later and more focussed efforts did not start until after the Second World War. The many years of work mean that it is now possible to present a fairly detailed picture of the geology of the whole of Greenland, both on land and offshore, which gives a solid foundation for exploration for natural resources.

This book is based on some 60 years of surveying work by the Geological Survey of Denmark and Greenland (GEUS*), supplemented by results from many other research institutions and commercial companies. Since the Second World War, the Survey has carried out field work in Greenland with the participation of on average 100 research and support personnel each season. The results of these efforts include a set of geological overview maps of the entire country at a scale of 1:500 000 together with 60 more detailed maps at 1:100 000 and a wide range of special geological and geophysical maps. Around 400 000 samples of rocks and fossils have been collected through the years and much of this material has been studied and analysed, often in a collaboration between GEUS and a large number of Earth Science research institutions in both Denmark and abroad. The results have been published in more than 2300 scientific papers and documented in many thousand reports and maps. The recognition of the many geological units in Greenland is based on the definition and description of around 1000 stratigraphic groups, formations and lower rank units, all of which consist of a rock unit that can be mapped regionally. Additionally, many thousand species of fossil plants and animals have been collected and described.

The book has been written for readers with an interest in the Earth Sciences, but special knowledge of geology is not a requirement. Importance has been attached to explaining geological processes and terms that are necessary for general understanding. The book is not a scientific treatise and consequently the detailed references normally used in scientific literature have not been included. To help readers who want a deeper and more specific knowledge of Greenland's geology, some of the most relevant background articles and suggestions for further reading are listed at the end of the book. Readers with a professional background can find a more detailed list of references on http://www.geus.dk/isbn211.html

The volume is structured around text, illustrations and 'boxes' containing explanations of geoscientific methods, technical terms and geological processes. It need not be read from cover to cover – it can readily be used selectively as a source of background information on specific topics. A list of many geological terms as well as an index at the end of the book will aid readers to find their way around geology.

The geological development of the Earth took place in three dimensions – length, breadth and depth – so geological structures must also be understood in three dimensions. Often only rocks at the surface can be seen and there is thus a tendency to view geological phenomena in two dimensions. Greenland's landscapes, of high mountains with steep, vegetation-free slopes down to valleys and fjords, give outstanding possibilities of visualising the three-dimensional reality; this is one of the country's greatest benefits for geological research. The fourth dimension – time – is of utmost importance in geological studies. This factor can be difficult to grasp in the perspective of our day-to-day existence, since most geological processes happen extremely slowly, over very long periods on the human scale. Nonetheless, small displacements over very long periods can add up to kilometres or even thousands of kilometres of movement when they take place over millions of years. Greenland contains traces of geological developments back to about 3800 million years ago. To relate to the ages of the various geological units discussed in this book, the reader may find that it helps to think in units of a million years (written as 'Ma' throughout the book).

The science of geology is important not just as a part of our culture, but because most of our raw materials and energy resources come from the Earth. A vital prerequisite to discovering these materials is knowing how and where they were formed. This is an important reason why there has been so much effort devoted to describing and understanding the geological evolution of Greenland and this endeavour continues to preoccupy both the public and the private sectors.

*The Geological Survey of Greenland (GGU) was amalgamated with the Danish Geological Survey (DGU) in 1995 to form the present Geological Survey of Denmark and Greenland (GEUS). Throughout the book the term GEUS has been used to include the activities of both GGU and GEUS.

Three main stages in the evolution of the Earth: The first stage shows a red-hot (about 1400°C), glowing Earth early in the Hadean. The second stage shows the Archean when the Earth's crust had differentiated into oceanic and various continental regions. The third stage depicts the present-day Earth, with the Moon in the background.

Satellite photo: NASA. Drawing: C.E. Thuesen, GEUS

GREENLAND'S GEOLOGICAL EVOLUTION

3800 million years of geological history

1

FROM THE DAWN OF TIME TO THE PRESENT

GREENLAND'S GEOLOGICAL EVOLUTION
Precambrian

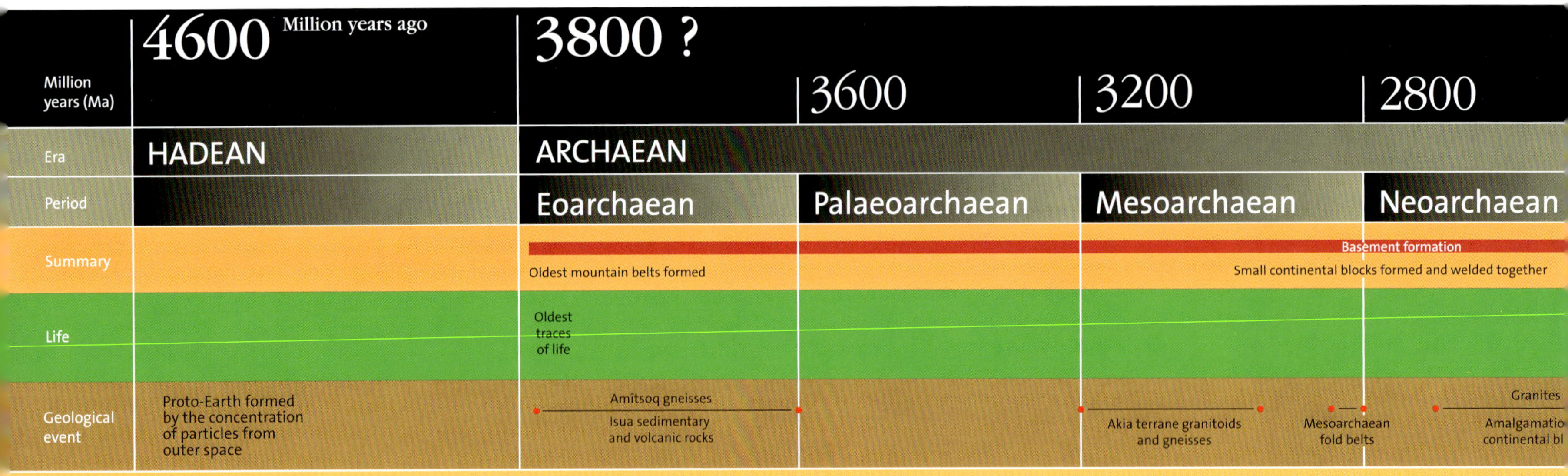

Million years (Ma)	4600 Million years ago	3800 ?	3600	3200	2800
Era	HADEAN	ARCHAEAN			
Period		Eoarchaean	Palaeoarchaean	Mesoarchaean	Neoarchaean
Summary		Oldest mountain belts formed		Small continental blocks formed and welded together	Basement formation
Life		Oldest traces of life			
Geological event	Proto-Earth formed by the concentration of particles from outer space	Amîtsoq gneisses / Isua sedimentary and volcanic rocks		Akia terrane granitoids and gneisses	Mesoarchaean fold belts / Granites / Amalgamation continental bl

Core, mantle and crust

The Sun and the planets began to form about 4560 Ma ago from concentrations of gases and small particles from space. Once cooling of the gas cloud around the Sun had begun, iron, nickel and silicate-bearing particles gathered into belts, and then particle accumulation led to the formation of the inner planets, including the Earth. At an early stage in the Earth's development, the densest part formed a nickel-iron core with less dense magnesium-rich silicate minerals forming a surrounding mantle. Subsequently, the Earth was characterised by violent volcanic activity with local melting and the differentiation of mantle material into discrete basaltic and granitic magmas that were forced upwards and emplaced near the surface to form parts of the Earth's crust.

No direct evidence of this earliest phase of Earth development has been found in Greenland.

The proto-Earth formed by the concentration of particles in space.

Formation of the earliest continents

Some of the Earth's oldest rocks are preserved in Greenland in the form of the about 3800 Ma old portion of the metamorphosed sediments and lavas of the Isua area. These sediments show that at this early point in the Earth's development there were already continents and oceans. The rocks formed in mountain belts about 3800–3600 Ma ago and their remains now occur in a narrow belt through Godthåbsfjord near Nuuk. The sediments were originally laid down at the Earth's surface and were later intruded by granite-like rocks, now gneisses. These rocks represent an early continent, with an accumulation of sediments that were derived by erosion of a still older continent and into which newly formed granite-like rocks from the Earth's mantle were emplaced.

Oldest basement

Banded iron-formation from Isua, a 3700 Ma old chemical sediment.

Older basement

The Nuuk region in West Greenland preserves three areas with rock formations that date from the Meso- and Neoarchaean. These areas form so-called terranes, which are disrupted parts of continents that originally were widely scattered, but that were brought together by tectonic processes (plate-tectonic movements). One of these terranes preserves the remains of a large feldspar-rich, layered intrusion associated with plutonic granitic rocks that is known as the Fiskenæsset anorthosite complex. Other parts of the Greenland basement also include Archaean gneisses, granites and metamorphosed sedimentary rocks that were subsequently caught up in Proterozoic folding events so that most of the original Archaean structures were obliterated and Proterozoic structures now dominate. Extensive unmodified Archaean basement is now only found in the southern part of Greenland.

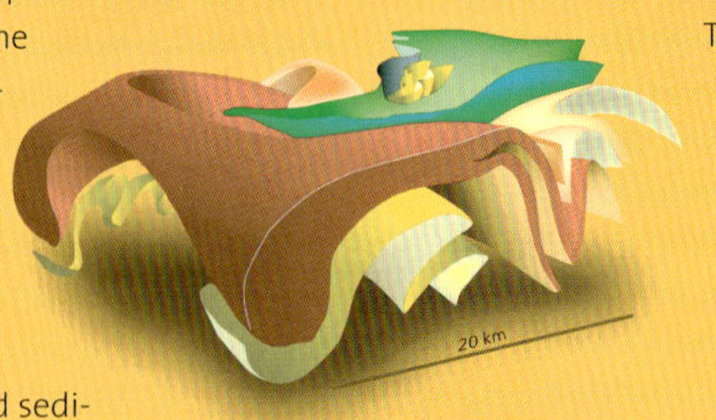

Continental block assembly

About 2720 Ma ago, the different small Archaean terranes of the Nuuk region were amalgamated into a larger continental mass. At the same time, other small Archaean continental blocks were welded onto its margins resulting in the formation of one large coherent Archaean block. The remnants of this continental mass are now found over much of West Greenland and are also present in South-East and East Greenland.

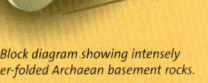
Archaean mountain belts

Block diagram showing intensely inter-folded Archaean basement rocks.

The Nuuk region showing amalgamated Archaean terrane blocks.

The red-hot proto-Earth.

Continents in the Archaean. Land areas surrounded by oceans.

GREENLAND'S GEOLOGICAL EVOLUTION

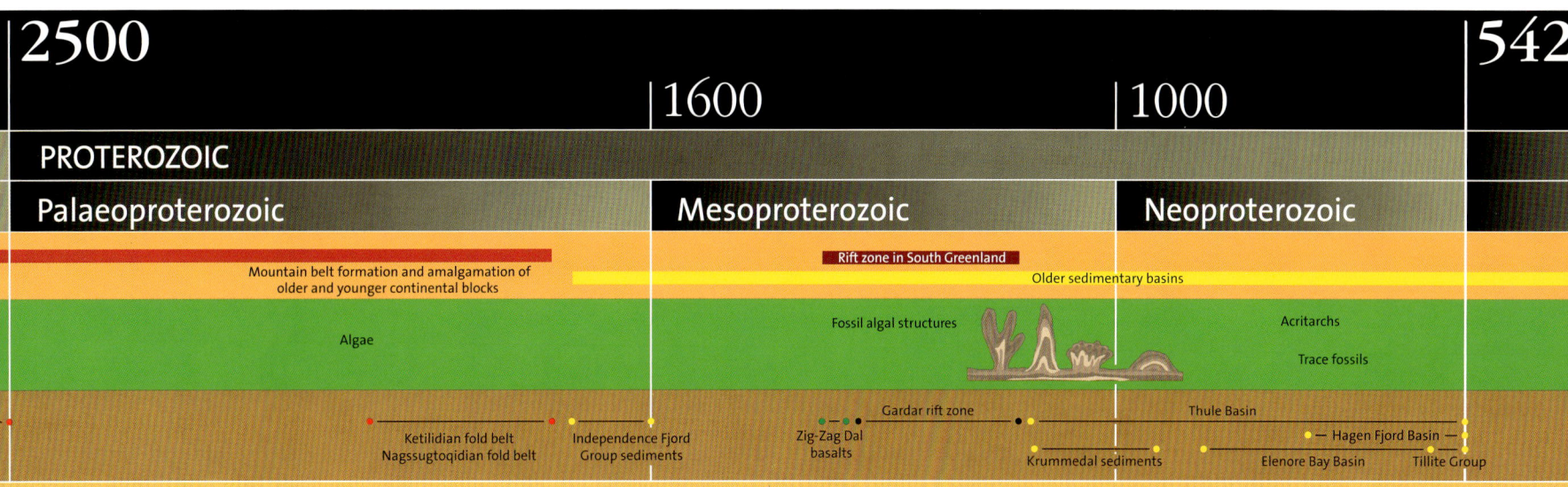

| 2500 | | 1600 | | 1000 | | 542 |

PROTEROZOIC

| Palaeoproterozoic | Mesoproterozoic | Neoproterozoic |

Mountain belt formation and amalgamation of older and younger continental blocks
Rift zone in South Greenland
Older sedimentary basins

Algae
Fossil algal structures
Acritarchs
Trace fossils

Ketilidian fold belt
Nagssugtoqidian fold belt
Independence Fjord Group sediments
Zig-Zag Dal basalts
Gardar rift zone
Krummedal sediments
Thule Basin
Hagen Fjord Basin
Elenore Bay Basin
Tillite Group

Younger basement

Between 2050 and 1750 Ma ago, a large part of the Archaean basement was involved in a series of new mountain building processes with deposition of younger supracrustal sequences and generation of granite-like plutons. The old Archaean continent was broken up by extension and ocean crust formation and then once again amalgamated into a new configuration including the new Palaeoproterozoic fold belts. These new mountain belts thus incorporate both reworked older material and younger, newly-formed rocks in the Earth crust.

The most widespread Palaeoproterozoic belt is the Nagssugtoqidian fold belt of West Greenland (about 1900–1800 Ma) and its continuation in the Rinkian area north of Disko Bugt. In South Greenland, the Ketilidian fold belt formed about 1850–1725 Ma ago as an independent mountain belt.

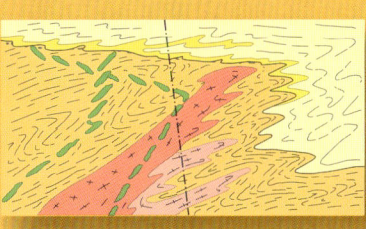
Cross-section through two inter-folded basement complexes.

Final development of the basement

Between 1800 and 1725 Ma ago, different parts of the Greenland crystalline basement were assembled into progressively larger, coherent continental blocks that eventually marked the establishment of the stable Greenland basement block (a craton). The later geological development of Greenland took place principally along

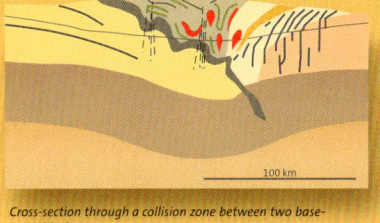
Cross-section through a collision zone between two basement blocks.

the margins of this stable craton. The stable block established at about 1750 Ma has escaped subsequent tectonic processes apart from local rift formation and limited block uplift and subsidence.

Early sedimentary basins and rift formation

Shortly after the stabilisation of the basement shield, the first major sedimentary basin (the Independence Fjord Basin) formed in eastern North Greenland and accumulated continental sand-rich sediments. In the same region, about 350 Ma later, a large volcanic province with plateau basalts (the Zig-Zag Dal basalts) developed. In North-West Greenland, another continental basin dominated by sandstones was formed (the Thule Basin).

In South Greenland, the Gardar period (1350–1125 Ma) was marked by the development of a complex NE–SW-trending rift zone within the basement. A continental basin formed within this rift zone in which sedimentary and volcanic basaltic sequences were deposited. After this came a series of plutonic intrusions that solidified as dykes or large coherent bodies within the crust.

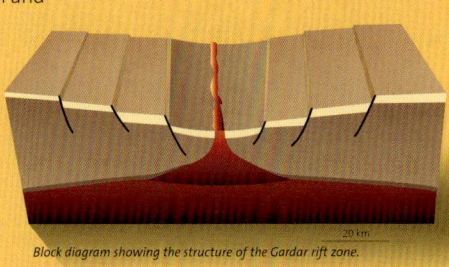
Block diagram showing the structure of the Gardar rift zone.

Late Proterozoic sedimentary basins in North Greenland

Between 1100 and 1000 Ma, a thick succession of sandy sediments (the Krummedal sediments) were deposited in a great basin that formed south-east of Greenland.

These sediments were reworked in the Grenvillian orogenic event at about 930 Ma.

Throughout the Proterozoic, several other large basins developed along the eastern margin of the basement. In eastern North Greenland, deposition of both nearshore, shallow-water sediments (Hagen Fjord Basin) as well as deeper water sediments (Hekla Sund Basin) took place. The largest basin developed offshore adjacent to North-East Greenland (the Eleonore Bay Basin) with the accumulation of about 14 km of sediments. The upper part of this succession comprises warm-water deposits, but a sudden climate change at the end of this period led to deposition of glacial sediments (the Tillite Group).

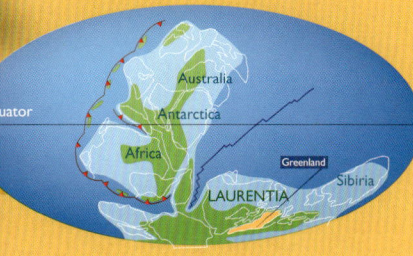
Position of Greenland at about 650 Ma.

GREENLAND'S GEOLOGICAL EVOLUTION

Phanerozoic

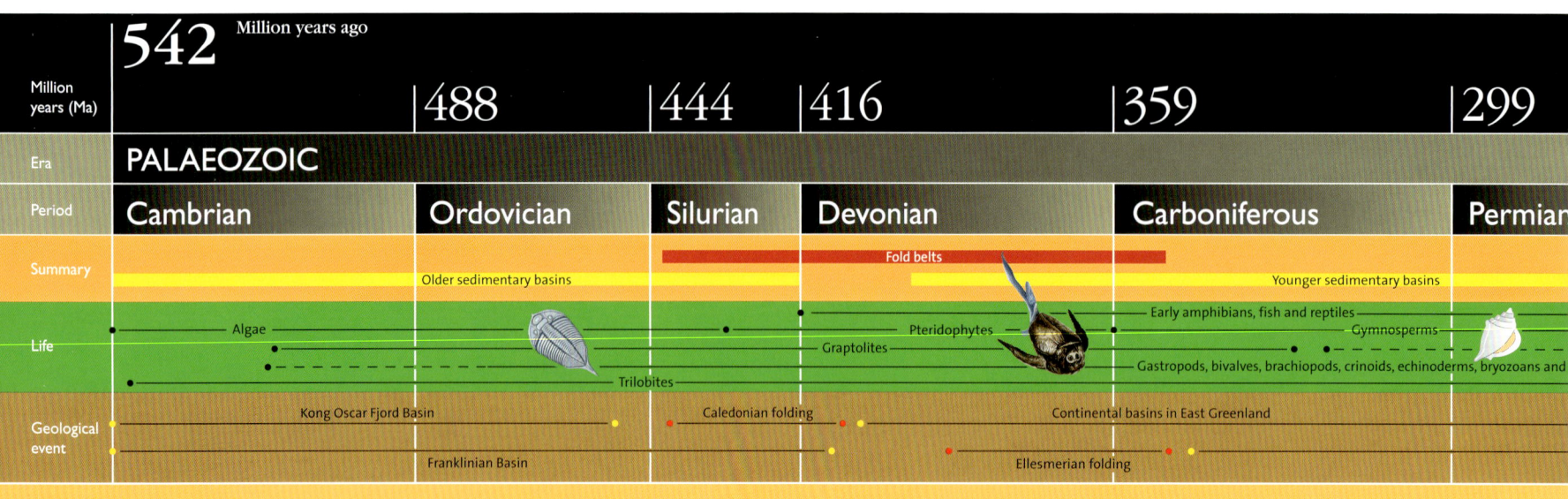

Lower Palaeozoic sediments

At the end of the Proterozoic era, at the transition to the Palaeozoic, a new ocean (Iapetus Ocean) formed between the continents of Laurentia (North America and Greenland) and Baltica (including part of Scandinavia). A succession of carbonate sediments about 4 km thick (Kong Oscar Fjord Group) was deposited under tropical conditions in the continental shelf region of East Greenland during the Cambrian and Ordovician periods.

Along the margin of North Greenland and through Arctic Canada, an east–west-trending depositional basin about 2000 km long (Franklinian Basin) developed. A broad, shallow shelf along the south side of the basin built up a carbonate succession about 4 km thick. North of this, a deep-water trough developed in which accumulated about 8 km of sands and muds – material from the adjacent land areas deposited from submarine currents.

Fold belts in North and North-East Greenland

At the end of the Ordovician, the Iapetus Ocean began to close and in the Silurian the continents of Laurentia and Baltica, formerly separated by an ocean, collided. This caused formation of a mountain belt with major upfolding of the compressed crust, melting of sedimentary rocks, granite formation, and major westwards thrusting. This 1300 km long orogenic zone forms the Caledonian fold belt in North-East Greenland.

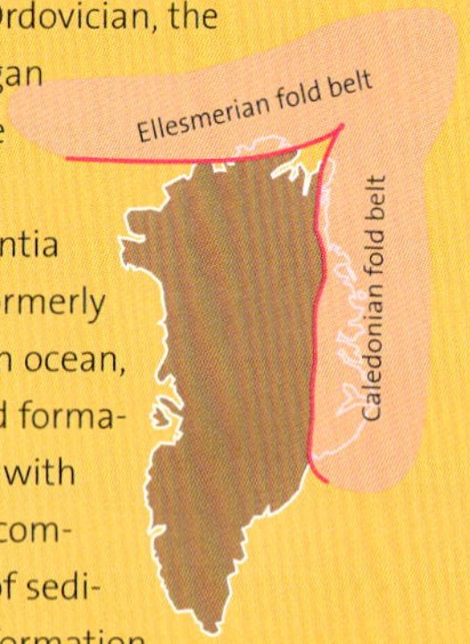

Another, slightly younger mountain belt, the Ellesmerian fold belt, developed late in the Devonian as an east–west-trending belt along the margin of North Greenland. Deformation here was less intense than in the earlier Caledonian fold belt.

Younger Palaeozoic basins in North-East Greenland

Formation of the Caledonian fold belt in North-East Greenland by E–W compression was followed by uplift, erosion, collapse and extension of the mountain belt.

Within the Caledonian belt, large continental fault-bounded basins developed in which more than 10 km of river, lake and dune sediments accumulated through the Devonian, Carboniferous and Early Permian. The Devonian deposits, dominated by sandstones, contain vertebrate fossils, amongst which occur transitional forms between fish and amphibians, the so-called 'quadruped fish'.

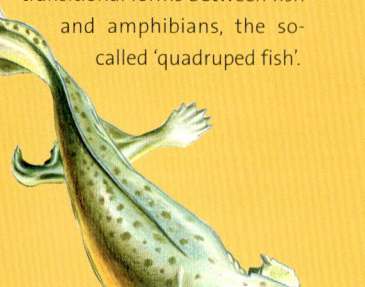
'Quadruped fish' from the Devonian.

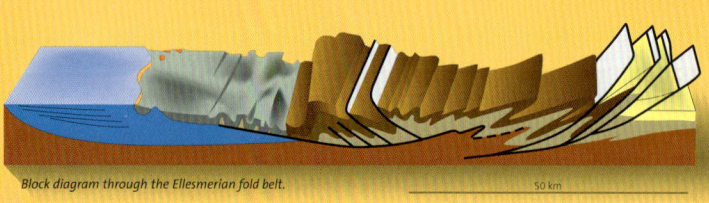
Folds in the Caledonian fold belt. 1 km

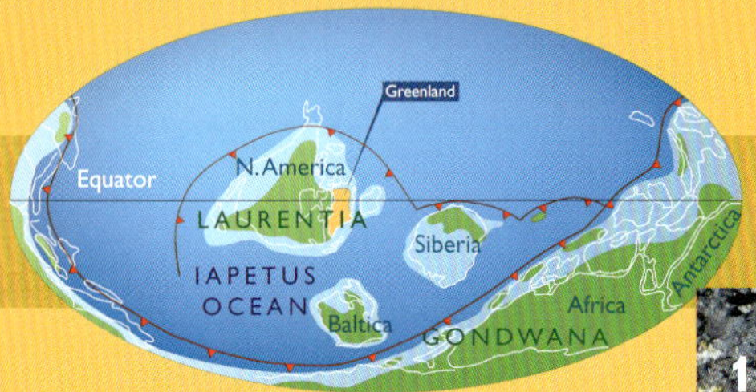

Block diagram through the Ellesmerian fold belt. 50 km

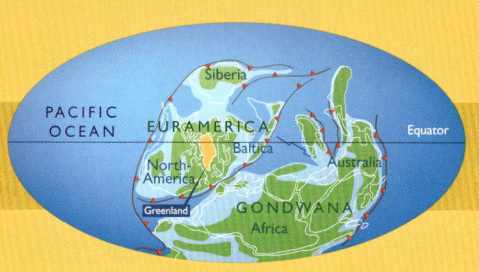

Position of Greenland in the Cambrian 514 Ma ago.

Distribution of the continents in the Devonian 390 Ma ago.

GREENLAND'S GEOLOGICAL EVOLUTION

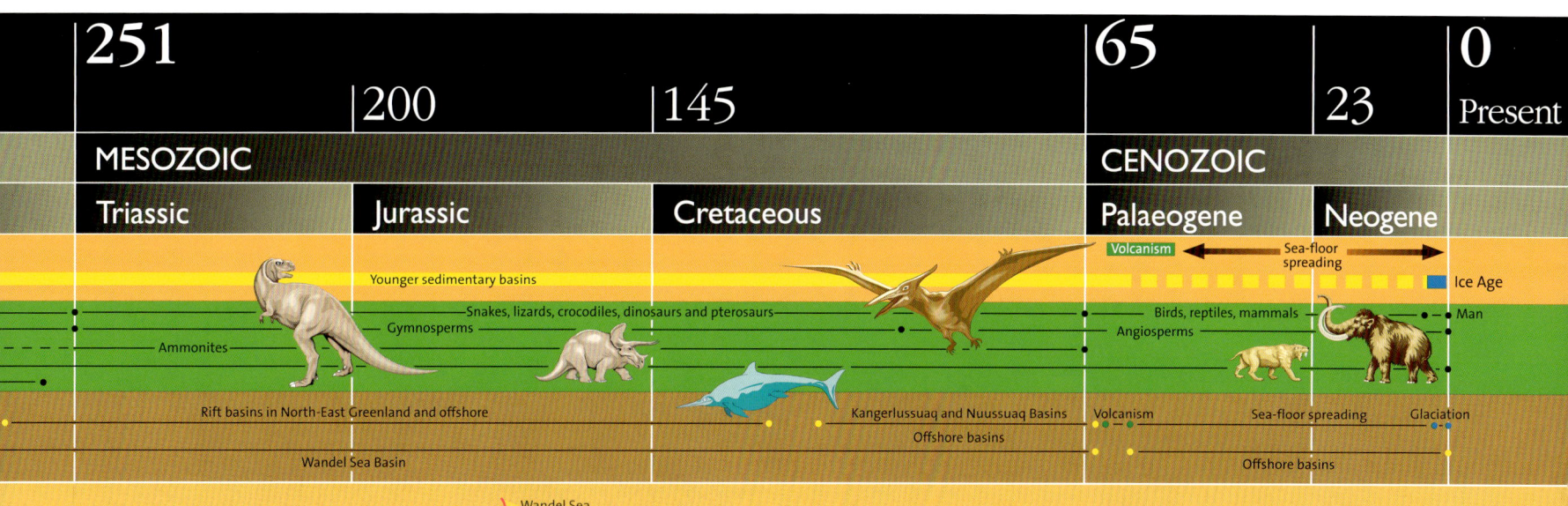

Rift basins in North-East Greenland

Following the continental collision that formed the Caledonian mountain belt, Greenland was joined to Laurasia and subsequently the Earth's continents were assembled into the super-continent Pangaea that existed from about 300 to about 200 Ma. Pangaea then split up due to rifting and sea-floor spreading, developing a sea connection along the east coast of Greenland between the Polar Sea in the north and the Tethys Ocean south of Laurasia and Baltica (Europe). The region between Greenland and Scandinavia formed a series of shallow-water, marine and continental rift basins in which thick sedimentary sequences accumulated.

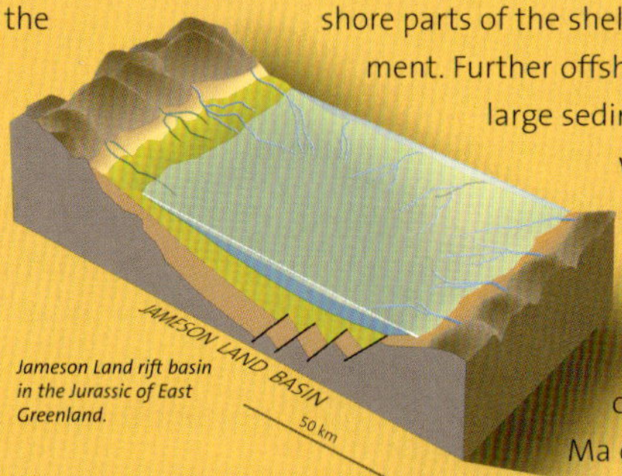

Jameson Land rift basin in the Jurassic of East Greenland.

Offshore areas

Greenland is surrounded by shallow-water shelves that have roughly twice the area of the ice-free land. Where the coastal region comprises Precambrian basement rocks, the near-shore parts of the shelf are also basement. Further offshore there are large sedimentary basins with kilometre-thick, principally Mesozoic sequences that are locally covered by 60–54 Ma old plateau basalts. Following the formation of the basalts was a thick sedimentary succession deposited offshore.

Palaeogene volcanism

Opening of the oceans around Greenland between 60 and 54 Ma ago led to formation of an enormous sequence of plateau basalts in the coastal regions. In West Greenland, a volcanic centre formed west of the Disko–Nuussuaq region and created a 4–10 km thick pile of lavas on the underlying sediments.

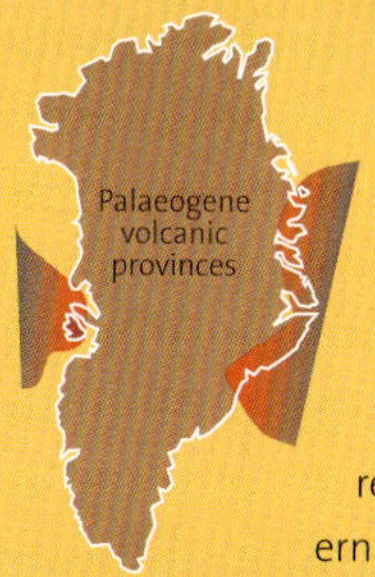

In the central parts of East Greenland, volcanic activity occurred in two phases forming a sequence of basalts up to 6 km thick. After the extrusion of the lavas, continued volcanism resulted in the emplacement of over 100 small intrusive complexes.

Volcanic fissure eruptions, similar to the Krafla fissure on Iceland, occurred in the Palaeogene in East Greenland.

The Ice Age

The Inland Ice is a relic of the extensive continental ice sheet that formerly covered a large region of the Northern Hemisphere. The Greenland ice sheet began to form about two million years ago, and the ice-free land areas bear witness to more widespread glaciation with erosional plains, glacial valleys and fjord systems. Glacial deposits include moraines and glacial river sediments in front of the ice margin both on land and offshore on the shelf up to 100 km away from the nearest coastline. Throughout the Ice Age, climate fluctuations between colder and warmer periods caused the Inland Ice to expand and contract. The latest glacial retreat began about 11 700 years ago.

Local basin formation

In addition to the large regional basins some local sedimentary basins were also formed. These included the Carboniferous to Palaeogene Wandel Sea Basin in eastern North Greenland, and the Cretaceous to Palaeogene Kangerlussuaq Basin in South-East Greenland and Nuussuaq Basin in West Greenland.

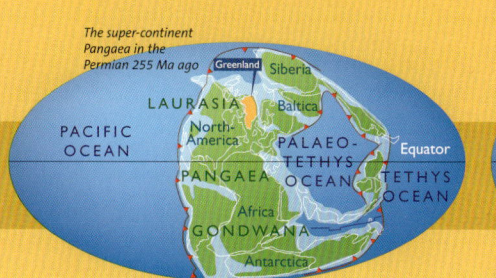

The super-continent Pangaea in the Permian 255 Ma ago.

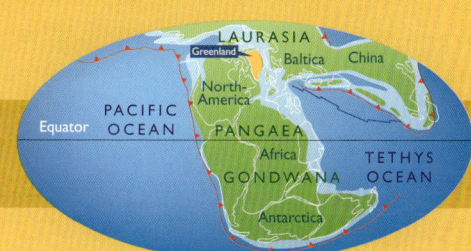

Greenland drifts towards the north. Position in the Jurassic 195 Ma ago.

Ice cover in the northern hemisphere about 100 000 years ago.

15

Meltwater deposits in Kronprins Christian Land.
Photo: J. Lautrup, GEUS

GEOLOGY IN GREENLAND

Geological research and exploration for natural resources

2

GREENLAND THROUGH MILLIONS OF YEARS

GEOLOGY IN GREENLAND

"Geo" is the Greek word for the Earth and "logos" means description or knowledge of - hence geology, the study of the composition, origin and changes of the solid Earth. The science consists of studies and descriptions of the materials and processes that have been involved in the shaping and development of the Earth during its 4600 million years of evolution.

The geological history of Greenland reaches back 3800 million years and the country contains a record of nearly all periods of the Earth evolution: from the earliest Precambrian to the Holocene. Greenland has been the focus of geological studies for several hundred years and today there is a fairly good and comprehensive knowledge of the geology of the whole ice-free land area and much of the continental shelf.

Regional geological mapping of Greenland on an overview scale was recently completed after a 40-year long dedicated programme. All of Greenland's ice-free land areas are now covered by a series of 14 coloured geological maps on a scale of 1:500 000. Research projects were carried out in parallel with the mapping, so today there is a sound understanding and description of the history of the entire country's geological formation and development. Geophysical and geological studies of the offshore areas carried out during the last few decades complement the mapping of the land areas and oil exploration wells have been drilled in West Greenland at six offshore sites and one onshore site. The results of these studies, both onshore and offshore, are summarised on a map (page 5) that shows the generalised geological structure of Greenland.

Part of a regional geological map of South Greenland. A geological map uses colours to show the occurrence and extent of the various geological units. On the map shown, the light yellow, orange and red colours show various gneisses and granites in the Ketilidian fold belt that are about 1800 million years old, while the dark brown, grey and blue colours indicate the younger, unfolded rocks of the 1350–1120 Ma old Gardar Province.

The size of Greenland compared to Europe

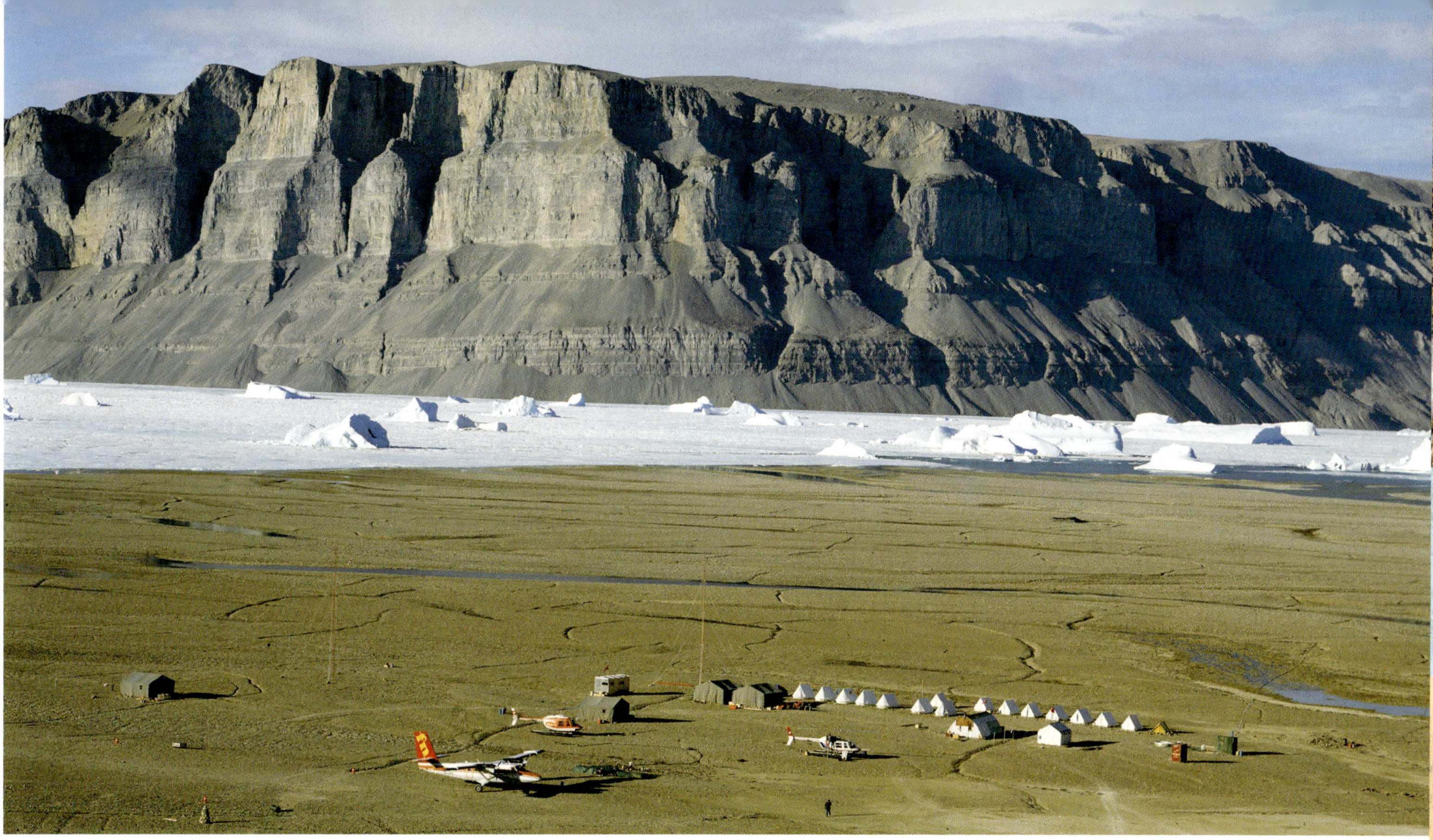

Photo: J. Lautrup, GEUS

Greenland's size and accessibility

Greenland is the largest island in the world with an area of more than 2 million square kilometres, of which roughly 80% is covered by ice. It extends 2675 km from north to south and 1250 km from east to west at its broadest part. The Inland Ice that covers the central part of the country, has a maximum thickness of about 3.4 km. The surrounding ice-free land area is up to 300 km wide and has a total area of about 410 000 km^2 nearly twice the size of the United Kingdom. Geologically, Greenland continues offshore under its continental shelves and adjacent oceanic areas, where the exclusive economic zone covers an area of 825 000 km^2.

As an island, Greenland is, of course, surrounded by water, but because it is situated in the Arctic, parts of this sea are frozen and covered by ice for much of the year. Only southern West Greenland has open water where ships can sail generally all year round. Most of the population and infrastructure are found

Tented base camp for geological investigations in unpopulated North Greenland. The field work was carried out by summer expeditions each comprising 30–40 participants who, supported by helicopters and a small fixed-wing aircraft, could map an area of about 150 x 300 km every summer. The picture is from Warming Land in central North Greenland where the rocks consist of unfolded Palaeozoic carbonates.

Photos: N. Henriksen, A.K. Higgins & S.A.S. Pedersen, GEUS

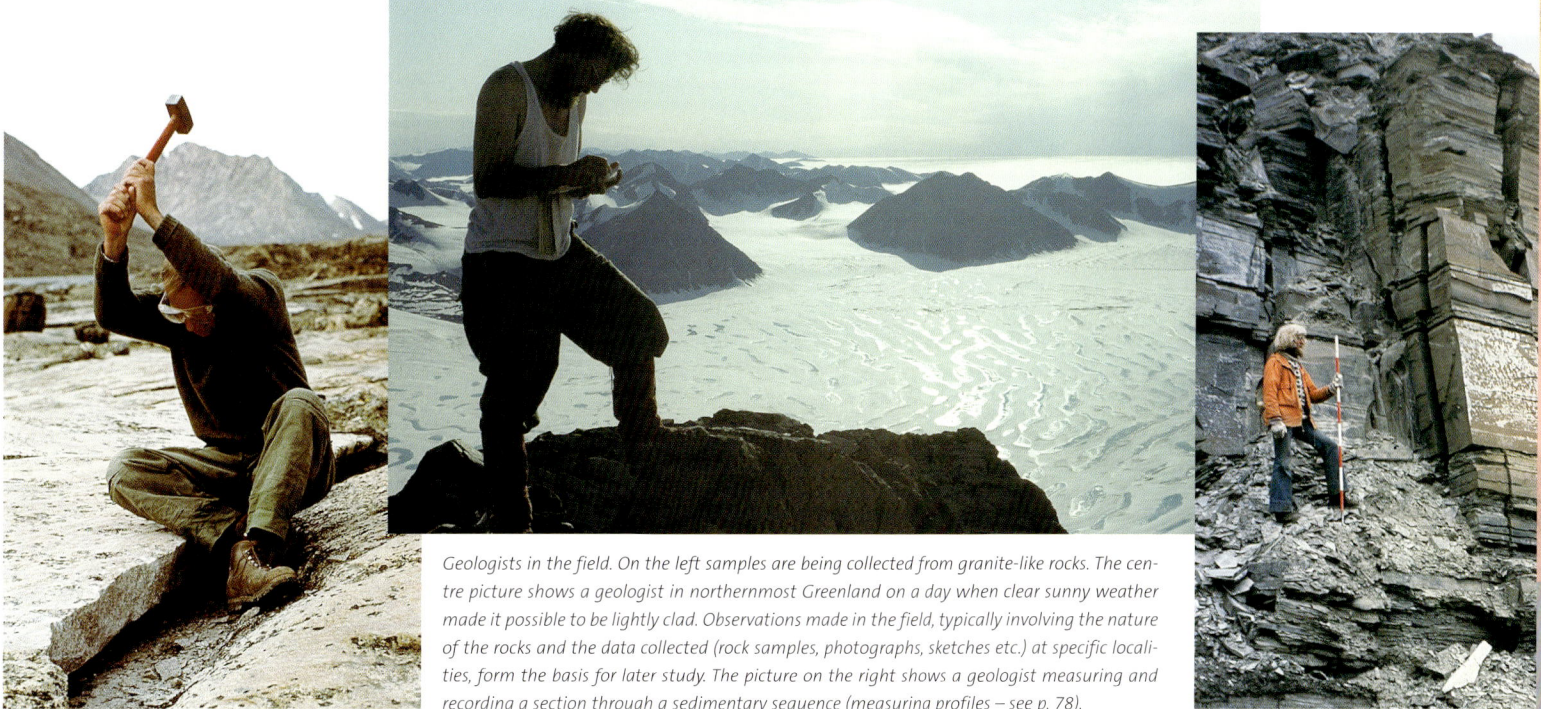

Geologists in the field. On the left samples are being collected from granite-like rocks. The centre picture shows a geologist in northernmost Greenland on a day when clear sunny weather made it possible to be lightly clad. Observations made in the field, typically involving the nature of the rocks and the data collected (rock samples, photographs, sketches etc.) at specific localities, form the basis for later study. The picture on the right shows a geologist measuring and recording a section through a sedimentary sequence (measuring profiles – see p. 78).

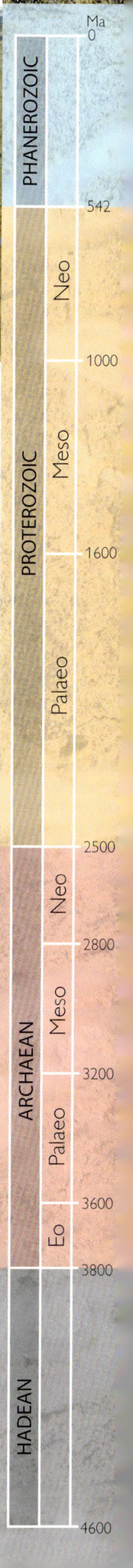

GEOLOGY IN GREENLAND

The material collected during field work is worked up in GEUS' laboratories where samples are examined under a microscope and analysed. One of the instruments used is the Scanning Electron Microscope (SEM) shown in the picture. It works by scanning a beam of electrons across the sample and collecting the reflected electrons. Computer processing is used to assemble a final picture. The SEM can enlarge up to 100 000 times and the instrument can also be used to make both qualitative and quantitative chemical analyses.

in this more accessible area of southern West Greenland. Only scattered communities are found in more remote areas, and large stretches of North and North-East Greenland, some thousand kilometres in extent, are devoid of permanent inhabitants. Geological research in Greenland has, consequently, always been logistically difficult. Field work is generally possible only during a few summer months when the country is more or less free of snow. Where there is no infrastructure, the work has been carried out by expeditions whose personnel and provisions have to be transported to and from Greenland. During the last 30–40 years, transport while in the field has been by means of specially chartered ships, helicopters and small aircraft.

Geological exploration

Geological surveying started during the first scientific voyages of discovery that reached West and East Greenland in the early nineteenth century and North Greenland at the end of that century. Systematic mapping began with the 'Danish Expeditions to East Greenland' (1926–1958) and in West Greenland in 1946 by the Geological Survey of Greenland (GGU, an acronym for the Danish name Grønlands Geologiske Undersøgelse). Since the late 1960s, the latter organisation has extended its work to cover the whole of Greenland. It was amalgamated with the Danish Geological Survey in 1995 to form the Geological Survey of Denmark and Greenland (GEUS) that today undertakes geological work in Greenland on behalf of the Danish and Greenland authorities.

Exploration of the sea areas around Greenland started in the 1970s when interest was awakened in the country's hydrocarbon potential. Much of the information available from the offshore areas originates from exploration by commercial companies, as well as from marine geological and geophysical studies by the Survey and international academic institutes.

Geological research

Traditional studies of the geological development of an area consist of field work, recording and mapping the presence and extent of different rocks and their structures, combined with observations of, for example, the presence of minerals and fossils. After completion of the field work, extensive analyses of the observations and studies of the sampled material are carried out in the laboratory, often involving a large amount of work. The result of all this is an interpretation of the processes and changes that took place during the development of an area and an understanding of how it formed.

Geological structures and their development must be interpreted in three dimensions. Although field observations are commonly only available from the

THE STRUCTURE OF THE EARTH

The Earth's radius is 6371 km. It consists predominantly of a core and a mantle; the boundary between them being at a depth of 2900 km. The entire surface is covered by the crust, a thin skin that varies in thickness from 3 to 70 km. Most visible geological processes take place within this surface layer.

The three parts have very different chemical compositions. The core (1) consists of iron and nickel (density c. 10–13.5 g/cm^3) and the mantle (2) is made of heavy silicate rocks of basic and ultrabasic composition (density 3.3–5.5 g/cm^3). The crust (3) is composed of lighter silicate rocks of two types – the oceanic areas that are dominated by basalts (density c. 3.0 g/cm^3), and the continents that are dominated by granitic rocks and sediments (density 2.5–2.8 g/cm^3).

The outer parts of the Earth form a 50–150 km thick rigid shell (A) – the lithosphere – which can move over an underlying, ductile layer (B) – the asthenosphere. The transition between the two occurs in the upper mantle, and is caused by a change in the mechanical behaviour of the mantle rocks. It is thus important to notice that the boundary between the lithosphere and the asthenosphere is not defined by differences in composition, but by increasing temperatures and pressures with depth.

The lithosphere is divided into a number of large plates that comprise both the crust's oceanic and continental rocks. The plates move relative to one another and are the cause of the dynamic development of the Earth (plate tectonics, see p. 53).

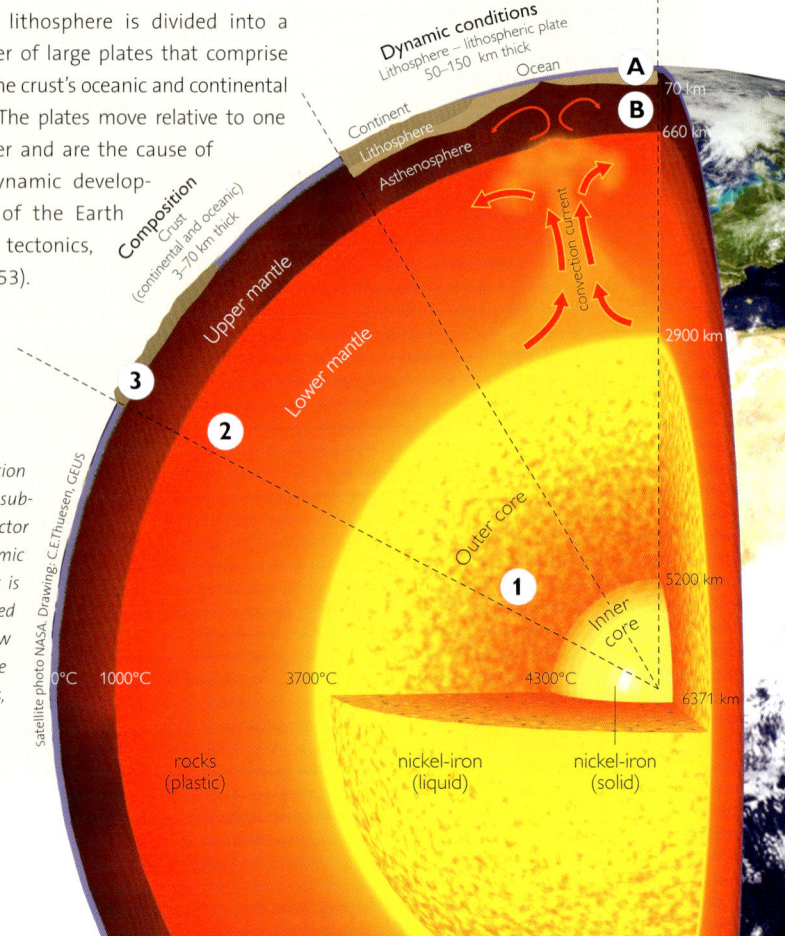

Cross-section of the Earth showing its structure, composition and dynamic conditions. The sector on the left shows the subdivision according to composition, while the right sector shows the division according to physical properties (dynamic conditions), i.e. how the layers react to deformation. It is important to note that the two types of divisions are based on entirely different concepts. The figure illustrates how various parts of the upper mantle – despite having the same composition – deform in entirely different ways, depending on their temperature and pressure regime.

Earth's surface, it is often possible to use them to establish models that interpret the deeper structures.

Understanding geological processes requires an appreciation of time – the fourth dimension. Due to the great age of the Earth, and the continuous dynamic conditions of its geological evolution, processes that act even extremely slowly can accumulate to have very large effects. For example, continents have moved thousands of kilometres relative to one another even though their speed of drift is only 2–10 cm per year. Sediment is normally deposited in the deep ocean at rates of only 1–20 cm per thousand years, but can, nonetheless, build accumulations several thousands of metres thick.

The construction of the geological time scales has evolved as the science has developed. The names of the divisions of geological time were devised when geologists recognised that stratified rocks could be divided into units that represent the natural divisions of time. The actual names are derived from parts of the world where rocks of a particular age were originally studied and described. The absolute time scale, defining ages in millions of years, has been worked out only during the last 50–75 years. The main divisions of geological time are shown in the figure on the left. The major divisions are divided into a large number of smaller segments that allow accuracy down to less than 1 million years, even for sedimentary rocks as old as 540 million years.

Glaciologists drilling into a local ice cap in North Greenland. The drilling was carried out with a hot-water drill that used water at a temperature of 70°C to melt a 280 m deep hole through the ice. The borehole was used to measure temperatures in the ice.

The divisions of geological time (geochronology) with the names of the units and their ages in million years (Ma). The right side of the diagram shows which main groups are represented in Greenland. The colours in this part of the diagram correspond to those used on page 5.

STUDY AREAS WITHIN GEOLOGY

The study of geology is so comprehensive and complex that several specialities of study within the subject have developed, each demanding its own special insight and knowledge. Geologists within each area have their own highly developed techniques and skills and the specialists generally do not have the ability to work at the same highly specialised level within another discipline. The main subject areas that have been used in Greenland are as follows:

Economic geology – the study of the occurrence, composition and structure of potentially economic minerals.

Geochemistry – the study of the chemical composition of rocks, minerals and water.

Geochronology – the study of geological time, particularly absolute age determinations (see p. 48).

Geophysics – the study of the Earth using quantitative physical methods to measure properties such as magnetism, gravity and reflection and refraction of seismic (sound) waves.

Geological mapping – mapping the presence and extent of different rock units together with studies of their age-relations and structure etc.

Glaciology – the study of ice, especially the Inland Ice and glaciers.

Mineralogy – the study of the occurrence, composition and structure of minerals.

Palaeontology – the study of life in the past using fossils. Macrofossils are visible with the naked eye, while microfossils must be studied using a microscope.

Petroleum geology – the study of the presence, chemistry, methods of formation, migration and entrapment of hydrocarbons (mineral oil and gas).

Petrology – the study of the occurrence, formation, composition, structure and development of rocks – especially magmatic and metamorphic rocks.

Sedimentology – the study of sedimentary rocks, their form, constitution, physical shapes and the processes forming them.

Stratigraphy – the study of layered successions of rocks, their age relationships, distribution and fossil content.

Structural geology – the study of how rocks deform, fold and break under stress by studying the three-dimensional structural patterns within them.

Caledonian mountain chain containing strongly-coloured Proterozoic sediments west of Kong Oscar Fjord (uppermost right).
Photo: J. Lautrup, GEUS

LANDSCAPES

Mountains worn down by ice reveal ancient geological formations

3

ICE-AGE EROSION CAN BE SEEN EVERYWHERE

3 LANDSCAPES

Landscape forms in the ice-free zone of Greenland mirror the underlying geology and the erosion styles that have affected it. During most of the Ice Age Greenland was entirely covered by ice that deeply eroded the country, planing it off and sculpturing the mountains. This erosion often followed existing lines of weakness and topographic elements such as fracture zones, river valleys and fjords that had precursors dating from earlier geological periods. The surface of existing high-level plateaux was planed off and the valleys were occupied by glaciers that gave them their present U-shaped cross-sections.

U-shaped glacial valley floored by meltwater deposits. A short way into the valley on the left the tongue of a little local glacier can be seen with a large terminal moraine in front of it. The mountains around the valley are up to 2000 metres high. The picture is from the Caledonian mountains in North-East Greenland.

Material ground off by the glaciers was redeposited as thin layers of boulder clay on the mountain plateaux and as marginal moraines along the flanks of the glaciers. Enormous amounts of eroded rock were deposited as meltwater sands and gravels in river valleys and in particular were shed offshore onto the continental shelves and into the deep seas around Greenland. These sediments can be several kilometres thick.

The topographic morphology is very varied and ranges from coastal plains near sea-level to alpine summits over 3000 m high. Near the margin of the Inland Ice, thousands of small ice-free areas – so-called nunataks – emerge like islands from the ice. The highest ice-free areas are commonly found in the nunatak zone; among them is Greenland's highest summit – Gunnbjørn Fjeld at 3693 m in South-East Greenland. The height of the Inland Ice increases gradually inland away from its margin and reaches over 3230 m at its central and highest point; its average height is about 2100 m. The inner regions

Map showing water depths where Scoresby Sund opens into the North Atlantic; depths are in metres. The depth contours show the shape of the large depositional fan of eroded material that river systems and glaciers have transported through the fjord system and deposited on the continental shelf out towards the deep sea. The fan is about 100 x 200 km in extent. Water depths in the inner fjord are 1000–1500 m and the mountains around it reach up to heights of a couple of thousand metres. Erosion has thus sculpted a topographic relief of over 3000 metres, since basalts were deposited in this area about 55 Ma ago and the North Atlantic started to open. The dashed line shows the boundary between continental and oceanic crust. Since the depositional fan clearly stretches a long way beyond this boundary, the deposition must have taken place since the start of oceanic spreading.

Photo: J. Lautrup, GEUS

Source: KMS photo

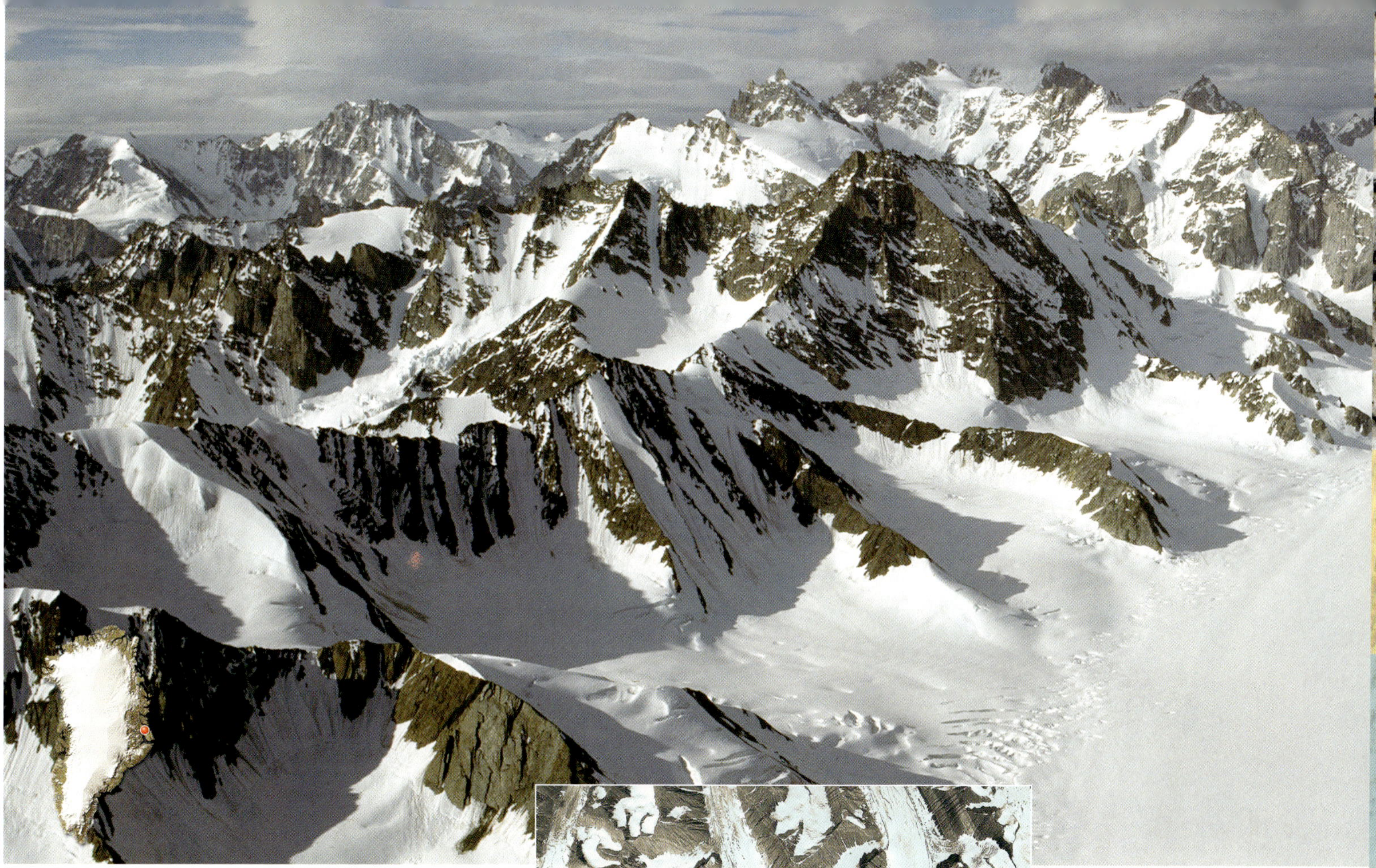

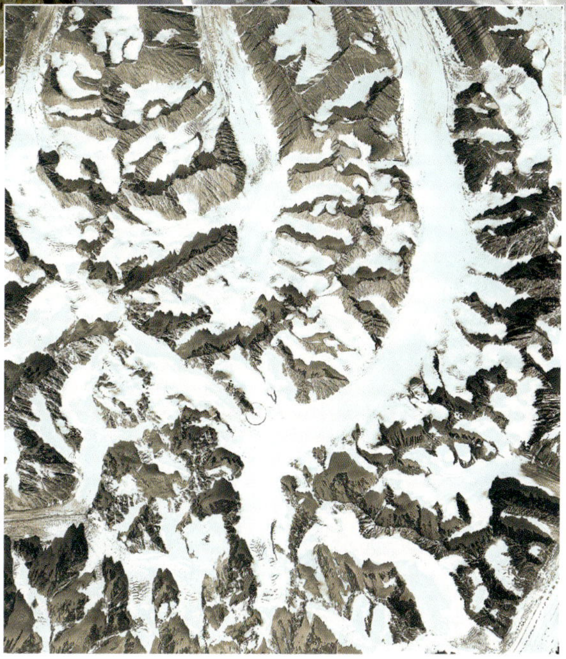

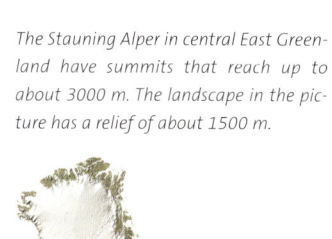

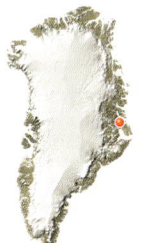

Vertical aerial photograph of the Stauning Alper in East Greenland. The picture shows how such an alpine landscape consists of isolated summits separated by a tight network of corrie glaciers (seen as light-coloured, branching patches). Compare with the upper photograph from the same area.

The Stauning Alper in central East Greenland have summits that reach up to about 3000 m. The landscape in the picture has a relief of about 1500 m.

rest on a bowl-shaped depression of the basement shield whose deepest parts lie several hundred metres below sea level.

The landforms depend on the area's geological background. The basement and fold belt regions are dominated by gneisses, metamorphic schists and granites that are hard and resistant to erosion. Erosion by glaciers leaves such rocks bare, standing out and well exposed. In contrast, the un-metamorphosed sediments disintegrate easily, resulting in landscapes dominated by gentle hills and valleys where the sediments themselves are commonly obscured by landslipped material. The basalt sequences consist mostly of hard rocks, but in places softer, more readily weathered layers (e.g. ash) are found between the lava flows, so the resulting landscape forms can be very variable or intermediate between the landscapes produced by the rocks in the basement and sedimentary areas.

In many landscapes, the uppermost parts of the mountains consist of a flat-lying erosional surface – a

Drawing: C.E.Thuesen, GEUS

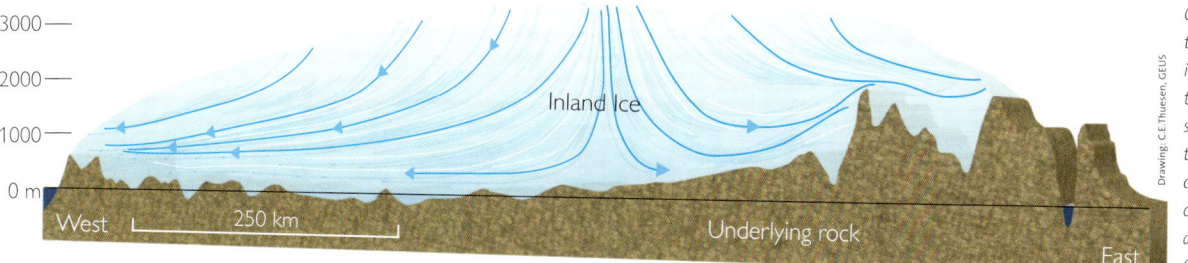

Cross-section through Greenland showing that the Inland Ice is more than 3 km thick in the centre. Note that the deepest part of the ice lies below sea level. The flow lines show how the ice moves from its centre out towards its margins. The rocks under the central and western parts consist primarily of Precambrian basement, while the high areas in the east consist of rocks in the Caledonian fold belt.

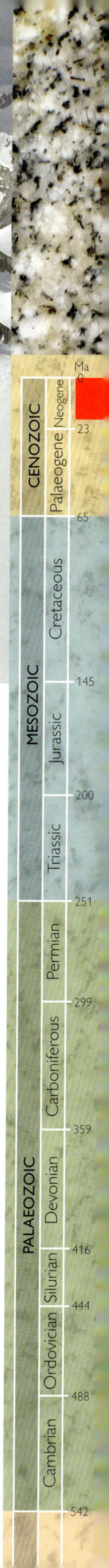

Photo: J. Lautrup, GEUS

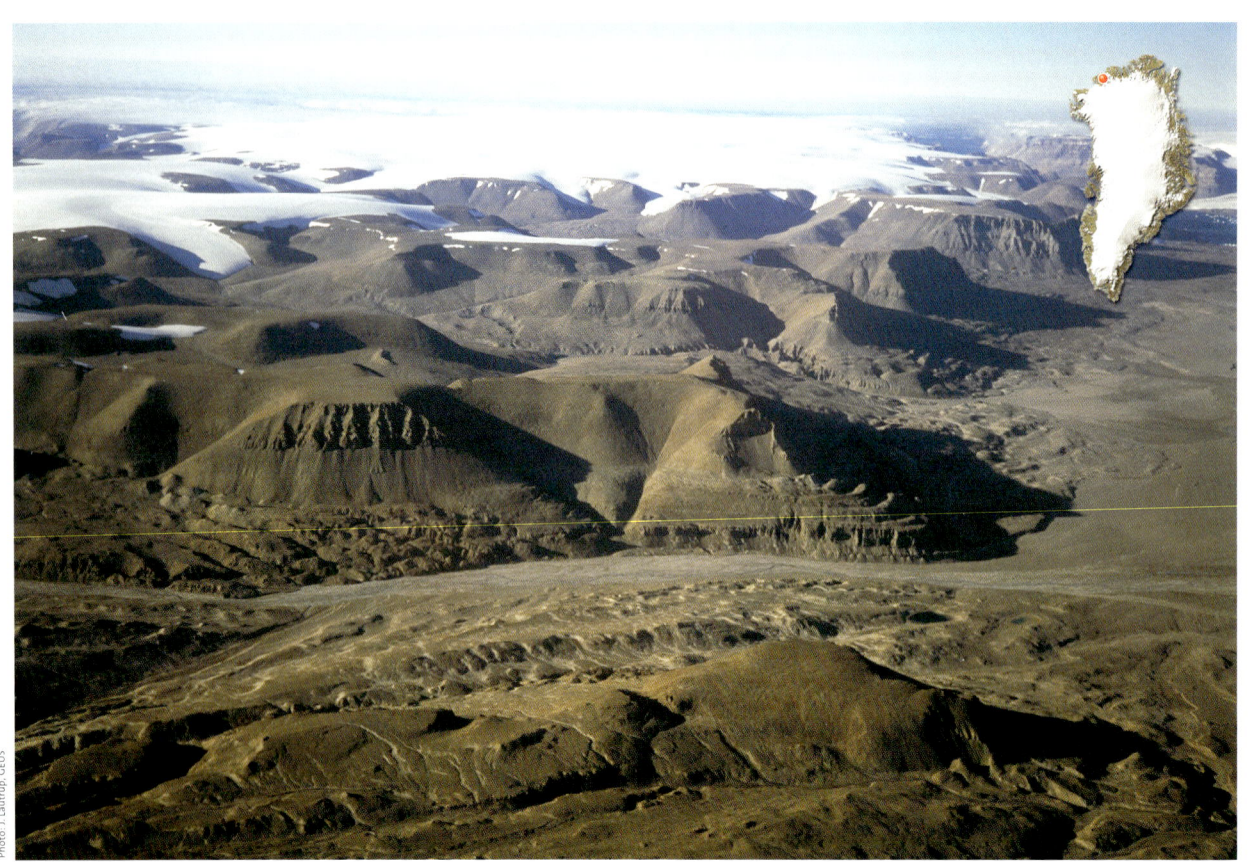

Typical North Greenland sedimentary area in the south-eastern part of Washington Land with flat-lying layers of carbonate rocks that weather comparatively easily, so the mountainsides disintegrate and the landscape develops gentle shapes. The top of the mountains is a flat plateau (at heights of around 1000 m) that is the remains of an old peneplain (a sea-level erosion plain uplifted to its present elevation). A local ice cap can be seen on top of the plateau.

Typical West Greenland basement landscape in the central part of Godthåbsfjord near Nuuk. The grey mountains consist mainly of granite-like rocks with scattered, darker, amphibolite bands. The mountain range, reaching heights of up to 1500–2000 m, is dissected by a branched fjord system with steep sides, yielding excellent opportunities for geologists to study geological structures in three dimensions.

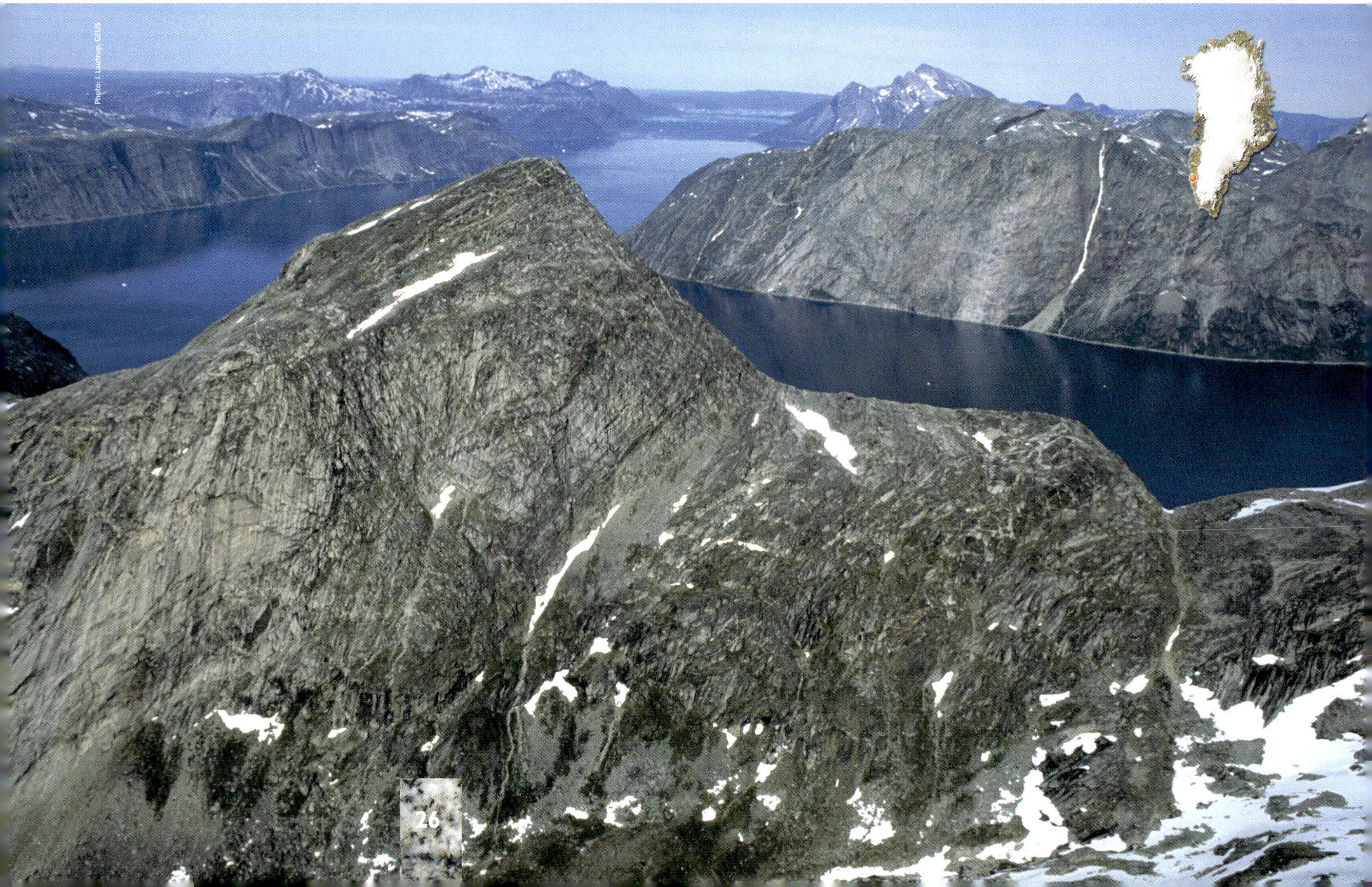

Photo: J. Lautrup, GEUS

so-called 'peneplain'. Such high-lying flat plateaux are old erosion surfaces, originally formed by slow erosion and shaping of the landscape by meandering rivers at low levels. These rivers finally removed the last variations in height and levelled the terrain to a coastal plain. The remains of these old peneplains are found today high up in the mountains – in places up to a couple of thousand metres above sea level – due to uplift that occurred after formation of this peneplain surface (see p. 179).

View over a mountainous landscape in central East Greenland, west of Kong Oscar Fjord. The area consists of large mountain massifs dissected by branching fjord systems and deep valleys. The mountain summits are at about 2000–2400 m and small local ice caps are found in many places. Despite resembling a rugged alpine landscape when seen from low altitude, the picture shows that the summits are, in fact, a flat plateau that can be seen on the picture as a level line along the horizon. This flat surface is an old peneplain.

UPLIFT OF THE LAND – ISOSTASY

The rocks we see exposed in the mountains at heights of up to 2 or 3 km were nearly all formed at much lower levels. Sediments were deposited near or below sea level and crystalline rocks in the basement and in fold belts have been formed at depths of 10–40 km in the Earth's crust. When such rocks are now exposed at much higher levels it is a testament to the uplift that took place after they were formed.

In places where the Earth's crust is being thickened, for example by deposition of large volumes of sediments or basalts, or when continents collide and their edges are superimposed during mountain building in fold belts, the immense additional weight forces the lithosphere to sink and its base is pressed downwards into the underlying upper mantle (the asthenosphere). The downwards depression is compensated for in the asthenosphere by lateral flowage of mantle material outwards and away from the bulge. The outwards flow results in a natural adjustment or balance whereby the total mass of all the rock units above a certain level (depth of compensation at about 110 km below the Earth's surface) is the same all around the earth. This theory of mass balance is called isostasy (from the Greek for 'equal weight'). According to this all large portions of the Earth's outer shell are floating on an underlying nearly molten denser layer (the asthenosphere). An analogue to this is an iceberg floating on water.

When a load on the crust is eased or removed, for example by erosion of mountains, the rocks below the terrain move upwards and asthenospheric material 'flows' inwards to compensate for the material removed from shallower levels. Hereby the mass balance is kept equal. During the late phases of mountain-building, the isostatic rebound restores the lithosphere and the rock units can rise tens of kilometres.

In Greenland a recent effect of isostasy is related to the ice cap, which has covered the land during the Ice Age. The weight of the ice has pressed the land down by several hundred metres, but after melting of the margins and the upper parts of the ice cap the land rose a little over a hundred metres. The effects of this are seen as raised shorelines in the coastal areas all around Greenland (see p. 190).

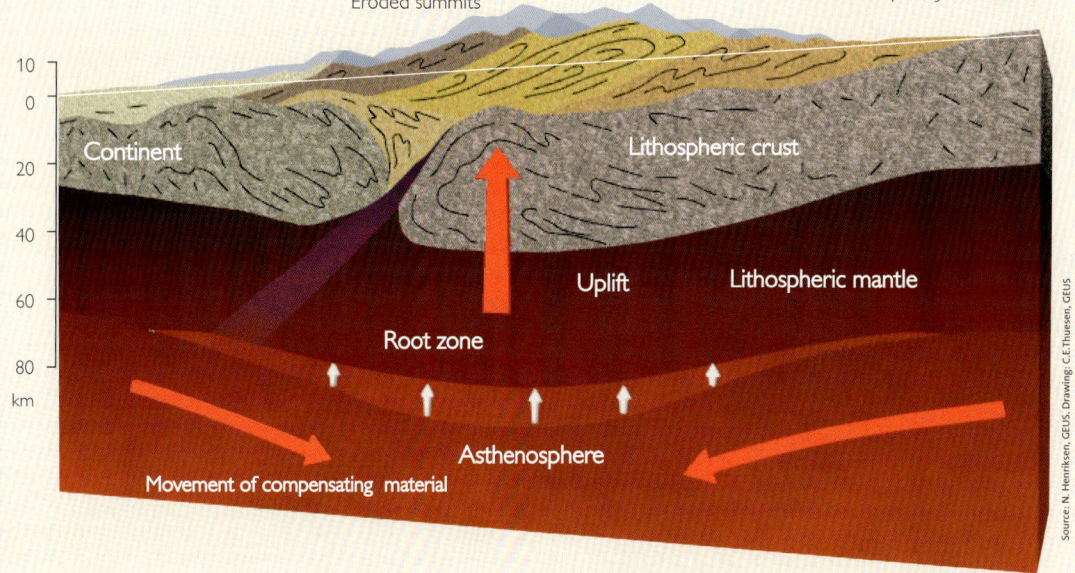

Cross-section showing how the base of the lithosphere is depressed during mountain building. As the tops of the mountains gradually erode, the load on the underlying asthenosphere decreases, so the remaining rocks in the mountains rise gradually upwards and material in the asthenosphere flows inwards.

Basement rocks near Godthåbsfjord, West Greenland.

Photo J. Lautrup, GEUS

THE CRYSTALLINE BASEMENT

Rocks formed in Precambrian fold belts

4

THE GEOLOGY OF THE GREENLAND SHIELD

THE CRYSTALLINE BASEMENT

By far the largest part of Greenland is composed of a Precambrian basement shield that is dominated by pale, folded gneissic and granitic rocks containing layers and inclusions of other rock types. This crystalline shield is composed of a series of different Archaean (3800–2500 Ma) and Palaeoproterozoic (2000–1750 Ma) mountain chains, which are welded together. Today these rocks crop out at the surface of the Earth, but were formed at depths of about 20–50 km in the crust where temperatures were 400–800°C. After their formation the mountain chains partially collapsed due to stretching (extension) of the crust, and the upper parts of the folded mountains were gradually eroded away. Over millions of years the deep-seated rock masses were brought to the surface, so that what is seen today are only the roots of the earlier mountain belts.

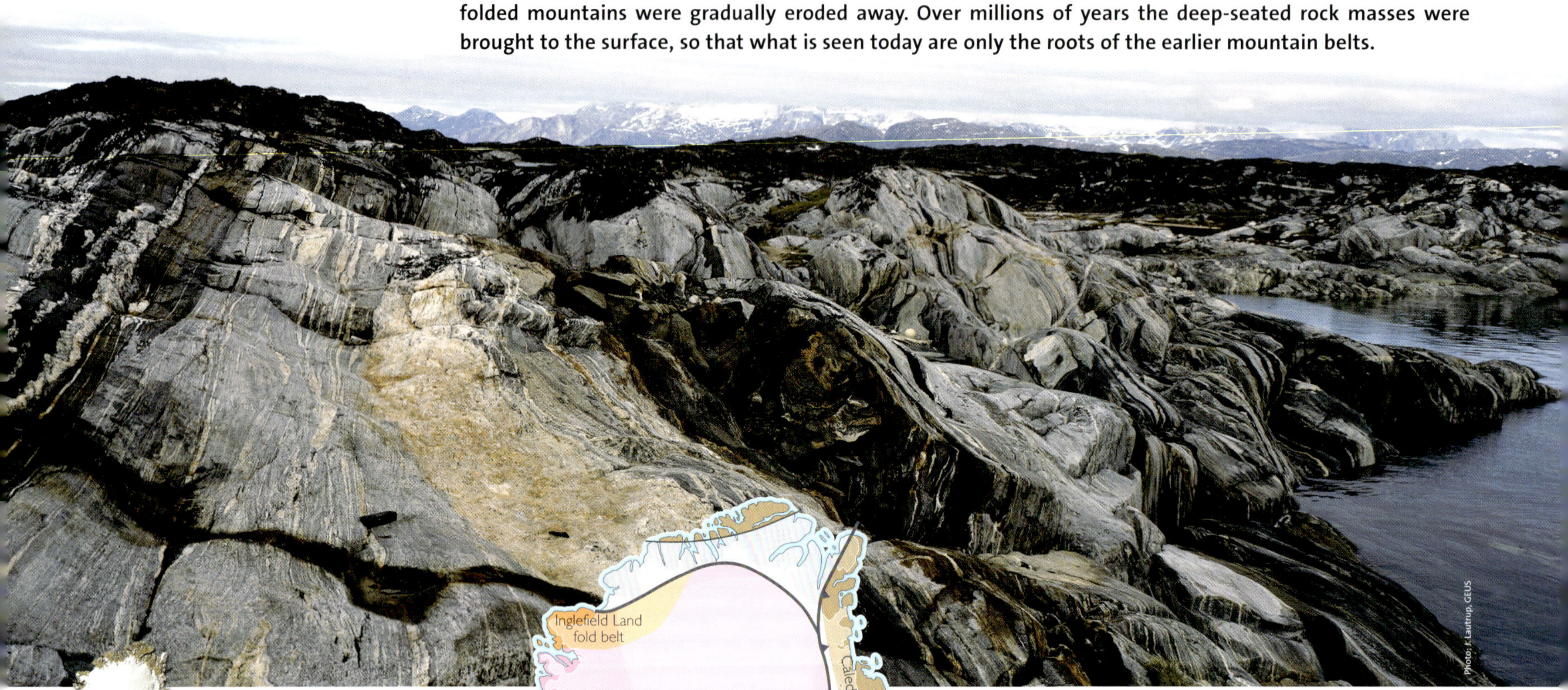

Typical basement rocks composed of grey gneisses with layers of black amphibolite. These are cut by paler, coarse-grained, quartz-feldspar sheets (pegmatites). This gneiss complex was deformed and transformed (metamorphosed) under amphibolite facies conditions, 20–40 km deep in the crust at temperatures of 450–750°C. Photograph taken on the coast of Godthåbsfjord in the Nuuk area. The coastal outcrop in the foreground is 10–15 m wide.

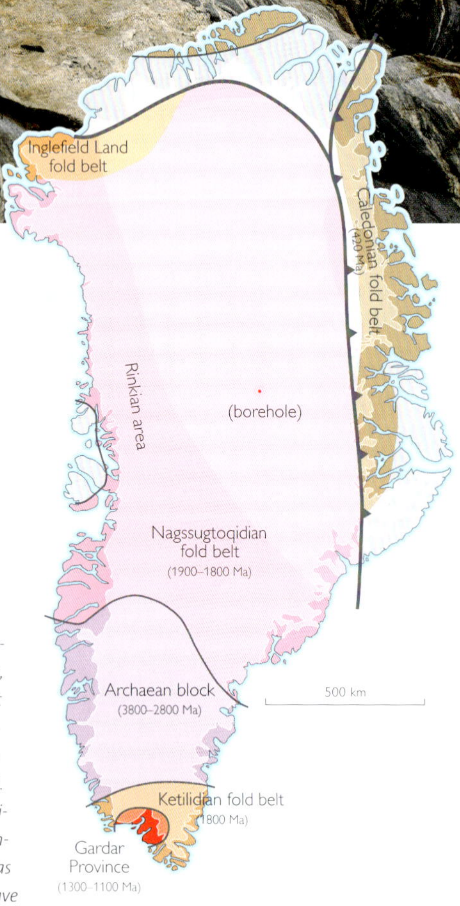

The basement gneisses that make up the Greenland shield consist of Precambrian fold belts welded together. The core region, originally formed in connection with Archaean mountain belt formation (3800–2500 Ma), but extensive regions were overprinted by Palaeoproterozoic events (2000–1750 Ma) in which old and newly formed crustal materials were often interleaved. The transformation was sometimes so intense that their original characteristics are totally obscured. In North-East Greenland, the 420 Ma old Caledonian fold belt contains large areas of Archaean and Palaeoproterozoic basement rocks which have been affected by the Caledonian deformation. Rock distribution below the Inland Ice is unknown, so the map here is based on assumptions.

Inglefield Land fold belt
Caledonian fold belt (420 Ma)
Rinkian area
(borehole)
Nagssugtoqidian fold belt (1900–1800 Ma)
Archaean block (3800–2800 Ma)
Ketilidian fold belt (1800 Ma)
Gardar Province (1300–1100 Ma)

The Precambrian shield can be studied throughout the whole of the ice-free parts of West and South-East Greenland, but the greater part is hidden by the ice cap. Formation of the shield took place in stages over a time span of some 2300 Ma. The oldest core region of the Archaean shield is bordered by younger mountain belts produced by plate tectonic processes. The different mountain belts originally formed as parts of an extensive series of global mountain building events that successively built up large continental areas. Parts of these were again broken up by continental drift, so that some of the components lost their original coherence. The typical development was that younger mountain chains were formed in belts along the margins of the earlier continents whereby the basement shield grew step-wise. The Greenland shield is a good example of this.

The Archaean basement in Greenland originally made up a very large portion of the present shield, but only the southern part escaped modification by

Banded gneisses with pale veins of quartz-feldspar pegmatite and thin, black amphibolites. These are called orthogneisses and were formed by deformation of an originally homogeneous granitic rock. The pale-coloured pegmatite bands were formed by the injection of melted material.

later orogenic events. The northerly parts of the Archaean shield were reworked in a new fold belt formed about 2000–1750 Ma ago. This means that their Archaean structures and metamorphic history have been largely obliterated and the rocks have lost their original features. There remain, however, a few, small, isolated areas where the Archaean rocks are unmodified and where relicts of their composition and structure are preserved.

The younger fold belts that make up the Greenland basement shield mainly developed between 2000 Ma and 1750 Ma ago, in the Palaeoproterozoic era (see p. 21). Each of these different fold belts has its own name and distinctive character. South of the Archaean block the Ketilidian mobile belt characterises South Greenland. North of the Archaean block in West Greenland the Nagssugtoqidian fold belt occurs, and still further north comes the Rinkian area. Other crystalline basement areas formed in Greenland at about the same time, but these fold belts are not so well defined.

Between 1750 Ma and 1600 Ma ago the crystalline basement in Greenland stabilised, and subsequent geological processes took place around the edges of this Precambrian shield. The greater part of the crystalline basement that we can observe today has a form and character unchanged for more than 1600 Ma. However, in many places the basement was broken up and intruded by swarms of dark, linear igneous intrusions called dykes. Where these dykes are undeformed and unaltered it can be concluded that the crystalline basement has not been substantially affected by later events.

Rocks in the crystalline basement

The crystalline basement is made up of different types of gneisses and deformed granite-like rocks, which have roughly the same appearance irrespective of their age or geological history. The gneisses are frequently banded with alternating paler layers, dominated by the minerals quartz and feldspar, and darker layers that contain a large proportion of minerals like hornblende and biotite. In the gneisses there are frequently dark layers of amphibolitic rocks, which principally comprise equal parts of plagioclase

Homogeneous, weakly banded granite. On the clean coastal surfaces in the foreground the rocks have a light, grey colour, which cannot be seen when the rock has been overgrown by lichen (algae and fungi). In the foreground, where the geologist is sitting, the lichens are dark grey. On the hillside in the background to the left the lichen cover is thick and nearly black.

THE CRYSTALLINE BASEMENT

feldspar and hornblende. These amphibolite layers are often metamorphosed basaltic rocks that were originally emplaced as magma into linear fissures in the host gneisses.

In addition to gneisses and granite-like rocks the crystalline basement may contain bands and zones of schists. These are former sedimentary and volcanic deposits laid down at the Earth's surface, and transformed by heat and pressure into schists. The schist units vary greatly in thickness from sequences several kilometres thick, down to very thin layers and bands. The schists and the surrounding gneisses are often deformed together, making it very difficult to establish the age relationships by observation. The gneisses could be oldest, and have formed a basement to the supracrustal rocks. Alternatively, the supracrustal rocks may have been deposited first, and the gneisses could represent granitic magmas that were intruded into the sedimentary and volcanic rocks when deep in the crust. In the latter case the granitic rocks, transformed by pressure and deformation into gneisses, are the youngest. Establishment of the correct age relationships may require costly laboratory analyses.

The structures in the basement

The structures in the basement were formed during mountain building processes (orogenies). Different types of folds and shear zones developed due to compression and heating at high pressures. Deformation patterns are revealed by the orientation of the different layers and their three-dimensional geometry. During fieldwork, geologists unravel these structural patterns by mapping the shape and extent of the layers, and through spatial geometrical analysis, work out how the rock sequence was folded and at which crustal levels the deformation took place. Structures formed deep in the crust tend to have plastic shapes with tight fold patterns whereas those formed closer to the surface develop simple, open fold patterns.

Pale, grey gneisses with two generations of dark (nearly black) amphibolitic rocks. The oldest generation occurs as thin, conformable layers in the gneisses. The younger generation occurs as the cross-cutting dyke, just above the sitting geologist, and was intruded into the gneisses after they were deformed and metamorphosed together with the first generation of amphibolite. The thick, cross-cutting dyke was originally intruded into a fissure as a basaltic magma. Its transformation into a metamorphic amphibolite can only have taken place in connection with a second phase of regional metamorphism (see box p. 35).

Typical metamorphic mica schists that comprise an alternation of greyish and red-brown layers made up of quartz, feldspar and micas. The numerous conformable, white to pale quartz-feldspar veins were formed by local partial melting of the host rock. The rock was originally a sediment with alternating sandy and muddy layers. During a subsequent folding event at high temperatures, these sediments were totally recrystallised (metamorphosed under amphibolite facies conditions - see box p. 35) and deformed by which part of their present banded character was formed. The red hammer is about 40 cm long.

THE CRYSTALLINE BASEMENT

STRUCTURES IN THE BASEMENT

The crystalline basement comprises mostly gneisses and granitic rock types that are composed of alternating layers with slightly varying composition; they may contain bands and layers of other rock types. The rocks have a dominant layered and striped character that indicates the three-dimensional structural patterns in the basement.

The structures reflect the deformation and recrystallisation the rocks have undergone during creation of fold belts (orogenesis). The banded or striped structure in the rocks may be the remains of an original layering but, in most cases, originates as a secondary phenomenon connected with the deformation. During metamorphism some components move by a form of diffusion and segregate into layers of different minerals.

The structural forms are very varied and depend upon where in the crust they developed. In the upper levels, where the temperatures are relatively low, simple open folds and crush zones occur along faults. In the deeper parts of the crust, where temperatures and pressures are higher, the rocks are deformed in a more plastic manner often resulting in complex fold patterns. Deformation zones where intense movements are concentrated result in sheared rocks with thinned, streaked out layering (ductile deformation).

OPEN FOLDS

Open folds in the basement often formed at a late stage in the deformation history. These folds frequently have steeply inclined symmetry planes (axial surfaces), and sometimes deform flat-lying earlier folds. Such refolding results in so-called interference patterns (see example on p. 39). Open folds can form both at moderate and deep levels in the crust.

PLASTIC FOLDS

Plastic folds in banded, light grey gneisses with pale veins. Folds of this type are often formed deep in the Earth's crust (10–40 km, 500–800°C), where the physical condition of the rocks is plastic and less brittle than higher up in the crust. The photograph shows a section with a height of about two metres.

BOUDINAGE / SAUSAGE STRUCTURE

A rock sequence composed of banded, grey metamorphic schists containing layers of a black, amphibolitic rock. Such amphibolitic bands are comparatively more rigid than the surrounding material and during deformation break up into a string of short, sausage-shaped pieces with the host rocks squeezed into gaps between them. This structure is called boudinage after the French word for a blood sausage. The hammer is about 40 cm long.

EYE STRUCTURES

Pale, grey gneisses with dark, amphibolite layers. The structure shown is a closed, near elliptical shape that is called an eye structure. This shape is a characteristic interference pattern, developed when an early set of folds is deformed by a second set with a different orientation. The hammer is about 40 cm long.

'GHOST' STRUCTURES

Migmatitic rocks with a heterogeneous nebulitic or ghost structure. These rocks are mixtures of two different components (see p. 34), formed by the break-up of an originally pale, banded gneissic rock by a network of diffuse veins and patches of granitic melt.

FLATTENING AND SHEARING

Very strongly flattened, finely banded gneisses with lenses of amphibolite, typical of zones with intense deformation. Some of the dark amphibolite layers have locally retained their original thickness, but as the photograph shows they can be stretched out into thin bands only 1/20th of the original. The rocks have been tightly folded both during and after the flattening event.

THE CRYSTALLINE BASEMENT

MAIN ROCK GROUPS

Rocks occur as aggregates of minerals with a grain size that spans from less than a millimetre up to several centimetres. Mineral grains are the smallest building block of the Earth's crust. The rocks are classified by their origin and their composition into the following main groups.

Rocks at the Earth's surface
- Sedimentary rocks
- Volcanic (extrusive) rocks

Rocks formed in the deeper parts of the crust
- Plutonic (magmatic) rocks
- Metamorphic rocks
- Migmatitic (partially melted) rocks

Examples of commonly occurring rock types from each group are shown on this page. The compositions of the rocks with their mineral components are described in the glossary (see p. 246-265).

SEDIMENTARY ROCKS

Siltstone — Fine-grained
Conglomerate — Coarse-grained
Sandstone — Medium-grained

Sediments are accumulations of mineral grains that are deposited by water or wind at the surface of the Earth. Examples are sands, gravels and conglomerates. Other types include rocks that are formed by chemical or biological precipitation in water such as rock salt and limestone.

VOLCANIC ROCKS

Vesicular basalt — Rock with holes from former gas bubbles
Ignimbrite — Rock formed from an incandescent ash cloud
Basalt — Solidified lava flow with flow structures

Extrusive volcanic rocks formed at the surface of the Earth by cooling of lavas (melted rock) or other eruptive products such as ash. Examples are basalt, pillow lava (magma erupted under water) and tuff (solidified ash).

PLUTONIC ROCKS

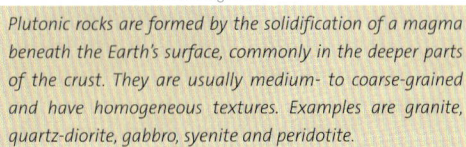

Granite — Reddish, medium-grained
Syenite — Pale, medium-grained 
Gabbro — Dark, medium-grained 

Plutonic rocks are formed by the solidification of a magma beneath the Earth's surface, commonly in the deeper parts of the crust. They are usually medium- to coarse-grained and have homogeneous textures. Examples are granite, quartz-diorite, gabbro, syenite and peridotite.

METAMORPHIC ROCKS

Gneiss — Grey, banded with pale veins 
Gneiss — Dark grey, folded
Augen gneiss — Gneiss with feldspar veins and 'eye' structures
Marble — White, weakly banded
Amphibolite — Grey-black, banded

Metamorphic rocks and crystalline schists are formed by the transformation of other rocks through the effects of pressure and temperature. These rocks form typically in connection with fold belt formation or heating associated with magmatic intrusions. Examples are gneiss, quartzite, mica schist, amphibolite and marble.

MIGMATITES (Mixed rocks)

Migmatitic gneiss
Migmatitic amphibolite

Migmatitic rocks contain two different components; an older one (the palaeosome) and a younger component (the neosome) that occurs as pale veins permeating the palaeosome. The neosome usually forms by partial melting of material in the rock itself or from rocks nearby. The melted material mainly comprises quartz and feldspar and has a composition close to that of granite. Migmatites are very widespread in the deeper parts of mountain belts where temperatures are sufficiently high to mobilise components and melt parts of the rock.

Photos: E. Schou Jensen, Geologisk Museum. & J. Lautrup, GEUS

FORMATION OF METAMORPHIC ROCKS

During the formation of fold belts parts of the crust are forced down to great depths where the rocks are subjected to greater pressures and higher temperatures than nearer the Earth's surface. Below the surface there is a gradual, downwards increase in pressure such that at a depth of about 60 km the pressure is typically 16–18 kBar (1 kBar = 1000 atmospheres) and the temperatures are 600–800°C.

Pressure and temperature at depth cause the minerals in the rocks to recrystallise and modify their composition, because the stability of a mineral depends upon the pressure and temperature conditions. The formation of metamorphic rocks can take place without melting, a process that is widespread in all orogenic belts. However, most rocks will melt if temperatures are high enough and there is access to hydrous fluids. Metamorphism (heat and pressure) can transform muddy sediments to mica schists, limestone to marble and basalt to amphibolite.

The crust is divided downwards into pressure and temperature zones at different depths, each of which has its own characteristic mineral assemblages. Geologists work with the following principle metamorphic facies:

Regional metamorphic facies	Pressure ~ Depth in km	c. temp. in °C
Greenschist	10–35	350–500
Amphibolite	12–40	450–750
Granulite	15–50	750–950
Eclogite	30–60	350–800

Metamorphic rocks form the majority of all rocks in fold belts. They include gneisses, different types of deformed granites, other recrystallised magmatic rocks and transformed sedimentary rocks (mica schists, marbles). Usually, the rocks are deformed at the same time as they are metamorphosed, which leads to the development of structures in the rocks in the form of layers, bands and folds.

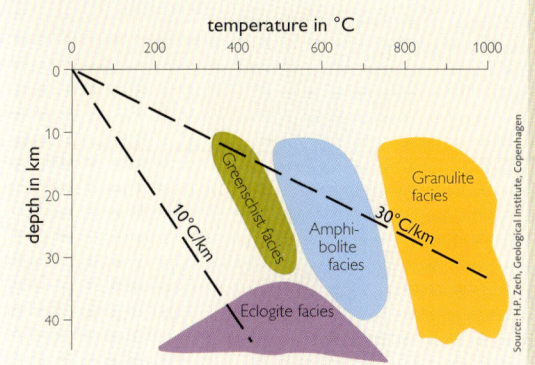

Diagram showing depth and temperature conditions for the various metamorphic facies regions. Black curves: geothermal gradient.

GREENSCHIST FACIES
Banded chlorite schist that is a transformed sedimentary or volcanic rock, recrystallised to chlorite, feldspar and quartz.

AMPHIBOLITE FACIES
Recrystallised, banded sediment composed of quartz and feldspar and with the minerals andalusite and staurolite.

GRANULITE FACIES
Folded gneiss with the minerals pyroxene, quartz, feldspar, garnet and sillimanite.

ECLOGITE FACIES
Green basic inclusion in gneiss with pyroxene (green) and reddish garnet. The inclusion is enveloped by grey-brown gneisses.

METAMORPHISM OF A BASIC MAGMATIC ROCK

Drawing of two alteration stages of the same doleritic material viewed under the microscope

Greenschist facies: Originally an intrusive dyke rock with yellow-green pyroxene and white striped plagioclase feldspar. Metamorphic recrystallisation begins with small, needle-shaped hornblende grains forming in the pyroxene, while the plagioclase remains unaltered.

Amphibolite facies: With greater heat and pressure the rock is completely transformed and the new minerals segregate into alternating bands. The original pyroxene grains recrystallise to hornblende (brownish colours) and the plagioclase is changed into an aggregate of smaller crystals.

THE CRYSTALLINE BASEMENT

Archaean – the first geological record of time

The Archaean basement

The Archaean is the name of the early geological period that includes the time of formation of the Earth's early crust 3800 to 2500 Ma ago. In Greenland this period is represented by rocks that belong to two of the oldest recognised periods of mountain belt formation 3800–3600 Ma ago and later 3000–2800 Ma ago. In the southern part of West Greenland and on the opposite side of the Inland Ice in South-East Greenland, Archaean rocks make up a large, coherent unit of stable basement, which is essentially unaffected by later geological events. Other parts of the old continental block were later involved in younger mountain belt formation, or were broken up and detached from the Greenland block.

Most of the Archaean basement comprises gneisses and granitic rocks that were formed in the later part of the period between 3100 and 2600 Ma ago. Only in the Nuuk region is there an older group of gneisses, called the Amîtsoq gneisses, made from a series of deformed and transformed granite-like intrusions, that were emplaced between 3800 and 3600 Ma ago. In the same region there also occurs an important group of rocks that comprise the Earth's oldest examples of well-preserved metamorphosed volcanic and sedimentary units.

The Archaean gneiss complex is cut by many dark, linear intrusions (dykes) that were formed from basaltic material. These dykes occur in swarms that were emplaced between 2200 and 2000 Ma ago and, because they are undeformed, it means that this part of the Archaean basement was not reworked by the 1850 Ma old mountain building (orogenic) events that affected large parts of the Greenland basement.

Geology of the Nuuk region

The Nuuk region comprises three different geological complexes, called terranes (see box p. 39), each of which has its own characteristic developmental history. The boundaries between the three terranes occur as zones of intense deformation, similar to those seen where continental blocks are brought together by plate tectonic displacements. Age dating shows that these movements occurred about 2720 Ma ago.

The oldest rock complexes occur within the Akulleq terrane that forms a 30–60 km wide zone running through Godthåbsfjord. Transformed sedimentary and volcanic rocks are found as layers within the surrounding slightly younger, but still very old Amîtsoq gneisses. The Earth's oldest substantial series of supracrustal rocks, the Isua supracrustal belt, comprises a sequence of metamorphosed and deformed volcanic and sedimentary rocks that were deposited about 3800–3700 Ma ago. The unit forms an up to 2–4 km wide belt that can be followed in an arc-shape for about 35 km at Isukasia, in the inner part of Godthåbsfjord. In places the units are so little

Geological sketch map of the Nuuk region. Different types of gneisses (yellow-orange and olive-green colours) dominate the area with scattered intrusive granitic bodies (red colours). The gneisses contain layers of different earlier supracrustal formations that are now mostly amphibolites (green colours). The different terranes are, for the most part, made up of the same range of rock types making it difficult to recognise the composite nature of the area. Map area outlined by red rectangle is shown on page 37.

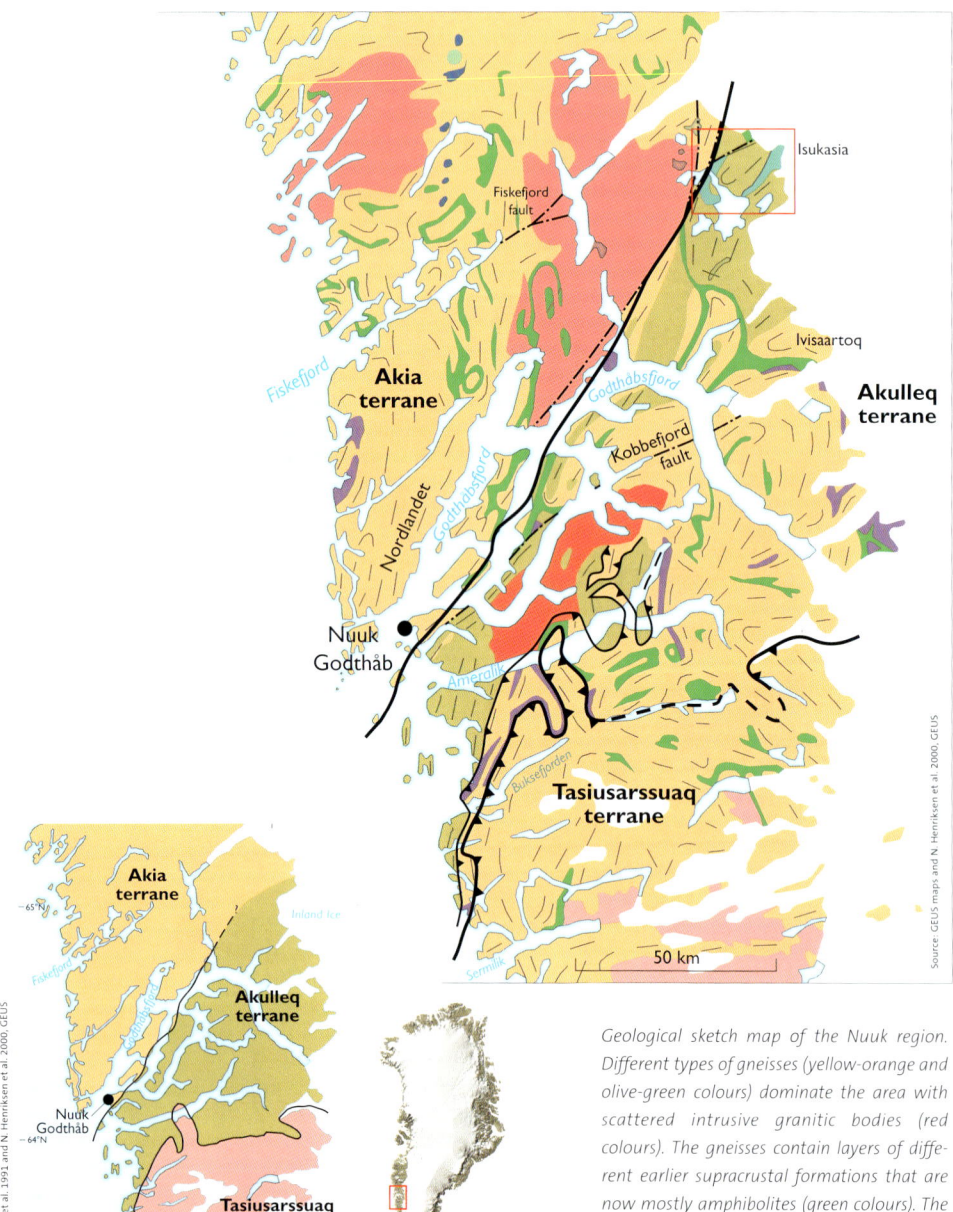

The Nuuk region was assembled by plate tectonic processes from three different continental blocks – so-called terranes.

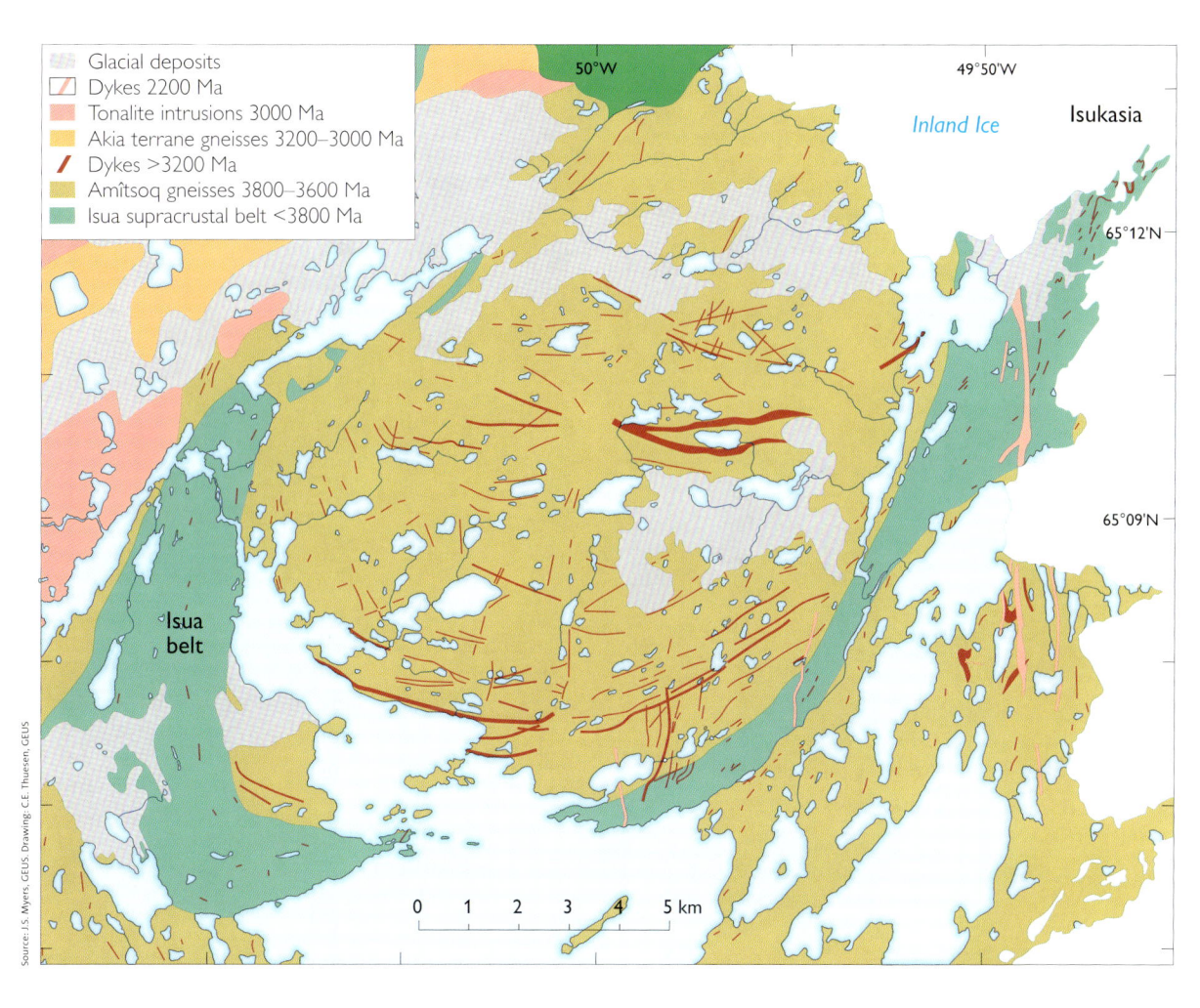

Geological sketch map of the Isukasia area in the inner part of Godthåbsfjord. The map shows the occurrence of the Earth's oldest, well-preserved sedimentary and volcanic rocks (green colour). These rocks formed part of the crust 3800–3700 Ma ago. The supracrustal rocks are surrounded by gneissic and granitic rocks, shown in olive-green, orange and reddish colours. The old interlayered volcanic and sedimentary rocks occur in a 2–4 km wide belt, which because of later deformation is folded together with the gneisses to form an arc-shaped structure in the basement.

deformed that their original structures can be recognised. Locally, the rocks contain carbon particles, which isotopic studies have shown were originally organic matter that very likely was derived from primitive planktonic organisms deposited in the layered muddy sediments at the bottom of the sea. These findings have great significance for the understanding and interpretation of the Earth's earliest geological evolution and the appearance of life (see box page 38). The Isukasia area is thus a study area of great international geological importance.

The Amîtsoq gneisses were originally formed between 3850–3600 Ma ago as a complex of different granite-like intrusions which were later cut by basic dykes. Subsequently, this complex was deformed and transformed by metamorphism and migmatisation, and the dense swarm of basic dykes is now in most places totally transformed and broken into lens-shaped inclusions within the gneisses.

The Late Archaean part of the basement

Outside the Nuuk region, the Archaean gneiss complex is mainly composed of deformed and transformed, originally deep-seated (plutonic) intrusive rocks. Only 10–20% of the outcrop comprises scattered supracrustal rocks including sedimentary and volcanic types. The rocks occur now as gneisses and granitic bodies often with interlayered metamorphosed schists, which can be followed as layers within the gneisses.

CHRONOLOGY OF THE NUUK REGION

Approx. age in Ma	Akia terrane	Akulleq terrane	Tasiusarssuaq terrane
2200	----------	basic dykes	----------
2550	granite sheets	Qôrqut granite	granite sheets
2720	terrane assembly and granite sheets in all terranes		
2810			Ilivertalik granite
2825		Ikkattoq gneisses	
2835		Younger sedimentary and volcanic rocks	
2840–2860			Regional gneisses
2850			Anorthosite intrusions
2920–2940			Early gneisses
2970	Late granites		
3000–3020	Regional gneisses		
3200	Dioritic gneisses		
3460–3510		Ameralik dykes	
3600–3850		Amîtsoq gneisses	
3700–3800		Isua sedimentary and volcanic rocks	
3800–3900		Earliest supracrustal rocks and gneisses	

THE CRYSTALLINE BASEMENT

3800-3600 Ma old Amîtsoq gneisses from the northerly part of Godthåbsfjord. The complex is constructed of granitic intrusions and cut through by dark dykes that were later deformed and transformed. The dykes are broken into pieces (boudinaged) and can be seen now as dark blocks within the gneisses. Hammer shaft is 45 cm long.

In the Fiskenæsset area, a part of the Tasiusarssuaq terrane south of Nuuk, a rock complex occurs dominated by the feldspar mineral plagioclase. This so-called anorthosite complex is the remains of an originally large, flat-lying intrusion that was emplaced into the crust about 2850 Ma ago. The anorthosite complex, which comprises up to 5% of the rocks in the area, now occurs as a series of layers that can be followed throughout the surrounding gneisses and granites. The layers are complexly folded and show that the rocks have suffered a minimum of three phases of Late Archaean folding.

The gneisses were originally various types of granite-like rocks that were intruded deep in the

THE OLDEST TRACES OF LIFE IN THE WORLD

Photograph of layered schists with well-preserved sedimentary layering of light-, and dark-coloured mud layers that contain very small, 1/200 mm sized, carbon particles. Isotopic analyses show that the carbon has an organic origin. This carbon is interpreted as derived from algae-like, planktonic organisms that sank to the sea bottom and were incorporated into the sediment. These analytical results bear witness to the existence of life for more than 3700 Ma.

North-east of the inner part of Godthåbsfjord, at Isukasia, an about 3 km wide arc-shaped belt of moderately to strongly deformed sedimentary and volcanic rocks occurs that is part of the oldest documented deposits on Earth. The Isua supracrustal belt is divided into two parts, formed about 3800 and 3700 Ma ago. These sequences contain different sedimentary deposits, originally muds, silts, sands and conglomerates, together with volcanic rocks. Some of the volcanic rocks preserve pillow structures, indicating that they were erupted in water. The composition and structure of the rocks show that they formed on a submarine shelf and demonstrate the existence of land and sea areas at this early time in the Earth's development. Some of the sediments were products of erosion on nearby land areas. These sediments include different fine-grained shale layers supposedly deposited from mudslides. Some of these rocks contain fine carbon particles that have been shown by isotopic investigations to be of organic derivation. The organic material probably originated from algae-like organisms in the seawater that sank to the bottom and became incorporated with the sediments on the seabed. The Isua volcanic and sedimentary rocks are folded and usually so transformed that their original character is lost. However, the finds of traces of Earth's earliest life have made the Isukasia area and Isua supracrustal belt, the target of a series of international research projects that have resulted in many detailed publications on the earliest phases of the Earth's geological development.

Gabbro anorthosite: A special intrusive plutonic rock with large light crystals of plagioclase and between which are domains dominated by pyroxene and hornblende. When the rock is almost exclusively made of plagioclase it is called anorthosite. Hammer shaft is 30 cm.

The Archaean basement

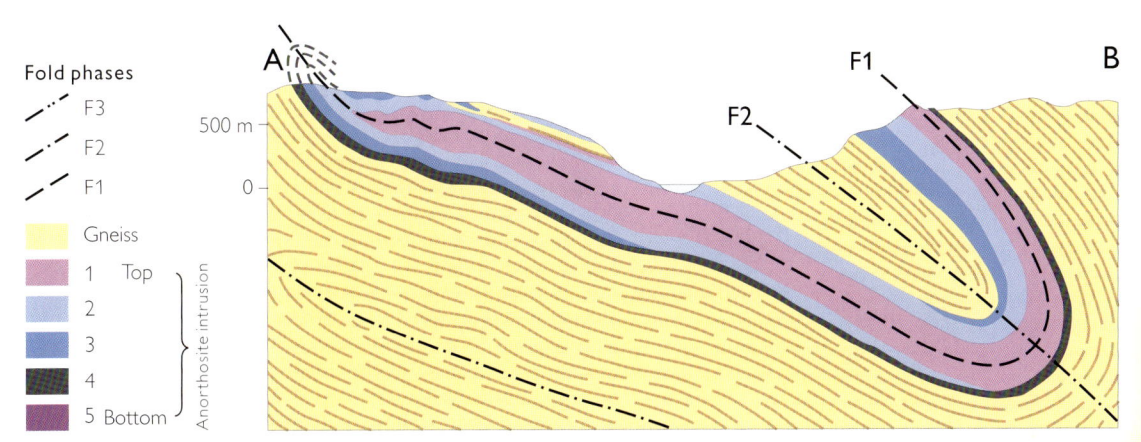

Geological sketch map and profile (A–B) showing how the Fiskenæsset intrusive layered anorthosite complex occurs as a triply folded structure within the surrounding Archaean gneisses. The original intrusion was layered (layers numbered 1 to 5) with denser, darker rocks at the bottom of the magma chamber and pale-coloured, less dense rocks, including anorthosite at the top. The layered series is repeated about the axial plane of the earliest fold phase (F1), as seen clearly in the cross-section. The first phase structures are then refolded by later, more open structures (F2). The axial surfaces of the third fold phase (F3) are parallel with the cross-section and therefore do not appear in the profile, but are shown on the map. This type of complex fold pattern, produced by successive fold phases, is typical of basement gneisses and the deeper parts of fold belts.

crust about 2800 Ma ago. Most were emplaced as flat-lying, sheet-like magma bodies into the pre-existing rock units that included the anorthosite complex in the Fiskenæsset region as well as metamorphosed supracrustal rocks originally laid down as sediments at the Earth's surface. The newly formed granite-like rocks were considerably more voluminous than the rocks they intruded, and dominate the present day outcrop in this part of the gneissic basement. In many places, high temperatures associated with the regional metamorphism have transformed the rocks into migmatites. These formed where granitic melt has permeated the host rock and is preserved in a fine network of veins and diffuse segregations.

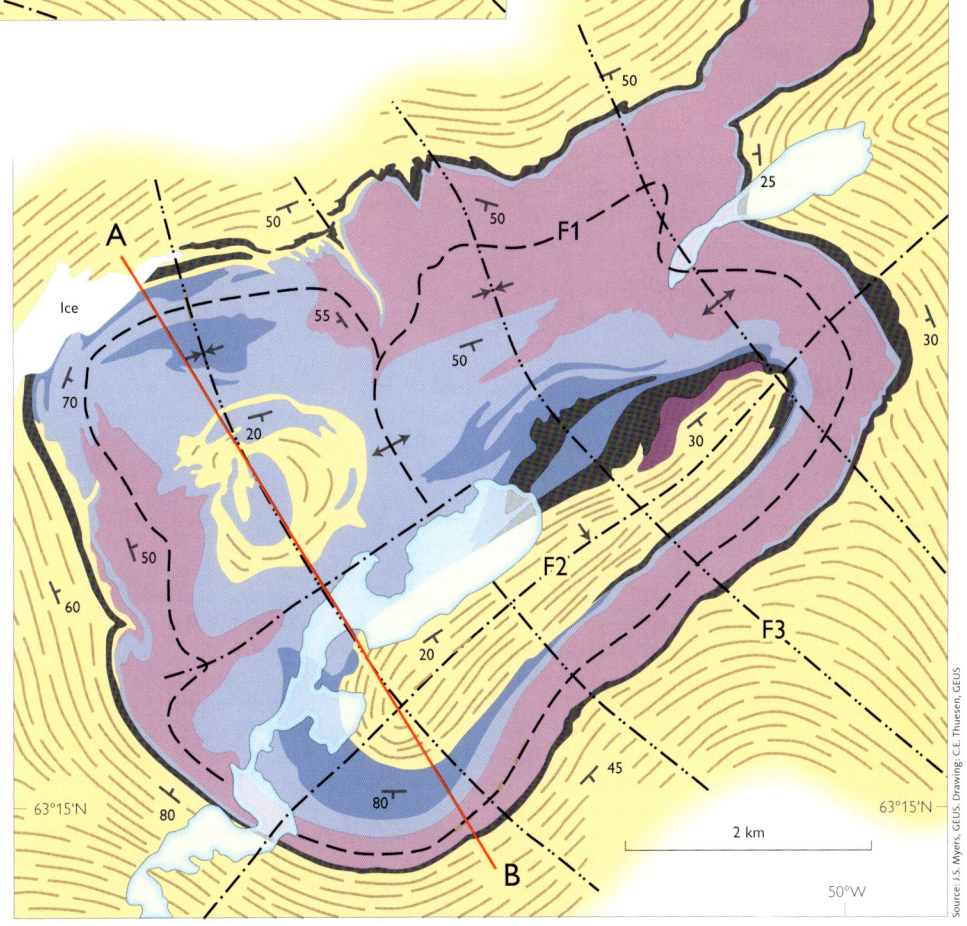

DETACHED FRAGMENTS OF OLD CONTINENTS – geological terranes

As a consequence of plate tectonic processes, parts of larger continental areas can become detached from their original coherent continental block and drift as independent crustal segments. Each of these may move independently many thousands of kilometres across the Earth's surface, until they collide and amalgamate with a continental block or another terrane, by involvement with an active orogenic belt and an associated subduction zone. Geological terranes are often found within younger mountain belts as distinct, foreign elements. They characteristically have clear boundaries against the surrounding rocks and possess their own separate stratigraphy and structural geological history. In some Phanerozoic fold belts, palaeomagnetic investigations have demonstrated that such terranes were sometimes formed many thousands of kilometres from the place of origin of the adjoining rocks. Figuratively speaking, terranes can be likened to 'ships' of crustal material that sail across an ocean until they 'dock' against another continental block.

In the beginning of the 1970s geologists recognised such foreign terrane elements within younger fold belts – principally in the western parts of the USA and Canada. Based on these discoveries, geologists began to look for the possible presence of terranes in older fold belts, such as the crystalline basement regions of Greenland. In the Nuuk region, three different terranes have now been recognised in the Archaean part of the basement. Each terrane has its own assemblage of rocks and a distinct structural and metamorphic history. The three terranes may originally have formed in different places on the Earth and then, in the late Archaean, were brought together and welded into a coherent part of the Archaean basement that now forms a part of the Greenland shield.

THE CRYSTALLINE BASEMENT

Enlargement and stabilisation of the basement

The younger part of the basement

The geological period that followed the Archaean is the Proterozoic (the name derives from a Greek word and means a period with faint traces of life, in contrast to the Archaean). It represents a very long time span that is subdivided into three main parts, namely the oldest part – Palaeoproterozoic (2500–1600 Ma), the Mesoproterozoic (1600–1000 Ma), and the youngest part, Neoproterozoic (1000–542 Ma). The greater part of the basement in Greenland reached its present shape as a consequence of formation of the Palaeoproterozoic fold belts.

The youngest rock-forming processes in the basement ceased at the end of the Palaeoproterozoic, after which the Precambrian basement shield stabilised. The later geological developments took place around the margins of the shield, partly in the form of younger fold belts and in part related to the development of sedimentary basins and volcanic provinces.

In Greenland, the principal part of the basement comprises Palaeoproterozoic gneiss complexes that commonly include material derived from earlier Archaean units as well as newly-formed granitic crustal material. The only exceptions to this are the Ketilidian in South Greenland and the Inglefield Land area in North Greenland that are dominated by newly-formed rocks derived from the Earth's mantle during the latest orogenic processes. Another large region with newly-formed crustal material occurs in the northern part of the Caledonian fold belt in North-East Greenland. However, most of this region is located east of the basement shield and is included in large thrust sheets that were pushed over the shield from even further east.

ROCKS IN THE YOUNGER BASEMENT

1. Atâ granite

An Archaean intrusive complex (about 2800 Ma old) that has largely escaped modification by later Proterozoic deformation events. The intrusive complex comprises a variety of different granitic rock types emplaced as magma injections where younger phases cut through earlier, solidified phases.

2. Metamorphosed Archaean sediments

A banded series of metamorphic rocks, originally a sequence of Archaean, volcanic and sedimentary rocks that were deformed during the formation of a Proterozoic fold belt. The banded sequence is cut by a later undeformed black dyke that contains abundant white feldspar crystals.

3. Proterozoic transformed older gneisses

Pale, banded orthogneisses, originally homogeneous granite-like rocks that developed banding during reworking by later deformation. Pale, migmatitic veins invade the gneisses.

4. Proterozoic metamorphosed sediments

Proterozoic marine sediments originally deposited on an Archaean basement. The metamorphosed siltstones and sandstones are part of a succession more than 2500 m thick.

Four examples from localities shown on the index map on the page opposite.

Proterozoic
- Sandstone and marble

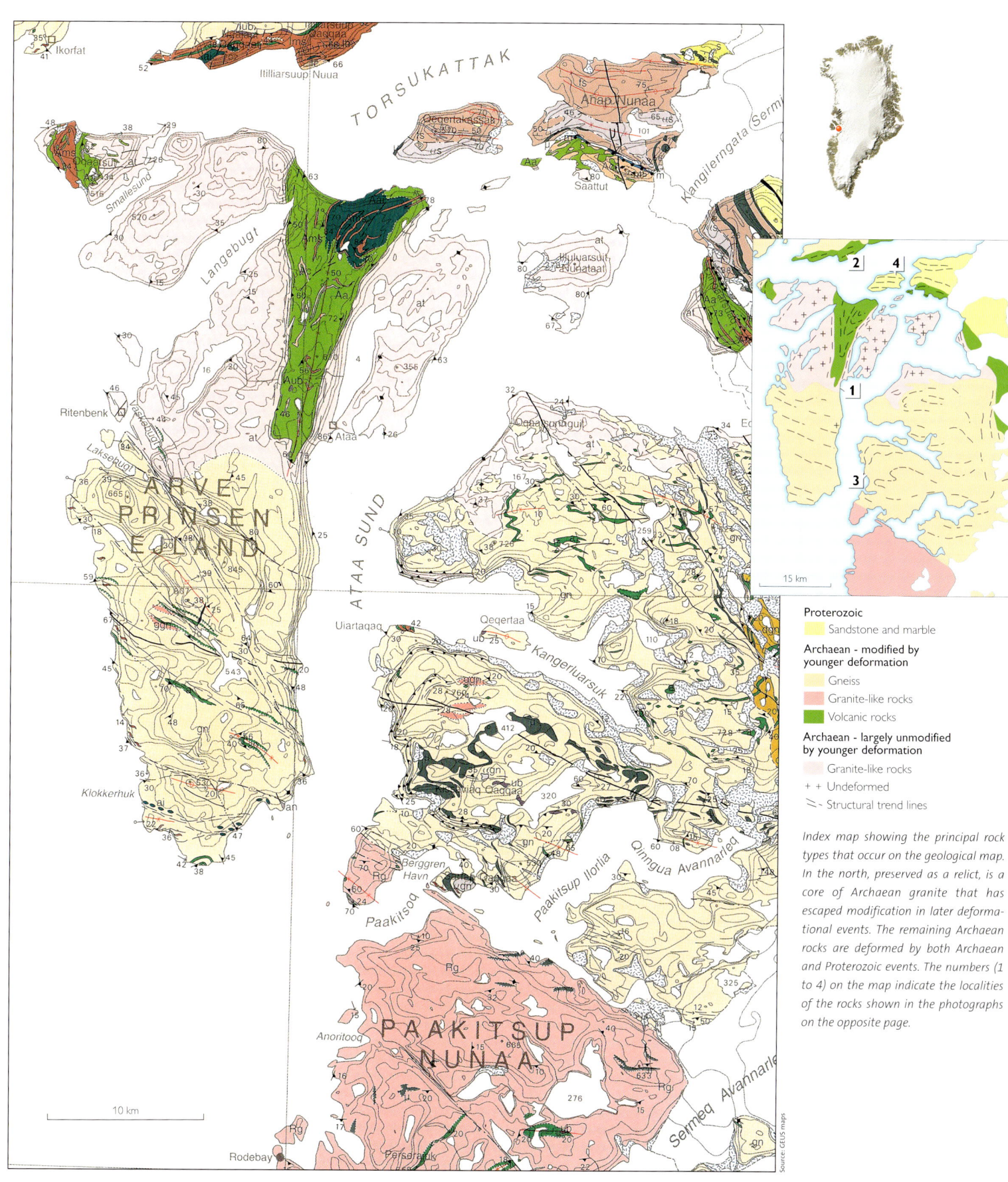

Archaean – modified by younger deformation
- Gneiss
- Granite-like rocks
- Volcanic rocks

Archaean – largely unmodified by younger deformation
- Granite-like rocks
- ++ Undeformed
- ~ Structural trend lines

Index map showing the principal rock types that occur on the geological map. In the north, preserved as a relict, is a core of Archaean granite that has escaped modification in later deformational events. The remaining Archaean rocks are deformed by both Archaean and Proterozoic events. The numbers (1 to 4) on the map indicate the localities of the rocks shown in the photographs on the opposite page.

Part of a geological map of the basement area on the eastern side of Disko Bugt, north of Ilulissat Isfjord. The map covers an area composed of both Archaean and Proterozoic rocks that were inter-folded during Palaeoproterozoic fold belt formation. Pale yellow colours show gneisses and the reddish colours are various granite-like rocks. Green and yellow colours represent respectively metamorphosed volcanic and sedimentary rocks.

THE CRYSTALLINE BASEMENT

Collision between rifted Archaean continents

The Nagssugtoqidian in southern West Greenland

The Nagssugtoqidian is an approximately 1900–1800 Ma old fold belt that formed in the collision zone between two parts of a previously broken up (rifted), large Archaean continent. In West Greenland it can be followed from its southern boundary near Kangerlussuaq (Søndre Strømfjord) to Disko Bugt, and continues farther north into the contemporary Rinkian area. To the east it disappears underneath the Inland Ice, but reappears in the Ammassalik area in South-East Greenland.

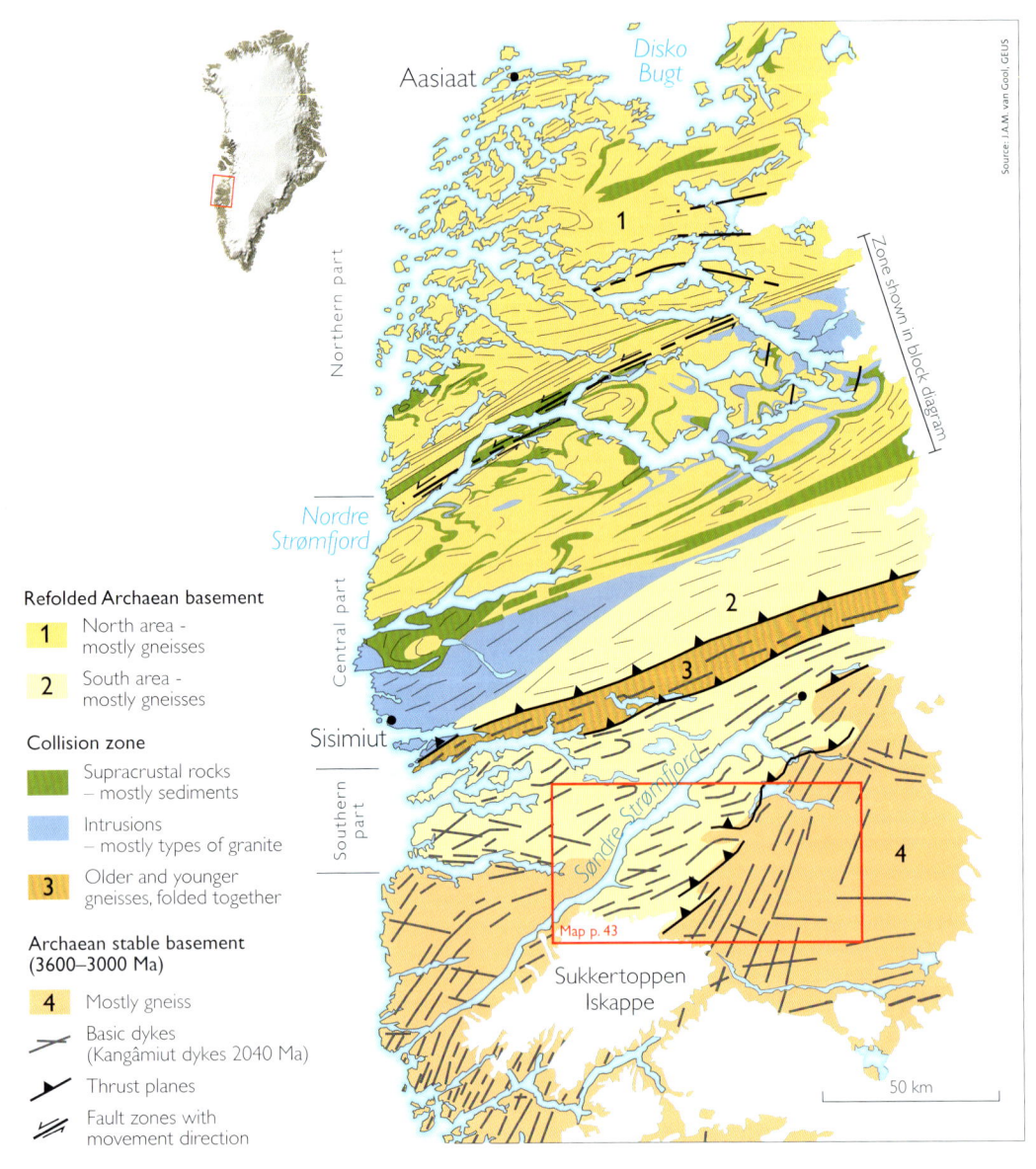

The Palaeoproterozoic Nagssugtoqidian fold belt is named after the fjord Nagssugtoq (Nordre Strømfjord) and it has its southern boundary with the remnant core of the Archaean basement. It extends northwards until it passes into the Rinkian area (see p. 46). It is mainly composed of modified and refolded basement rocks from an old Archaean continent. The rocks are dominantly gneisses and granites, with remnants of metamorphosed interlayered Archaean sedimentary and volcanic rocks. During the collisional events that produced the fold belt, a series of younger sediments were deposited on the old shield and different granitoid rocks were intruded. The fold belt thus has a complicated construction with slices of Archaean and Palaeoproterozoic rocks intensely

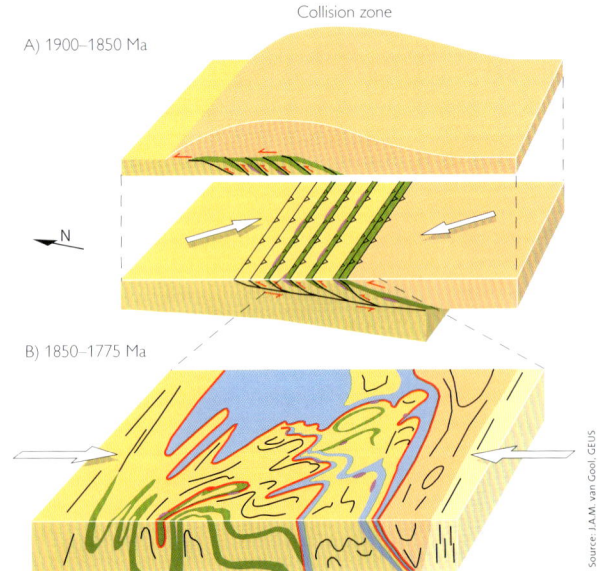

Refolded Archaean basement
1. North area – mostly gneisses
2. South area – mostly gneisses

Collision zone
- Supracrustal rocks – mostly sediments
- Intrusions – mostly types of granite
3. Older and younger gneisses, folded together

Archaean stable basement (3600–3000 Ma)
4. Mostly gneiss
- Basic dykes (Kangâmiut dykes 2040 Ma)
- Thrust planes
- Fault zones with movement direction

Geological map of the Nagssugtoqidian fold belt between Kangerlussuaq (Søndre Strømfjord) and Disko Bugt, also showing the boundary region with the stable Archaean basement in the south. The development of the fold belt is interpreted to be the result of a Palaeoproterozoic plate tectonic cycle involving continental rifting, ocean-floor formation, island-arc volcanism and sedimentation and, finally, continental collision with interfolding of the rocks in the collision zone (see figure p. 44).

Block diagram showing a two-stage development of the orogenic belt. (A) The early interleaving of northerly and southerly blocks along a wide dislocation zone in the crust. A series of thin, similarly constructed, slices are separated by thrusts (flat-lying movement planes) to form a stack. Each slice contains interfolded older gneisses, younger sedimentary rocks and granitic intrusions. (B) The following phase refolds the earlier thrust planes producing a fold pattern related to N–S-oriented compression. The green and blue colours represent Nagssugtoqidian sedimentary rocks and intrusions, with the yellow and orange colours depicting basement rock of, respectively, the former northern and southern continental blocks. Arrows indicate compression directions.

folded together, such that it is often only possible to distinguish them by radiometric age dating.

The geological map on p. 42 shows how the fold belt of West Greenland is divided into southern, central and northern parts, where each has its own development and structure. The newly formed Palaeoproterozoic rocks, including sedimentary rocks and large granite-like intrusions, occur in the central part where, as shown in the block diagram, thrusts and later deformation have led to intense interfolding with Archaean rocks. The rocks and structures in the central zone are interpreted as the remains of a collision zone between two continents. The ocean that once separated the continents has now totally disappeared because it descended beneath the continental crust into the lithospheric mantle by subduction (see p. 44).

Cliff (about 150 m high) in Nordre Strømfjord showing a flat-lying sequence of Nagssugtoqidian meta-sedimentary schists and intrusive granites overlying much older Archaean gneisses. The younger rocks (about 2000–1900 Ma old) are a coherent package that has been displaced over the underlying approximately 2800 Ma old gneisses along a movement (thrust) plane. This has caused rocks of very different ages to have been brought into contact with each other.

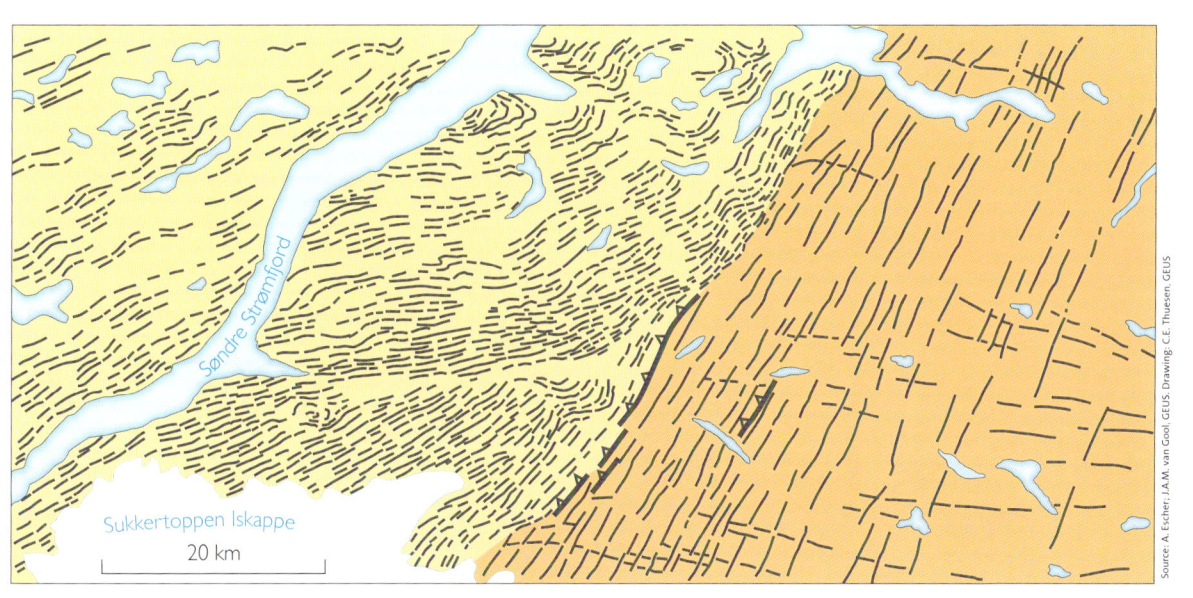

The boundary between the stable Archaean block in the south-east (orange) and Archaean gneisses deformed by the Nagssugtoqidian orogeny in the north-west (yellow). Both regions contain swarms of linear basic intrusions, the approximately 2040 Ma old Kangâmiut dykes (black lines). The map depicts clearly how the approximately 1850 Ma old Nagssugtoqidian event deforms and breaks up the formerly linear dykes, so that both the dykes and their host rocks develop new, modified orientations.

THE CRYSTALLINE BASEMENT

The southern boundary zone

The boundary between the unmodified Archaean basement in the south and the Nagssugtoqidian fold belt to the north occurs as an approximately 40 km broad transition zone where the Archaean rocks, from south to north, show an increasing intensity of the younger deformation. The deformation can be seen very clearly through its influence on a dense swarm of intrusive basic dykes, the Kangâmiut dykes, that were emplaced into the Archaean basement about 2040 Ma ago. Farthest to the south the dykes are completely unaltered and their original doleritic (dolerite: a gabbro-like rock) composition can be seen. However, towards the north the dykes become progressively changed into amphibolites by metamorphic processes, and at the same time are broken up (boudinaged) so that they end up as a line of lens-shaped pods within the gneiss.

The plate tectonic model

The Nagssugtoqidian fold belt is interpreted to have formed through a series of typical plate tectonic developmental phases. The earliest phase was the rifting of an Archaean continent about 2040 Ma ago

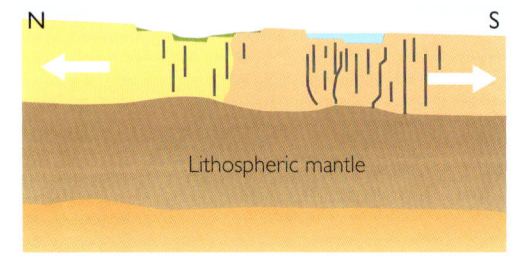

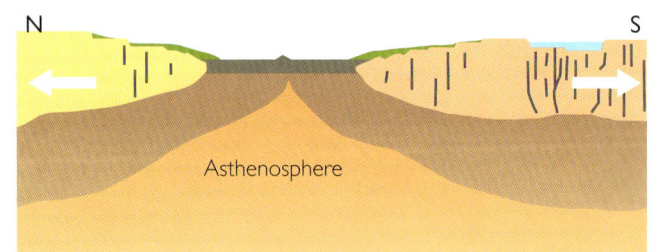

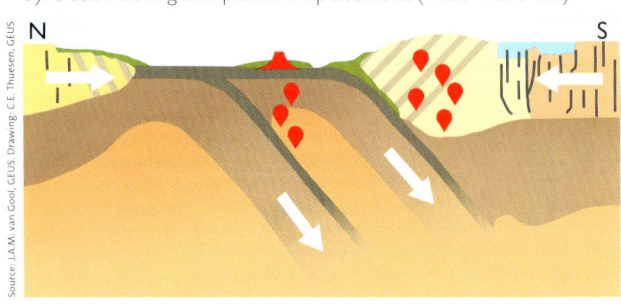

Remains of Kangâmiut basic dykes (dark) and host Archaean gneisses (light colour) in an area on the south side of Kangerlussuaq (Søndre Strømfjord) where they are reworked by Nagssugtoqidian deformation. Many of the dykes have been broken into a series of lenticular fragments. This type of structure is called boudinage (see box p. 33), and is characteristic of the deformation of layered sequences with rocks of different stiffness (competence); in this case the dykes are more rigid, and the gneisses less rigid and more easily deformed. In this example there has been compression towards the plane of the dykes and an extension at right angles along them. The dykes in the centre of the photo are about 30 m across.

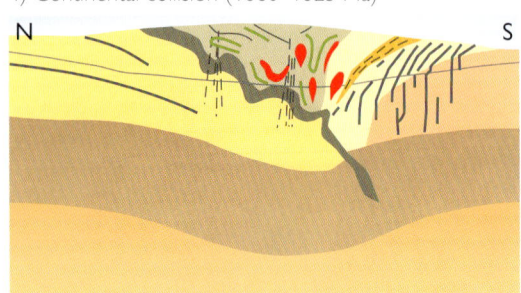

Schematic model for the formation of the Nagssugtoqidian fold belt between Kangerlussuaq (Søndre Strømfjord) in the south and Disko Bugt in the north. The four stages shown are:
1) *Initial rifting of an old Archaean continent and intrusion of a swarm of basic dykes (Kangâmiut dykes).*
2) *The two separated continental blocks drift away from each other and new oceanic crust forms between them.*
3) *The continents move towards each other and the ocean floor (dark olive-green) on one side is displaced downwards beneath the encroaching continent, a process called subduction. At the same time island-arc volcanism occurs over the subduction zone.*
4) *Collision between the two continents, during which intense deformation and interleaving of old (Archaean) and younger (Palaeoproterozoic) rocks occur. The effects of the Nagssugtoqidian deformation and metamorphism extend for a considerable distance away from the collision zone into the adjoining Archaean basement. The rocks in the northern and southern Nagssugtoqidian area are shown with yellow colours whilst the stable Archaean block is shown in orange. The collision zone itself is shown in light brown tones and the supracrustal rocks in green. Intrusive rocks are red. Ocean-floor rocks are shown as grey-brown and the Kangâmiut dyke swarm is depicted by broken black lines.*

and the emplacement of basaltic material in a dense dyke swarm – the so-called Kangâmiut dykes. A new ocean developed in the rift between the two continental blocks, and different, thick sedimentary sequences were deposited in the coastal regions about 2000–1920 Ma ago. When ocean-floor spreading stopped, a new phase started in which the continents began to move back towards each other, and the oceanic crust in between was lost through subduction. This phase included, about 1920–1870 Ma ago, the development of a volcanic island-arc system with formation of granite-like magmas at depth. The following collision phase led to the interfolding of the remains of the old Archaean continent with the newly deposited sediments and the island-arc rocks, and simultaneously (about 1860–1825 Ma ago) the whole fold belt was subjected to intense metamorphism under high temperatures and pressures. The folding and later deformation continued until the final phase of development, which ended about 1775 Ma ago.

This plate tectonic interpretation of the region is based on detailed geological mapping of the Nordre Strømfjord area.

AIRBORNE MAGNETIC SURVEYING

A twin-engined aircraft (here a Britten-Norman Islander) equipped for aeromagnetic surveying. The aircraft is fitted with a magnetic sensor that is mounted in the long boom attached to the tail. A set of measuring and recording instruments is installed in the cabin.

Many rocks contain variable amounts of magnetic minerals, such that different rock types possess characteristic magnetic properties. The magnetic field strength of rocks can be measured with sensitive instruments (magnetometer) installed in an aircraft that typically is flown over the project area in a systematic grid pattern with an altitude of some hundred meters above the mountains. The variations in the rock magnetic field strength are recorded, together with the varying background strength of the Earth's magnetism, and information on altitude, the absolute position (latitude and longitude) and the time together with different instrument data. All the data collected are combined with a topographic map of the project area to produce a so-called magnetic anomaly map (see example). This map reflects the distribution of the local magnetic intensity from the different rocks, and often shows a close resemblance to the structural pattern of a geological map of the same area. However, the measured magnetic intensity may also reflect strongly magnetic rocks at depths of 5–10 km in the crust. Anomaly patterns can be interpreted using computer modelling.

The anomaly map shown covers the southern part of the Nagssugtoqidian fold belt, centred on Kangerlussuaq (Søndre Strømfjord); the coastline is marked with a white line. The colours reflect falling magnetic intensity from violet through reds and yellows to greens and blues. The map extract shows clearly the structural contrast across the boundary between the ENE-striking rocks in the Nagssugtoqidian fold belt and the slightly irregular, north-west trends of the Archaean basement south-east of the dashed black line. Compare this with the figure showing the distribution of the Kangâmiut dykes (p. 43).

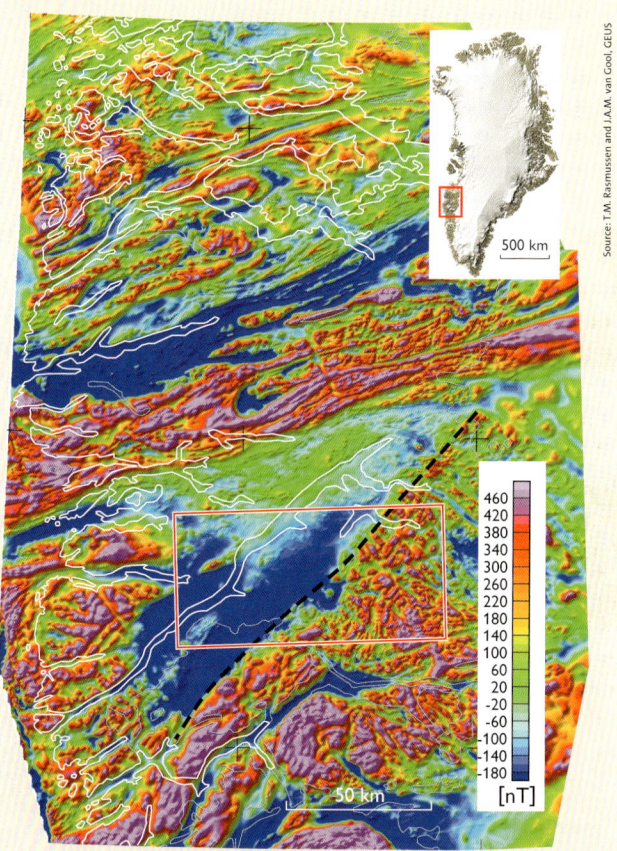

Aeromagnetic map that shows the variation of magnetic intensity over a large region as measured from an aircraft that has overflown the area in a series of closely spaced, parallel tracks. In the aircraft a sensitive magnetometer and various recording instruments are installed that accumulate magnetic data during the flight. After digital data reduction a magnetic anomaly map is made that shows how the magnetic field varies throughout the area. Red colours indicate high intensities, yellow colours intermediate values and greens and blues show the lowest intensities. The map depicts the area centred on Søndre Strømfjord (Kangerlussuaq), the box indicating the same area as shown on p. 43. The boundary between the Archaean basement in the south and the Nagssugtoqidian belt is seen as a marked change in the intensity (colours) and in the structural pattern of the anomalies.

THE CRYSTALLINE BASEMENT

Younger sedimentary rocks and older gneisses interfolded in a new orogeny

Fold belts in central and northern West Greenland

The Palaeoproterozoic fold belts in the central and northern parts of West Greenland formed at about the same time approximately 2000–1800 Ma ago. These include the Rinkian area between Disko Bugt and Qaanaaq/ Thule and the Inglefield Land area, north of Thule. The Rinkian area is similar to the Nagssugtoqidian as it comprises interfolded Archaean gneisses and younger overlying sedimentary rocks. The Inglefield Land fold belt is formed only of Palaeoproterozoic rocks. The relationships between the two fold belts are partly unknown, because their boundary zone in the Thule region is covered by the Inland Ice.

The Rinkian area

The basement between Ilulissat Isfjord (Jakobshavn Isfjord) in Disko Bugt and Thule is designated the 'Rinkian area'. This is the root zone of a fold belt formed about 1850 Ma ago that comprises slices of Archaean basement rocks interfolded with an overlying thick sequence of sedimentary rocks dominated by marine sandstones and mudstones that, during orogenesis, were transformed into psammites and schists. In the central part of the region different Proterozoic granitic intrusions occur which form large plutonic bodies (batholiths) that may reach more than 100 km across.

Archaean pillow lava from a greenstone belt in the basement north of Disko Bugt. The pillow structure forms when basaltic lava erupts under water causing rapid cooling of the magma that draws it into pillow-shaped lumps. Each of the pillows has a distinct, fine-grained, chilled margin against the outside, and the space between the pillows quickly fills with loose fragments or sedimentary deposits (in the photograph the dark areas are the inter-pillow space). In greenschist belts the rocks are completely metamorphically transformed and recrystallised, but some of the rocks preserve primary structures, e.g. pillow structures. The photograph shows an ice-scoured and polished surface with distinct glacial striae. The hammer shaft is about 45 cm long.

Newly formed pillow lava.

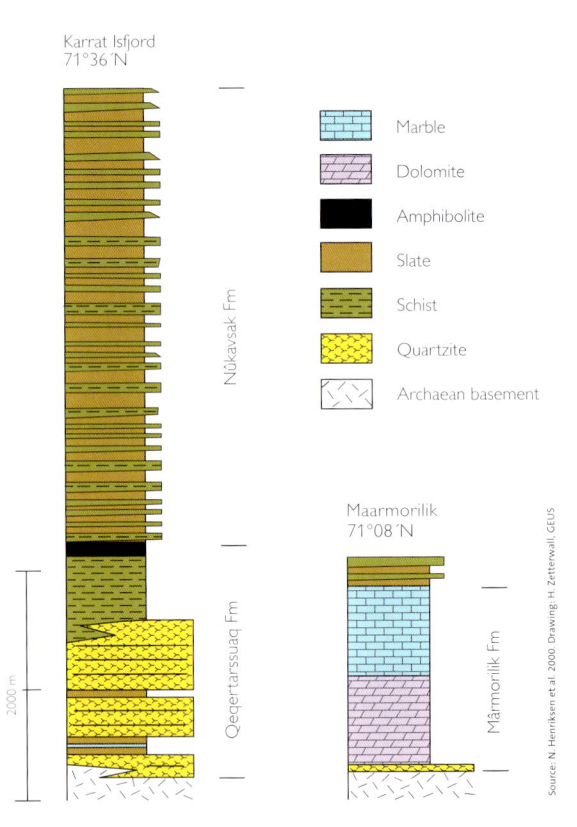

Stratigraphic column showing the composition of the approximately 8.5 km thick Karrat Group of sedimentary rocks deposited between 2000 and 1900 Ma. They were deposited on an Archaean gneissic basement with which they were interfolded about 1850 Ma ago. The sedimentary sequence is divided into named formations.

The Archaean basement is locally very little modified by the Proterozoic deformation, and especially north-east of Disko Bugt there is an area with well-preserved Archaean textures and structures. Here there are old volcanic units comprising a so-called greenstone belt that is cut by approximately 2800 Ma old granite intrusions. After the stabilisation of these rocks as part of an older fold belt, it was followed several hundred million years later by the deposition of a different, thinner sedimentary sequence on the basement. The whole complex was then deformed by a younger folding event about 1850 Ma ago (see map p. 41).

The main occurrence of Palaeoproterozoic sedimentary rocks consists of the Karrat Group, a sequence with a thickness of about 8.5 km, that is widespread east of Svartenhuk Halvø and north of Upernavik. The sequence includes three units with marble lowest, sandstone in the middle and uppermost a finely layered series of alternating sandstones and mudstones, so-called turbidites. The latter type reflects formation through supply of turbulent pulses of sediment where, as the current wanes, the sediment is sorted with the coarse sands separating from the finer muds. Each pulse forms an up to one metre thick bed and the whole series comprises thousands

Large recumbent fold in the basement in the Rinkian area, north of Uummannaq. The structures deform both the approximately 3000–2800 Ma old Archaean gneisses and the overlying approximately 2000 Ma old transformed Palaeoproterozoic sedimentary rocks. These two rock complexes were intensely interfolded and have been welded together so that they now behave as a coherent, stable basement block. The structures in the area are spectacularly displayed in the steep mountainsides along the fjords. The photograph shows a folded sequence with a marked, dark grey, amphibolite band that clearly defines the outline of the recumbent fold. In the core of the fold there is a thin, white, marble layer and outside the amphibolite the rocks pass into old, grey gneisses. Height of the cliff is about 800 m.

An approximately 1 m high section of the several kilometres thick Karrat Group, showing metamorphosed sand and mud sediments. The sequence is principally made up of sands and muds deposited in deep water from pulses of sedimentary material as turbidity currents. Each comprises a mixture of mud and sand in water such that the denser mass flows down the continental slope out over the seabed until the current stops and all the sediment is deposited. It is characteristic that the material in a single layer is sorted according to grain sizes, with the coarser sand grains at the bottom and the finer muds at the top. On the photograph three approximately 30–40 cm thick layers are shown where the pale-coloured portion of each bed equates with the finer-grained part of the sediment.

THE CRYSTALLINE BASEMENT

The basement in Inglefield Land comprises mainly gneissic rocks that have a dark, brownish colour and are comparatively strongly weathered and crumbly. The rock surfaces are often covered with dark lichen obscuring the true paler colour and structure of the rocks, which is the main difficulty when mapping an area with this type of exposure. Geologists must, therefore, frequently knock off small pieces with a hammer to tell what sort of rock they are on.

of uniform sedimentary beds that total a thickness of several kilometres, a style of deposition that is typical of deep water. These sediments were deposited in the period from about 2000 to 1900 Ma ago. The whole of the sedimentary sequence was metamorphosed during orogenesis, and today the Karrat Group sedimentary sequence occurs as a series of dolomites, marbles, slates, mica schists and quartzites.

The Karrat Group with its underlying basement gneisses was intensely interfolded during the formation of the fold belt. The earliest deformation resulted in compression in a north-easterly direction but later the transport direction changed to more westerly. In this phase the Archaean basement with the overlying Karrat Group sedimentary rocks were thrust over one another. At the same time a series of very large recumbent folds were developed, and finally the whole of this folded sequence was refolded by a series of large-scale, upright, open folds. The structures are seen very clearly in the clean, steep sides of the fjords around Karrat Isfjord, north of Uummannaq.

Inglefield Land fold belt

Inglefield Land is an isolated area in the north-westerly part of West Greenland, bounded by the Inland Ice and the sea. It is found north of the Archaean/Proterozoic basement area in the Thule region but,

ABSOLUTE (RADIOMETRIC) AGE DETERMINATIONS

The age of many igneous and metamorphic rocks can be determined by using a naturally existing clock that occurs within different minerals. The method is based upon certain radioactive elements that naturally transform into other, non-radioactive elements. The decay, or change, of one radioisotope into another isotope takes place over an absolute given time, a characteristic constant called the 'half-life'. For any elemental isotope this is the length of time it takes for half of the original amount of the isotope to change into its decay product.

One of the most used elements is uranium (U), two different isotopes of which the ^{235}U and ^{238}U decay into two different stable lead (Pb) isotopes, respectively ^{207}Pb and ^{206}Pb, with half-lives of 4500 and 700 Ma. By measuring the ratios between the isotopes, one can calculate how much lead has formed and therefore one has a measurement of how long a time has elapsed since the decay process started and thus a determination of the age of the mineral that originally contained the uranium isotopes.

The method is technically very demanding and requires that one can analyse atomic components in a mass spectrometer whereby the individual isotopes can be separated and measured. Interpretation of the results of the measurements is also often difficult. Some minerals are very resistant and are able to survive later high temperature reworking, so there is no impression of the last metamorphic phase. On the other hand some minerals and isotopic systems react to even low temperature reworking, and so there is a record of the last regional temperature regime. By combining different isotopic methods and mineral associations, an expression of the thermal history of the development of a rock unit can be obtained.

In some instances a mineral can be formed by stepwise growth. A mineral that commonly shows this is zircon, which may be made from an old core that comes from an earlier event and a younger rim that grew around the core during a later metamorphic event. If a measurement of the whole crystal is made, a meaningless average result is obtained. However, if a laser or ion beam is used to release the isotopic components from the individual parts of the crystal and these are measured separately, a more accurate picture of the ages of the different parts of the crystal can be obtained.

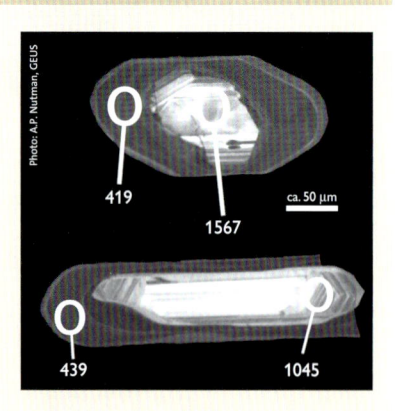

Two zircon crystals, separated from their rocks, upon which spot radiometric age determinations, using the so-called SHRIMP method, have been made. Each crystal is made up of an inner, old core (respectively 1567 Ma and 1045 Ma) which was derived from an old basement and deposited as grains in a sediment about 1000 Ma ago. During a later metamorphic event, new zircon shells (respectively 439 Ma and 419 Ma) grew around the old cores.

Fold belts in central and northern West Greenland

because of its isolation, relationships with the Rinkian area are partly unknown.

The Inglefield Land belt is composed of two different rock groups. An older group containing metamorphosed sedimentary rocks originally deposited as limestones with mudstones and sandstones, together with different volcanic formations (the Etah Group). This sequence is cut by a series of younger plutonic intrusive rocks that include different types of juvenile granites and syenites. All of the rocks are metamorphosed and intensely deformed by several phases of folding, and occur today as crystalline schists and gneisses. After a Palaeoproterozoic orogeny the area became stabilised into a coherent basement that was later overlain by undeformed younger Proterozoic and Palaeozoic sedimentary rocks.

Pegmatite banded and small-folded, dark orthogneisses from the north coast of Littleton Ø, west of Inglefjeld Land. The hammer is about 50 cm long.

RELATIVE CHRONOLOGY IN THE BASEMENT

The complicated construction with interfolding of younger supracrustal and plutonic rocks, together with reworking by one or more orogenic episodes, are reflected by the structure and relative age differences between the different rock units. In the basement, there are no direct age indicators, e.g. fossils in the rocks. Therefore, investigations of the relative age of the components, in order to uncover the geological development step by step, are an important part of field work. Later, the absolute ages of single rocks can be measured in the laboratory through radiometric age determination (see p. 48).

The principle in unravelling the relative chronological succession is that the geologist seeks to uncover the relations, older or younger, between different rocks and structures that adjoin each other. By sequential older/younger comparisons the complete sequence of events that represent the geological development of a given area can be elucidated. Figures A and B show schematically two developmental stages in the construction of a typical basement area that has undergone two orogenic episodes. The figures are drawn as a vertical section through the upper part of the Earth's crust.

Figure A is a section through an older, folded gneissic area (1) that contains an intrusive granite sheet (2), which, after interfolding with the gneisses, was cut by a larger granitic body (3). After the end of this orogenic phase and uplift of the rocks to a higher crustal level, two generations of mafic dykes (4 and 5) were emplaced. In connection with later continental rifting and ocean-floor formation, a sea area was formed on the edge of the old continent in which was deposited an enormous sedimentary sequence (6). These sandstones (yellow) with siltstones and mudstones (pale yellow) were laid down on the gneissic basement with an angular unconformity.

Figure B shows the same area as A, but the region has now undergone a new, younger folding event (7) in which the sedimentary rocks and their basement were folded together; the older structures in the basement also reflect this younger deformation. During this new orogenic event the whole of the region was pushed down to a deeper crustal level and all of the rocks were strongly transformed (metamorphosed), so that the sedimentary rocks were converted into metamorphic schists and the mafic dykes changed into amphibolites. The last stage in the development was the formation of a fault zone (8) that occurred due to horizontal crustal movements after the fold processes had stopped and the rocks brought up to a high crustal level.

Theoretically, this development could reflect a 3000 Ma Archaean basement (1–3) that was traversed by 2200 Ma mafic dykes (4, 5) emplaced in connection with continental rifting. On the edge of the newly formed ocean a many kilometres thick sedimentary sequence was deposited about 2000 Ma ago (6). After this the whole region was subjected to a new orogenic episode about 1800 Ma ago (7) and later faulted (8).

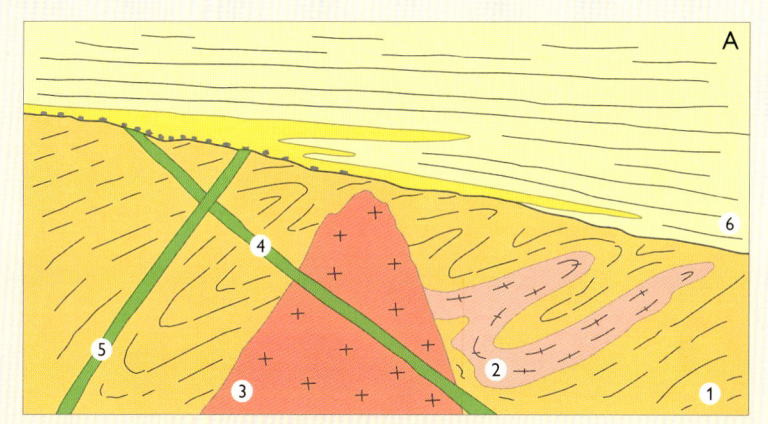

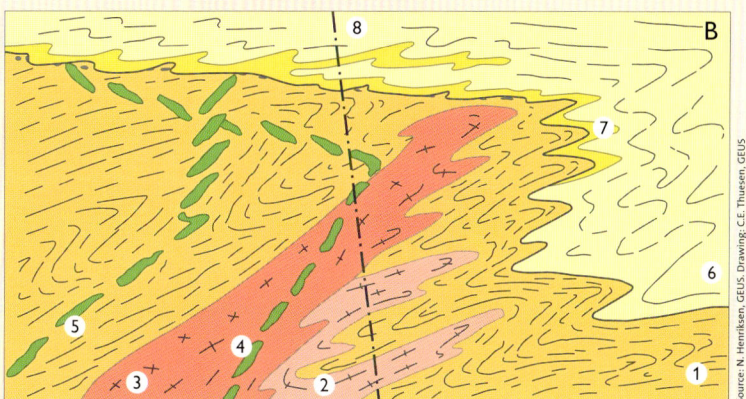

Sketch profiles of two theoretical developmental stages, A and B, of a basement area that was subjected to two orogenic episodes. See box text for details.

THE CRYSTALLINE BASEMENT

1800 million years old plate tectonics

The Ketilidian fold belt in South Greenland

The basement terrain of South Greenland formed from the roots of a NE–SW-trending fold belt developed along the southern margin of the Archaean block. It is the best example in Greenland of a fold belt generated by plate tectonic processes in Precambrian basement. The development took place from about 1850 to 1725 Ma ago. A cross-section through the fold belt traverses from a foreland in the north, across a boundary zone, through a central granite zone into two zones of metamorphosed sedimentary rocks in the south, which were formed from erosion of the granite zone late in the development of the fold belt.

Ice-polished surface of an outcrop of the Julianehåb batholith. The outcrop is made up of grey, granite-like rocks with large feldspar crystals (porphyritic minerals). This rock is dissected by thin, pale quartz-feldspar veins (pegmatites) that were formed at a late stage of the crystallisation history of the granite. Hammer shaft is about 45 cm long. Locality is east of the town Qaqortoq (Julianehåb).

The Ketilidian fold belt began with the development of an active plate margin between the stable Archaean basement block to the north and a newly formed ocean to the south. The oceanic plate was pushed down beneath the margin of the continental block (subducted) resulting in a down warping in the continental block, which allowed the deposition of a series of Ketilidian sedimentary and volcanic rocks. In connection with the subduction, parts of the mantle above the sinking plate melted, and the magmas generated ascended to form a volcanic island arc south of the continent, while at lower levels the roots of this arc crystallised as granite-like bodies. The granite roots were subsequently exhumed to the surface of the crust, where they were eroded producing sediments that were deposited as a thick sequence of sandstones and mudstones

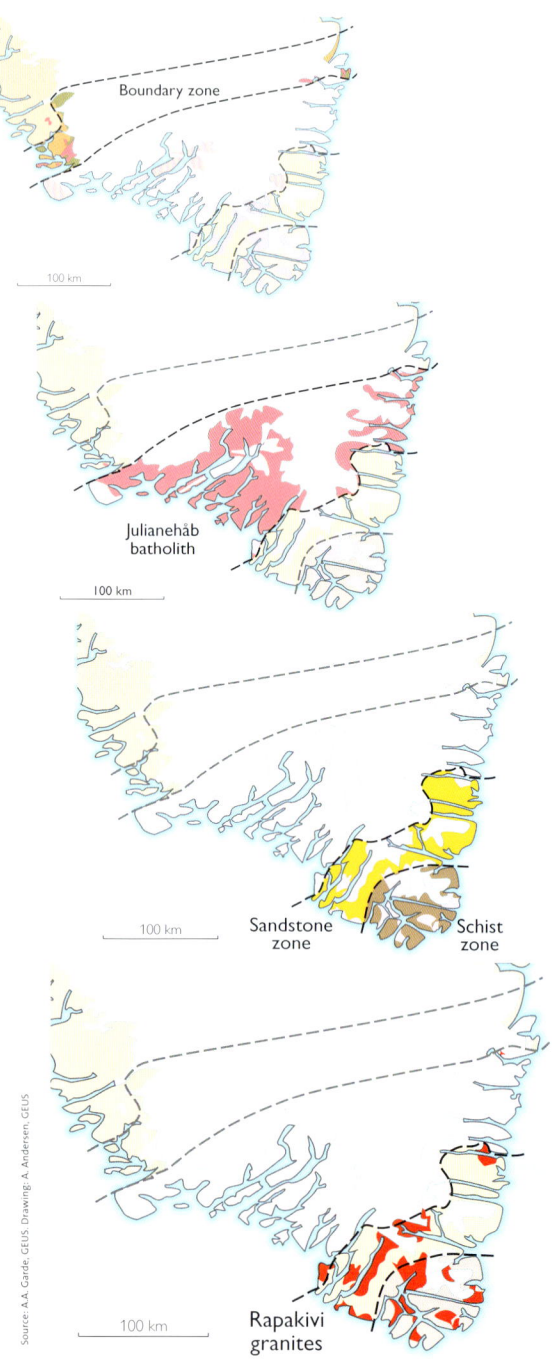

Map of the Ketilidian fold belt in South Greenland showing the division into four zones and the location of late granitic intrusions (rapakivi granites).

on the margin of the ocean. The development of the fold belt continued with intense deformation and metamorphism of all the rock units, until near the end of the orogeny numerous cross-cutting granitic bodies were emplaced in the southernmost areas.

The four zones of the fold belt

The boundary zone with the unaffected Archaean rocks occurs as an approximately 50 km wide zone of gneisses about 2800 Ma old, upon which the 1900–1850 Ma Ketilidian sedimentary and volcanic rocks were deposited prior to the start of deformation in the fold belt. These supracrustal rocks are now preserved in the Ivittuut region. The granite zone is characterised by the Julianehåb batholith. This developed over a period of about 60 Ma as a series of large, lens-formed granitoid intrusions (batholiths) that now form one coherent mass. The granite zone is up to 150 km across and makes up the predominant part of the fold belt at the present-day exposure on the surface. The sandstone zone is 30–40 km wide, and was formed immediately south of the batholith by deposition of coarse erosion products from rivers. The sediments include cobbles, gravels and sands that are now transformed into metamorphic rocks. The schist zone is found farthest south and comprises transformed, finer grained deposits that were laid down at greater water depths, beyond the sandstone zone.

Deformation, metamorphism and late granites

After deposition of the sediments south of the batholith about 1800 Ma ago, the whole of the fold belt was subjected to intense deformation and folding that took place in several phases. At the same

Large-scale migmatite formation in the schist zone in the southernmost part of the Ketilidian fold belt. The dark, red-brown host rocks comprise mica schists that originally were sands and muds. These schists are cut by a network of slightly younger, pale, granitic veins formed by partial melting during metamorphism. Both the schists and the intrusive granite sheets are strongly deformed and interfolded. Mountainside is about 500 m high.

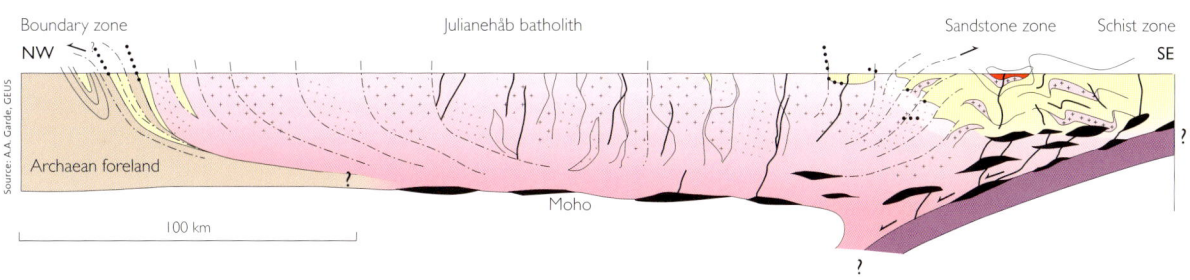

A schematic cross-section through the Ketilidian fold belt showing an interpretation of its plate tectonic structure. The interpreted subduction zone is situated at the assumed boundary with oceanic crust.

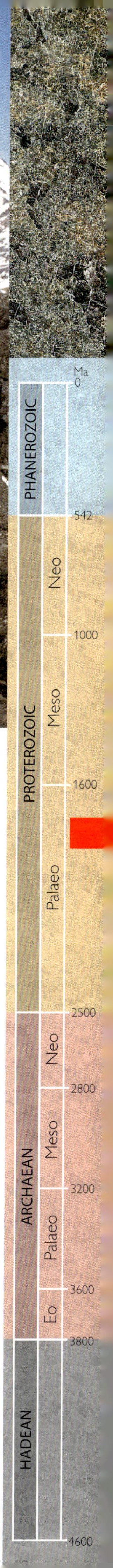

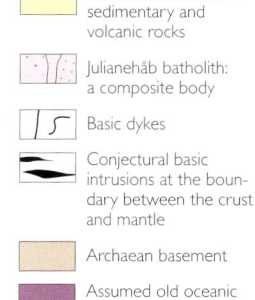

THE CRYSTALLINE BASEMENT

Outcrop of metamorphosed sandstone in the sandstone zone. Danell Fjord can be seen in front of the 2500 m high, alpine mountains at the edge of the Inland Ice.

Geochemical map showing the distribution of gold in the Ketilidian fold belt as detected through analyses of systematically collected stream sediment samples. Red colours show high anomalies, while dark blue colours show areas without gold.

time the sedimentary rocks were metamorphosed and, in the deeper parts of the schist zone, locally melted to form migmatites. The latest group of rocks in the Ketilidian fold belt is a series of late granites (rapakivi granites) that were emplaced about 1750–1725 Ma ago into the southern part of the fold belt, for the most part after deformation had ceased.

Gold in the Ketilidian fold belt

Mineralisation of gold and other metals is often associated with metamorphism of sedimentary and volcanic rocks in basement terrains and it was therefore natural to search for indications of such minerals in South Greenland. During the early 1990s the whole of South Greenland was prospected using a combination of geological, geophysical and geochemical techniques. As a result it became clear that the Ketilidian fold belt contained gold mineralisation, which later led to the discovery of the Nalunaq gold occurrence (see p. 212).

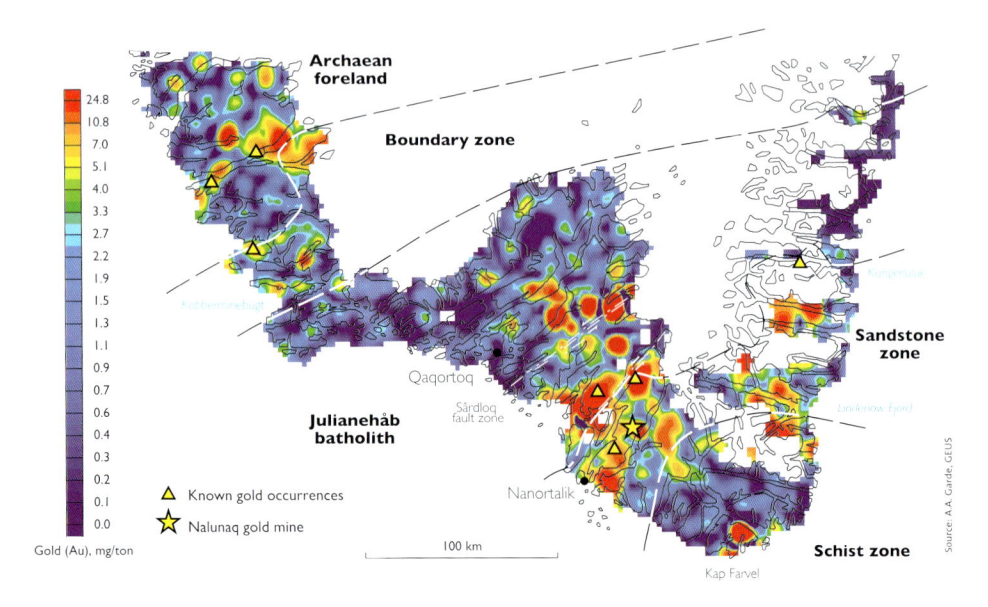

In the sandstone zone metamorphosed sandstones and volcanic rocks overlie the Julianehåb batholith. Gold mineralisation occurs in thin quartz veins that vary from a few centimetres to approximately one metre in thickness. These quartz veins are often concentrated in thrust and crush zones that dissect the volcanic and granite-like rocks. Locally, gold-bearing veins can contain visible gold grains up to 2 mm across, but usually they are too small to be seen with the naked eye.

The mineralised quartz veins formed by precipitation from percolating, hot hydrous fluids (hydrothermal fluids) at temperatures of 250–350°C and at low pressures. These fluids are assumed to be derived from seawater with a high salinity that was heated by the still hot rocks of the Julianehåb batholith. During circulation through the rocks the hot fluids were able to dissolve gold, for example from the volcanic rocks, which have only small amounts of very finely disseminated gold. After dissolution the gold was transported away in the fluids, and later precipitated with quartz in the veins at lower temperatures and pressures, or due to a change in chemical composition.

The Ketilidian fold belt in South Greenland

PLATE TECTONICS

Plate tectonics is the name for a concept, a geotectonic model, explaining large-scale geological phenomena such as fold belts, earthquake zones, ocean-floor formation and continental drift. The model is based on the interpretation of a series of geological and geophysical observations that have led to the following assumptions:

1) The Earth's outer shell, the lithosphere (see p. 20), functions as a rigid plate that rests on and can move across an underlying, more plastic part of the upper mantle, the asthenosphere.

2) The lithosphere is divided into a series of large plates that change their size and shape with time. The plates are in constant motion and their position on the surface of the Earth is continually very slowly changing.

The lithospheric plates contain both oceanic and continental crust. The mutual boundaries between plates are of two main types. Constructive plate boundaries form at mid-oceanic ridges (spreading zones) where new volcanic material is added from the mantle. Destructive plate margins occur where oceanic crust is forced underneath the adjacent plate (subduction) and partially melts.

Mountain belts form along subduction zones, where crustal material can become squeezed between two plates, and the rocks become deformed and folded. Frequently, fold belts occur where opposing continents, each on its own plate, collide and newly deposited sedimentary rocks between the continents are folded up and transported in over the margins of the adjacent, older continent.

While old oceanic crust is continually descending into subduction zones, the less dense continents 'float' above, and progressively increase in size by addition of new mountain belts. There is thus a gradual increase in the amount of continental crust preserved at the surface of the Earth. Consequently, the oldest oceanic floor is only about 200 Ma old, but many continents have a core that is more than 3000 Ma old.

At the plate margins, along the mid-oceanic ridges, an extension of the ocean floor occurs which is described as sea-floor spreading. Here, basaltic magma from the mantle forces its way into the oceanic crust, and both sides of the ocean floor are forced apart. This spreading pattern can be traced back about 200 Ma. Due to the Earth's changing magnetic polarity with alternating north–south pole directions (see p. 113), these magnetic changes are preserved in the newly formed volcanic rocks and can be detected by the pattern of the magnetic anomalies on the ocean floor. The rates of spreading at the ridges are normally from 1 to 2 cm per year, but may be up to about 15 cm per year.

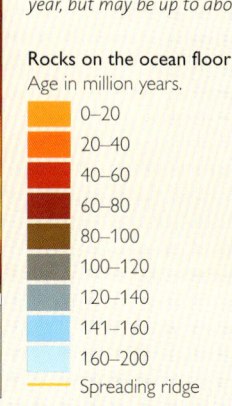

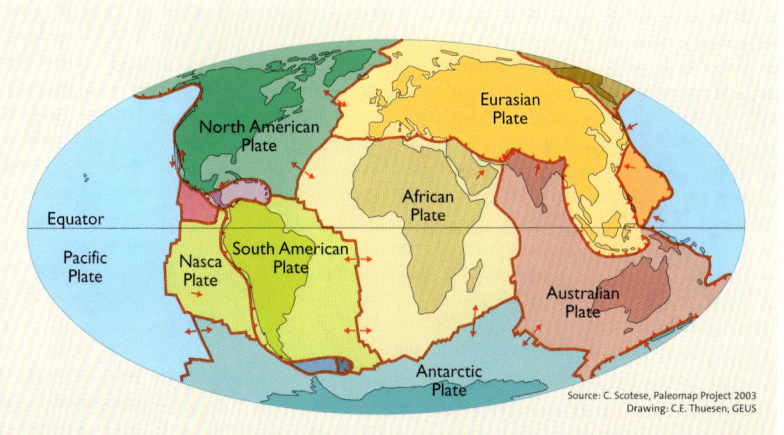

Rocks on the ocean floor
Age in million years.
- 0–20
- 20–40
- 40–60
- 60–80
- 80–100
- 100–120
- 120–140
- 141–160
- 160–200
- Spreading ridge

Map showing the structure of the Earth's surface with seven large and a series of smaller lithospheric plates. Note that the single plates comprise both continental and oceanic crust.

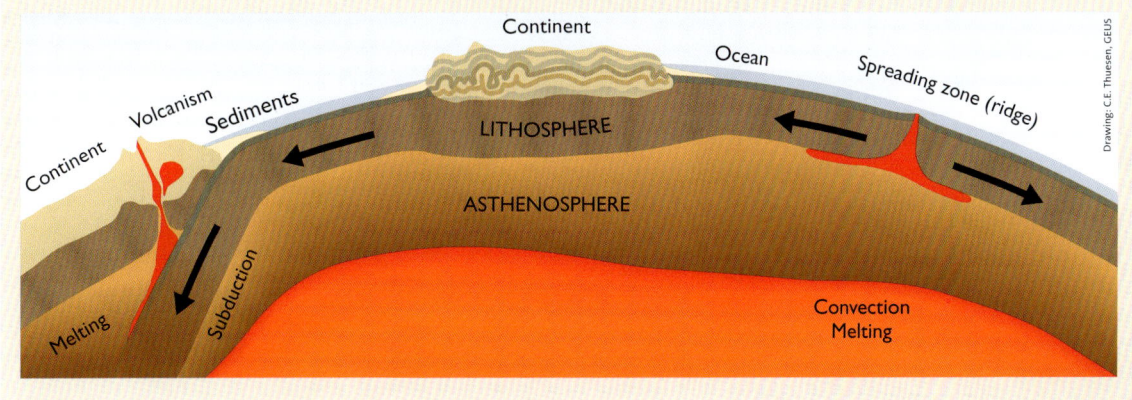

The principles of plate tectonics: In spreading zones (ridges), newly formed oceanic crust develops over a rising convection cell from the mantle. A lithospheric plate with oceanic crust sinks into the mantle by subduction underneath less dense continental crust. By this process, partial melting of the subducted lithosphere and overlying rocks causes volcanism at the surface and granitic intrusions at depth. The continent in the middle is part of the lithospheric plate that is moving left. Eventually, the two continents will collide, and a mountain belt will be formed in the collision zone.

Motzfeldt Sø (the lake in the left foreground of the photograph) is surrounded by high mountains comprising rocks of the Motzfeldt intrusion in the Gardar Province. Mountains are about 1600 m high.

Photo: T. Tukiainen, GEUS

MINERALS IN THE GARDAR PROVINCE

Several of the Gardar Province intrusions contain large quantities of rare elements and a correspondingly large number of rare minerals. Particularly important are the Ilímaussaq intrusion near Narsaq and the cryolite occurrence near Ivittuut. More than 225 different minerals were found in the Ilímaussaq intrusion and 30 of these are new minerals that were discovered and described for the first time. Similarly at the cryolite locality, there are more than 100 different minerals of which 16 were discovered and first described from here. Examples of well-developed, peculiar minerals from the province are shown.

AEGIRINE

A comparatively common mineral in alkaline intrusive rocks, which is named after the Norse god of the sea, Ægir. The mineral is found in many places in the world, but is particularly beautifully developed in the Ilímaussaq intrusion, often as rosettes of needle-shaped crystals – aegirine suns.

TUGTUPITE

First discovered in 1957 in the Ilímaussaq intrusion near Narsaq, this was recognised internationally as a new mineral in 1965. The name Tuttu (reindeer) refers to a place in the mountains where a girl, according to Greenlandic legend, gave birth to a child in secret. The mineral occurs most commonly as dense, red masses and only occasionally as crystals. It is considered today as Greenland's national gemstone.

SØRENSENITE

This was observed for the first time in 1962 at Kvanefjeld, near Narsaq, in the Ilímaussaq intrusion. In 1964 it was described and named after the Danish geologist Henning Sørensen who has studied the minerals of the intrusion for many years. Sørensenite is only known from this locality in South Greenland.

PACHNOLITE

This mineral is only known from the Ivittuut cryolite locality where, together with a few other minerals, it occurs as a secondary alteration product of cryolite. Pachnolite has a characteristic crystal form with very uniform prisms that have pyramidal terminations. Pachnolite occurred mainly near the surface of the cryolite body. It was principally collected in material excavated from the top of the body during the early mining activities.

AMAZONITE

This mineral is a turquoise green variety of the common feldspar microcline. The name was given for a locality on the River Amazon. In Greenland the mineral mostly occurs in pegmatites, often as centimetre-large crystals, especially in one of the Gardar intrusions on the south side of Kobberminebugt, 35 km south of Ivittuut. Amazonite can be cut and polished and its silvery sheen makes it a sought-after gemstone.

CRYOLITE

The mineral's name stems from the Greek words for ice (kryos) and stone (lithos), i.e. 'ice stone'. Its refraction properties and its ice-like appearance have given the mineral its name. The mineral was first identified in 1802 and is normally white, although brown and black varieties are also seen in Ivittuut. The mineral is very rare and the deposit in Greenland was the only large mineable occurrence in the world. Cryolite has been used for more that 120 years in different industrial applications, but primarily in connection with the production of aluminium as a flux in the electrolytic smelting process. Following the closure of the mine at Ivittuut, artificial cryolite has replaced natural cryolite in industrial processes.

Lower part of the Thule Supergroup comprising layered sandstones and a red-brown weathered basaltic layer. The mountain, west of Qaanaaq, is about 1000 m high.

Photo: P.R. Dawes, GEUS

THE GARDAR PROVINCE

A 1350–1125 Ma old rift valley with sedimentary deposits, basalts and intrusive magmatic rocks

5

FISSURES IN THE EARTH'S CRUST

THE GARDAR PROVINCE

In the later part of the Mesoproterozoic the South Greenland basement shield was affected by stretching of the Earth's crust with formation of a NE–SW-trending fracture zone in the continent. In this so-called rift zone, faulted blocks sank and, at the surface, gave rise to the formation of sedimentary basins. Some faults extended down into magma chambers in the upper mantle from where basaltic magmas rose along these fissures to erupt onto the surface as lavas. Later, various magma types were emplaced as large plutonic intrusions at higher crustal levels below the land surface.

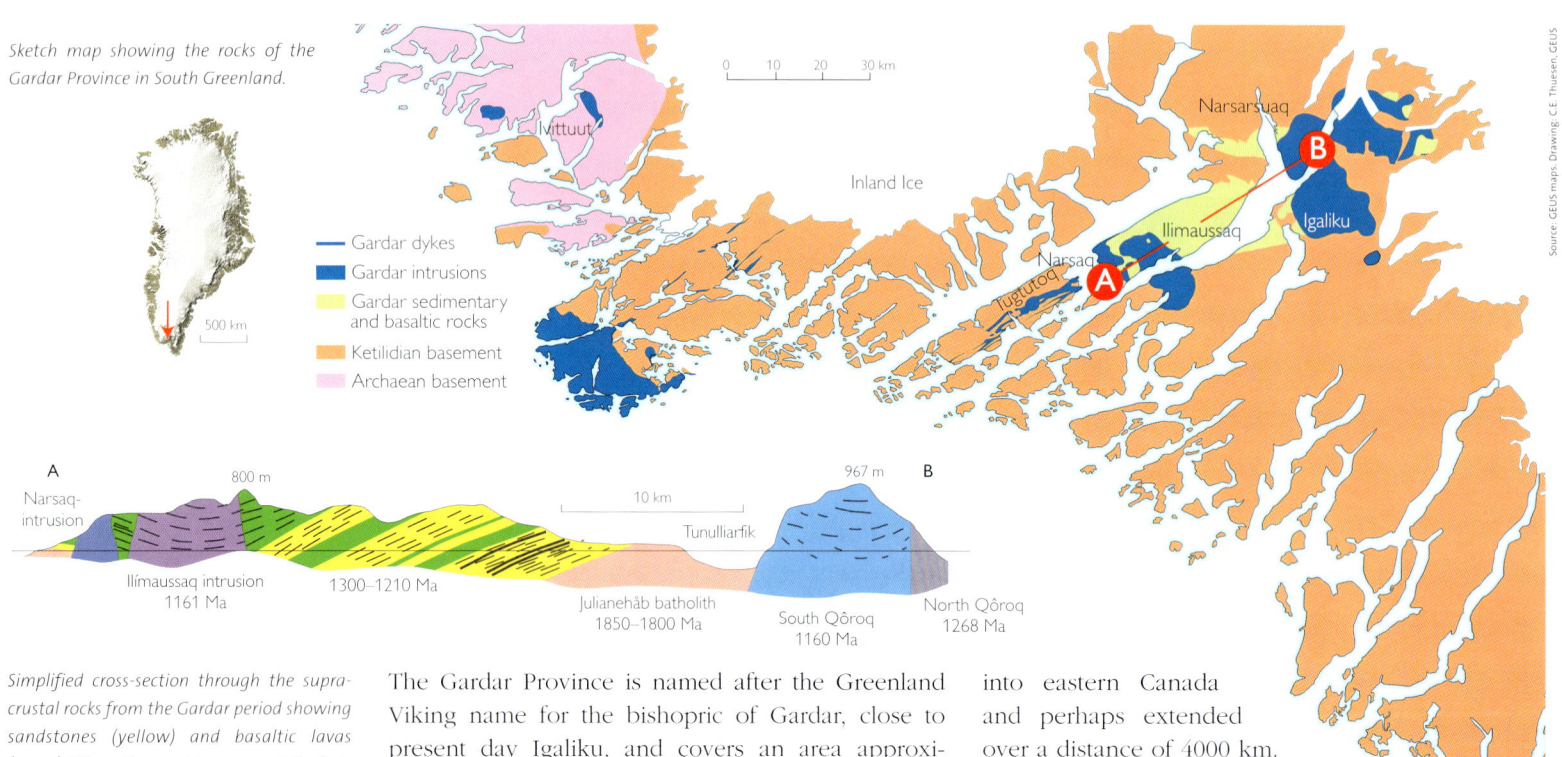

Sketch map showing the rocks of the Gardar Province in South Greenland.

Simplified cross-section through the supracrustal rocks from the Gardar period showing sandstones (yellow) and basaltic lavas (green). Where the sandstones are at the bottom of the sequence they are intruded by flat-lying basaltic intrusions (black). The position of the profile (A–B) is shown on the sketch map above. The supracrustal rocks rest on granites of the Julianehåb batholith (pink) and are cut by different, large Gardar-age plutonic intrusions. The section is drawn with a vertical exaggeration of four times, so that the actual inclination of the layers is much less than shown on the figure. The true dip of the sandstone layers is only 5–10°.

The Gardar Province is named after the Greenland Viking name for the bishopric of Gardar, close to present day Igaliku, and covers an area approximately 200 km along the rift zone and 100 km across. The development of the Gardar Province spans more than 200 Ma and comprises three periods:

1) The earliest phase (from about 1350 to 1260 Ma) with sedimentary deposits and the accumulation of lavas at the surface, and a few large intrusions at a high crustal level.
2) The middle phase (from about 1250 to 1200 Ma) mainly comprising emplacement of basic dykes and two smaller intrusions in the northern foreland.
3) The late phase (from about 1185 to 1120 Ma) in which the principal part of the Gardar Province rocks formed. These comprise basic dykes and high-level magmatic intrusions that originally crystallised in magma chambers 2–5 km beneath the surface.

Formation and early development of the Gardar rift

The rift development in South Greenland forms part of a major NW–SW-trending system that can be traced into eastern Canada and perhaps extended over a distance of 4000 km. Extension of the crust took place perpendicular to the length of the rift and the first development in Greenland led to the formation of a series of faults and subsidence of the blocks between the faults. In these depressions a sequence of sandstones and basaltic lavas more than 3600 m thick was deposited. The sandstones are the oldest, and comprise erosion material from adjacent land areas deposited from an old river system that flowed parallel to the rift valley. Higher up in the sequence, the first of the basaltic horizons occurs, followed by alternating sandstones and basalts. Subsequently, the whole of this sequence was covered by an outpouring of extrusive basaltic lavas. Today remnants of the sequence are only preserved in the 20 km broad central part of the rift zone, but probably it originally covered a much larger area.

The earliest intrusions comprise a syenite complex (6 × 4 km) at Grønnedal (near Ivittuut) in the foreland north of the rift zone and two large syenitic intrusive centres (10 × 15 km and 7 × 4 km) on both sides of the Tunulliarfik fjord near Narsarsuaq.

Finely banded, red Igaliko Sandstone with thin, white layers and spots formed by bleaching of certain zones and horizons by circulating water. The sequence comprises alternating silt and sand layers that have been lithified into the resistant rock quartzite. In some places the sandstones are intensely red coloured (as in the photograph) due to the red iron oxide, hematite, filling the spaces between the sand grains. This red colour is often characteristic of deposits formed under desert conditions.

The middle period

A widespread swarm of basic dykes were intruded into a fissure system that formed parallel to the trend of the rift zone. The period was characterised by marked stretching and fissuring of the rigid, upper part of the crust which made space for the emplacement of many tabular dykes that make up some 10–15% of the rocks in the rift zone. At least four different generations of dykes were intruded successively with slightly different directions. This implies that the orientation of the spreading direction, at right angles to the rift, also changed slightly with time.

Also during this period, two small, characteristic intrusions with unusual compositions formed in the northern foreland to the rift zone. One of these is the Ivigtut granite (at Ivittuut) with its associated cryolite occurrence that was mined between 1856 and 1987 (see section on ore minerals, p. 205).

Main development of the Gardar Province – the late phase

By far the largest volume of rocks in the Gardar Province formed in the final phase of development from about 1185 to 1120 Ma. Extensive tectonic and magmatic activity can be traced throughout the whole of the province, and includes both a series of 'giant' basic dykes with an ENE trend as well as a series of very large intrusions; the largest measures more than 25 × 45 km. Some of the intrusions from this period are enriched in rare elements, which led to the formation of special rock types and an abun-

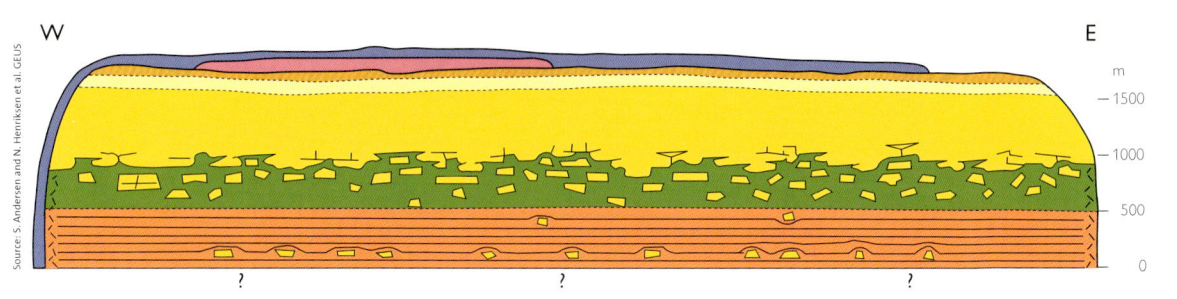

Diagrammatic cross-section through the 1161 Ma old Ilimaussaq intrusion that shows the layered construction. The rocks comprise a series of unusual nepheline syenites (rocks that are dominated by alkali feldspar and nepheline with various amounts of dark minerals). The intrusion represents the interior of a large magma chamber in which, as the magma crystallised, different rock types gradually separated from the magma. The least dense minerals were carried up to the top of the magma chamber and here gradually solidified downwards in a series of rare rock types such as pulaskite, foyaite, sodalite foyaite and naujaite. The denser minerals sank to the bottom and solidified upwards as layered kakortokites. The final magma residue crystallised in the middle as layers of lujavrite. A large number of rare minerals occur in the intrusion and are concentrated in the late crystallising phases; the lujavrites and kakortokites in particular are enriched, for example in uranium and zirconium.

THE GARDAR PROVINCE

Steep slope exposing layered kakortokites in the lower levels of the southern part of the Ilímaussaq intrusion. The magmatic layering was formed in a magma chamber with periodic addition of new melt in pulses. After each pulse the intrusive magma gradually crystallised and magmatic differentiation took place (see box p. 152). The denser minerals sank to the bottom and were overlain by lighter minerals. This process was repeated with the injection of each new magma pulse. The mountain in the right background is made of granite outside the boundary of the intrusion, and is about 900 m high.

dance of many different rare minerals. From the Ilímaussaq intrusion more than 225 different minerals have been described.

Seven different intrusions have been mapped, that include syenites and granite-syenite bodies. These intrusions vary in size from small (3 × 4 km) through medium sized (10 × 20 km) to very large (25 × 45 km) bodies. Each intrusion is typically made up of a number of different rock types that represent a differentiation series from the parent magma. The best known is Ilímaussaq, where the rocks occur as a stratified sequence in a high-level magma chamber. At an early stage, after formation of the magma chamber, part of the melt began to solidify around the margins against the country rocks, while the magma in the interior remained liquid and slowly separated into various types by magmatic differentiation (see box p. 152). During this process, less dense crystals were separated and floated upwards whilst the denser minerals sank to the bottom. Different rock types thus formed at the top and bottom of the solidifying intrusion. The final phases of the remaining melt are enriched with rare elements and it is these rocks that contain interesting and unusual mineral occurrences.

GEOCHEMICAL INVESTIGATIONS

The chemical composition of a rock can be determined through the analysis of its different elements. There are normally between 10 and 12 major elements and more than 30 other elements that occur only in small amounts, known as trace elements. A geochemical survey begins with the collection of rock samples, followed by laboratory analyses. The samples are crushed into a very fine homogeneous powder that is representative of the rock and they are then analysed geochemically. It is also possible to let nature help with the homogenisation by collecting and analysing sand and silt samples from small streams that are representative of local areas such as a small drainage basin. When more detailed geochemical studies are undertaken, such sand and silt samples can be taken approximately every 5 km^2, while reconnaissance studies may be based on one sample for every 20–30 km^2.

By systematic collection and analysis of stream sediments a record of how the different elements are distributed in the investigated area can be obtained. Such surveys show that the distribution pattern closely follows the occurrence of the different rock types in the field, and that different geochemical provinces are distinguishable. The results are entered into a geochemical database from which a geochemical map for each investigated element can be made. By undertaking an area-based smoothing of the more heterogeneous primary data, an anomaly map that shows the variation of occurrence of a single element throughout the investigated region can be prepared. An example of such an anomaly map of South Greenland is shown here for the element niobium. Together with unusual elements such as uranium, tantalum, zirconium and other rare elements, niobium characterises the intrusive rocks of the Gardar Province. The anomaly map shows that the distribution of niobium reflects the occurrence of the Gardar intrusions shown on the geological sketch map (p. 56).

Geochemical investigations have been carried out throughout the whole of West and South Greenland, as well as over large parts of North and North-East Greenland. The results can be used as the basis for indicating areas where a mineralisation potential for certain elements may exist. General geological research and mapping studies also use geochemical data to distinguish between different rock units, separate fold belt provinces and to identify genetically coherent regions with intrusive and extrusive volcanic rocks.

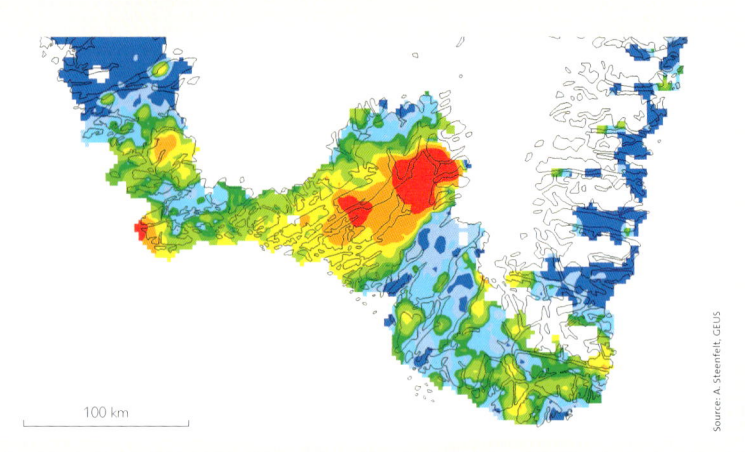

Geochemical anomaly map, based on a systematic collection of stream sediment samples, showing the distribution of the element niobium in South Greenland. The niobium content, from highest to lowest, is indicated by a colour scale from red, through yellow and green to blue. The occurrence of niobium in South Greenland reflects very clearly the distribution of the Gardar intrusions (compare with figure on p. 56), and this is a good example of how geochemical investigations can contribute to revealing the geological make-up of a region.

BASIN DEPOSITION

Sedimentary basins – depressions in which mud, sand, gravel and calcareous deposits accumulated

6

SUBSIDENCE AND DEPOSITION

6 BASIN DEPOSITION

After formation and stabilisation of the Greenland Precambrian shield about 1700–1600 Ma ago, later geological developments primarily took place along the margins of the shield. At different times various so-called sedimentary basins developed here by gradual subsidence of the Earth's crust. At the same time sediments were carried into and deposited in the basins, the rate of deposition keeping pace with the subsidence. The formation and development of such basins can take place over periods from tens to hundreds of million years. The sedimentary basins can extend for more than 1000 km along the shield margins and the accumulated deposits can vary from a few kilometres up to more than 20 km in thickness. Most of the basins formed in the sea marginal to the old continent, but locally, fresh water basins on land were filled by deposits laid down from rivers and in lakes. A special depositional type was the formation of layered volcanic rocks (plateau basalts) laid down in kilometre thick units deposited on land or in the sea.

Sedimentary successions form by the deposition of the erosion products of rocks (e.g. muds, sands and gravels), chemical precipitates (e.g. salt) or the remains of organisms (e.g. calcareous shells) in a sedimentary basin (see box p. 64). Supply to the basin is via transport in water (by rivers), through the air (by wind) or in ice (during glaciation). Deposits laid down in water immediately sink to the bottom and initially the entire mass is loose and water saturated. The weight of accumulating, overlying sediment squeezes the water out and, as temperature increases with depth in the basin, the individual mineral grains gradually become cemented together. However, the original sedimentary character, clearly seen as variations in both compositional layering and primary structures formed during deposition (e.g. wave and current ripples) are preserved. As the lithification process progresses sands are changed into sandstones, muds into mudstones and later shales, and calcareous muds into limestones.

A modern example of a sedimentary basin is the North Sea where, during a period of some hundred million years, deposition of sediments has taken place. This basin contains transported erosion products derived from the surrounding north European landmass as well as calcareous deposits. In Greenland the offshore shelves bordering the continent are the sites of recent sedimentary basins. Deposition here has taken place over several hundred million years, mainly during the Jurassic, Cretaceous and Palaeogene.

Sediment types

Distinction can be made between three main types of sediments formed by different processes. They comprise clastic, chemical and biogenic sediments.

Clastic sediments are generally formed from the erosion of older rocks, for example, of an old basement area. Erosion results in the formation of different grain-size fractions that, during transport, are broken into smaller fragments, sorted and polished. On deposition the coarsest fractions (pebbles) are laid down nearest the source and the middle fraction (sands) are transported further away. The finest fractions (silts and muds) are moved the furthest out into the basin and deposited in areas where the transporting currents have effectively ceased. The sediments can be well or poorly sorted; short transport distances mean the grains are generally angular while longer transport usually results in rounded grains. Where more than half of the deposit comprises granules and pebbles the sediment is called a conglomerate. The matrix between the coarser clasts in such a deposit normally comprises sand grains.

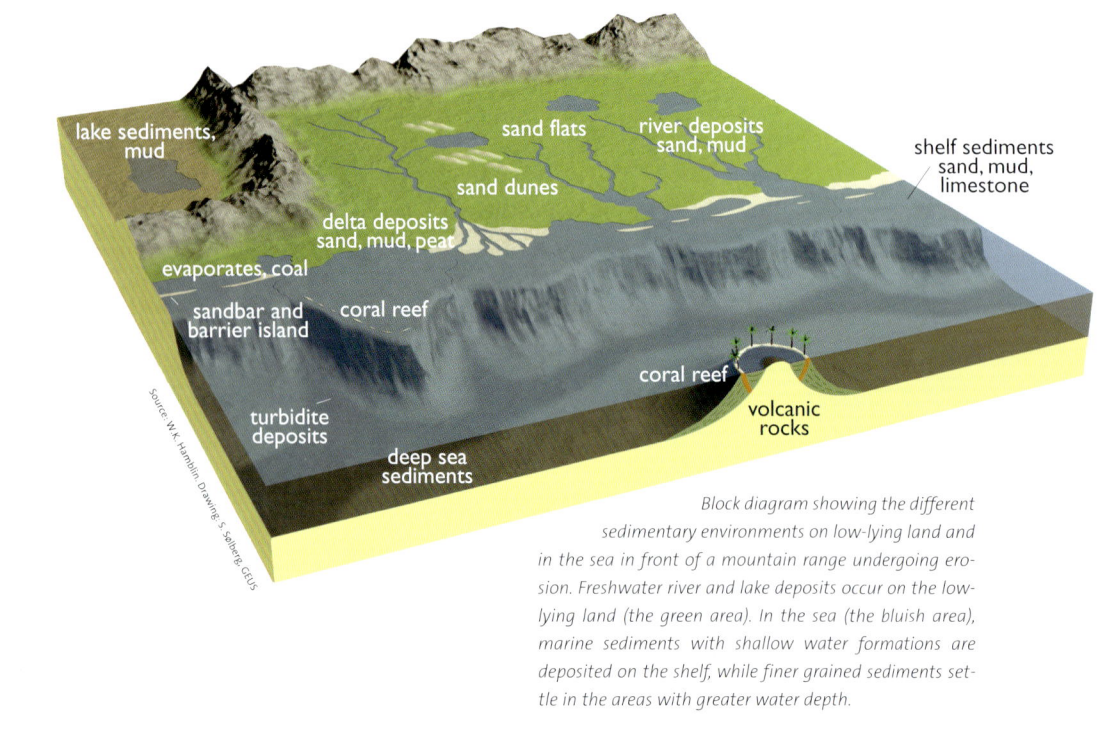

Block diagram showing the different sedimentary environments on low-lying land and in the sea in front of a mountain range undergoing erosion. Freshwater river and lake deposits occur on the low-lying land (the green area). In the sea (the bluish area), marine sediments with shallow water formations are deposited on the shelf, while finer grained sediments settle in the areas with greater water depth.

Chemical sedimentary deposits occur when dissolved components become supersaturated and precipitate from solutions. This process takes place when either a surplus of dissolved material is supplied to a basin, or the water in a basin evaporates and is not replaced by input from outside. Examples of such chemical sediments include many Precambrian limestones and evaporitic deposits such as salt and gypsum.

Biogenic sediments are widely distributed in the geological record, typically as limestones and dolomites, but did not form significant deposits before the early Cambrian evolutionary explosion of life about 540 Ma ago. This type of sediment is fre-

Horizon of dark, fine-grained, muddy limestone between two layers of pale, brecciated limestones; Middle Cambrian shelf limestones of the Brønlund Fjord Group, south-western Peary Land.

Conglomerate with well-rounded, small pebbles, principally of quartzitic rocks, in a matrix of pale, yellow-brown sandstone; Thule Supergroup, North-West Greenland. Field of view is about 50 cm wide.

Layered, flat-lying sedimentary sequence of Cambrian age. Dolomitic limestones at the bottom are overlain by dark sandstones and mudstones, with a pale limestone sequence at the top. Photograph shows a 600 m high section on the north side of Jørgen Brønlund Fjord, southern part of Peary Land, North Greenland.

SEDIMENTS – TYPES AND CLASSIFICATION

Sediments that are developed and then preserved at the same place (e.g. limestone) are called autochthonous, while sediments that are deposited after transport to a different place (e.g. sandstone) are called allochthonous.

Autochthonous sediments

Autochthonous sediments include chemical precipitates with evaporitic deposits, such as gypsum, rock salt and anhydrite, as well as biogenic deposits such as coal, limestone, dolomite and silica-rich sediments (chert). Of these types, limestones and dolomites are common in Greenland. They occur in Cambrian and younger deposits, and were primarily formed from an accumulation of organic remains from calcareous-shelled organisms. Limestones are made up of three components – grains, matrix and cement. The grains comprise carbonate sand, skeletal remains or small precipitated spheres (ooids). The matrix is usually lime mud that fills in the space between the grains, while the cement is calcite that bonds the components together. Any remaining open space in a rock, not filled by matrix or cement, is known as the pore space.

Allochthonous sediments

Allochthonous sediments are mainly the products of mechanical weathering and erosion derived from a land area (clastic sediments). The deposits are divided according to grain size:
1) muds (comprises mainly clay)
2) silts
3) sands

The coarser fractions include clasts or rock fragments of increasing size:
4) granules
5) pebbles and cobbles

Sedimentary grain sizes (diameter) are:
Clay < 0.002 mm
Silt 0.06–0.002 mm
Sand 0.06–2.0 mm
Granules 2–4 mm
Pebbles 4–64 mm
Cobbles > 64 mm

Within some of the grades fine, medium and coarse divisions may be recognised. The composition of a sediment can also be used as a method of classification; for example, a quartzitic sandstone comprises > 90% quartz grains, while an arkose comprises a mixture of quartz and feldspar grains. Another common type of rock is a greywacke that has a large proportion of sand-grade quartz and rock fragments in a muddy matrix.

Transport and deposition of sediments

During transport sediments are usually sorted according to grain size, primarily due to variations in the current strength. Strong currents can transport both small, light grains and heavier material like pebbles, while a weak current only carries finer material such as silt and mud. Thus in the upper reaches of a river, gravels will be deposited, while further downstream sands are deposited, perhaps in a delta. Sands and silts settle out in the near-shore area, whereas mud particles only settle out and sink to the seabed far offshore. How well the sediments are sorted may often vary so that, for example, a mixture of sand and silt can be deposited in the same layer. A distinction can be made between well-sorted, moderately sorted and poorly sorted types of deposits. An example of a poorly sorted deposit is a conglomerate that is a mixture of pebbles, granules and sand. Conglomerates are a widespread type of deposit in the lower parts of large sedimentary basins and often indicate the base of a new sedimentary cycle.

Allochthonous sediments produced by erosion of a hinterland can be deposited in nearby basins, or be carried very long distances before deposition occurs (see box p. 76). There are many examples of material being transported thousands of kilometres away from the place of origin before being deposited in a sedimentary basin.

Present-day sandy deposits (sediments) from a stream running through a rocky valley. The sand deposits are distinctly layered and at the top of the photograph the surface structure of a single layer is seen. The stones in the foreground were deposited at an earlier stage when currents were strong, while the sand was deposited at a later stage with weak currents.

Cobbles > 64 mm	Pebbles 4–64 mm	Granules 2–4 mm
Coarse sand	2.0 – 0.6 mm	
Medium sand	0.6 – 0.2 mm	
Fine sand	0.2 – 0.06 mm	

Silt 0.06–0.002 mm   Mud < 0.002 mm

Grain size of different sediment types. Pebbles and granules usually comprise rock fragments while sand can be a mixture of both rock fragments and single mineral grains. In clastic sediments sand and silt typically comprise quartz grains and skeletal fragments, or calcite grains in carbonate rocks. Muds are dominated by clay minerals that have characteristics that distinguish them from other minerals.

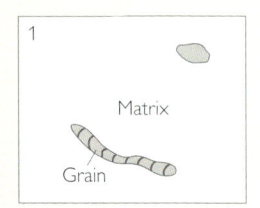

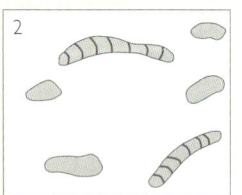

Limestone has a simple chemical composition and is mainly composed of calcium carbonate ($CaCO_3$). Nonetheless, its structure is very variable, as limestone comprises different amounts of grains, matrix and cement. The grains include skeletal remains from small invertebrates and small, globular carbonate lumps. The matrix comprises mud-grade calcite and the cement is calcite that binds the other parts together by occupying some of the pore space. Depending upon the different amounts of the three components, distinction may be made between:
1) *mudstone, with < 10% grains*
2) *wackestone, with > 10% grains that are contained within the matrix*
3) *packstone, where the grains touch each other and the matrix fills the spaces in between*
4) *grainstone, lacking mud-grade material*
5) *boundstone, where the original components have grown together, e.g. by formation of reefs.*

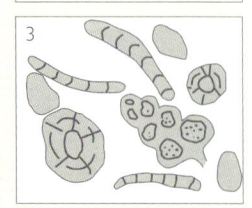

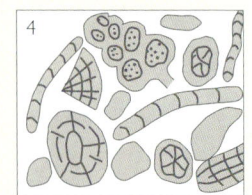

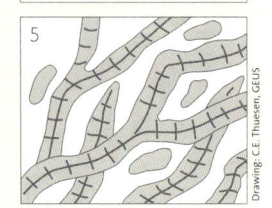

quently made of fine calcareous shell fragments from both plankton (microfossils) and from fossils that can be seen with the naked eye (macrofossils). Complete micro- and macrofossils are often well preserved in a matrix of finer components and form rocks such as coral, bryozoan and foraminiferal limestones. Some fossils have silica shells, and their accumulations can lead to development of siliceous sediments, such as deep-sea sediments from various planktonic, unicellular organisms (radiolarians and diatoms). Peat and coal deposits, comprising the remains of plants are also biogenic sediments.

Depositional environments

The depositional environments of sediments can be deduced from the types of rocks that are present and the sedimentary structures they exhibit. Mapping and analysis of the succession of different layers allow a geologist to visualise the sedimentary environment under which they accumulated and the direction of supply can be determined from the sedimentary structures. The content of characteristic indicator minerals can provide clues to the provenance of the sedimentary detritus transported into a sedimentary basin.

The principal sedimentary environments include:

Continental environments
- Sediments laid down from river systems on land (alluvial deposits)
- Sediments laid down in lakes (lacustrine deposits)
- Sediments laid down on land by wind (aeolian deposits)
- Sediments laid down in a delta system at the transition between land and sea.

Near-coast and shallow water marine environments
- Clastic sediments deposited near the coast in shallow water (marine sands, lagoons, tidal areas)
- Marine salt deposits and clastic sediments laid down in shallow water
- Clastic sediments deposited in shallow water (siliciclastic sediments)
- Marine carbonate deposits formed in shallow water (limestones).

Deep-water marine sediments
- Clastic deep-water sediments (muds and silts, turbiditic layered sequences)
- Oceanic sediments formed by deep-water muds and oozes (pelagic sediments).

Glacial deposits
- Moraines and fluvioglacial sediments (ice-age deposits).

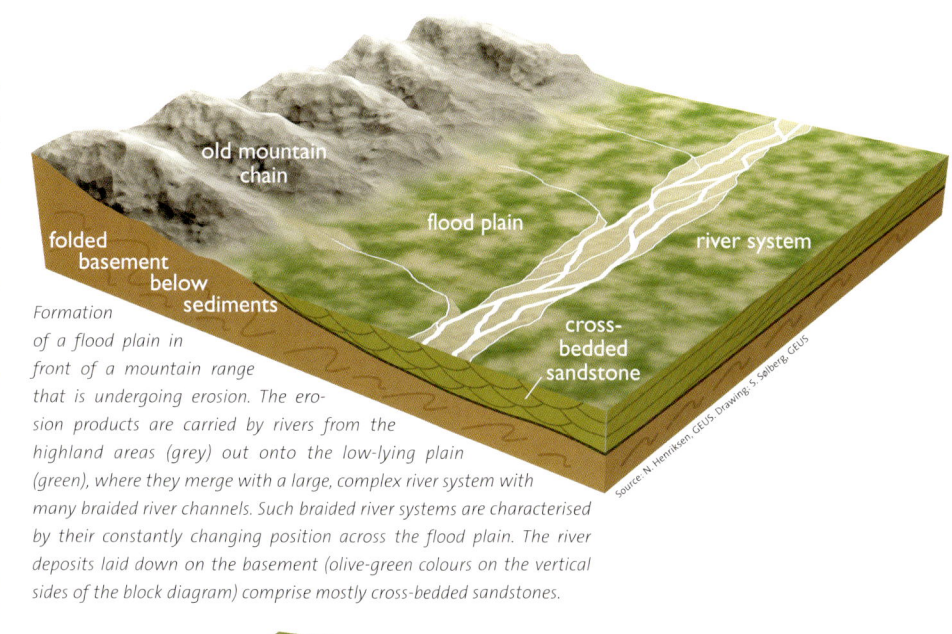

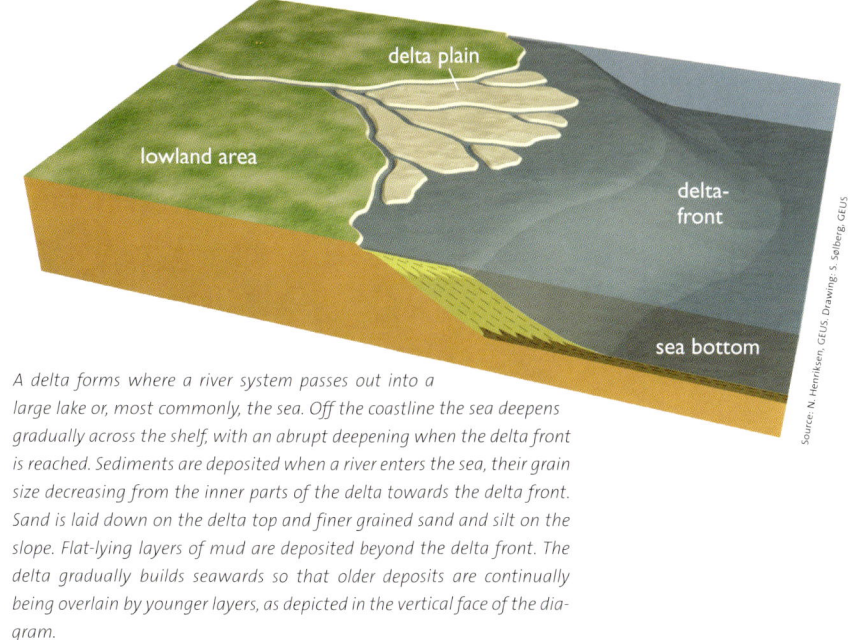

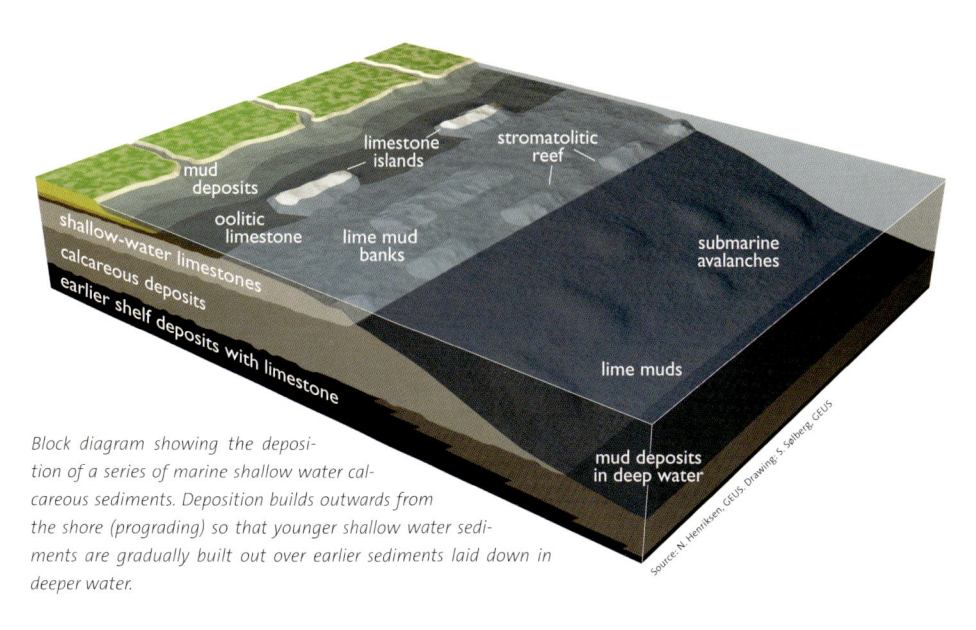

Formation of a flood plain in front of a mountain range that is undergoing erosion. The erosion products are carried by rivers from the highland areas (grey) out onto the low-lying plain (green), where they merge with a large, complex river system with many braided river channels. Such braided river systems are characterised by their constantly changing position across the flood plain. The river deposits laid down on the basement (olive-green colours on the vertical sides of the block diagram) comprise mostly cross-bedded sandstones.

A delta forms where a river system passes out into a large lake or, most commonly, the sea. Off the coastline the sea deepens gradually across the shelf, with an abrupt deepening when the delta front is reached. Sediments are deposited when a river enters the sea, their grain size decreasing from the inner parts of the delta towards the delta front. Sand is laid down on the delta top and finer grained sand and silt on the slope. Flat-lying layers of mud are deposited beyond the delta front. The delta gradually builds seawards so that older deposits are continually being overlain by younger layers, as depicted in the vertical face of the diagram.

Block diagram showing the deposition of a series of marine shallow water calcareous sediments. Deposition builds outwards from the shore (prograding) so that younger shallow water sediments are gradually built out over earlier sediments laid down in deeper water.

BASIN DEPOSITION

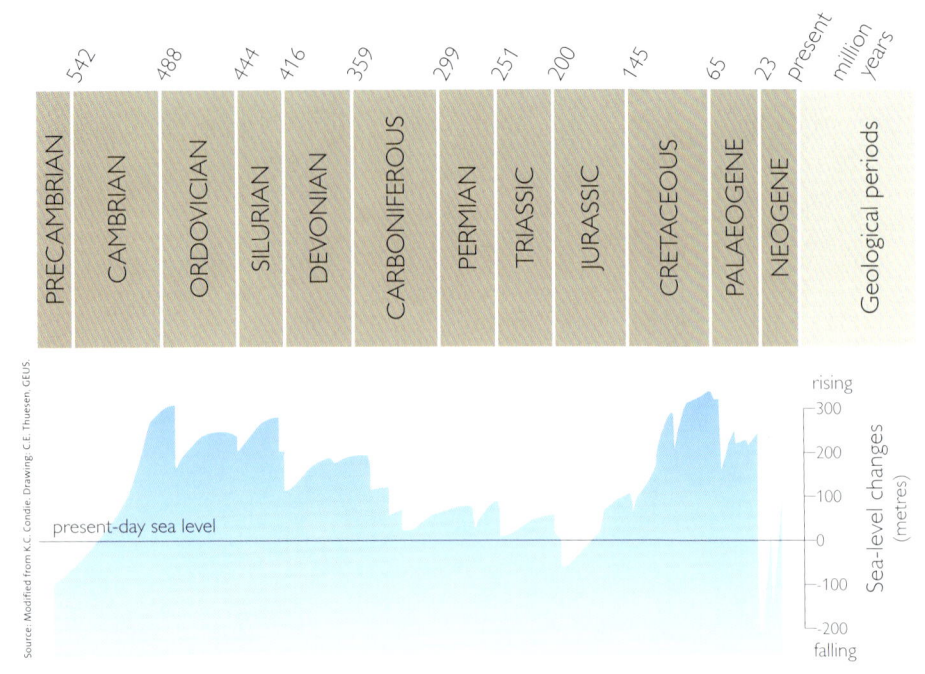

Global sea level changes during the last 550 Ma. The global sea level depends on the total volume of water in the oceans, as well as on how much frozen water (ice) is contained in ice sheets. The spatial size of the oceans depends on plate tectonic processes; newly formed mid-oceanic ridges and deep-sea trenches respectively decrease or increase the ocean volume. During ice ages a large amount of frozen water is contained in ice caps, so that the sea level falls. As shown in the diagram, the range of sea level change has been about 500 m.

Rates of sedimentation are very variable and depend upon the depositional conditions. In a deep-water basin in North Greenland, marine sand and mud deposits laid down by turbidity currents accumulated several hundred metres of sediment per million years, while in periods with low sediment supply (starved basin conditions) only a few metres per million years were formed in the same basin. Bordering shallow water platform areas may accumulate calcareous deposits at rates of 25–50 m per million years. On land and in near shore areas sedimentation rates in fluviatile and deltaic systems are high, while they are generally much lower in lacustrine environments. However, there can be large local differences depending upon the rates of subsidence and the location within the basin. The highest sedimentation rates occur adjacent to steep faults that form the boundaries between elevated land areas and rapidly subsiding basins.

The sea level in the World's oceans has varied considerably with time, just as the altitude of the continental land masses has varied. Sea level rise leads to flooding of low-lying land depositing sedi-

DOLOMITIC LIMESTONE

Dolomite is a carbonate mineral ($CaMg(CO_3)_2$) described chemically as calcite ($CaCO_3$) with substitution of some calcium by magnesium (Mg). The mineral dolomite occurs in dolomitic limestones or makes up the bulk of the rock dolomite. Most carbonate deposits are primarily limestones (an almost pure $CaCO_3$ rock), but by later, secondary alteration (diagenesis and dolomitisation), magnesium can be added to the rock, replacing calcium and the rock changes from limestone to dolomite. The photograph shows a dark limestone (about 50 cm wide) with pale dolomitic parts.

Ymer Ø

ment over an earlier land surface. Conversely, a sea level fall can expose deposited sediments as dry land. Any sedimentary basin can thus be characterised by alternating periods of active sedimentation and times of non-deposition. A time gap in a layered sequence, when no sedimentation occurs, is called a hiatus (see figure p. 84). The duration of such a hiatus can vary from a few million to several hundred million years. However, a hiatus in a sedimentary sequence may be unrecognisable unless the character of the sediments has changed significantly or crustal movements during the hiatus have led to tilting or folding introducing a discordance between the two sides of the time gap.

Sedimentary structures

During deposition 'primary' sedimentary structures are frequently formed. These must not be confused with 'secondary' structures that relate to subsequent deformation (e.g. orogenesis). The primary sedimentary structures reflect the conditions at the time of deposition. The most important sedimentary structure is the primary bedding, which is nearly always roughly horizontal or, in general approximately flat lying. Bedding reflects natural variations in the sediment supply leading, e.g. to alternating sand and mud layers. These variations can relate to changing conditions within the basin, coarser sediments being deposited when there are strong currents and finer material being laid down when currents are weaker. The variations may also reflect variations in precipitation in the source area of the sediments, which may depend upon short-term seasonal alterations or longer-term climatic changes.

Sedimentary structures form essentially at the same time as the deposition, but even within this very narrow time frame three different types can be distinguished. The earliest of these (predepositional) are characterised by small-scale erosion in a single layer before the overlying layer is deposited. The second type forms structures concurrent with the depositional processes themselves (syndepositional). The third type forms after the sediments were

Numerous small current ripples overprinting larger, linear ripples. Both sets of ripples have a long, shallow-dipping slope and a shorter, steep slope. This asymmetry shows that deposition was from a current flowing from left to right. Yellow-grey sandstone from the Thule Supergroup, North Greenland. Field of view is about 2 m across. Beneath the photograph is a schematic profile through the structures seen.

A cyclic (repeating) sedimentary sequence in which the grain size decreases upwards in each cycle. Yellow-white sandstone layers grade upwards (to the right) into darker coloured, silt- and clay-bearing mudstones. The build-up of each cycle is the same, and shows that the depositional environment has repeatedly oscillated from near-shore sedimentation (the coarse-grained sandstones) to outer shelf sedimentation (fine-grained mudstones). This cyclic structure between the marked boundaries (dashed white lines) reflects a repeated, large-scale change of relative sea level. The profile illustrates a 2 km sequence in the upper Eleonore Bay Supergroup on Ymer Ø, North-East Greenland.

BASIN DEPOSITION

Trace fossils in red, fluviatile Upper Triassic sandstones in the Jameson Land Basin in East Greenland. Trace fossils are preserved sedimentary structures that are created by the activity of animals. Examples of trace fossils include the burrows of worms, snails and bivalves, movement trails of different animals, footprints of a wide range of animals from insects to dinosaurs, feeding trails of animals eating sediments, and their excreta. The photograph shows burrows made by small insects – the drawing pin in the lower right-hand corner shows the scale.

deposited, but while they were still soft and water saturated (postdepositional). These structures are usually triggered by the weight of overlying accumulating sediments (compaction), but may also involve tectonic disturbances. Examples of the three types:

- Erosional structures – Predepositional:
  Small-scale erosion channels, removal of loose material (scours), stream erosion marks (flutes), drag marks caused by moving objects (grooves)
- Depositional structures – Syndepositional:
  Cross-bedding (small- and large-scale), lamination, cross-lamination, current ripples
- Load structures – Postdepositional:
  Collapse structures (slumps), slide structures, convoluted lamination, dewatering structures (e.g. flame structures).

Other structures found in sediments include different traces of biological activity such as burrows formed in a soft substrate, as well as the mixing of material from different layers due to animal activity (bioturbation). A family of structures can occur due to exposure and drying out of a sediment surface raised above water level (e.g. desiccation cracks in muddy sediments and break-up of surface layers to form loose flakes), or subsurface erosion by running water (karst landscapes and caves in limestones).

Most sedimentary structures are in the range of a few centimetres to a few metres in size, but both larger and smaller examples can occur. The various structures are important indicators of the conditions prevailing during deposition. Their association with each other and variations at different levels in a sequence also provide valuable information on the changing depositional environments.

Older and younger basins in Greenland

Between 420 and 350 Ma ago two major, coast-parallel mountain belts formed in Greenland, namely the Caledonian fold belt of North-East Greenland and the Ellesmerian fold belt that traverses North Greenland (see map p. 92). The orogenic deformation (mountain building) disturbed the sedimentary rocks laid down in (older) sedimentary basins prior to the formation of the fold belt, while those sediments deposited in (younger) sedimentary basins, formed after the folding, and were obviously not deformed. It is appropriate when describing the geological development of Greenland to distinguish between the series of older depositional basins that formed in the period from 1750 to 400 Ma and the depositional basins younger than 420–350 Ma. The older and younger basins are treated separately in the chapters that follow, although in many respects they reflect quite similar geological processes. The main differences between the basins are determined by their age and their different exposure to crustal movements, continental displacements, sea level changes and climatic variations.

At the present time the older depositional basins occur almost exclusively in North and North-East Greenland, while the younger basins are mainly found in the coastal areas of central East and central West Greenland, as well as in the continental offshore regions that surround most of Greenland.

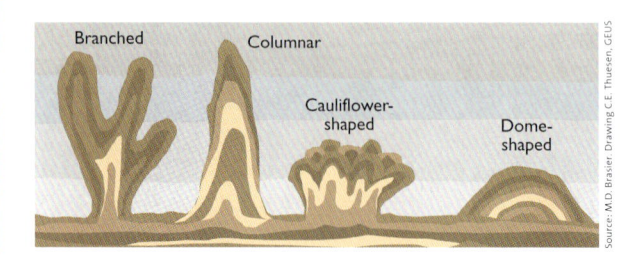

Stromatolites are calcareous structures that form from an interplay between the growth of organisms and the precipitation of calcareous sediment. The organisms are usually primitive types of blue-green algae and bacteria, whose biological activity leads to the precipitation of calcareous mud that becomes trapped. The photograph shows a top view of several cauliflower-shaped depositional structures with a diameter of 30–40 cm (penknife gives scale). Many stromatolites occur in late Precambrian sedimentary deposits. Example is from the Thule Supergroup, northern North-West Greenland.

Different stromatolite types shown in vertical section. The structures represent fossilised algal and bacterial mats (sediment adheres to the surface of the organism) that are principally formed from blue-green algae (Cyanophyta). Each millimetre-scale layer represents a single depositional event that might reflect daily changes such as tidal variations. Modern algal mats form in both brackish and fully marine environments, frequently in warm water and typically where there are few bottom living animals to disturb the delicate layers.

STRATIGRAPHY

The word 'stratigraphy' is derived from stratum, Latin for a layer, and is used for the geological discipline that describes the construction and age relations of a layered sequence. Sedimentary deposits show great variation in rock types, and in the conditions under which they were deposited and the fossils they may contain. The same sediment type, e.g. sandstone, can be deposited under different conditions. Conversely, different types of sediment, e.g. sandstone, siltstone and mudstone, may be deposited at the same time, but in different places within a basin. The fossil content within a sediment may also vary at different places in the depositional basin. Because of these environmental variations it is necessary to distinguish between different stratigraphic relationships, depending upon whether rocks (lithostratigraphy), fossils (biostratigraphy) or time (chronostratigraphy) are used as the basis for subdivision.

Sedimentary sequence with layers of different composition reflected in the differing colours of the sediment. Part of the Neoproterozoic Rivieradal Group in eastern North Greenland; older, dark-coloured siltstones in the lower part of the cliffs are overlain by younger, pale-coloured sandstones and conglomerates. The cliff is about 700 m high.

Lithostratigraphy

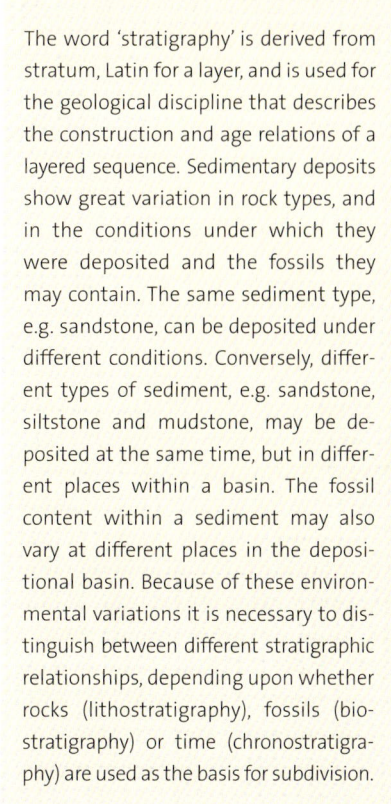

Geological mapping in Greenland is largely based on lithostratigraphic divisions with recognition of units that can be visually distinguished from each other in the field, and whose distribution and thickness variations can be recorded Thus a pale-coloured sandstone unit or sequence can be directly distinguished from a dark-coloured silt- and mudstone unit or sequence. The smallest lithostratigraphic unit is a 'Bed' that often has dimensions from a few decimetres to metres in thickness. The next division is a 'Member' that comprises a number of beds. A 'Formation' is the fundamental, formal unit, which can be several hundred metres thick and must exist as a coherent body traceable over a large distance. Lithostratigraphic divisions as members and formations can immediately be used as mapping units. The next division is a 'Group' that usually contains several Formations, and normally occurs as layered successions from several hundred metres to several kilometres thick. A number of successive Groups may be placed together as a 'Supergroup'. Lithostratigraphic divisions are defined and named from a specific geographical area (a type locality).

Biostratigraphy

Biostratigraphic divisions are based on the diversity and occurrence of fossil assemblages present in a unit. The basic unit is the 'biozone' or 'zone' that is a part of a sequence characterised by a particular fossil or fossil assemblage that distinguishes it from adjacent parts. This usually requires that a specific faunal assemblage is present, with one or more species occurring in association in this zone. An individual biozone can span over several of the smaller lithostratigraphic divisions. A biozone is usually named after one or more of the characteristic fossils that occur within the zone; e.g. the *Paraglossograptus tentaculatus* Zone named after an Ordovician graptolite that occurs in North Greenland.

Chronostratigraphy

Chronostratigraphic divisions are employed for rock sequences that formed during a specific geological time interval. Distinction is made here between the designation of a depositional sequence formed in a certain time and the designation for a time period. The smallest sedimentary sequence deposited in a certain time is called a 'Stage', corresponding in time to an 'Age' (e.g. Oxfordian). The next division is a 'Series' that in time units equates to an 'Epoch' (e.g. Upper Jurassic). Then comes a 'System', equal in time to a 'Period' (e.g. Jurassic), and above this is the 'Erathem', equivalent to an 'Era' (e.g. Mesozoic). The largest division is the 'Eonothem' that in time corresponds to the 'Eon' (e.g. Phanerozoic). Formal rules govern subdivisions for both depositional and time periods. Thus, a deposit of Lower Cambrian limestone is said to have formed in Early Cambrian time, and an Upper Jurassic sandstone to have been deposited in Late Jurassic time. Many people are confused by this – but the difference can be remembered, if one bears in mind that the upper part of the Eiffel Tower was built in the late 19th century (- not late Eiffel Tower built in the upper 19th century).

LITHOSTRATIGRAPHIC UNITS (SHOWN IN ORDER OF RANK)

Unit	Examples
Supergroup	Eleonore Bay Supergroup
Group	Lyell Land Group
Formation	Skjoldungebræ Formation
Member	Aggersborg Member
Bed	Gråklint Beds

CHRONOSTRATIGRAPHIC UNITS (SHOWN IN ORDER OF RANKS)

Applied to rock sequences	Applied to time intervals	Examples
Eonothem	Eon	Phanerozoic
Erathem	Era	Mesozoic
System	Period	Jurassic
Series	Epoch	Upper Jurassic
Stage	Age	Oxfordian

BASIN DEPOSITION

Very well-preserved Proterozoic basin deposits

Older continental basins in North and North-West Greenland

After consolidation of the basement shield, two separate subsiding areas developed in North Greenland leading to basin formation adjacent to the northern continental margin. The oldest basin developed in eastern North Greenland around Independence Fjord about 1750 Ma ago, with development of widespread sandstone deposits. On the eastern edge of this basin a volcanic province developed and lavas were interlayered with the sandstones. After an interval of some hundred million years without deposition (a hiatus), a new volcanic province developed in eastern North Greenland about 1380 Ma ago, and a thick succession of lava flows was laid down on the earlier sandstones and volcanic rocks.

In the Thule region, in northern West Greenland, a younger basin developed. During the period 1270–650 Ma a more complex succession of continental and marine sedimentary deposits together with occasional volcanic lavas and intrusive rocks accumulated.

Except where involved in Caledonian deformation in North-East Greenland the deposits are not deformed, and the preservation is comparable to corresponding, much younger successions elsewhere in Greenland.

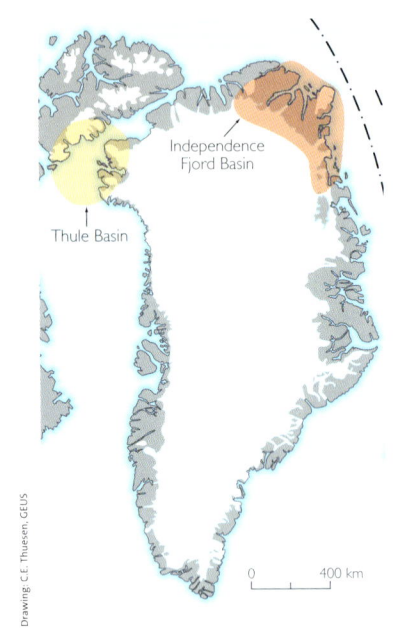

Older continental basins of Proterozoic age in North Greenland. In the Independence Fjord Basin, sandstone sequences were deposited first, forming the Independence Fjord Group. Some hundred million years later they were overlain by a series of plateau basalts – the Zig-Zag Dal Basalt Formation. Deposition in the Thule Basin took place over a 600 Ma time period. These deposits, the Thule Supergroup, were dominated by sandstones with scattered basaltic units.

The oldest basins in North Greenland (1750–1200 Ma)

The earliest sedimentary basin was a continental, inland basin that developed adjacent to the north-east margin of the Greenland basement shield. Here a more than 2 km thick sandstone sequence, the Independence Fjord Group, was deposited. These sediments are found today in eastern North Greenland, mainly as unfolded layered sequences, in the areas west and south of the Palaeozoic fold belts. The basin originally extended farther east and in Kronprins Christian Land the sediments were affected by deformation in the Caledonian fold belt. The rocks underlying the sandstones are not exposed but indirect evidence suggests that they rest on a gneissic basement. The sandstones were originally lake sediments, and include some beds of wind-borne material. Alternating with the sandstones are characteristic, silty beds that preserve indications of deposition in a salt lake. It is estimated that the entire succession accumulated under a climate characterised by low precipitation with, at times, desert-like conditions. The Independence Fjord Group crops out over an area of more than 80 000 km². Towards the eastern margin of the basin the sedimentation was disrupted by faulting and the extrusion of volcanic lava flows, which suggest the rocks here were probably laid down near the boundary between the continent and an old ocean. The age of the Independence Fjord Group sandstones is poorly constrained, but some of the volcanic rocks associated with the sandstones in the Caledonian fold belt in the east are about 1740 Ma

Mountain wall on the north side of Ingolf Fjord in Kronprins Christian Land that shows Independence Fjord Group sandstones cut by dark, volcanic intrusions. The rocks here are intensively deformed by the Caledonian orogeny, which resulted in steeply inclined structures. The original angles between the bedding planes of the sandstones and the cross-cutting intrusions have been almost obliterated, so they are now nearly parallel. The mountainside is about 1000 m high.

Cliffs on the south side of Independence Fjord in North Greenland showing a section through unfolded sandstones of the Independence Fjord Group (pale, yellowish-grey colours) cut by a network of dark volcanic intrusions that form dykes and flat-lying sills (Midsommersø Dolerites). The dykes and sills are feeder channels to the overlying basaltic rocks (Zig-Zag Dal Basalt Formation), seen at the upper left of the section. Cliffs are about 800 m high.

old, indicating that the Independence Fjord Group sandstones are of similar age.

After a protracted interval of several hundred million years, without evidence of geological events, a basaltic lava sequence, at least 1350 m thick was laid down over a very large part of eastern North Greenland. These plateau basalts are known as the Zig-Zag Dal Basalt Formation and comprise more than 50 individual flows. In the lower part of the sequence, pillow lava structures show that some flows were erupted under water, while the upper part of the sequence was clearly extruded onto a land surface. The volcanic activity occurred either as the beginning of ocean floor spreading north of a Greenland–Canadian continent, or due to the passage of the continent over a volcanic centre in the Earth's mantle (a hotspot). The basalts originate from a deep-seated magma chamber from which eruptions reached the surface through a network of dykes and flat-lying intrusions (sills). This dyke-sill network,

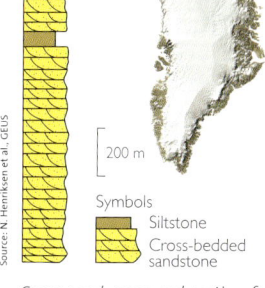

Compound measured section for the about 1750 Ma old continental deposits of the Independence Fjord Group in an area south of Peary Land, North Greenland.

BASIN FORMATION ON CONTINENTAL AREAS

Sedimentary basins generally form as slowly subsiding, bowl-shaped depressions in the continental crust (intra-continental basins), or are developed at the boundary between a continent and an ocean (epi-continental basins). The intra-continental basins can be up to several thousand kilometres across and usually the deposits are thickest in the middle, tapering off towards the edges. Basin formation can take place as a steady development through several hundred million years with slow subsidence closely related to the rates of sedimentation, so that deposition takes place in shallow to moderate water depths at all times. During this, often long time span, the sediments may accumulate to many kilometres in thickness.

The basins in a continental area can include both fresh water deposits from rivers and lakes, and marine sediments that form if the sea breaks through and floods an otherwise isolated basin. This means that geographically basins often form large, partially closed arms of the sea or bays, rather than completely land-locked inland lakes. Modern examples of such marine basins in a continental area are Hudson Bay in Canada and the Baltic Sea in Scandinavia.

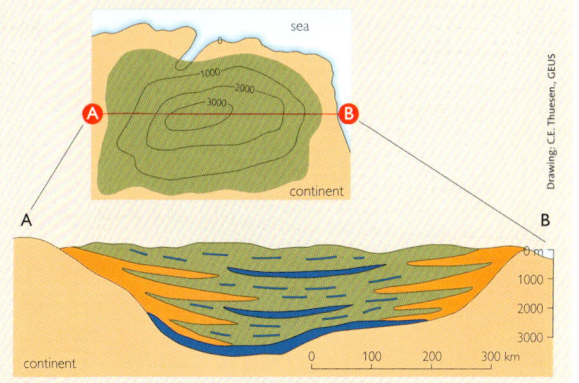

Sketch map and section through a model basin with continental deposits formed by subsidence of an inland continental area. Blue – limestone; orange – sandstone; green – fine-grained mudstone. The profile is shown with a large vertical exaggeration.

BASIN DEPOSITION

Part of the Zig-Zag Dal Basalt Formation in eastern North Greenland. The individual layers are flat lying, like the sediments in a basin. Each layer represents a volcanic eruption where the lava has flowed out over an extensive flat land surface. The lava was very fluid and able to flow tens of kilometres away from the eruption centre. The profile is about 500 m high.

known as the Midsommersø Dolerites, comprises numerous intrusions, some up to 100 m thick, which dissect the undeformed Independence Fjord Group sandstones. The volcanic sequence formed over a short time period and eruption activity was often very violent. Some lava flows are over 100 m thick and cover very large areas, and probably had volumes of many hundreds of cubic kilometres. The dolerite dykes, that are feeders to the basalts, are about 1380 Ma old. Away from the Caledonian fold belt the old sandstones and basalts are all preserved in an unmodified state comparable with that of much younger sedimentary and volcanic formations, e.g. the Palaeogene volcanic province (see p. 138).

The Thule Supergroup sandstones and basalts of North-West Greenland (1270–650 Ma)

In the northerly part of the Greenland-Canadian basement shield a large subsiding area formed, which lasted for more than 600 Ma. A sedimentary basin more than 300 km across developed covering the area from Thule in Greenland to the coastal areas of Ellesmere Island in Canada. A layered sequence was deposited in

PLATEAU BASALTS

Basalt magma can form by partial melting of ultrabasic rocks within the Earth's mantle at relatively shallow depths under the ocean floor (6–10 km), but at greater depths under the continental crust (50–100 km; see p. 138 and box p. 152). The melts accumulate in magma chambers at depth and from there they can reach the surface as volcanic eruptions with different forms, e.g. a central volcanic cone or a large, widespread eruption such as a plateau lava sequence called flood or plateau basalts. Such basalts occur in different places and times in Greenland. The oldest of these eruptive successions make up the 1380 Ma old Zig-Zag Dal Basalt Formation in the eastern part of North Greenland and the youngest occur in the central parts of East and West Greenland, where they form parts of the extensive eruptions related to the North Atlantic volcanic province about 60–55 Ma ago (see p. 138).

Plateau basalts often form in the early stages of continental spreading and oceanic crust formation, and most commonly along continental margins, where the basalt lavas flow out over the underlying crust.

When the basalt magmas reach the surface they have a temperature of about 1200°C and a very low viscosity. The very fluid magmas form large, coherent lava flows that build up into layers 5 to 50 m thick. An individual flow can spread out over many hundreds of square kilometres and forms a nearly horizontal body. Each flow follows the terrain, infilling irregularities or hollows in the landscape. The result is the build up of a flat-lying sequence of basalts with each layer succeeding the next, just like the layers in a sedimentary basin. However, while sediments have a comparatively low depositional rate that is measured in tens of metres per million years, basalts may accumulate considerably faster, at up to several kilometres in one million years.

The magma from deep-seated magma chambers finds its way to the surface through feeder channels, usually part of a large fissure system that cuts through the crust below the basalts. Where the melts cut through a sedimentary sequence they often form an irregular network of narrow basalt intrusions that are emplaced in steeply inclined fissures (dykes) or as flat lying sheets (sills). A good example of such a dyke and sill system is the Midsommersø Dolerites (see p. 71) in eastern North Greenland. It is generally difficult to find traces of feeder dykes in the plateau basalts themselves.

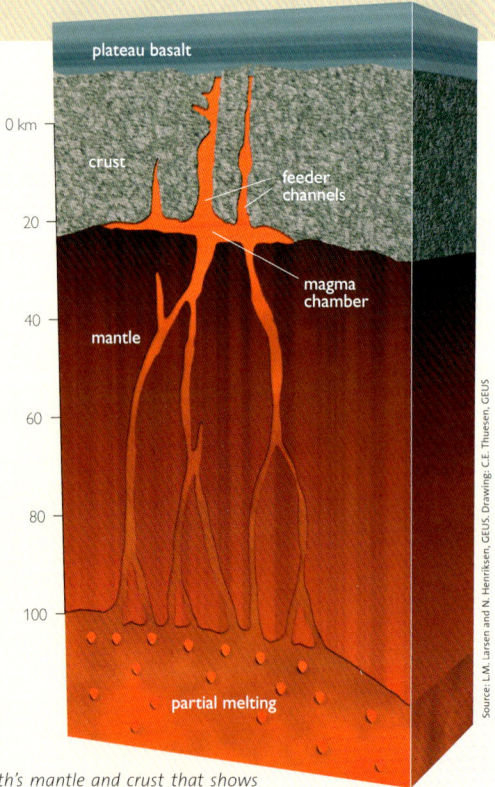

Section through the upper part of the Earth's mantle and crust that shows how the source material, for the very large sequences of plateau basalts, forms through partial melting of the underlying portion of the mantle. The melted magmas travel up towards the surface through a branching network of channels. At different levels local magma chambers can be formed from which the fluid basalt magma moves upwards to the surface as pulses of magma that make up an individual eruption.

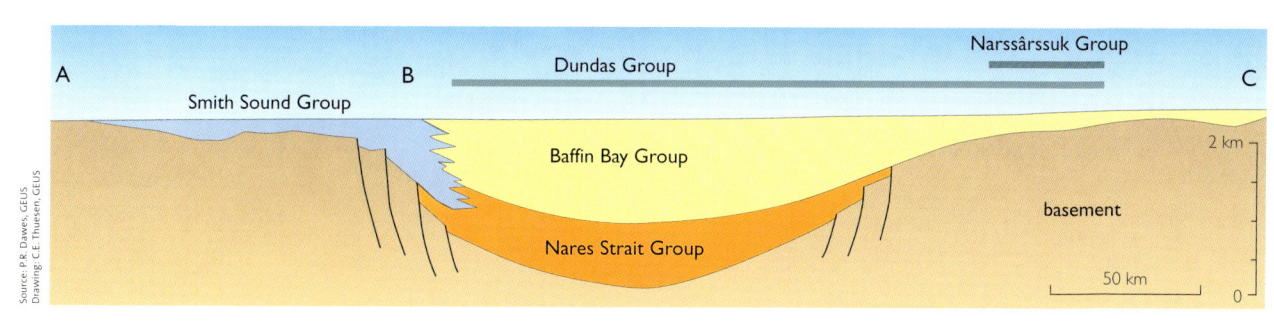

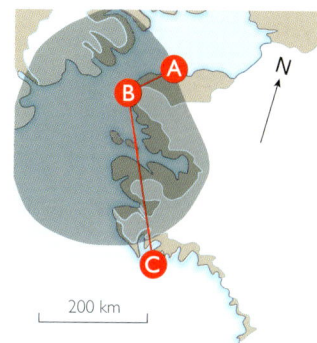

Older continental basins in North and North-West Greenland

this basin that in Greenland reaches a thickness of 6–8 km, but could be even thicker in the sea-covered portion of the basin.

The Thule Supergroup comprises a variety of unfolded often strongly coloured sedimentary rocks with a marked stratified appearance that even from a great distance can be recognised as basin deposition. Lowest in the sequence, sandstones and shales dominate with subordinate units of limestone and layers of basaltic lava that were extruded onto a land surface. The upper part of the sequence includes alternating units of sandstone, mudstone and limestone with local salt deposits. The whole succession reflects deposition in a large shallow-water sedimentary basin in a continental area, and includes intercalations of wind-blown deposits. Most sediments are river and lake deposits, but some intervals include different shallow water marine deposits laid down in a tidal environment. This indicates that the sea periodically broke into the basin and locally covered it with salt water. In the upper part of the succession, many beds with algal structures occur together with cyclic deposits as well as different sediments precipitated by evaporation. The environment for these deposits can be compared with a coastline near a desert area such as that around the Persian Gulf today.

The Thule Supergroup deposits preserve different traces of biological activity. In addition to the algal constructions (stromatolites) there are primitive forms of microfossils known as acritarchs, as well as trace fossils formed in soft sediment by the life activities of organisms (e.g. fossilised tracks, burrows, borings etc. made by animals moving, feeding, digging etc.). The fossils can, to a limited extent, constrain the age of the basin's development.

North to south section through the Thule Basin, with an inland basin developed on the Precambrian basement. The Thule Supergroup comprises five groups shown on the figure with different colours. The two upper and youngest groups are shown schematically as bars above the main part of the figure. There is a vertical exaggeration of 25 times relative to the horizontal scale, and the unfolded succession is in reality flat-lying.

Section through the lower part of the Thule Supergroup in the central part of the Thule area, North-West Greenland. Dark volcanic horizons alternating with different sedimentary deposits occur at the base and are succeeded by reddish mudstones, pale yellowish sandstones and at the top reddish sandstones. The mountainside is about 700 m high.

BASIN DEPOSITION

The thickest basin succession in Greenland is about 20 km thick

Older marine basins in North-East Greenland

The southern part of the Caledonian mountain belt in North-East Greenland incorporates two very thick, folded, sedimentary successions, found within Caledonian thrust sheets. In many places the sediments are only partially modified by deformation and metamorphism, so it is possible to recognise their original character and determine the conditions of their formation within large sedimentary basins.

The oldest sediments were deposited in a basin that was formed about 1000 Ma ago and are known as the Krummedal supracrustal sequence; the sedimentary material was supplied mainly from source areas outside Greenland. A younger sedimentary succession (Eleonore Bay Supergroup, Tillite Group and Kong Oscar Fjord Group) formed in another basin over a long period from about 900 Ma to 450 Ma, and it is well exposed in the central fjord zone (72–75°N) of present-day East Greenland.

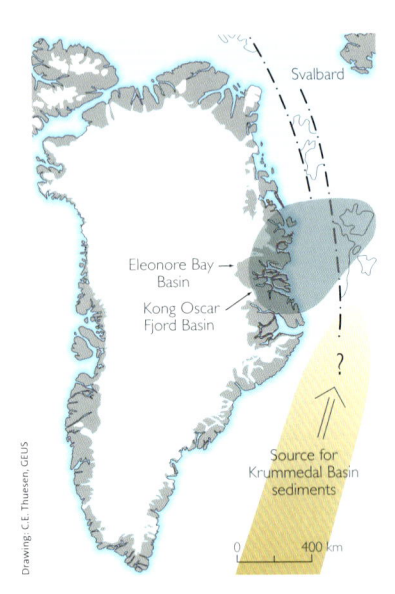

Sketch map showing the distribution and extent of the older marine basins in North-East Greenland. The 1000 Ma old Krummedal sequence sediments were derived from a source area with basement rock units from parts of the North American and European continents in a position south-east of South Greenland. The present occurrence of the sediments in North-East Greenland was due to long-distance transport of the sediments, and later tectonic displacements (see p. 96).

Almost identical Neoproterozoic – Silurian successions were laid down in both North-East Greenland and Svalbard. In Greenland the succession comprises the Eleonore Bay Supergroup, the Tillite Group and the Kong Oscar Fjord Group. The sequences were originally deposited in the same extensive sedimentary basin that was split up and disrupted along crustal-scale faults with large horizontal displacements. This means that a part of present day Svalbard was originally positioned adjacent to the central part of East Greenland, as shown on the map.

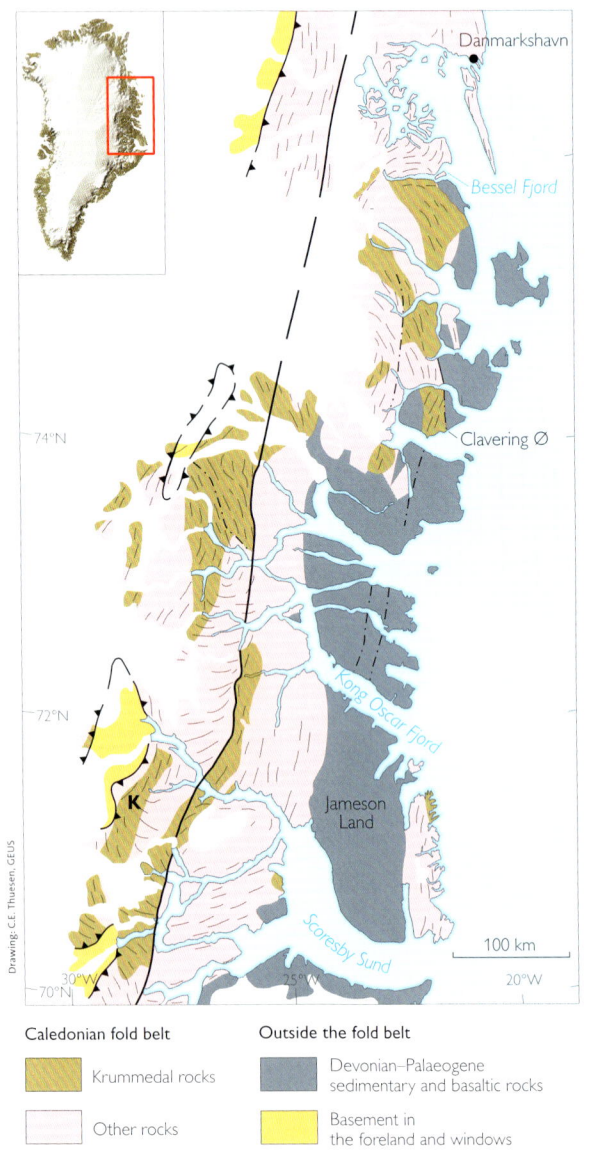

Caledonian fold belt
- Krummedal rocks
- Other rocks

Outside the fold belt
- Devonian–Palaeogene sedimentary and basaltic rocks
- Basement in the foreland and windows

The distribution of the Mesoproterozoic Krummedal sedimentation in the Caledonian fold belt in East Greenland. The sedimentary sequence is named after a key locality – Krummedal (K) – in the inner part of the Scoresby Sund region. The rocks were metamorphosed into schists, but in Krummedal many original sedimentary features are still preserved.

The deposits of the Eleonore Bay Basin – Kong Oscar Fjord Basin contain a succession with a total thickness of up to 18.5 km. It comprises exclusively sediments that were deposited on a shallow-water shelf with a slow, steady subsidence over several hundred million years. On the Svalbard archipelago in the North Atlantic, there is an almost identical sedimentary sequence of the same age. It is therefore likely that the original sedimentary basin had a much larger extent than indicated by the occurrences in North-East Greenland alone.

Krummedal sequence sedimentation in North-East Greenland (about 1000 Ma old)

The southern part of the Caledonian fold belt in North-East Greenland comprises some major thrust sheets several kilometres thick, which have transported rock units hundreds of kilometres westwards across the basement shield (see p. 100). The two major thrust sheets have been mapped between the Scoresby Sund area (70°N) and Bessel Fjord (76°N), and include very characteristical, metamorphosed and folded sedimentary successions. These sedimentary rocks are referred to as the Krummedal supracrustal sequence, named after a locality in the Scoresby Sund area where they were first recognised as a separate unit. Radiometric age determinations (see box p. 48) show that the deposits were laid down in a relatively short time period of 50–100 Ma about 1000 Ma ago.

The Krummedal sequence comprises a 2.5–8 km thick series of mud- to sand-grade sediments with subordinate local, calcareous units. The rocks have everywhere been metamorphosed into schists, showing they were transformed at pressures and temperatures corresponding to amphibolite facies conditions (see box p. 35). Furthermore, within much of the upper thrust sheet the sediments have been subject to local melting and granite formation (migmatisation) during two distinct mountain-building events.

Despite the heating and deformation, the rocks in many places preserve some of their original sedimentary characteristics in the form of distinct layering and sedimentary features, such as current ripples and cross-bedding. There is a considerable variation in the composition of the sequence both horizontally and vertically which, together with the metamorphic transformation, makes it difficult to unravel the original structure of the basin. The dominant sedimentary components can, however, be interpreted to have originated in a rift-like, deep-water marine basin more than 700 km from north to south and up to 300 km wide.

Unravelling the origin of the Krummedal sedimentary rocks required a series of very detailed radiometric age determinations, where the mineral zircon is used as an age indicator (see box p. 76). Zircon crystals in the sediments mainly fall into two different age groups, one from 1700–1450 Ma, and the other from 1400–1100 Ma. This spread of ages makes it clear that the zircons represent transported grains (sand grains) of different ages derived from a geologically varied hinterland and carried into a younger basin. These age groups of zircons are not represented in the adjacent Greenland basement areas, so the source area must be located in another part of the North Atlantic area, away from Greenland. This could indicate that the older group of zircons might have been carried into the Krummedal Basin from the Baltic basement in Scandinavia, while the younger zircon group could have been derived from the eastern part of North America. After deposition in

Weakly modified, folded Krummedal sequence sediments that are composed of layers of quartzitic sandstone with thin layers of fine-grained material.

Small-folded, finely banded, quartzitic sandstones containing thin silt layers. The rocks have been metamorphosed under low amphibolite facies conditions (see p. 35), but they have not been partially melted (migmatised, as the rock shown on the adjacent photo). The notebook (centre) is about 15 cm long. These rocks occur near the type locality at Krummedal (indicated on the map p. 74) in the inner part of the Scoresby Sund region in East Greenland.

Strongly metamorphosed, dark grey sandstones of the Krummedal sequence with veins of whitish to rusty-coloured, granitic rocks that were formed by mobilisation (melting) of components with low melting temperatures in the sediments. The metamorphic grade equates to upper amphibolite facies conditions. The hammer shaft is about 40 cm long. These rocks occur in the central part of the Scoresby Sund region in East Greenland.

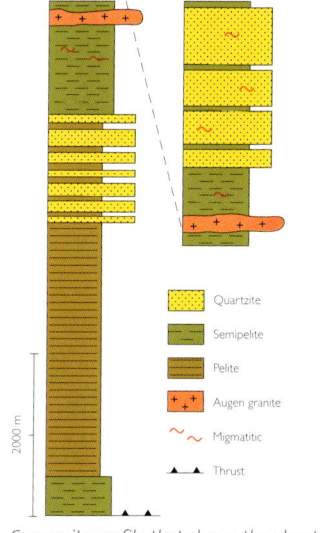

Composite profile that shows the about 8 km thick succession of the 1000 Ma old Krummedal sequence in the southern part of the Caledonian fold belt in the Scoresby Sund area. The rocks were originally laid down in a sedimentary basin south-east of Greenland, as sandy and muddy sediments that today occur in a transformed state because they were subjected to high temperature and pressure conditions, during the Caledonian folding.

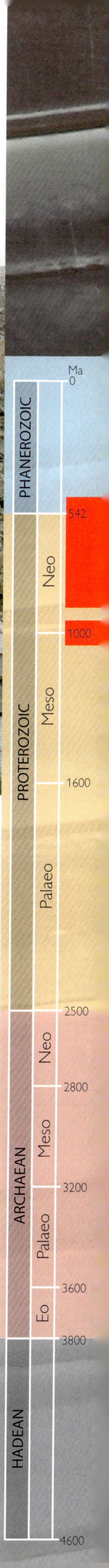

BASIN DEPOSITION

WHERE DO SEDIMENTS COME FROM?

Deposits laid down in sedimentary basins originate either as erosion products of an adjacent land area, or as chemical or biological precipitates formed within the basin. The transported material, called detritus, comprises mainly the mechanically eroded parts of older rocks. The products that accumulated in the basins range from large composite blocks of rock fragments down to very small grains of single crystals. When the transport distance is short the deposits are usually poorly sorted and the individual clasts or grains have a sharp, angular shape. Such sediments are called immature. Conversely, if the material has been transported over a long distance prior to deposition, the material is usually sorted into different grain-size fractions and the individual grains have a rounded shape; this type of sediment is called texturally mature. Deposits with immature sediments are nearly always derived from the immediate surroundings, for example, from a continent bordering the basin. Mature sediments on the other hand, may be transported up to several thousand kilometres before they become deposited. A basin with mature deposits can therefore be situated far from the area from which its deposits derive, and thus reflect a geological history completely different from that of its present location.

Many detrital sedimentary deposits are made of different mineral grains, some of which are very resistant to mechanical and chemical breakdown. One of these is the mineral zircon, single grains of which can be used to determine radiometric age (see box p. 48). By comparing these zircon ages with the ages of basement rocks in the adjacent hinterland, it may indirectly be possible to locate the region from which the zircon grain originated. An example of this method to estimate where material came from (provenance studies) was used on the 1000 Ma old Krummedal sequence sediments of the southern part of North-East Greenland (see p. 74). Age determinations on many zircon grains from these deposits yielded a spread of ages between approximately 1700 and 1100 Ma. Since rock units of this time interval are completely unknown in the extensive Precambrian basement complexes of Greenland, it is concluded that the sediments that make up the Krummedal sequence were derived from a land area not part of Greenland, and a long way away from their present day occurrence. The Krummedal sequence rocks are found exclusively in major thrust sheets in the Caledonian mountain belt in North-East Greenland. The sedimentary material was transported away from its original place of formation in two events: first, by long distance transport into the sedimentary basin and second, by tectonic thrusting of the whole sequence during the formation of the Caledonian fold belt.

Provenance studies of the components of sedimentary basins are today routinely carried out, both in connection with older and younger basins. They provide important information on how a basin development has taken place, and in younger basins help in the understanding of the potential for oil and gas resources.

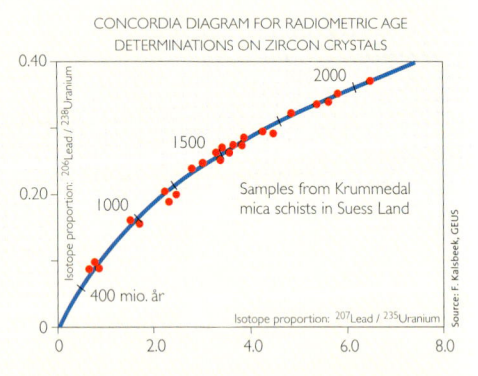

Diagram illustrating the spread of age determinations made on single zircon crystals from one individual rock sample (method described on p. 48). The blue line (shown with ages along the curve) illustrates the development with time of the isotopic components (uranium and lead) depending upon the age of the minerals. The red dots show the age determinations on different zircon grains. It is evident from the diagram that the zircon crystals have very different ages showing they were not formed at the same time. The largest groups of zircon crystals give ages from 1700 to 1100 Ma, which show that the zircons derive from erosion of different older basement areas. A few small groups of zircons have ages of around 930 and 435 Ma. These younger zircons reflect that the sediments from the Krummedal sequence have undergone metamorphism during mountain belt formation in two younger periods, when new zircon growth took place.

Grain shapes with different degrees of rounding. The shapes shown for sedimentary grains reflect transport over shorter or longer distances. The angular, rough grains were deposited close to their origin, whilst the smooth, rounded grains have travelled farther. Sediments that are dominated by one or the other of the grain types are called immature or mature.

the basin the sediments were first affected by mountain belt formation in the early Neoproterozoic about 950 Ma ago, and then later by the Caledonian orogeny about 420 Ma ago. During both of these events metamorphism transformed the mineral assemblages, and renewed zircon growth occurred.

Fjord zone basins, North-East Greenland (about 900–450 Ma)

The most distinctive basin succession in Greenland is the Neoproterozoic–Ordovician sequence preserved in the uppermost Caledonian thrust sheet of the central fjord zone (72–75°N). The up to 18.5 km thick succession reflects almost continuous deposition for about 450 Ma in a sedimentary basin that extended for approximately 500 km from north to south. The original basin must have formed far to the east of where the rocks occur today, as the sequence is preserved in a very large Caledonian thrust sheet (see p. 100) that has transported them several hundred kilometres westwards. Nearly identical rocks are found in the Norwegian archipelago of Svalbard, showing that the basin originally had a much greater extent. Part of the Svalbard archipelago was originally adjacent to East Greenland, and was later moved laterally towards the north. The original size of the basin is estimated from the proposed original location of the relevant part of Svalbard. This East Greenland basin was originally bounded to the west by an approxi-

Older marine basins in North-East Greenland

mately N–S-trending coastline, while the other boundaries are unknown. In North-East Greenland the sequence is divided into the Neoproterozoic Eleonore Bay Supergroup, the overlying Tillite Group, and finally the Cambrian–Ordovician Kong Oscar Fjord Group.

Eleonore Bay Supergroup sediments (about 900–600 Ma)

This very conspicuous sedimentary sequence comprises a series of weakly to moderately folded almost non-metamorphic rocks. Towards the base of the very thick sequence, tectonic modification and metamorphic transformations increase. The formation of the basin took place at about the same time as the development of the Hagen Fjord and Hekla Sund basins in the northern part of North-East Greenland, but the basin areas were not connected and the sedimentary sequences are generally very different.

The total thickness of the Eleonore Bay Supergroup sequence is up to 14 km. It is divided into a lower sandstone-dominated part (Nathorst Land Group) with a thickness of 9 km and an upper carbonate-dominated part (Lyell Land, Ymer Ø and Andrée Land groups) with a combined thickness of up to 5 km.

The Nathorst Land Group sandstones have sporadic intercalations of siltstone and mudstone and a single 250 m thick unit of limestone. In detail, the sequence is very variable and the thickness of a single unit changes both along and across the basin.

Panorama of the upper part of the Eleonore Bay Supergroup in the region west of Mestersvig in North-East Greenland. The 1900 m high cliffs show the multi-coloured sediments that include fine-grained quartzites, mud rocks and different limestones and dolomites. They were all deposited on a shallow-water shelf, approximately 700–620 Ma years ago. In the right foreground a tight fold in yellow and brown, striped dolomites is clearly shown. Several major faults, marked by gullies, cut up the layered succession into different fault blocks.

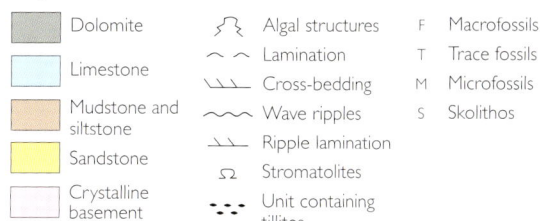

Legend

- Dolomite
- Limestone
- Mudstone and siltstone
- Sandstone
- Crystalline basement
- Algal structures
- Lamination
- Cross-bedding
- Wave ripples
- Ripple lamination
- Stromatolites
- Unit containing tillites
- F Macrofossils
- T Trace fossils
- M Microfossils
- S Skolithos

Composite sedimentary column that shows the main divisions of the about 18 km thick, layered sequence, laid down in the period from early in the Neoproterozoic (about 900 Ma) to the Middle Ordovician (about 460 Ma). The rocks were involved in Caledonian folding and thrusting, and occur in major thrust sheets (see p. 101) that have been transported westwards an estimated 300–500 km from their original place of deposition, east of the present coastline of North-East Greenland.

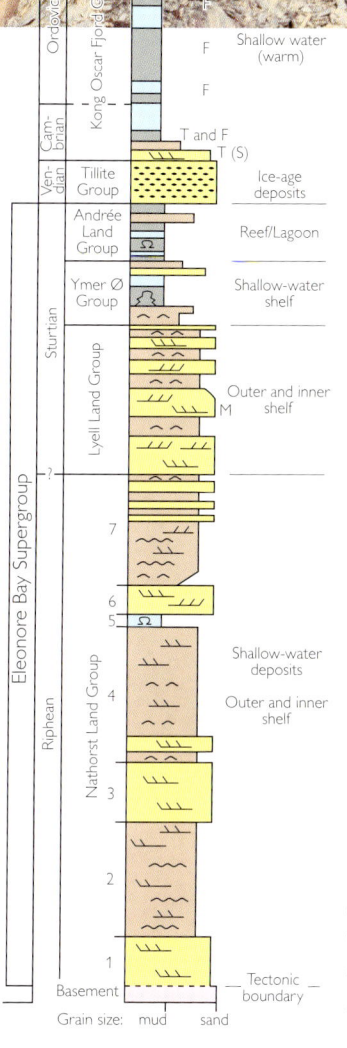

MEASURING SEDIMENTARY PROFILES – SECTION LOGS

Analysis of the development of large sedimentary basins is based on a study of the rocks making up the sequence, the sedimentary structures and any fossils they contain. Such studies are undertaken primarily by measuring sections in the field (see p. 19) through the succession – profile measurements – that include recording the occurrence of various rock types in cross-sections, up or down, through the succession, perpendicular to the bedding.

Such cross-sections may be drawn at different scales, depending upon the thickness of the sequence to be studied and the detail required. In detailed studies layers down to a few decimetres thick may be recorded, but in large, more regionally oriented studies the chosen scale for the measured units can be up to many metres in thickness. Typical observations recorded include the rock types making up the individual layers, their thickness and any significant grain-size variations. It is important to record the relationship between the beds, e.g. if they have erosional or transitional contacts. Sedimentary structures such as current indicators or wave ripples are recorded, together with any fossils present, and samples may be collected for later studies.

When measuring sedimentary profiles, fossils are collected when they occur. Geologists spend much time systematically extricating the fossils from the rocks using a hammer.

A section log

The measurements and observations recorded are often represented as a section log, which in geological terms is called a profile. The data are recorded using standard symbols and ornaments to produce a vertical column, the width of which is proportional to grain size, thus giving an overview of the rocks and structures present as well as any information on fossil content. The illustration is a section log through a part of the lower Palaeozoic sedimentary succession at the base of the Kong Oscar Fjord Group on Ella Ø, in the southern part of North-East Greenland. Profile measurement is not only utilised in the study of sedimentary basins, but can also be used in other layered sequences with flat-lying units, e.g. in areas with plateau basalts.

Within a basin area, a series of different profiles can be measured forming the basis for comparison and correlation between different parts of the depositional area. From the local geology and geography, geologists will carefully select where a profile is to be measured; it is here very important to make rational assumptions and judgements because each profile measured in the field can take one or two days to complete.

Basin analysis

After a large number of well-distributed, measured profiles have been made through a sedimentary basin succession, comparison and correlation can be undertaken. It becomes possible to evaluate how the depositional conditions in the basin area have developed, and it is often possible to determine sedimentation rates and determine from where the sediment originated. Indications of sea-level changes, climatic conditions and development of animal life can also be revealed. This information forms the basis for a basin analysis, which is not simply a description of the rock sequence, but an unravelling of the dynamic origin of the basin formation in terms of its overall geological context.

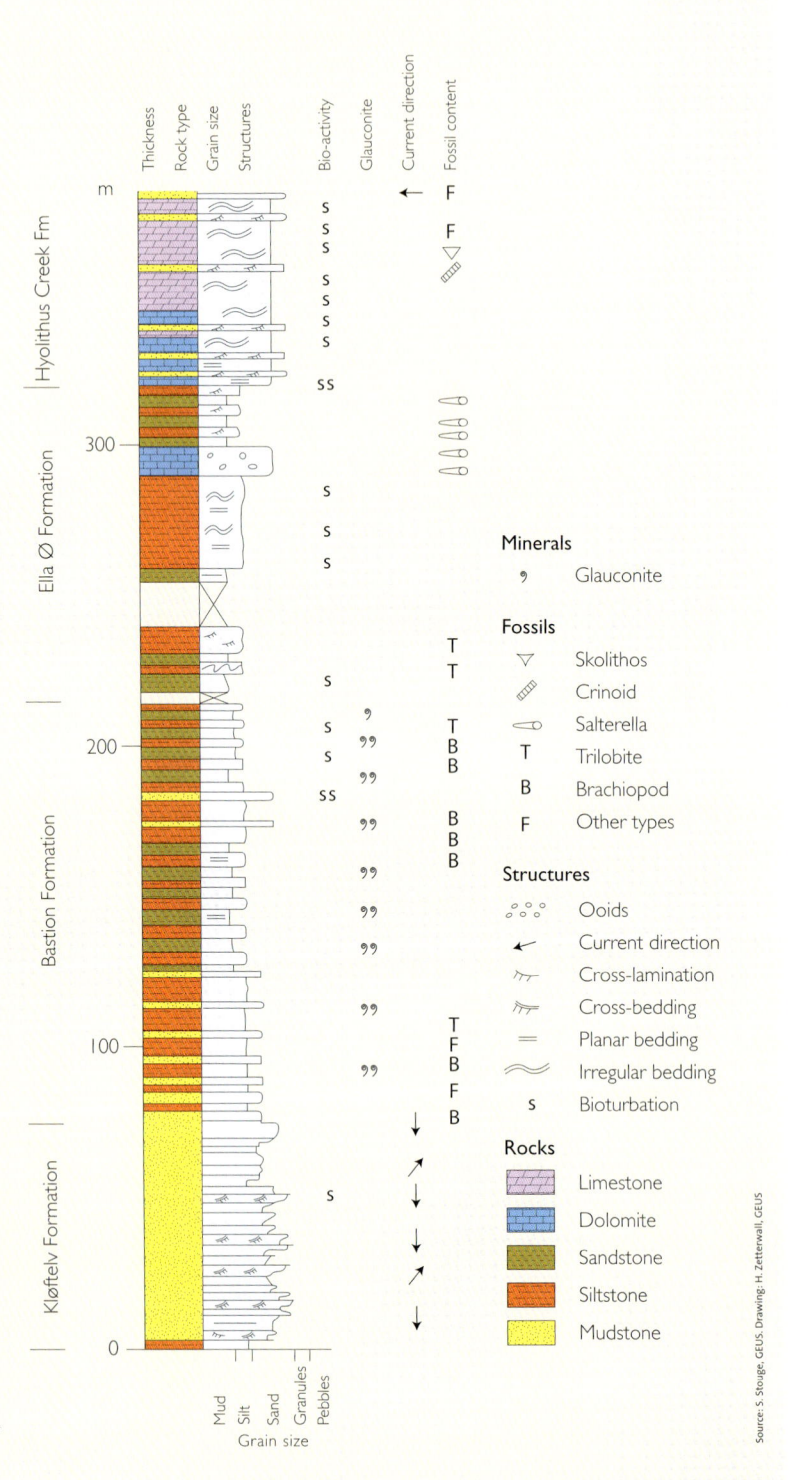

Example of a measured section with a record of the data in standard format. The measured section spans a sequence about 400 m thick in the lower Cambrian part of the Kong Oscar Fjord Group in the southern part of North-East Greenland. The section is divided into four named formations, and the coloured, left hand column shows the rock types. At the side of the column the grain size of the sediments and their structures are given. Information on the fossil content and other features are given on the right. The presence of the mineral glauconite is noted because it is an indicator of sediments laid down under marine conditions.

However, there is a general tendency for the sediment thickness to increase from the western margin towards the outer, eastern part of the basin. Throughout the whole of the 9 km thickness of the lower sedimentary sequence, sedimentary structures such as cross bedding, wave ripples, current features and load structures frequently occur. Algal structures and desiccation cracks are present in the limestone unit. These structures together with the types and characters of the sedimentary rocks show that the depositionary environment was a near-coast, shallow-water, marine area. The sediments represent deposition in a zone that spans across the outer shelf where, during storms, waves disturbed the sediments on the sea bottom, to the innermost shallow-water beach areas that were subject to tidal reworking and at times were left dry.

In the southern part of the basin, the upper carbonate-dominated part of the Eleonore Bay Supergroup sediments lies directly on top of the Nathorst Land Group, while in the northern part it lies on rocks that equate to the middle section of the lower sandstone-dominated sequence. This demonstrates that there was a break (hiatus) in the sedimentation in the northern areas, with erosion (removal) of some of the already deposited sediments. The sediments in the upper part of the Eleonore Bay Supergroup are characterised by a very stable, uniform development. This shows that the depositional conditions in the basin were extremely quiet and without any tectonic disturbance over a very long period of time, perhaps more than two hundred million years. The upper part of the supergroup is divided into three parts, each ranked as a group. The lowest group comprises pale sandstones and dark mudstones (Lyell Land Group), of similar type to those in the Nathorst Land Group sediments beneath. The following middle division (Ymer Ø Group) is a very characteristic, multi-coloured succession that comprises alternating limestones, dolomites and muddy sediments. The uppermost and youngest division comprises pale grey limestones and dolomites (Andrée Land Group). All of the upper part of the supergroup was deposited in a shallow-water, marine area and detailed studies of the deposits in the Lyell Land and the Andrée Land groups show that the sediments were transported from a land area in the west and south-west, out into near-shore areas in the sea to the north and east. The studies also show that there have been repeated changes in the relative sea level. The Andrée Land Group contains many reefs of algal deposits and shows depositional structures that can be compared to those found in modern coral islands in the tropics. Therefore, there are grounds to suggest that the younger parts of the Eleonore Bay Basin were situated in a tropical climate zone.

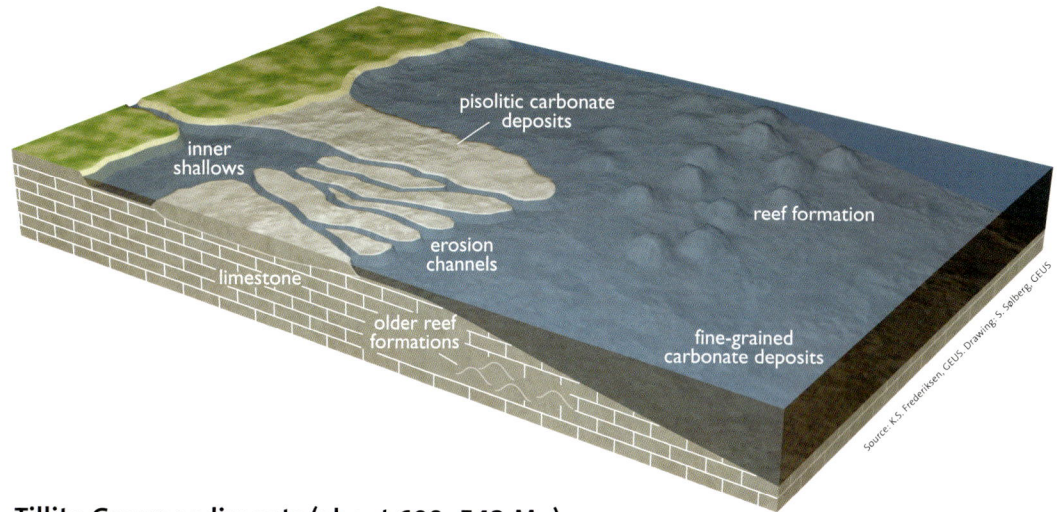

Block diagram showing the environment conditions along the coast of the Eleonore Bay Basin during deposition of the limestone-dominated Andrée Land Group, in the latest part of the Neoproterozoic. Offshore, in the shallow water, pisolitic limestones were laid down and were cut through by erosion channels from a river outfall. In slightly deeper water, banks formed with carbonate (limestone) reefs and farthest out on the outer slope, fine-grained limestone was deposited.

Tillite Group sediments (about 600–542 Ma)

Tillite is the name for a petrified ice-age deposit (glacial deposit). It is a poorly sorted sedimentary rock that comprises a variety of boulders and stones that are contained in a matrix of mud- and sand-grade material. The sediments comprise glacially transported material deposited either on land or in water. The sea drift ice and icebergs may carry glacial material with blocks of stone far away from their source, and when the ice melts, the blocks sink (drop stones) and are incorporated into the bottom sediment. Tillite deposits are indications of earlier ice ages and, therefore, important for the interpretation of past climatic conditions.

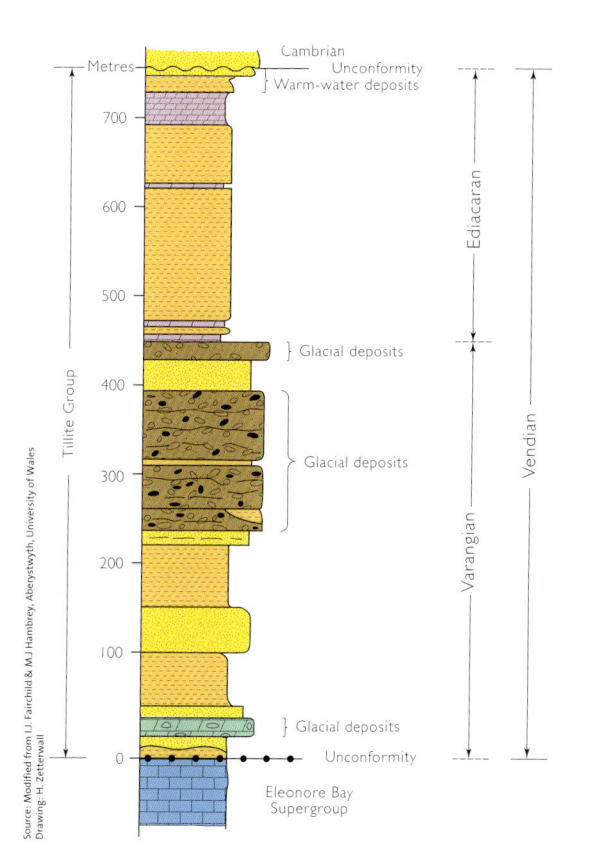

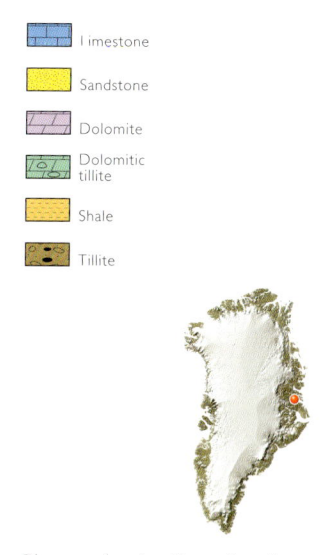

Diagram showing the sedimentary succession in the Tillite Group in the southern part of the Caledonian fold belt in North-East Greenland. The deposition spans a period of about 50 Ma, immediately prior to the beginning of the Cambrian. The sequence contains ice-age deposits (tillitic rocks) that clearly demonstrate that the region was glaciated on at least two occasions between 600 and 550 Ma ago.

BASIN DEPOSITION

POSITION OF GREENLAND 650 Ma AGO

Late in the Neoproterozoic, Greenland lay far to the south in the southern hemisphere, together with the adjacent part of North America (Laurentia) and in contact with continental blocks containing the future South America and Siberia. The oceanic areas formed an unbroken entity and several of the continents that today are found in the southern hemisphere then lay close to or north of the equator, in a large assembled block (a supercontinent).

A 600 Ma old ice-age deposit from the Caledonian fold belt in North-East Greenland. The rock is called a tillite, a lithified, glacial deposit that comprises a dark groundmass of sand and mud in which large sub-rounded stones and blocks occur. In the photograph most of the blocks are paler in colour than the groundmass. The largest blocks can be more than a few metres in diameter, but most are less than 1/2 m across. The many different rock types that occur in the tillites were transported and deposited by glaciers that have planed off irregularities in the old landscape. The tillite deposits are heterogeneous and unsorted, consisting of a mixture of the material that was contained within the glacier and deposited when the ice melted.

The Tillite Group sediments in North-East Greenland comprise a 700–800 m thick sequence that rests on the underlying Eleonore Bay Supergroup sediments, locally with an erosional contact (unconformity) between the two sequences. The Tillite Group is divided into five formations, of which only two contain proper tillite deposits. These two formations contain blocks of various rock types, showing that the material in them comes from different source areas. The formation between the two tillite units comprises sandstones and shales. Above the youngest tillite unit a mud- and silt-dominated unit occurs, which is directly overlain by a formation that indicates a sudden change to a considerably warmer climate. In this youngest formation imprints of salt crystals occur that can only form in sediments laid down in tropical or sub-tropical regions, such as evaporite deposits.

This sudden climatic change, from an ice age into a much warmer period, is very difficult to explain using current geological climate models. The sudden change is, however, a widespread event recorded over large parts of the World in the latest part of the Neoproterozoic between about 750 and 550 Ma ago. Tillite occurrences of this age are found on nearly all the continents, and some of these occurrences were formed at a time when they were located at or near the equator in a tropical or sub-tropical climate zone (between 30°N and 30°S). These occurrences are the basis for a theory that the whole world was, for a short time, completely ice-covered (called 'Snowball Earth'). At many places tillite deposits are directly followed by warm-water sediments, which demonstrate an extremely rapid melting of the ice. One explanation of these phenomena proposes that, at the beginning of a past ice age, a large number of the continents were concentrated in a belt around the equator, while at the same time the polar regions became ice-covered. This positioning of the continents may have resulted in increased precipitation and a dilution of the heat-insulating content of carbon dioxide (CO_2) in the atmosphere. As a consequence the global average temperature fell leading to additional ice formation in the oceans. As ice has a high thermal reflectivity, this development may have had a self-reinforcing effect leading to a rapid, total ice cover of the entire Earth. The subsequent melting of the ice, and return to a more normal climate, is thought to relate to a sudden 'greenhouse effect' due to a buildup of the CO_2, perhaps linked to volcanic activity. The Earth's average temperature during this period may have changed by more than 50°C in less than 5 Ma.

Kong Oscar Fjord Group sediments (542–460 Ma)

In the latest part of the Neoproterozoic most of the Earth's continents were assembled into a superconti-

Older marine basins in North-East Greenland

nent called Pannotia. This supercontinent included North America with Greenland, the Baltic and Siberian shields, part of South America, Africa and Antarctica. About 600 Ma ago these amalgamated continents began to split up, and new oceans and oceanic crust were formed at the same time. In the area between Laurentia (North America with Greenland) and Baltica (northern Europe and the Baltic shield) a new ocean, called Iapetus, formed. Greenland at this time, 575 Ma ago, was located approximately 30° south of the equator and had a warm to tropical climate.

On the western margin of Iapetus, along the east margin of the Laurentian coast, deposition occurred in an extensive basin in what is now offshore the central part of East Greenland; an up to 4 km thick sedimentary sequence of shallow-water deposits was laid down under quiet conditions. The lowest few hundred metres of the sequence comprise sandstones, but, the overlying deposits are nearly exclusively limestones and dolomites. The sequence spans in age from the Lower Cambrian (542 Ma) up into the Middle Ordovician (about 460 Ma). In the Ordovician part of the succession alone, about 3 km of limestones and dolomites were laid down. It is possible that the sedimentation originally extended up into the Silurian, as in North Greenland (see p. 88), but, if so, these younger sediments were eroded away in East Greenland prior to the onset of the Caledonian deformation. However, the East Greenland part of the basin could also have developed differently from that in North Greenland, as the Iapetus Ocean was already beginning to close again from the beginning of the Ordovician (at about 485 Ma).

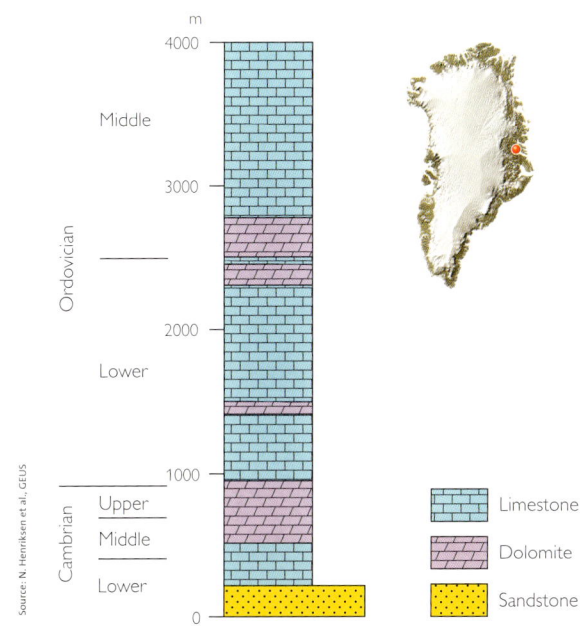

Composite stratigraphic column showing the Cambrian to Ordovician (Kong Oscar Fjord Group) succession in the southern part of the Caledonian fold belt in North-East Greenland.

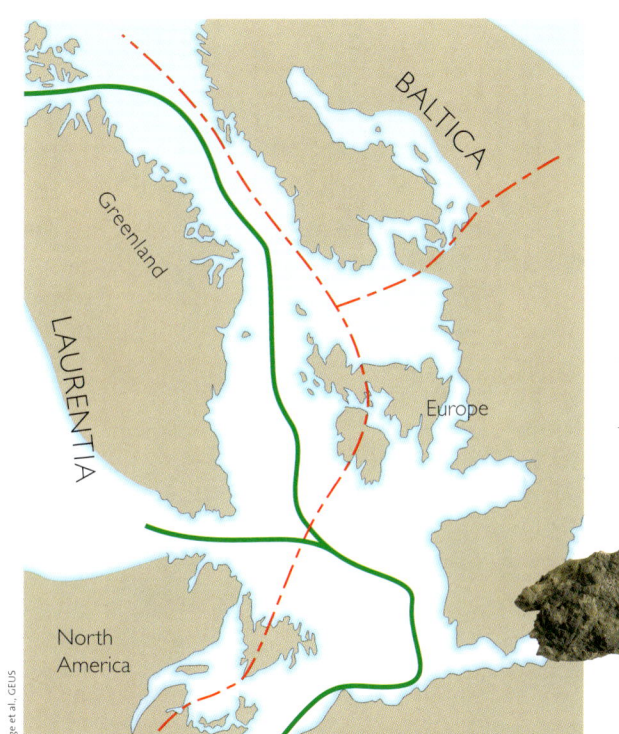

The Cambrian–Ordovician sediments are rich in fossils, in marked contrast to the earlier Precambrian sedimentary successions laid down in the basin. The sandstones in the lowest, about 100 m thick, part of the Cambrian deposits contain only a few traces of biological activity, such as animal tracks and burrows, but a little higher up the first real Cambrian body fossils are found. While only a few species occur in the lowest layers, rapid evolution of many different types of animals resulted in the appearance of more than 150 species through the sequence. The invertebrate fauna includes trilobites, brachiopods, gastropods, corals, bryozoans, bivalves and cephalopods. The fauna as a whole comprises animal species that are characteristic of the western side of the Iapetus Ocean – a fauna type, that belongs to the so-called Pacific Province. This fauna type was different from another faunal assemblage that developed at the same time on the eastern side of the Ocean in the so-called Atlantic Province. The Iapetus Ocean was too wide for the two assemblages to mix and therefore each fauna type was confined to its respective side.

With the gradual closing of Iapetus by continental drift during the Ordovician and Silurian, the East Greenland basin at first narrowed, and finally ceased to exist. At the end of the Silurian period the Iapetus Ocean, that had earlier separated the two continents of Laurentia and Baltica, closed and collision led to development of the Caledonian fold belt (see p. 92).

The present day North Atlantic Ocean developed gradually by continental spreading between Europe and North America/Greenland during a period of approximately 55 Ma. The figure shows how the two continental masses were situated in relation to each other before the opening began along the Atlantic spreading axis (green line). A similar oceanic development had taken place about 550 Ma earlier when the Iapetus Ocean opened along a different spreading axis (dashed red line). After this older opening, the continents drifted together again and collided about 420 Ma ago whereby the Caledonian mountain belt was formed (see p. 96).

Conical fossil shell of an extinct cephalopod –orthoceratite– from the Lower Palaeozoic in North Greenland. The shell comprises a series of living chambers (each compartment is a single chamber) that developed successively as the animal grew larger. The fossil illustrated is an internal cast, the result of sediment filling the empty living space after the animal's death. The fossil is about 50 cm long.

Cambrian trilobite (Olenellus svalbardensis) from North Greenland. Trilobites are an extinct, independant group of arthropods, which have segmented body armour made of thin layers of calcareous material. This armour, or the impression of it, may be preserved as a fossil. A trilobite's body is divided into three parts, a rigid head shield (cephalon), a segmented body section (thorax) with legs, and a rigid tail section (pygidium). Trilobites are found in deposits ranging from the Cambrian to the Permian. The majority of species are less than 10 cm long.

BASIN DEPOSITION

Among these is Greenland's largest sedimentary basin

Older marine basins in North Greenland

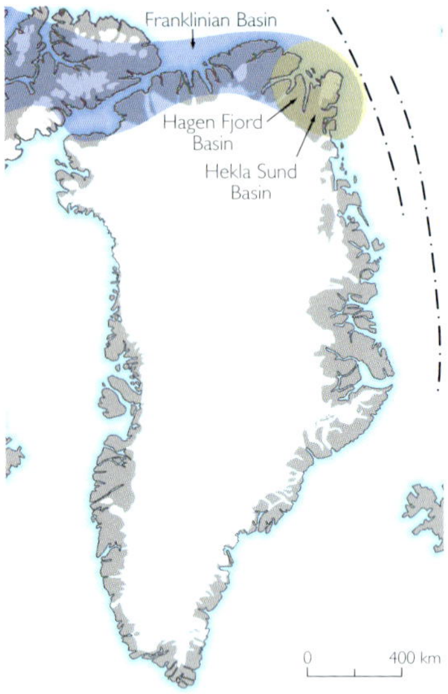

Two different sedimentary basins developed between 800 and 410 Ma ago in North Greenland. The earlier developed in eastern North Greenland at the transition between continent and ocean. It included a shallow-water, continental to marine area close to the land (Hagen Fjord Basin) and a deeper depositionary basin (Hekla Sund Basin) farther seaward. These basins were confined to eastern North Greenland.

The later, and much larger Franklinian Basin, developed during the Cambrian with deposition continuing to earliest Devonian time (542–410 Ma); the basin extended for more than 2000 km along the continental margin of North Greenland and into north-eastern Canada. The Franklinian Basin succession is several kilometres thick, and developed with three sedimentary environments; a shallow-water, dominantly carbonate shelf nearest the continent, an outer shelf and slope with variable sediments laid down in outwards increasing water depth, and outside this a deep-water trough in which a thick flysch succession accumulated. The northern parts of the basin deposits were deformed in Devonian to Carboniferous time during the Ellesmerian orogeny. A later deformational event affected the northernmost part of the sequence late in the Cretaceous.

The older marine basins in North Greenland contain deposits from the Proterozoic and Lower Palaeozoic. In the Hagen Fjord and Hekla Sund Basins sequences of continental and marine sediments were deposited.

The largest sedimentary basin in Greenland, the Lower Palaeozoic Franklinian Basin, extends across North Greenland and into Arctic Canada. The Basin contains a several kilometres thick sequence that includes shallow-water, marine carbonate sediments and a thick sequence of deep-water, marine sediments.

The Hekla Sund and Hagen Fjord Basins in eastern North Greenland (about 800– 542 Ma)

After the eruption of the Zig-Zag Dal Basalt Formation a period of about 400 Ma elapsed in which no deposits are preserved in eastern North Greenland. Late in the Neoproterozoic, deposition resumed in a two-part basin covering the north-easternmost part of Greenland. Farthest east a fault-bounded, rapidly subsiding basin (Hekla Sund Basin) accumulated a thick sequence of dominantly sandy sediments (Rivieradal Group). This was followed by quietly deposited, shallow-water sediments (Hagen Fjord Group) that began in the west, and whose upper units spread eastwards to overlap the basin containing the Rivieradal Group sediments. These basin developments took place at the transition between a stable continental area in the west and a future ocean basin in the east; i.e. prior to the initial opening of the Iapetus Ocean about 600 Ma ago (p. 81), which developed into a major ocean between the North American continent and Baltica.

The Rivieradal Group of the Hekla Sund Basin deposits crop out today in folded thrust sheets in the Caledonian fold belt in Kronprins Christian Land, while the main exposures of the Hagen Fjord Basin make up unfolded sequences in the foreland west of the fold belt.

Neoproterozoic basin development in eastern North Greenland began 800–700 Ma ago with the formation of a so-called half-graben. This basin was at least 200 kilometres from north to south and 50 kilometres wide, with a faulted western margin against the continental margin. Subsidence of the basin was probably associated with movements on the marginal fault system. Easterly flowing river systems draining the high land to the west transported enormous quantities of sediment into the basin. Near the land, coarse conglomerates and sandstones were laid down, while in the deeper parts of the basin towards the east, the sediments are fine-grained silt- and mudstones with local carbonate horizons. In places several thousand metres of turbidites were deposited, comprising repeated sequences of layered sand–silt sediments with an upwards decrease in grain size. These deposits were laid down by intermittent pulses of sediment such as mudslides, probably triggered by

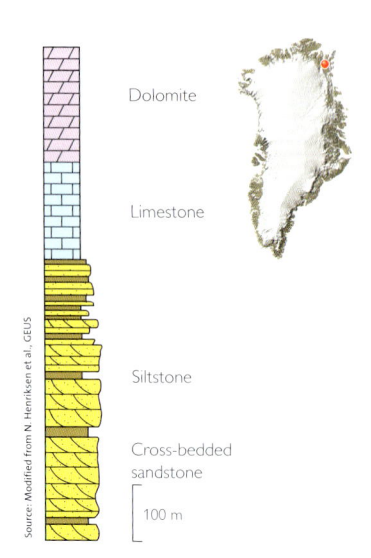

Composite sedimentary log for the deposits of the Hagen Fjord Group, measured in the area south of Peary Land, North Greenland. The sequence reflects development from a continental flood plain into a marine carbonate platform.

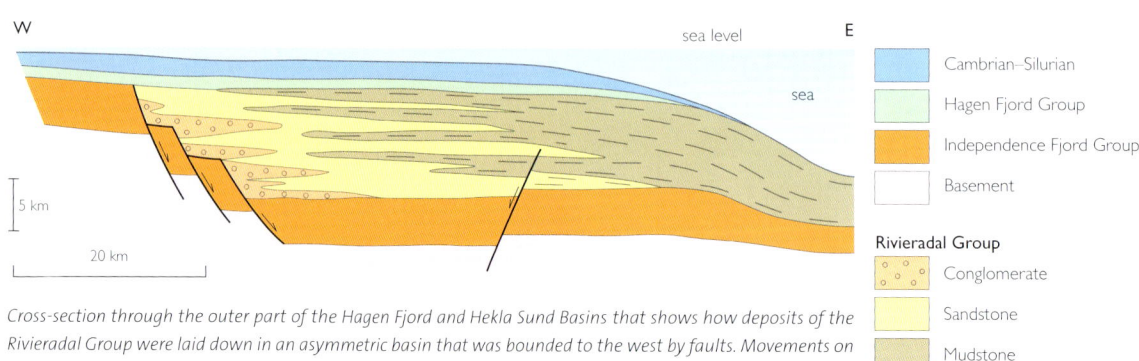

Cross-section through the outer part of the Hagen Fjord and Hekla Sund Basins that shows how deposits of the Rivieradal Group were laid down in an asymmetric basin that was bounded to the west by faults. Movements on these faults occurred at the same time as a gradual subsidence of the crust underneath the basin, so deposition of sediment kept pace with the subsidence.

minor earth tremors due to fault movements along the basin margin. At the western margin of the basin the distribution of conglomerate units suggests deposition from three distinct delta systems that were formed at the mouths of large rivers flowing into the basin. The total thickness of the Rivieradal Group sediments is estimated to be between 7.5 and 10 km.

The Hagen Fjord Group sediments crop out over a region more than 300 km across in eastern North Greenland. The succession is over 1 km thick and is divided into four formations. The oldest formation, more than 500 m thick, comprises sandstones deposited by streams (fluvial) that gradually pass upwards into shallow-water, marine sediments deposited in a tidal zone. The second formation comprises sand- and siltstones, with calcareous units containing algal structures, deposited in a tidal environment. The third formation is a distinctive, red-brown limestone unit (about 400 m thick), overlain by another characteristic yellow-weathering, dolomitic sequence (Fyns Sø Formation) containing widespread developments of algal structures (stromatolites). The Hagen Fjord Group basin has thus evolved from a continental flood plain into a marine shelf with limestones. The three uppermost formations extend across the site of the Hekla Sund Basin, where they directly overlie the Rivieradal Group. This shows that the rapid subsidence in the fault-bounded Hekla Sund Basin was superseded by a marine platform succession laid down under quiet conditions with slow, steady subsidence.

After an interval of about 50 Ma the region of the Hagen Fjord Basin was overlain by Lower Cambrian sandstones. During this long hiatus (see p. 84) the Fyns Sø Formation dolomites were exposed to erosion and karst structures and caves were eroded by circulating groundwater. The earliest Cambrian sandstones infilled the caves and the karst erosion features of the dolomite surfaces; the boundary between the dolomites and overlying sandstones is often highly irregular. Although there was a time interval of about 50 Ma between deposition of the two units, the boundary between them is frequently knife-sharp and it is only the regional, geological history together with the biological age indicators that reveal the existence of a considerable time gap (hiatus).

Franklinian Basin in North Greenland (about 500–410 Ma)

The largest sedimentary basin in Greenland developed along the almost 2000 kilometres long E–W-trending continental margin of present-day northern Greenland and Canada and accumulated sediments of early Cambrian to early Devonian age. The elongated Franklinian Basin through most of its lifetime was characterised by a shallow-water shelf in the

POSITION OF GREENLAND 514 Ma AGO

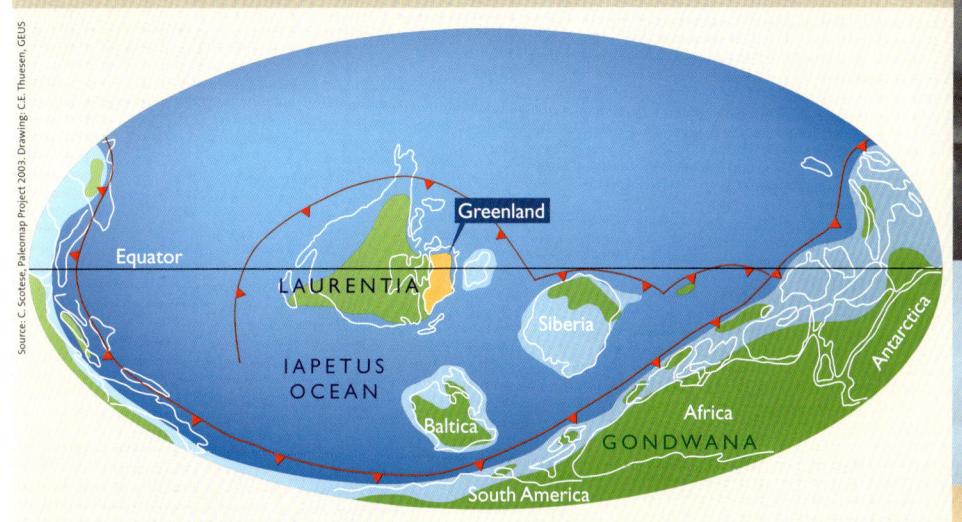

The position of the continents in the Middle Cambrian, 514 Ma ago. The North American continent, 'Laurentia', with Greenland along its eastern part, was at that time located over the equator, separated from the other continents by wide oceans.

Valley side in Kronprins Christian Land, eastern North Greenland, exposing a section through the Rivieradal Group sandstones, appearing in the lower, scree-covered slopes. These are overlain by more massive sandstones of the lower part of the Hagen Fjord Group. The mountainside is about 600 m high.

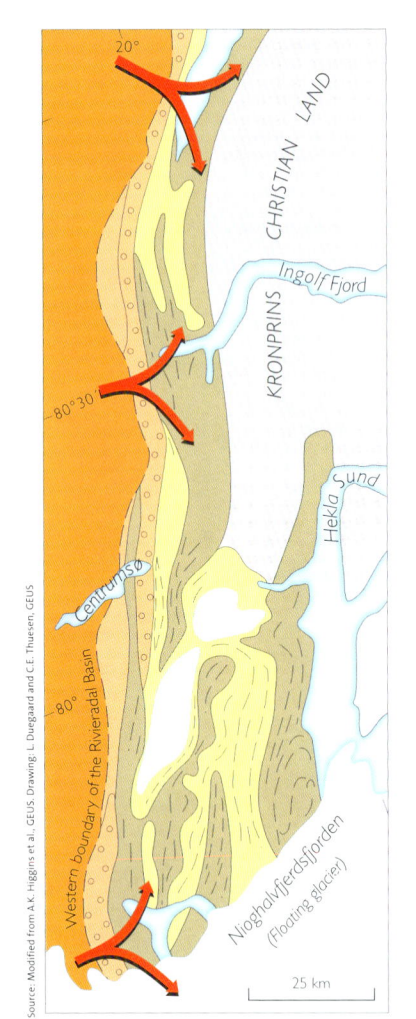

BASIN DEPOSITION

The Rivieradal sediments were deposited from a series of distinct river systems draining from a continent to the west, into a coast-parallel basin along the boundary of the adjacent sea. The local occurrence of conglomeratic fans and the overall sedimentary patterns indicate the location of the mouths of the rivers (red arrows). The deposits within the basin are sorted, with the coarse-grained material laid down nearest the land, while the finer-grained silt and muds were transported further away from the coast and laid down in the deeper parts of the basin.

Small dome-shaped sedimentary structures that are formed by growth of micro-organisms – probably algae or bacteria – which have trapped thin layers of fine-grained carbonate during their growth. These structures are called stromatolites, and can occur in a number of different forms (see p. 68). The sequence depicted covers the transition between the Precambrian and Cambrian in the southern part of Peary Land, North Greenland.

Rock sample that contains a hiatus representing a time gap of approximately 120 Ma. The lower part of the sample comprises a 600 Ma old, pale dolomite (Fyns Sø Formation), the surface of which was exposed and eroded during the hiatus so that it became irregular and cut by fissures. After a time interval of about 120 Ma the dolomite was overlain by Lower Ordovician, dark limestone (Wandel Valley Formation, about 480 Ma old) that infilled the uneven, older, eroded surface. The sample was collected in Kronprins Christian Land, eastern North Greenland.

south nearest to the continent, and a deep-water trough to the north (see p. 86). In the Cambrian, the Franklinian Basin was located far from its present position, as the whole of the North American continent, including Greenland, was located near the equator. This geographical position is clearly reflected in the warm, shallow-water, sedimentary deposits and associated fossils, interpreted to have been laid down under tropical to sub-tropical conditions.

In eastern North Greenland, the Cambrian tropical to subtropical carbonates with algal reefs or smaller banks with stromatolitic structures, overlie a thin tillite horizon (lithified ice-age deposits). This sequence with ice-age deposits, here known as the Morænesø Formation, is of the same type and age as the ice-age deposits of the Tillite Group in the southern part of North-East Greenland (see p. 79). This places it stratigraphically in the youngest Precambrian (Vendian), and immediately prior to the development of the Franklinian Basin.

The deposits of the Franklinian Basin can be traced for more than 900 km across the whole of North Greenland, and to the west continue along strike for a further 1000 km into Arctic Canada. In total a sequence approximately 4 km thick accumulated on the shallow-water shelf closest to the continent, while the deposits in the deep-water basin or trough are up to 8 km thick. Development of the basin through its 140 Ma time span can be divided into a series of phases. These are primarily linked to changes of the boundary between the shallow- and deep-water parts of the basin and to fluctuations in the sediment supply. Both shallow-water and deep-water sediments preserve a rich fauna. The shelf limestones include fossils that principally lived on the sea floor (benthonic fauna) such as trilobites, gastropods, bivalves, corals, brachiopods, different soft-bodied animals and micro-organisms. In the deep-water deposits, the fossils were free-swimming invertebrate animals (planktonic fauna) such as graptolites that on their death sank to the bottom and were incorporated into the sediment.

The development of the basin comprises seven stages (see p. 88) that from the oldest to the youngest are:

1) Earliest Cambrian (540–530 Ma). Deposition of sands, muds and dolomites on a marine shelf.

2) Early Cambrian (530–520 Ma). Division of the basin into a southern shelf and a northern deep-water trough with a broad intervening slope. Calcareous sediments accumulated throughout the basin, but at the transition to the deep-water part a steep slope (escarpment) developed, and further offshore sedimentation was partly composed of

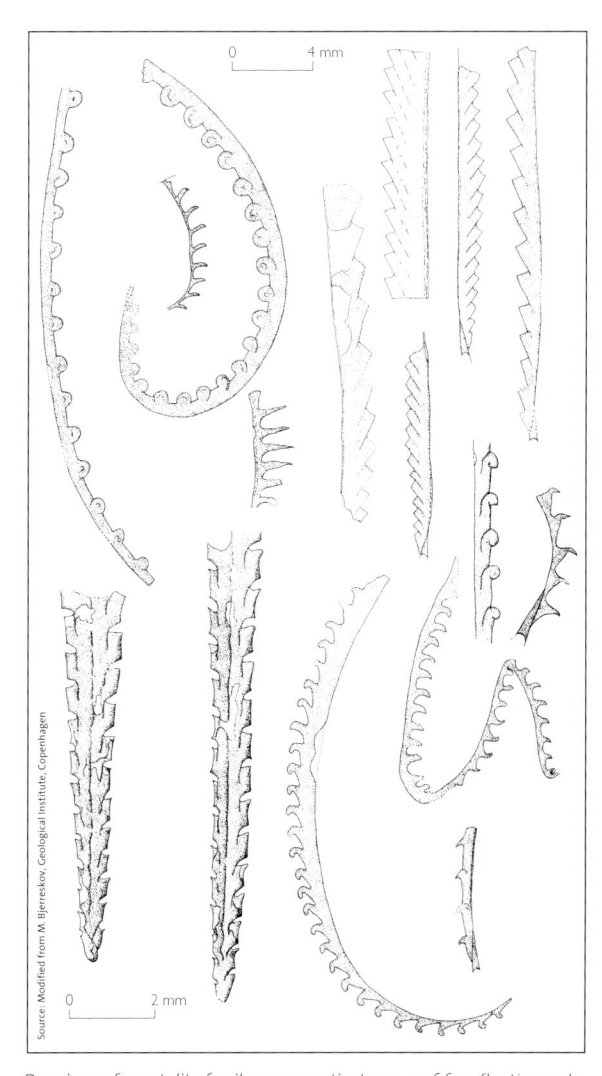

Drawings of graptolite fossils, a now extinct group of free-floating, colonial, soft-bodied animals. They are principally known from their impressions in deep-water, muddy sediments, and due to their shape variation with evolution they can be used as zone fossils for part of the Lower Palaeozoic. The examples shown come from Silurian sediments in Washington Land, North Greenland, and are about 430–420 Ma old. The examples on the left are drawn at double the scale of those on the right.

BASINS AT THE TRANSITION BETWEEN CONTINENT AND OCEAN

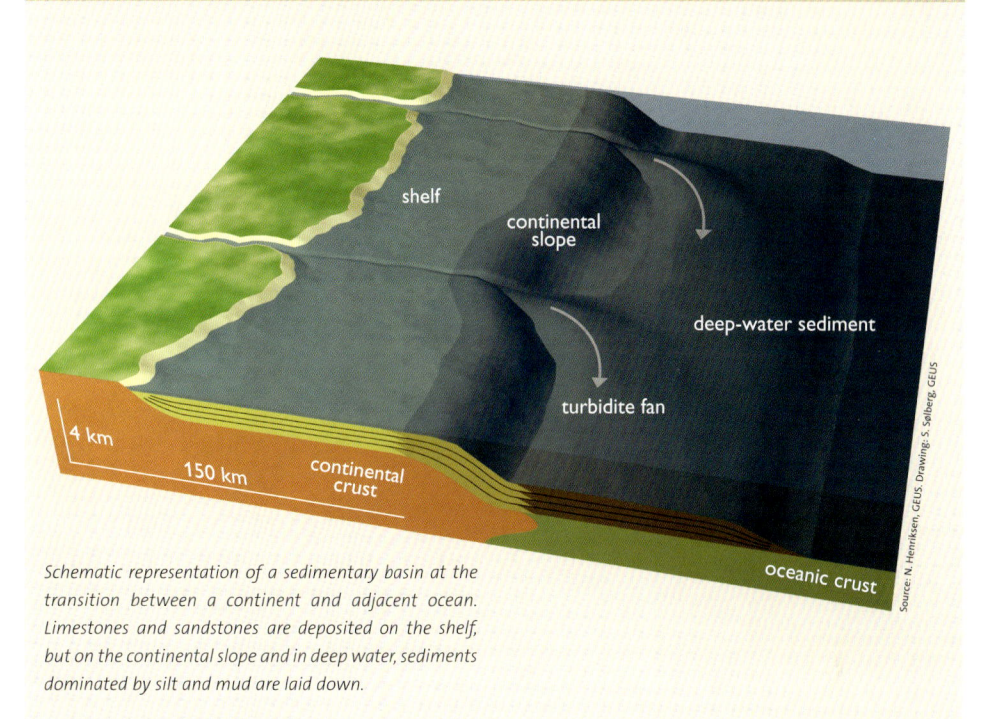

Schematic representation of a sedimentary basin at the transition between a continent and adjacent ocean. Limestones and sandstones are deposited on the shelf, but on the continental slope and in deep water, sediments dominated by silt and mud are laid down.

Large, elongate, sedimentary basins are often formed along the continental margin with the ocean. Such basins may be up to several hundred kilometres across, and generally formed in connection with continental drift and sea-floor spreading. The break-up of an old, coherent continent often begins along a tectonic rift zone that widens as the two continental blocks move away from each other. In the earliest phases of basin formation, an intra-continental, fresh-water basin forms, but with wider separation between the blocks the sea transgresses into the basin and marine sediments are laid down. With still wider opening between the two now separated continents, basaltic oceanic crust develops between them. The former, single basin now comprises, two sedimentary basins, each associated with the adjacent continental block, with oceanic crust between them.

The subsequent development of the two sedimentary basins often shows differences in types of sedimentation and the ages of deposition. Different sedimentary sequences may be deposited, depending upon the tectonic development, sediment supply and climatic conditions. On a tectonically inactive, so-called passive margin, basin development may be marked by slow subsidence rates, with sediments accumulating on a broad, shallow-water shelf. Different types of sand and silt sediments as well as limestones may be deposited, if the sediment supply and climate conditions are suitable. Beyond the shelf, on the continental slope where the water is deeper, fine-grained, deep-water sediments are laid down.

Where a continent–ocean boundary is characterised by active tectonic conditions, for example, by convergent plate-tectonic movements, the sedimentary basin evolution will be strongly affected by the tectonic development. Different sedimentary zones may occur with an inner, shallow-water shelf and an outer, deep-water trough dominated by more fine-grained sediments. In many cases volcanic island arcs may form parallel to the length of the basin. These can divide the basin further into a part landward of the arc, and a part on the oceanic side. Each of these parts may exhibit different sedimentary patterns and can also include volcanic material, either as intrusions or as sediments derived from the erosion of already deposited volcanic rocks.

Sedimentary basins formed at the boundaries between continental and oceanic areas can have very different lengths of existence. In most cases the whole development spans a few hundred million years, but in some basins this period is much shorter. The accumulated sedimentary successions also vary greatly from 5 to 20 km in thickness.

BASIN DEPOSITION

erosion debris, including conglomerates derived from rocks of the shelf.

3) Early Cambrian (520–510 Ma). In connection with a global rise in sea level, a change of depositionary environment was initiated and, at the same time, the subsidence rate in the trough increased. The boundary between the shelf and the trough moved towards the south, and sandstones were laid down in the shallow-water area; a considerable amount of sediments were carried out into the deep-water trough by, massive mud flows, forming graded beds (turbidites). At the outer part of the slope, fortuitous depositional environments locally allowed the preservation of a group of soft-bodied animals. The soft-bodied fauna from Sirius Passet in North Greenland has been compared to the celebrated Burgess Shale fauna of Canada.

Part of the Lower and Middle Cambrian sequence in the south-western corner of Peary Land. The easily weathered grey-black shales and sandstones of the Buen Formation (stage 3, see text) are overlain by a prominent crag of banded limestones from the Brønlund Fjord Group (stage 4, shelf facies). The mountain is about 800 m high.

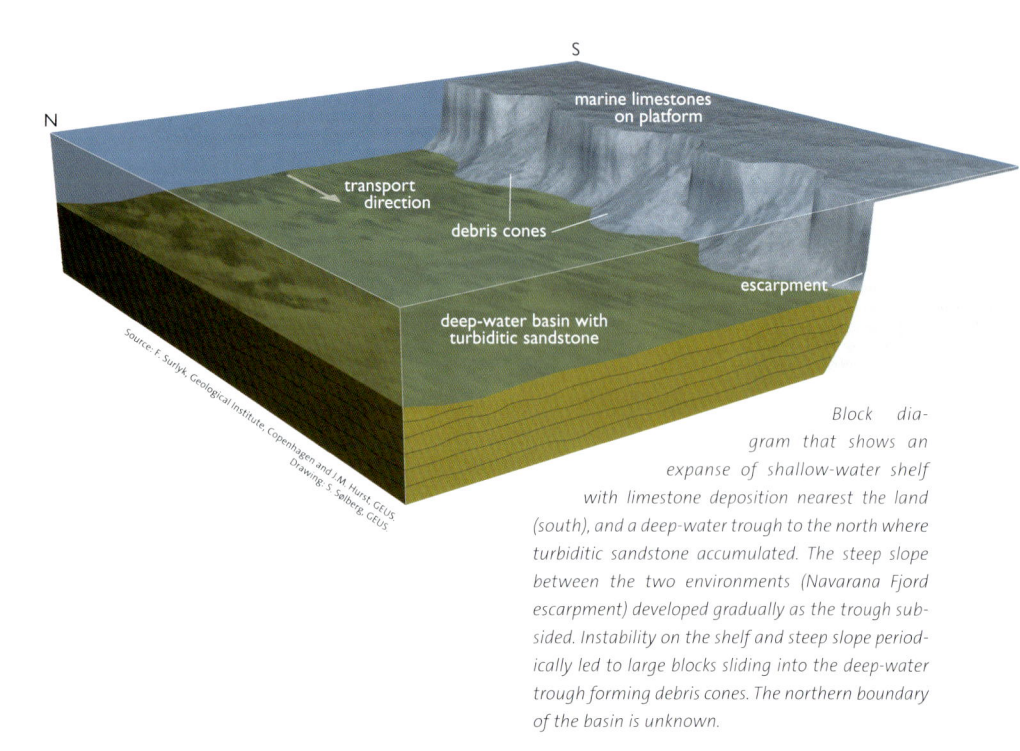

Simplified representation of the Cambrian–Silurian development of the Franklinian Basin of North Greenland

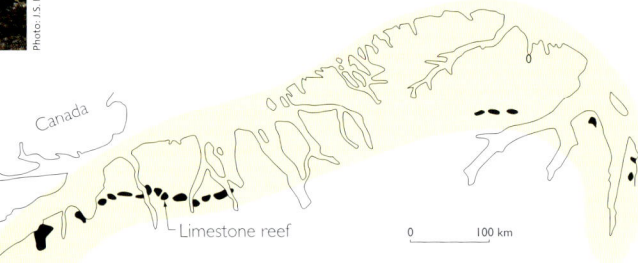

Stage 3: *Late Silurian (about 425 Ma). The deep-water trough was filled with sandy, turbiditic deposits from erosion of the Caledonian mountain chain in North-East Greenland. In the final phase these deposits covered the carbonate shelf with a thick succession of sandstones.*

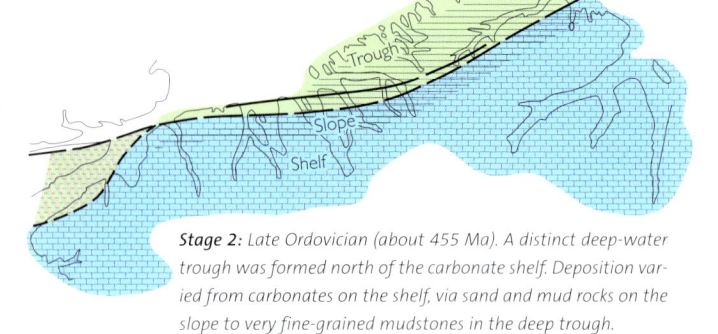

Stage 2: *Late Ordovician (about 455 Ma). A distinct deep-water trough was formed north of the carbonate shelf. Deposition varied from carbonates on the shelf, via sand and mud rocks on the slope to very fine-grained mudstones in the deep trough.*

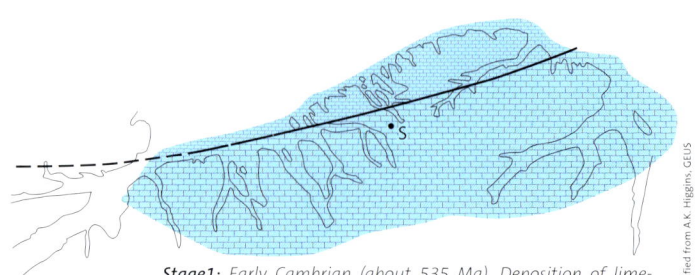

Block diagram that shows an expanse of shallow-water shelf with limestone deposition nearest the land (south), and a deep-water trough to the north where turbiditic sandstone accumulated. The steep slope between the two environments (Navarana Fjord escarpment) developed gradually as the trough subsided. Instability on the shelf and steep slope periodically led to large blocks sliding into the deep-water trough forming debris cones. The northern boundary of the basin is unknown.

Stage 1: *Early Cambrian (about 535 Ma). Deposition of limestones on a broad, shallow-water shelf which extends far to the north where the deep-water slope began. S: Sirius Passet.*

Older marine basins in North Greenland

4) Early Cambrian – Middle Ordovician (510–470 Ma). On the shelf in the south, calcareous deposition continued in the first part of this stage, but the area in the eastern part of North Greenland was gradually uplifted, while sedimentation remained unchanged in the west. Later, renewed subsidence of the shelf area in the east occurred, and deposition resumed following the established pattern. The supply of material was drastically reduced in the deep-sea trough region, and only a very thin, condensed sequence accumulated.

5) Middle Ordovician – Early Silurian (470–440 Ma). The pattern of the previous stages continued, but a steep submarine slope (escarpment) evolved with a depth change of about 1300 m between the shelf and the trough. Fault displacements appear to have triggered large submarine avalanches of calcareous material which slid off the shelf across the slope and out into deep water, where they formed lines of carbonate debris cones.

6) Early Silurian (440–430 Ma). This stage was marked by a dramatic change in the development of the basin that can be linked to the early stages of the Caledonian fold belt formation in North-East Greenland with large thrust displacements (see p. 96). Erosion of the Caledonian mountains to the southeast of the Franklinian Basin produced an enormous quantity of sand and silt that was washed out by rivers into the deep-water trough and then flowed westwards along the axis of the trough. Loading from the large quantity of sediment led to increased subsidence in the trough, and on the limestone shelf. The deep-water trough filled rapidly with sediment deposited from massive submarine turbidite flows, which are among the largest known examples on Earth. With a trough length of 1500–2000 km and a width of about 200 km, it has been estimated that more than two million cubic kilometres of sediments were deposited. On the shelf, limestone accumulation continued, and locally along the margin landslides of calcareous material periodically flowed northwards as debris cones.

7) Early Silurian – earliest Devonian (430–410 Ma). After the trough was filled by sandstones, and in combination with a global rise in sea level, the former shelf areas became covered by deep-water sediments. The shelf limestones were thus overlain by clastic sediments that comprise turbidites with layers of sand, silt and mud. Along the margins of the former shelf, limestone accumulation persisted in a series of up to 300 m high reefs that can now be traced as a 800 km long belt from Kronprins Christian Land in the east to Hall Land in the west.

Cliff section at the head of Navarana Fjord showing the 1300 m high, submarine slope (escarpment) that formed the north edge of the Silurian shallow-water limestone shelf. Dark-coloured sand and mud deposits of the deep-water trough laid down in stages 6 and 7 lap against the limestone escarpment. The Navarana Fjord escarpment can be traced as a distinct geological lineament from eastern Peary Land westwards to Nyeboe Land, a distance of more than 500 km.

Sandy and muddy turbiditic sediments, part of the fill of the Silurian deep-water trough in Peary Land. Turbidite is the name for a type of sediment deposited from sediment-laden mudflows that travel out into the deep-water basin and form graded beds as they settle out. The sandy sediments occur at the base of each bed, and silt- and mud-dominated material form the upper parts of each graded unit. The thickness of each graded layer can vary between a few metres up to about 20 m. The individual pulses of sediment may be triggered by small earth tremors or landslides in the hinterland, and the flows can transport the sediment several hundred kilometres along the axis of the basin.

BASIN DEPOSITION

Sedimentation continued with sandy and turbiditic deposits. Towards the top of the trough sequence, two very distinctive 30–200 m thick, conglomerate units occur, comprising exclusively pebbles and cobbles (clasts) of a flint-like rock (chert). These clasts appear to be derived from erosion of Ordovician deposits that are 30–40 Ma older than the conglomerate layers in which they are now found. This demonstrates that Ordovician rocks in eastern North Greenland were elevated in connection with Caledonian orogenic events to the east of the Franklinian Basin, and subsequently eroded. The conglomerates can be traced as continuous horizons over a distance of more than 600 km, and are interpreted as catastrophic submarine debris slides with sand, cobbles and boulders. Deposition in the Franklinian Basin ended with a unit of sandy and silty sediments.

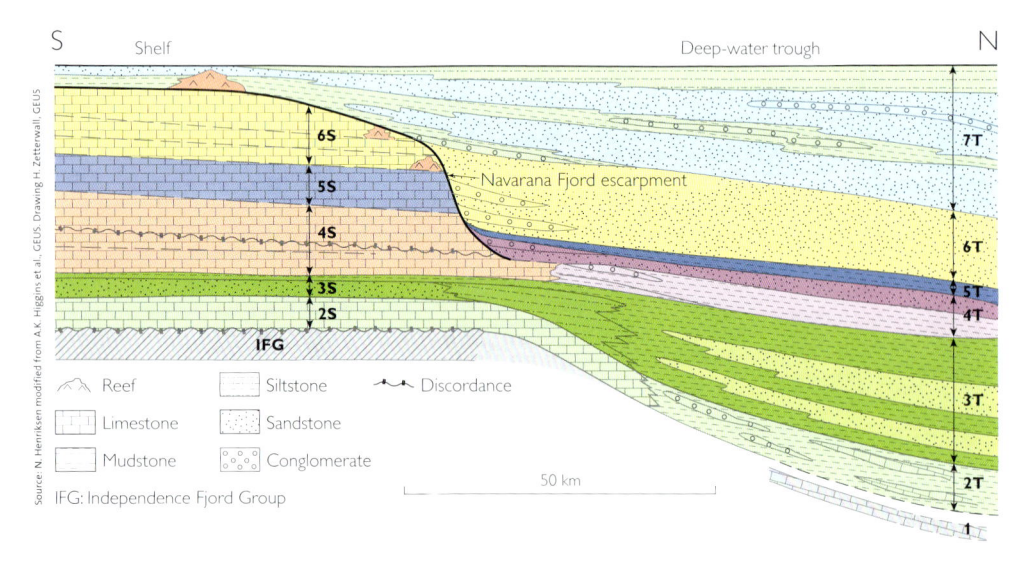

Flat-lying sequence of organically formed limestones, overlain by a domed series of reef limestones. From stages 6 and 7 (see sedimentary column on this page). Cliff is about 700 m high, Hall Land, North Greenland.

Halysites. Silurian chain coral from the Franklinian Basin, North Greenland.

South–north cross-section of the Franklinian Basin showing the transition from shallow-water carbonate shelf to deep-water, sand- and silt-dominated deposits in the trough. The succession is continuous from the earliest Cambrian (542 Ma) to the earliest Devonian (about 410 Ma). The shallow-water sediments of the carbonate shelf pass northwards into a much thicker sequence of clastic, deep-water sediments. The boundary between the shelf and the deep-water trough is a steep escarpment. In the Silurian, deposition of the turbiditic sediments was very rapid, the trough was filled, and turbidites spread out over the shelf areas. The development of the basin can be divided into 7 stages (see text); S indicates shelf deposits and T trough deposits. Sediment thickness on the shelf was 3–4 km and in the trough about 8 km. The profile is drawn with a vertical exaggeration.

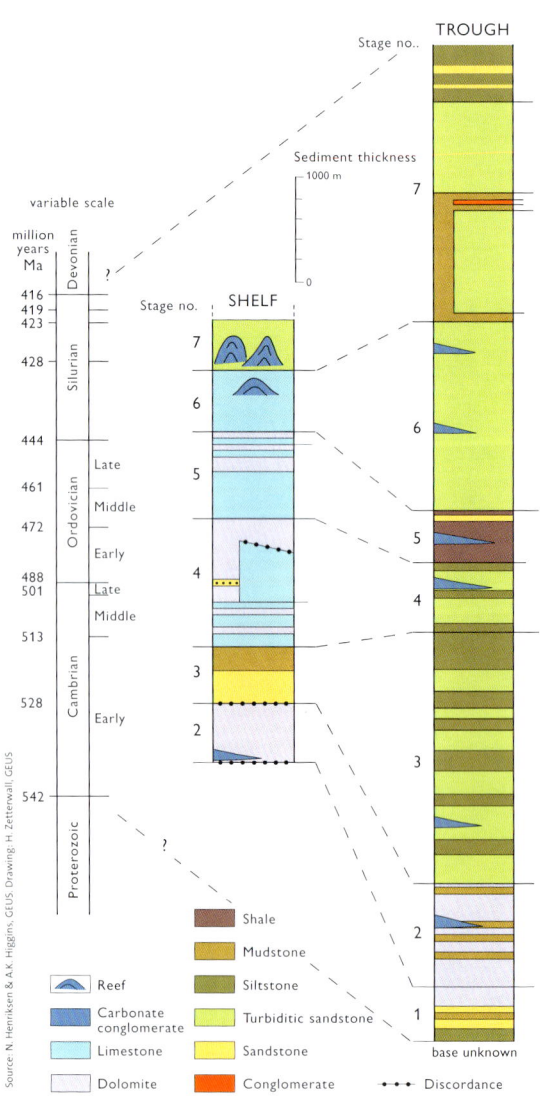

Composite sedimentary logs of the Franklinian Basin succession in North Greenland, showing the carbonate-shelf and corresponding deep-water trough developments.

TRACES OF PAST ANIMAL LIFE IN SEDIMENTS

Deposits from the beginning of the Cambrian period, about 540 Ma ago, show a sudden increase in the fossilised impressions of a large number of various animals, unknown from earlier periods. The traces of animal life found prior to the Cambrian (the Precambrian) are of comparatively primitive forms, e.g. bacteria and algae. At the beginning of the Cambrian a great variety of highly developed types of invertebrate animals appear. Many of these have hard external shells that form the impressions that are now found as fossils. The sudden appearance of this diversity of animals in the Cambrian is often referred to as the 'Cambrian explosion'. Since then, animal life has intermittently changed, with newly formed animal groups, families and species gradually developing (evolving), as described in Darwin's 'On the Origin of Species', by means of natural selection.

Traces of animal life in sedimentary rocks (fossils) are utilised by geologists to determine the age of the deposits in which they are found. Every species has adapted itself to the external conditions at the time it occurred. With changing environmental conditions, new and better-adapted species developed, whilst the older forms became extinct. Each species generally has a limited lifetime in geological history, and fossils can be used to divide sedimentary deposits into detailed age divisions (see p. 69). The special field where geologists work with fossils and traces of animal life is called palaeontology while the unravelling of geological history is called stratigraphy.

The development of early animal life comprises exclusively invertebrates, with both vertebrates and plants appearing much later. However, it is almost exclusively invertebrate fossils that are used for age determinations. There is a basic division into large fossils (macrofossils) that can be seen with the naked eye, and microfossils that are so small that a microscope is needed to determine what they are. In field work macrofossils are obviously the most important, because they can be identified and used directly to indicate the age of a rock.

The occurrence and distribution of fossils has, through the last couple of hundred years, been the subject of detailed palaeontological and stratigraphical research. As a result, the occurrence of individual animal species is now known in great detail, and can be listed as a tabulation of the occurrence of single species in chronological order. The whole subdivision of the geological timescale, from the Cambrian to the present, is based upon fossils. Therefore, they are an invaluable key to unravelling a considerable part of Earth's history. With developments in radiometric age determinations (see p. 48), it has become possible, especially during the last fifty years, to determine the absolute ages in millions of years on the time scale. Even though radiometric methods are important, the study of fossils is still the primary tool for geological age determinations in the period after the Precambrian. Classification of the units determined by fossils is so finely developed that a bed that is about 400 Ma old, for example, can be dated with an uncertainty of only one million years (i.e. plus or minus 1/4%).

Lower Cambrian fossil trilobite from 530 Ma old sandy shale from Peary Land. The animal has a segmented calcium carbonate armour, comprising a head shield, a segmented body with legs and a tail segment.

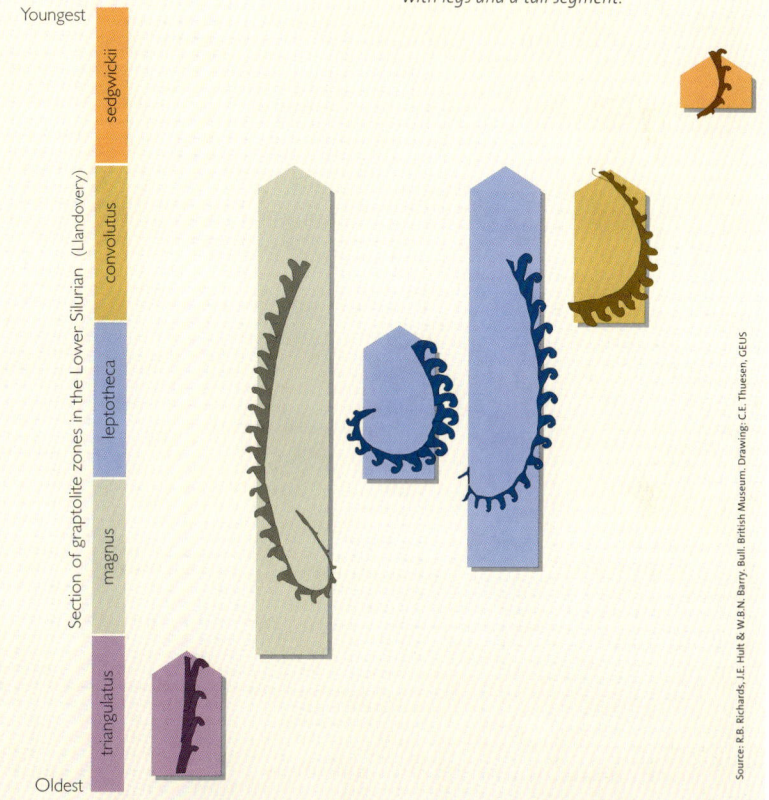

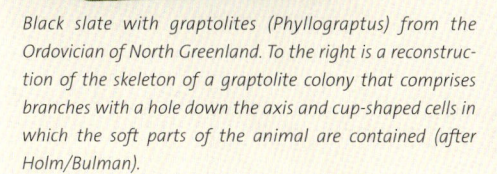

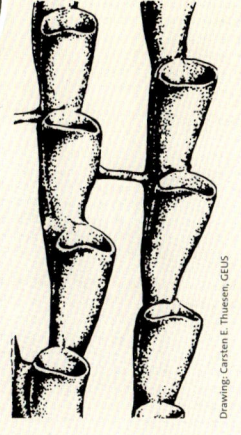

Black slate with graptolites (Phyllograptus) from the Ordovician of North Greenland. To the right is a reconstruction of the skeleton of a graptolite colony that comprises branches with a hole down the axis and cup-shaped cells in which the soft parts of the animal are contained (after Holm/Bulman).

An example of the use of fossils in a biostratigraphical division of a sedimentary sequence. The figure shows sketches of graptolites, an extinct group of colonial, soft-bodied animals, which lived in the sea. The animals lived in a thin, branched chitinous skeleton that may be found as impressions in mud rocks. Each of the illustrated fossils represents a different species within the family Monograptus. Each species existed only for a specific section of Earth's history, after which it became extinct and new species developed. On this basis a biostratigraphical division of time (see p. 69) can be made, in the form of graptolite zones, where each is characterised by the occurrence of an assemblage of certain species. The figure shows a continuum between six species and the time span in which they occurred in the lower part of the Silurian (Llandovery) that covered the period from 443 to 428 Ma and can be divided into 15 graptolite zones. On average a graptolite zone spans about 1 Ma and, related to a time span of 400 Ma, this is equivalent to a subdivision of about 0,25%, which is a remarkably fine resolution. This is equivalent to the possibility of dividing finds from the Viking age into intervals of 2 1/2 years.

Tightly compressed folds in a strongly banded, light-coloured gneiss with dark lenses and slivers of amphibolite. The rocks form part of an intensely deformed shear zone in the crystalline basement of the Caledonian fold belt. Locality near Danmarkshavn, North-East Greenland.

Photo: N. Henriksen, GEUS

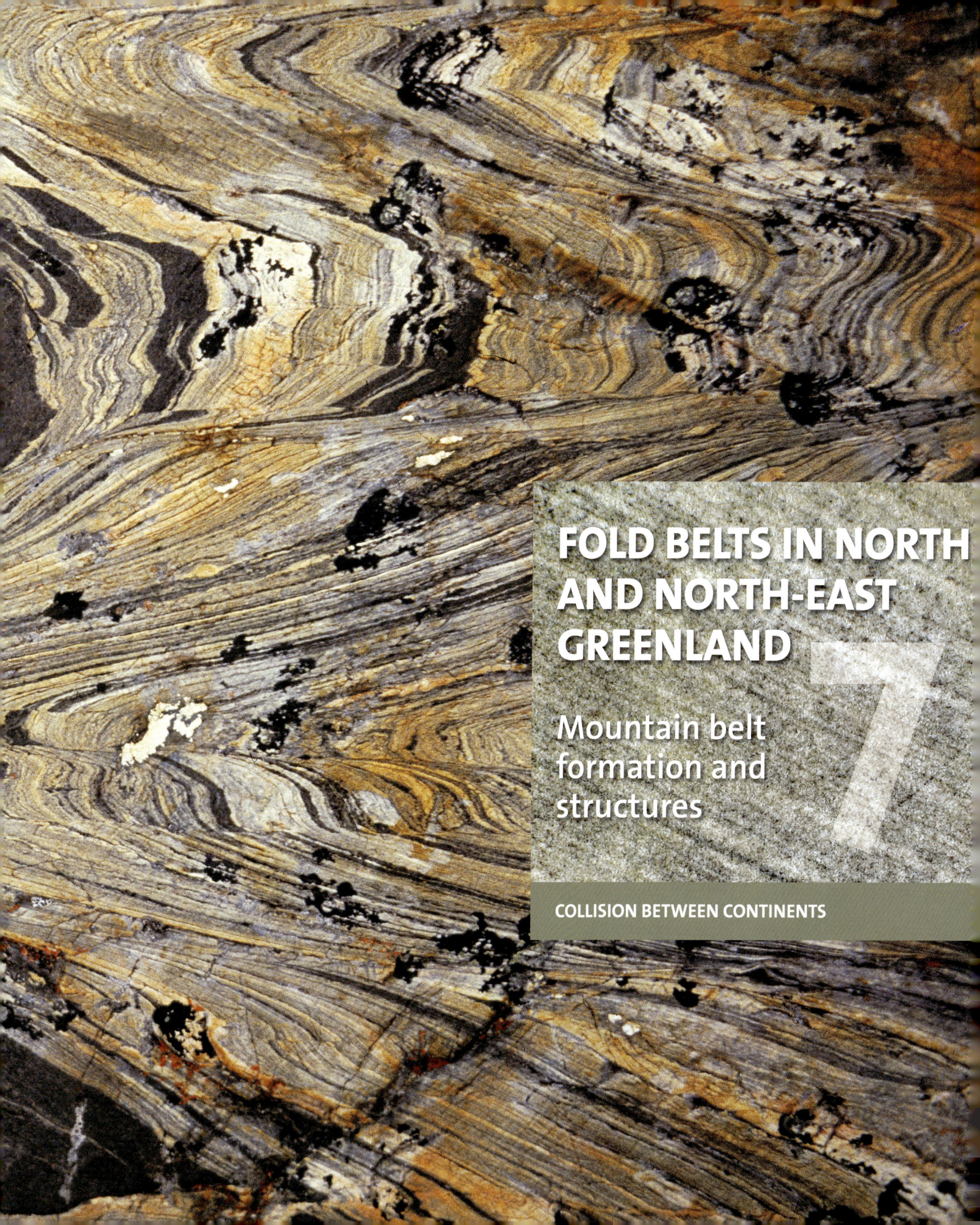

FOLD BELTS IN NORTH AND NORTH-EAST GREENLAND

Mountain belt formation and structures

7

COLLISION BETWEEN CONTINENTS

7 FOLD BELTS IN NORTH AND NORTH-EAST GREENLAND

Two Middle Palaeozoic, coast-parallel mountain belts occur in North and North-East Greenland. The older of the two is the Caledonian fold belt that occurs as a N–S-oriented belt in North-East Greenland between the Scoresby Sund area in the south (70°N) and Kronprins Christian Land in the north (81°30'N). This fold belt developed as a result of a collision between a North American continent, including Greenland, and Baltica about 430–410 Ma ago. The somewhat younger Ellesmerian fold belt in North Greenland occurs as an E–W-trending belt along the coast between Peary Land in the east (20°W) and Hall Land in the west (60°W) and continues farther west into Arctic Canada. This fold belt takes its name from Ellesmere Island in Canada that adjoins North Greenland. Folding and deformation occurred late in the Devonian, about 370–355 Ma ago, due to collision between a North American continent including Canada and Greenland and an unknown continent to the north.

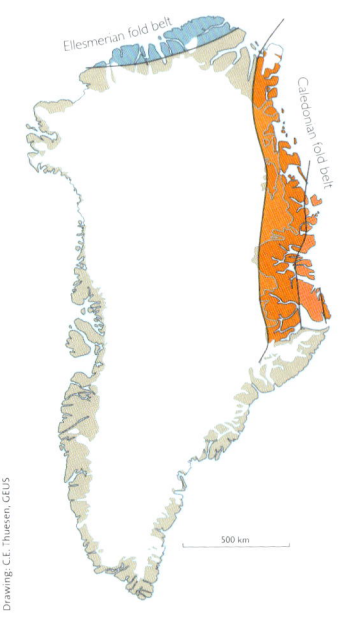

The Palaeozoic mountain belts in North and North-East Greenland. Both orogenic belts are parallel to the coastline, and on the landward side they border the stable basement shield with its overlying, flat-lying sedimentary sequences. In the coastal zone and beneath the sea beyond, the folded rocks are covered by successions of younger sedimentary and volcanic rocks.

Although these two fold belts in Greenland are almost identical in terms of age, and both were formed by comparable plate tectonic processes, their structural architectures are very different. The Caledonides are complexly constructed and incorporate components of older Precambrian basement that were refolded and reworked in the much younger fold belt. The fold belt is built up of major thrust sheets transported from the east across the margin of the Greenlandic basement shield in the west, forming a pile of kilometre-thick crustal slices stacked on top of each other. Some of the rock units have mineral assemblages developed at depths of 60–100 km in the Earth's crust. The Ellesmerian fold belt is, conversely, of a simple construction and developed almost exclusively within the sedimentary deposits of the Franklinian Basin (see p. 83). The transformations due to metamorphism and deformation are characteristic of processes that took place in the upper 10–30 km of the crust and only modified the original sedimentary structures to a limited extent.

The fold belts in Greenland are parts of larger mountain chains

The orogenic belts in North and North-East Greenland make up only small parts of much larger and more widespread mountain belts that can be followed as once coherent belts over thousands of kilometres through several continents. These major, curved or linear mountain belts formed as a conse-

MAPPING OF FOLDED SEQUENCES

The architecture of a mountain belt is primarily determined by the drawing up of a geological map on which the occurrence and extent of the different rock types are recorded. The map depicts the rocks exposed at the surface, but it can also be used to give a three-dimensional overview of the geology of an area. When mapping, the boundaries of the various units are followed and drawn onto a topographical base map, and the spatial orientations of the rocks and folds that deform them are recorded. Flat-lying, non-folded rock sequences show an outcrop pattern on the map that follows the height contours. In areas with folded rocks the outcrop pattern is very different. Here the spatial orientation of the layers varies over the map area and can, over short distances, change from flat-lying to vertical. Consequently the boundaries between individual layers delineate geometrical patterns with irregular waveforms. The deformation patterns in a mountain belt are constrained by structural geological principles and systems and, with the help of mapping data, three-dimensional models can be constructed to explain the spatial distribution of the rocks and predict the architecture and composition of the underground geology.

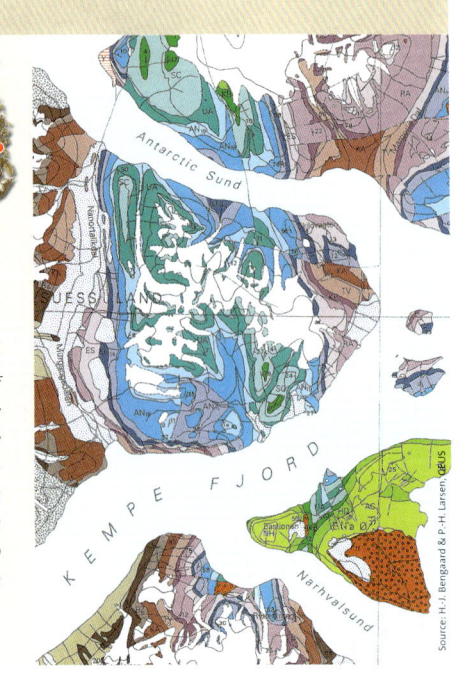

Segment of a geological map of the southern part of the Caledonian fold belt in North-East Greenland, showing folded rocks of the Eleonore Bay Supergroup, Tillite Group and Kong Oscar Fjord Group (Neoproterozoic to Ordovician; see p. 76). Each colour depicts the extent of a mapped rock unit. The curvature of the bands reflects the interplay between folding and the topography where the units occur. To an experienced geologist the coloured patterns show that the rocks are affected by a series of large, open folds. The map here covers an area about 50 km from north to south.

quence of plate tectonic movements when continents collided (see box p. 53) or when oceanic floor was subducted beneath a continent. Throughout the Earth's geological history major mountain belt formation has occurred regularly at intervals of several hundred million years.

Each cycle in a typical mountain belt event comprises three phases:

1. The splitting up of a block comprising several continents welded together (super-continent) along curved or linear zones of weakness.
2. Separation of the divided blocks of the continental mass by continental drift and formation of new oceanic floor in the expanding, intervening rift zone.
3. Opposite-directed continental drift, leading to the closure of an intervening ocean and formation of a mountain belt in the collision zone along the margins of the old continents.

Such a cycle (a Wilson cycle) typically spans several hundred million years. The youngest major mountain belt forming event was the Alpine Orogeny forming the Alpine and Himalayan mountain chains about 50–30 Ma ago. The next youngest, global mountain belt formation took place between 450 and 350 Ma ago and created the Variscan and Caledonian fold belts. Parts of these mountain belts are found on nearly all of the Earth's present-day continents, and the elements that can be studied in Greenland are fragments of such much larger systems. It is exactly because the mountain belts are parts of a global system that the segments found in Greenland are of great interest, also to geologists in other parts of the world. The exceptionally good exposures in Greenland illustrate new and established principles for the interpretation of mountain belt processes, and facilitate the investigation and interpretation of orogenic belts on other continents.

Deformation in fold belts is a consequence of lateral compression of crustal rock sequences that results in a horizontal shortening and a vertical thickening of the rock mass, but the volume is retained unaltered. This substantially increases the thickness of the crust during mountain belt formation, such

A 2000 m high cliff showing major, synformal (downwards facing) and antiformal (upwards facing) folds in the Caledonides of the inner part of the Scoresby Sund area. The folds deform a sequence of Archaean gneisses (about 3000 Ma old) overlain by younger, red-brown schists from the approximately 1000 Ma old Krummedal sequence preserved in the core of the synform. In the boundary zone between the two units strong flattening of the rocks (due to movements between them) is seen as an intense banding.

THRUSTS, FOLDS, TECTONIC WINDOWS AND FAULTS

Folding and deformation of rocks occur when a series of layers with different compositions and rigidity are subjected to lateral compression. At the surface of the Earth, most rocks are rigid and hard and are not easily deformed or folded. Conversely, if rocks are buried deep in the crust during mountain belt formation, they are subjected to high temperatures and pressures so that they become pliable and can be more easily deformed and folded; a significant factor is that such deformation can take place over very long periods of time, which can span thousands to millions of years.

Folds

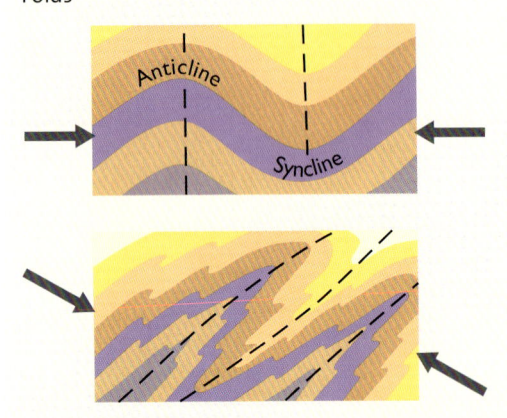

Sketches of profiles through:
Open folds (top) – seen principally in the upper parts of the crust. The sequence shown is youngest at the top.
Tightly compressed folds (bottom) – seen principally at deeper crustal levels. In this sketch the limbs of the folds both dip the same way and are known as overturned folds.

By lateral compression of flat-lying layered sequences, sediments in depositional basins may be deformed into simple, wave-like folds with a limited horizontal shortening of the sequence. Increasing compression leads to open folds with shallowly dipping limbs, and in more strongly compressed areas tight folds with steeply inclined limbs develop. The dimensions of the folds may vary from small-scale structures measured in centimetres to very large structures that are kilometres across. Large- and small-scale structures often form at the same time and have the same geometrical orientation. The longitudinal extension of the folds (hinge lines: line following the culmination of the folds) is perpendicular to the direction of compression.

Thrusts

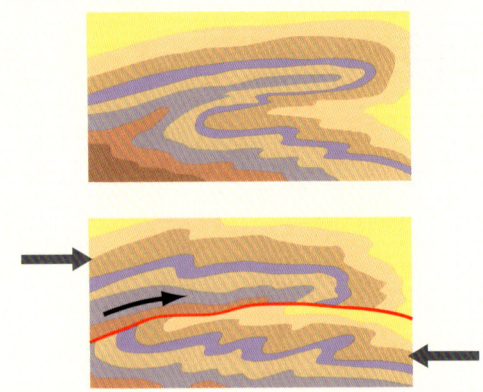

Sketches that show the development of a thrusted structure. Earliest stage (top) shows a recumbent fold (limbs roughly horizontal). Later stage (bottom) where further compression has caused a disruption of the structures and a flat-lying thrust plane (red) is developed where the overlying rock mass is displaced across the structures in the underlying block.

With continued compression, a tightly folded, banded succession may break up into large, flat-lying slices that can slide over each other. The contact planes (thrusts) between the slices (thrust sheets) may be initiated along zones of weakness, due perhaps to variations in the rock types. The accumulation of thrust sheets, one over the other, produces a thrust stack, and the boundaries between the slices are defined as thrust planes. The structures in these closely spaced stacks are usually dominated by flat-lying structural forms, and frequently the rocks adjacent to the thrust planes are highly strained and streaked out. Thrusts can develop from tightly compressed folds that, with further compression, gradually tilt over to become flatter lying and, with still further compression, one limb tears apart so that the overlying part is thrusted over the underlying part.

The distance a thrust moves varies greatly. The smallest can be on a metre-scale, while the displacements along thrust planes at the base of major thrust sheets can be measured in terms of hundreds of kilometres. Through thrust movements a characteristic, structural-geological phenomenon frequently occurs such that older units are found resting on top of younger units. This abnormal stratigraphic relationship may, for example, result in strongly transformed units from deeper crustal levels being found resting on top of unmodified sequences that formed near the surface. In compressional, structural scenarios a thrust plane, which is itself nearly horizontal, always cuts up through the sequence with continued compression, so that older, deeper level rocks are displaced upwards over the younger layers. In both the Ellesmerian and Caledonian Greenland fold belts numerous thrusts occur, and in the Caledonian fold belt the largest displacements of the overlying units are several hundred kilometres.

Thrust tectonics is typical of nearly all fold belts, both young and old. Usually, the transport direction of thrust sheets is from the oceanic margin towards the continent, so that a thrust sheet is pushed over the rigid substratum of the adjacent continent. Later erosion of part of a thrust sheet can occur, exposing the rock units below in a so-called tectonic window. If a part of the thrust sheet becomes detached from the remainder by erosion, the isolated block is called a 'klippe'.

Faults

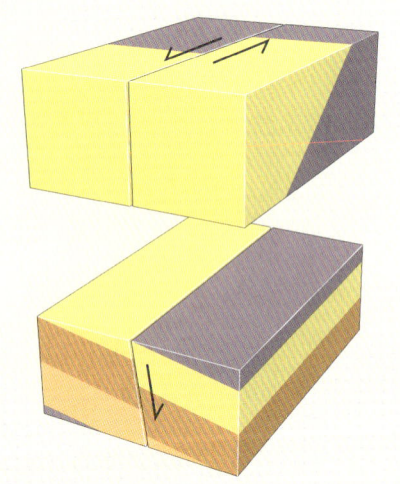

Three-dimensional sketches illustrating fault blocks.
Vertical fault with lateral strike-slip movement (top).
Inclined normal fault in which the dip-side block is downthrown (bottom).

In connection with deformation in the rigid upper part of the Earth's crust, movements along break lines (faults) can occur. Horizontal movements along steeply inclined fractures are called strike-slip faults and are caused by lateral forces, while movements in a vertical sense along dipping fault planes are named after the direction of the movement, either normal (down) or reverse (up), on the dipping fault plane. Normal faults (extensional faults) frequently form during the late phases of movement in the mountain belt formation, where there is a relaxation and an extension of the crust in the opposite direction to the former compression.

FOLD BELTS IN NORTH AND NORTH-EAST GREENLAND

that in some instances it may be up to 60–80 km thick. However, exhumation and thinning, together with erosion often mean that crustal thicknesses under old, eroded mountain belts are reduced to 30–50 km (see box p. 27).

Recent seismic investigations of the deeper parts of the Greenland basement have shown that the boundary between the crust and the upper part of the mantle (Mohorovičić discontinuity) occurs at a depth of about 30–40 km. Along the outer coast of North-East Greenland the crust is somewhat thinner, only 23–30 km, mainly due to continental break-up related to the opening of the North Atlantic.

Dark grey, folded gneisses with pale, quartz-feldspar veins. The structures were produced by several phases of deformation, but the pale pegmatites are only affected by the latest phase of deformation which moulds open folds. Such conditions are typical of areas characterised by thick-skinned tectonics.

THIN- AND THICK-SKINNED TECTONICS

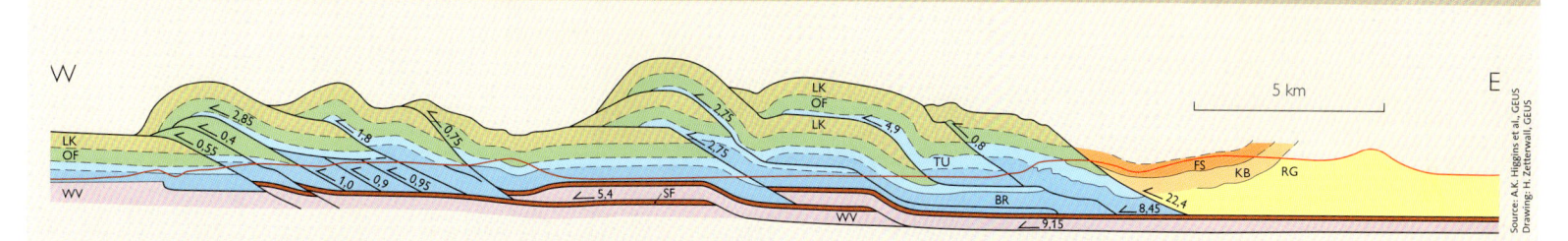

During mountain belt formation the rocks in the crust are deformed and folded in different ways depending on their depth of burial in the Earth's crust (the skin). In the upper parts of the crust temperatures are relatively low and the structures simple, while in the deeper parts more complicated deformation patterns form due to increased temperatures and pressures. Simple structures form down to depths of about 10–15 km, while in the deeper parts more intense deformation, typically with a more plastic style, is found all the way down to the root zone of the mountain belt at 50–100 km depth.

Fold belts are frequently formed along the edge of older, stable continents. These fold belts often comprise a marginal zone nearest to the continent where the rocks are formed by simply folded rock sequences that have slid over a more rigid block of an older continent or crystalline basement rocks. The structural pattern is analogous to sliding a tablecloth over a tabletop where the cloth folds, but the table is unaffected. Such structural style is called 'thin-skinned tectonics'. Belts where this type of structure occurs are often 50–100 km wide and can frequently be followed along a marginal zone between the mountain belt and the continent. In Greenland such structures are common along the margins of the Caledonian and Ellesmerian fold belts.

In the more central, inner and deeper parts of a fold belt, in contrast to the thin-skinned tectonics, the structures are more complex and characterised by so-called 'thick-skinned tectonics'. Such deep-level structures, developed under high temperatures and pressures, increase in intensity and complexity downward. Successive phases of deformation produce refolded folds, and thrust sheets involve both crystalline basement and little to moderately altered rocks from the cover sequences. In Greenland thick-skinned tectonics is widespread in all the Precambrian basement areas, as well as the inner and central parts of the Caledonian fold belt in North-East Greenland.

Profile through part of a thin-skinned fold/thrust zone in the western margin of the Caledonian fold belt in Kronprins Christian Land, northernmost North-East Greenland. The section shows stacks of small, thrust slices that were displaced westwards over one another producing a tiled pattern. The lower boundary is a slide surface (floor thrust) that follows a specific stratigraphic horizon, with underlying rock sequences unaffected by thrusts and folds. The section shows sedimentary rocks from the Neoproterozoic (RG, KB and FS), the Ordovician (WV, SF, BR and TU) and the Silurian (OF and LK). The numbers indicate the estimated westward displacement on each thrust slice measured in kilometres.

Deformation patterns in a region marked by thick-skinned tectonics. The structure is characterised by plastic folding that occurred under amphibolite facies conditions at 15–30 km depth with temperatures of 500–700°C. Such patterns are typical of the Archaean part of the basement and the deeper structural levels of the Caledonian fold belt. The profile is drawn from a fjord wall in the inner, north-western part of the Scoresby Sund region.

Legend:
- Pale granitic gneisses
- Amphibolite bands and lenses
- Dark hornblende gneisses
- Pegmatite

The Caledonian fold belt in North-East Greenland

Formed by collision between Greenland and Scandinavia about 420 Ma ago

At the present day the remnants of the Caledonian fold belt are found on both sides of the Atlantic Ocean. This orogenic belt was created in the lower Palaeozoic by the collision between the continent of North America with Greenland (Laurentia) in the west and a part of Europe (Baltica) in the east. At the same time the Iapetus Ocean that formerly separated the two continents was closed by continental drift processes. The now amalgamated continents formed about 300 Ma ago part of a supercontinent called Pangaea that included nearly all of the Earth's continents. In the Mesozoic (200 Ma ago) this supercontinent began to split up leading to the opening of the Atlantic Ocean and disruption of the Palaeozoic fold belts such that today parts of these fold belts are found in Europe–North Africa and in North America–Greenland.

A major overturned isoclinal fold delineated by an about 100 m thick pale, granitic layer doubled over itself. The fold nose (hinge) is clearly seen. The isoclinal fold may originally have been flat-lying but now dips steeply on the right side of the photograph because of later Caledonian folding. Inner part of the Caledonian fold belt at Grandjean Fjord comprising deformed gneissic and granitic rocks. Fjord wall is about 1200 m high.

In North-East Greenland the Caledonian fold belt comprises a series of thrusted and interfolded rock units derived from the margin of the old Laurentian continent together with overlying sedimentary basin deposits. The fold belt in Greenland is made of a series of very large thrust sheets that were displaced westwards over the continental margin, and stacked in a layer-cake fashion. The thrust pile probably reached a thickness of 30–40 km, and displacements on the major thrust sheets probably totalled more than 300 km. The original source of the thrust sheets was a long way east of the present coastline, and thrust displacements have transported coherent slices of crust from a location below the present shelf region westwards

Map of the Palaeozoic fold belts around the Atlantic with the continents shown in their original relative position about 300 Ma ago, before the sea-floor spreading that created the present-day Atlantic Ocean. The Caledonian fold belt in the north, shown in orange, was formed by continental collision between Laurentia and Baltica in the Silurian. The other Palaeozoic fold belts, depicted in blue, reflect the collisions of Laurentia with various microcontinents and Africa in tectonic events lasting into the Carboniferous. The stable continental areas are buff coloured.

onto the land area, where they occur today. Some of the individual thrust sheets comprise both older crystalline basement and overlying younger sedimentary sequences.

In the central part of North-East Greenland two major thrust sheet units are recognised. The lower one comprises exclusively crystalline and sedimentary rock complexes derived from the Greenland basement shield, whereas the upper thrust sheet unit is composed of a lower segment of similar basement rocks together with a higher segment of 900–450 Ma old sediments derived from the 'Fjord Zone Basin' (see p. 76). The basement rocks in the lower thrust sheet have a complex development reflecting formation during three different Precambrian orogenic episodes. The oldest of these resulted in a 3000–2700 Ma old Archaean gneiss complex. After this followed a very widespread, early Proterozoic episode that formed a gneiss complex 2000–1750 Ma ago, while the youngest episode is represented by metamorphism and deformation of an extensive sediment succession accompanied by granite intrusions about 1200–920 Ma ago. The eastern margin of the Laurentian shield represented in Greenland has been repeatedly broken up and re-assembled during Precambrian plate tectonic episodes, prior to involvement in the Caledonian fold belt.

Basement to the fold belt

Before the formation of the Caledonian fold belt, the margin of Laurentia in East Greenland was part of the Greenland basement shield, dominated by interfolded Precambrian gneisses and metamorphic schists. The eastern margin of the shield then extended several hundred kilometres east of the present coastline. In the Neoproterozoic–Lower Palaeozoic several large sedimentary basins developed along the margin of the shield, accompanied by a gradual subsidence of the basement underneath the basins. Where subsidence was greatest the basin deposits attained a thickness of

BASEMENT TO THE CALEDONIDES
Pre-Caledonian orogenies
highest

Grenville/Sveconorwegian orogeny
1200–920 Ma
deformation, migmatite and granite formation; deposition of the Krummedal sedimentary sequence

Palaeoproterozoic orogeny
2000–1750 Ma
deformation, extensive formation of new granitic rocks; sedimentary and volcanic rocks

Archaean orogenies
3000–2700 Ma
deformation, formation of gneissic and migmatitic rocks; sedimentary and volcanic rocks

lowest

POSITION OF GREENLAND 425 Ma AGO

The distribution of the continents in the Middle Silurian, after the closure of the Iapetus Ocean and the collision of North America/Greenland (Laurentia) and Scandinavia/Europe (Baltica) to form the Caledonian fold belt. In the southern hemisphere all the continents were assembled into a supercontinent, Gondwana, separated from the continents around the equator by the Rheic Ocean.

Pale, Archaean gneisses (3000–2700 Ma) cut by younger, dark metamorphosed basic dykes. The relationships between the pale gneisses and dark, cross-cutting dykes show that the area was influenced by at least two different orogenic episodes; the youngest deformation refolds the older structures and new metamorphic modifications have partially recrystallised the older rocks. The hammer is 40 cm long.

FOLD BELTS IN NORTH AND NORTH-EAST GREENLAND

18–20 km (Eleonore Bay Supergroup, Tillite Group and Cambrian–Ordovician sediments, see p. 76). The depth of the basin shallowed dramatically towards the continent in the west and, close to the margin of the basin; the equivalent Neoproterozoic–Ordovician sequence of sediments was only a few hundred metres thick.

The oldest Archaean basement rocks are found in the Scoresby Sund area and the southern part of the central fjord zone (70–73°N). They include pale-coloured gneisses with bands and lenses of dark amphibolitic rocks and metamorphosed sedimentary deposits. A characteristic feature of the Archaean gneiss complexes is the occurrence of transformed and folded thin, sheet-like basaltic intrusions called dykes. These dykes cut through the banding and fold patterns within the gneisses, and were thus intruded into the basement gneisses after the deformation episode between 3000 and 2700 Ma ago. As the dykes are themselves both metamorphosed and folded, the region in which they occur must have been subjected to another orogenic event, before the Caledonian orogeny.

The most widespread rocks making up the basement of the Caledonian fold belt in East Greenland comprise Palaeoproterozoic granitic gneisses with bands of metamorphosed volcanic and sedimentary rocks. The formation of these rocks is attributed to a major orogenic episode that occurred between 2000 and 1750 Ma ago. The exposures of these rocks are widely distributed between the southern part of the central fjord zone (about 73°N) and eastern Kronprins Christian Land (82°N). These gneisses were formed from granitic intrusions that were produced by melting and differentiation of material above a subduction zone formed during a Palaeoproterozoic orogeny. The granitic material represents new additions to the crust – juvenile crustal material, which contributes to the growth of the continent. Shortly after the emplacement of the granitic rocks, they were deformed by crustal movements related to orogenesis, that converted the originally homogeneous granitic rocks into banded, folded and veined gneisses.

Typical grey gneisses in the crystalline basement of the Caledonian fold belt. These gneisses originally formed about 1900 Ma ago as new granitic crust. This was subsequently deformed, and much later refolded and recrystallised during the Caledonian orogeny. These rocks occur in the central part of the roots of the mountain belt in the area north of Daneborg (about 75°N). The hammer is 40 cm long.

Folded and metamorphosed, pale sandstone and thin, darker mudstone layers of the Krummedal sedimentary sequence (about 1100–1000 Ma) in the southern part of the Caledonian fold belt. The original sedimentary character of the rocks is still preserved, although they have been recrystallised under low amphibolite facies conditions (see p. 35). Inner part of the Scoresby Sund area (72°N). The hammer is 40 cm long.

The Caledonian fold belt in North-East Greenland

Cliff section through the Målebjerg window (73° 40´N) in the inner fjord zone in the southern part of the Caledonian fold belt. The flat-lying thrust (NST) separates metamorphic sediments of the Krummedal sequence above from the foreland rocks below. The rocks above the thrust are the about 950 Ma old Krummedal sequence forming part of the Niggli Spids thrust sheet. Below the thrust (NST) pale, reddish-yellow lower Cambrian sandstones (SF) overlie about 1900 Ma old grey gneisses (G). The high-grade metamorphic Krummedal sequence rocks have been thrust westwards 100–200 km, and now overlie lower grade Cambrian sediments and their underlying gneissic basement rocks. Målebjerg at upper right of the photograph is about 1500 m high.

About 1300–900 Ma ago a supercontinent, called Rodinia, was formed. In the eastern part of the North American continent, a compound orogenic belt (the Grenville) developed, and a little later a similar Sveconorwegian fold belt formed in the southern part of the Scandinavian basement shield. In Greenland there is evidence that a similar 'Grenville/Sveconorwegian' orogeny affected parts of the basement complexes in the Caledonian fold belt 1100–900 Ma ago. Rocks included in this orogeny in North-East Greenland comprise a several kilometre-thick sequence of sandy and muddy sedimentary rocks called the Krummedal sedimentary sequence (see p. 74), parts of which were melted to produce migmatites and granitic intrusions. Like the two earlier Archaean and Palaeoproterozoic gneiss complexes forming the basement to the Caledonian fold belt, the remains of this 1100–900 Ma old mountain belt are today exclusively found as parts of major thrust sheets in the southern part of the Caledonian fold belt between 70°N and 76°N. These major thrust sheets of crustal material were derived from regions up to several hundred kilometres further to the east of where they are found today.

Deformation and transformation

With the collision of Laurentia (with Greenland) and Baltica (including Norway), the Iapetus Ocean closed, and the rocks along both sides of the collision zone were folded. In Norway, the rocks were displaced eastwards over the rigid Baltic shield, and in Greenland the rocks were transported in the opposite direction, westwards, over the Greenland Precambrian shield. Compression involved considerable shortening of the crust perpendicular to the length of the fold belt, and it is probable that the original width of the marginal, deformed zones was more than halved. Shortening occurred generally through stacking of major thrust sheets and by folding of the upper crustal rock sequences. Analysis of

Distribution of the continents 1200–1000 Ma ago, after the supercontinent Rodinia (light and dark brown colours) had formed from an assembly of land areas surrounded by a shallow-water continental shelf. The continents were assembled on one side of the Earth, while the other side was covered by ocean. The supercontinent existed in the period 1300–800 Ma. In the period 1100–900 Ma, it was reworked by an orogenic episode that is called the Grenville fold belt in North America, while in Scandinavia it is called the Sveconorwegian fold belt. This orogenic episode also affected the southern part of North-East Greenland and parts of Svalbard.

FOLD BELTS IN NORTH AND NORTH-EAST GREENLAND

the deformation has unravelled a complex sequence of structural events, where younger structures have overprinted and refolded older structural patterns, some of which were formed prior to the Caledonian deformation.

During the Caledonian folding in North-East Greenland, different structural patterns occurred in the upper and lower levels of the fold belt. In a western marginal belt characterised by thin-skinned tectonics (see p. 95), thrust sheets glided over an essentially unmodified basement of foreland rocks. In the central part of the fold belt, characterised by thick-skinned tectonics, the older crystalline basement is caught up in the deformation and Caledonian structures extend deep into the crust.

Examples of near-surface structural patterns occur in the northern part of the fold belt in western Kronprins Christian Land (80–82°N), where thrust sheets derived from the east displace older Proterozoic sedimentary rocks above younger Lower Palaeozoic sedimentary rocks on the Caledonian foreland (see p. 95). In the westernmost part of the marginal zone, in front of the major thrust sheets, a series of minor thrust segments affect Ordovician and Silurian sedimentary rocks. By constructing cross-sections, and restoring thrust displacements, it can be shown that a major thrust sheet has transported rocks more than 40 km towards the west, while folding and minor thrusts in the marginal zone have reduced the width of the folded zone by about 45%.

In the southern part of North-East Greenland the Caledonian deformation produced open fold patterns formed in upper crustal levels (down to 15–20 km depth). At greater depths a more intense and plastic deformation style with tight folding was accompanied by new mineral growth due to increased temperatures (metamorphism). The structural architecture comprises three principle zones or levels:

1. an unmodified or only slightly altered foreland that is also exposed in tectonic windows eroded through thrust sheets
2. a lower thrust sheet (Niggli Spids thrust sheet)
3. an upper thrust sheet (Hagar Bjerg thrust sheet) composed of
    a: a lower crystalline segment (the Hagar Bjerg unit), and above this
    b: an upper part with well preserved sediments (the Franz Joseph unit).

The names given to the thrust sheets are derived from place names in the area. These thrust sheets

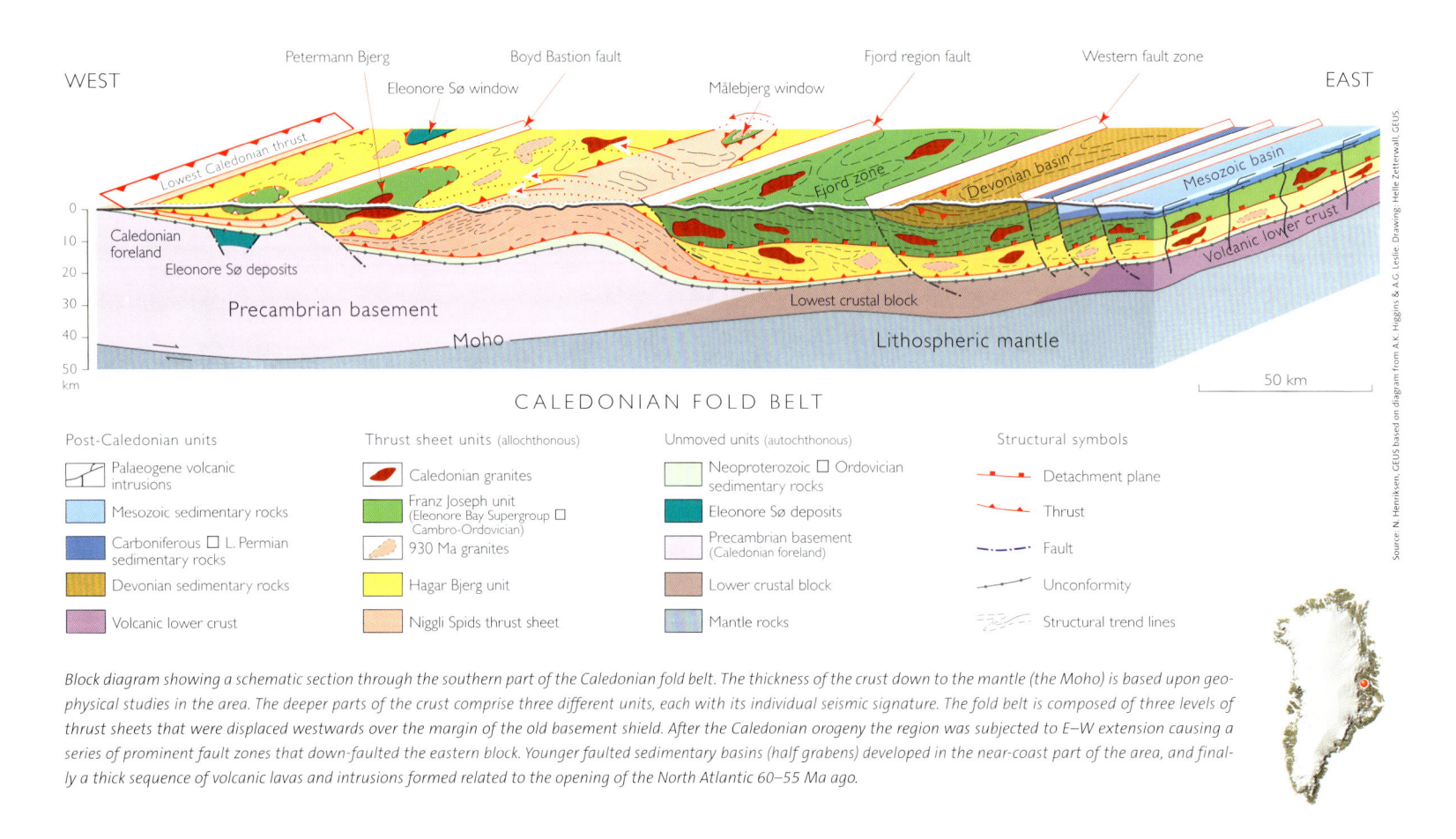

Block diagram showing a schematic section through the southern part of the Caledonian fold belt. The thickness of the crust down to the mantle (the Moho) is based upon geophysical studies in the area. The deeper parts of the crust comprise three different units, each with its individual seismic signature. The fold belt is composed of three levels of thrust sheets that were displaced westwards over the margin of the old basement shield. After the Caledonian orogeny the region was subjected to E–W extension causing a series of prominent fault zones that down-faulted the eastern block. Younger faulted sedimentary basins (half grabens) developed in the near-coast part of the area, and finally a thick sequence of volcanic lavas and intrusions formed related to the opening of the North Atlantic 60–55 Ma ago.

The Caledonian fold belt in North-East Greenland

both comprise a characteristic sequence of rock types that was originally derived from different areas. The Niggli Spids thrust sheet was derived from an area more than 100–200 km east of its present position, while the overlying Hagar Bjerg thrust sheet comprises rocks that came from an additional 100–200 km farther to the east. The Franz Joseph unit appears to have been carried westwards as part of the Hagar Bjerg thrust sheet. Thus, the 300–400 km broad zone that makes up the fold belt today represents an original 500–700 km wide segment of the margin to the Laurentian continent.

The composition of the thrust sheets is shown schematically to the right.

It is noteworthy that the major Caledonian thrust sheets were emplaced comparatively late in the evolution of the fold belt. The thrust structures clearly cut through the regional metamorphic patterns, resulting in high-grade metamorphic rocks locally overlying rocks with a lower metamorphic grade. The same type of juxtaposition of different crustal levels is seen in the tectonic windows, where older, crystalline, high-grade metasedimentary rocks of the Niggli Spids thrust sheet overlie younger, fossiliferous, Lower Palaeozoic sedimentary rocks of the foreland (see p. 99).

In collision zones between continents, some rocks become pushed down into the crust and other high-level rock units form thrust sheets that are stacked up on top of each other. This leads to thickening of the crust, with the rocks at depth being subjected to higher temperatures and pressures. In the Caledonian fold belt in North-East Greenland rocks range from completely unmodified types that have remained high in the crust, to rocks that belong to the high-pressure eclogite facies, reflecting alterations taking place at depths down to 60–80 km in the crust; all of the known metamorphic facies are thus represented (see p. 35).

When common rock types are subjected to temperatures of about 700°C at moderate pressures the quartz–feldspar component can begin to melt and locally pools to form granitic liquids. Melting is facilitated if water-bearing minerals occur in the rocks, and clay-rich sediments will provide compositions that are most likely to melt. Local partial melting can permeate the rock as a web-like texture of thin veins and diffuse pods of granitic melt, forming the mixed rock migmatite (see p. 34). Melts can also accumulate into large magma bodies that can move long distances up through the crust, so that granite dykes and bodies can cross-cut and intrude both the parent

Thrust sheets in the southern part of the Caledonian fold belt in North-East Greenland

highest

Hagar Bjerg thrust sheet

Franz Joseph unit

Caledonian granites (445–425 Ma). Sedimentary rocks: Eleonore Bay Supergroup, Tillite Group and Cambrian–Ordovician, with a total thickness of up to 18.5 km (see p. 76). The succession was deposited between 900 and 450 Ma ago.

— detachment —

Hagar Bjerg unit

Older granites (950–920 Ma) and Caledonian granites (445–425 Ma). Krummedal sequence (metamorphosed sedimentary rocks) (1200–950 Ma). Basement gneiss complex (2000–1750 Ma).

— major thrust —

Niggli Spids thrust sheet

Krummedal sequence (metamorphosed sedimentary rocks) (1200–950 Ma). Basement gneiss complexes of two ages (3000–2700 and 2000–1750 Ma).

— major thrust —

lowest

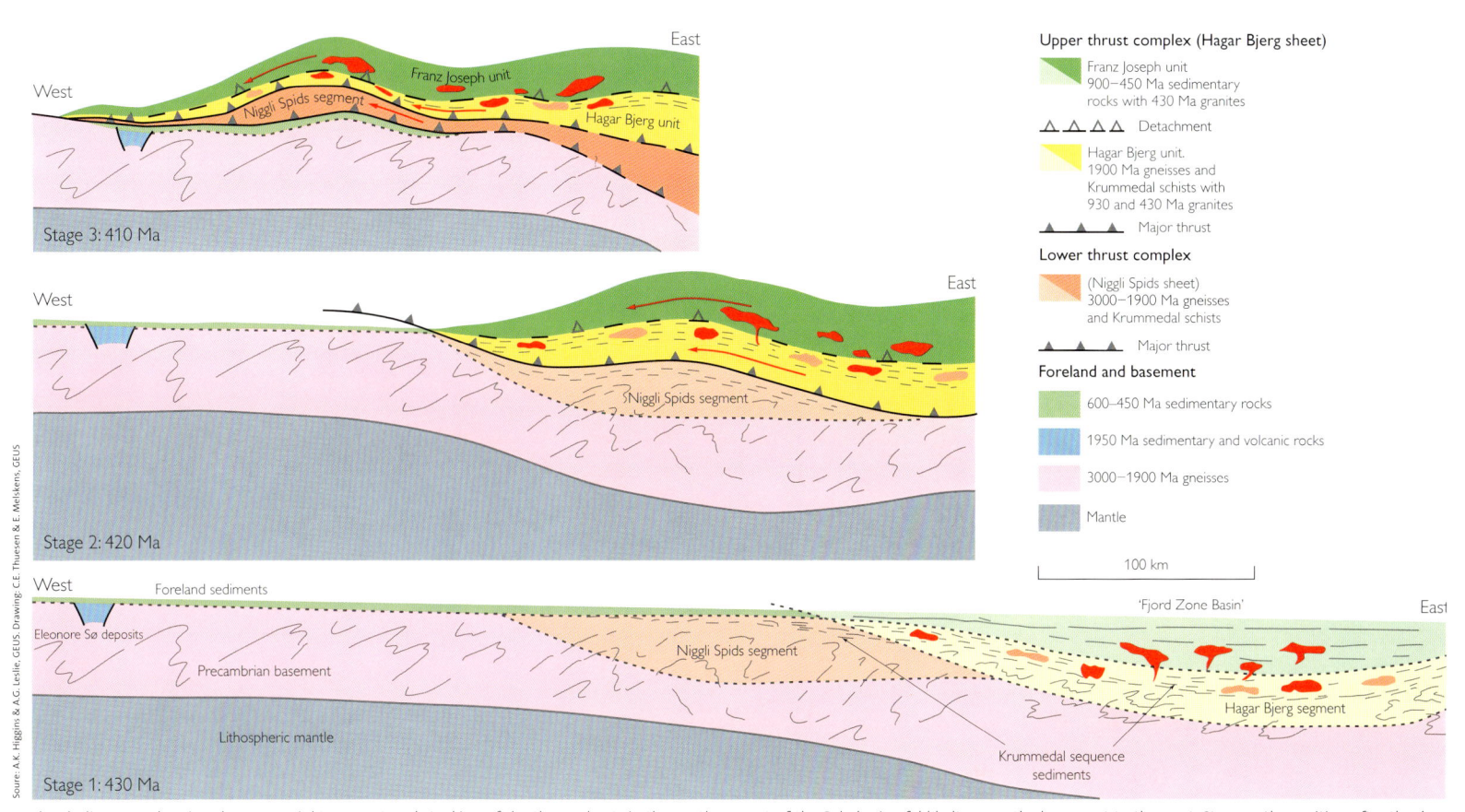

Sketch diagrams showing the sequential transport and stacking of the thrust sheets in the southern part of the Caledonian fold belt across the basement to the west. Stage 1: the position after the deposition of the sediments in the 'Fjord Zone Basin'. Stage 2: the Hagar Bjerg thrust sheet with the overlying sedimentary rocks of the Franz Joseph unit riding on the crystalline basement below. The entire thrust sheet was pushed across the western margin of the continent, which at that time also contained the Niggli Spids segment. Stage 3: the Niggli Spids segment, on the margin of the continent was detached from its basement and transported westwards together with the upper thrust complex (Hagar Bjerg thrust sheet) riding on top of it.

FOLD BELTS IN NORTH AND NORTH-EAST GREENLAND

Intrusive light grey Caledonian granite cutting across flat-lying sediments of the Eleonore Bay Supergroup. The network of granite sheets largely follows the sedimentary layering, and the large sedimentary xenoliths are in many places completely enclosed by granite. The outcrop shown in the photograph is about 1000 m high. Locality is in the innermost part of the fjord zone at about 73°20′N.

Deformed augen granite, comprising lens-shaped feldspar aggregates in a finer-grained foliated granitic groundmass. The rock was emplaced as a granite intrusion about 950 Ma ago, and developed its foliated structure during subsequent deformation. Locality in the nunatak region of North-East Greenland at about 74°20′N. The 15 cm long pen gives the scale.

rocks and overlying sequences. In the East Greenland Caledonian fold belt, the melting of rocks of the Krummedal sedimentary sequence is a very important source of granitic magma, for the generation of both migmatites and granites. These sedimentary rocks have been melted to produce widespread granitic intrusions during both the mountain belt formation about 950 Ma ago, and the Caledonian mountain building event 465–425 Ma ago.

Granites

In the southern part of the Caledonian mountain belt two sets of granites of nearly identical composition but very different ages occur. They are easily mistaken for each other, and can only be distinguished by their relative age with respect to the surrounding rocks and structures, or by using radiometric dating. The oldest generation of granites was formed about 950–920 Ma ago by partial melting of the metamorphosed Krummedal sediments. The younger Caledonian granites were also generated from partial melting of the Krummedal sequence rocks, but formed 445–425 Ma ago. Both types of granite occur as components of the Hagar Bjerg thrust sheet and were generated in the same region of the thrust sheets, prior to their westwards transport. The oldest generation of granites are found exclusively in the lower segment of the Hagar Bjerg thrust sheet (Hagar Bjerg unit), while the Caledonian granites occur in both the lower and the upper units (Hagar Bjerg unit and Franz Joseph units).

The granite bodies have many different forms, from small, thin veins, through medium-sized, flat-lying sheets, to large, coherent lens-shaped intrusions that are more than ten kilometres across. The latter bodies may contain angular inclusions of the wall rocks that, together with the pattern of the intrusions, show that they were forced into the surrounding rocks, either by simply pushing their way up through the overlying rocks following cracks and fissures, or by partial dislodgement of the overlying roof of country rocks from which blocks were broken off and sank down into the magma. The Caledonian granites cut through the open folds that deform the sedimentary rocks in the highest Franz Joseph thrust unit, demonstrating that they were generated relatively late in the evolution of the fold belt – after the beginning of the compressive phase, but prior to the displacement of the major thrust sheets.

Some of the granites of the oldest generation were emplaced at an early stage of the fold formation; they have been subjected to deformation and metamorphism, giving them their streaked gneissic appearance. Many of these granites are characterised by large oblong feldspar crystals, transformed into lenticular shapes. They are called augen (eye) granites.

The late development

Following collision of the continents and the build up of mountain belts on both sides of the collision zone, the large horizontal compression stopped. Subsequently, the over-thickened crust beneath the mountain belts that had descended into the lithosphere, began to rise because its density was less than the surrounding rocks in the mantle (see isostasy, p. 27). At the same time, relaxation of the horizontal pressures across the mountain belts led to an extension of the crust in the opposite direction to the earlier compression. This extensional movement and the uplift, or exhumation, of the depressed root zone of the fold belt are both effects of the cessation of compression. The extension phase typically leads to formation of a series of prominent, normal faults that may have vertical movements of 5–10 km. Uplift was very slow, about 1 km per million years, and as the rocks from lower crustal levels were successively uncovered they were worn away by erosion.

Erosion of the Caledonian fold belt in southern North-East Greenland resulted in deposition of sandstone-dominated successions throughout the Devonian, Carboniferous and Lower Permian, with a total thickness of up to 10 km (see p. 114). The erosion levels currently exposed in North-East Greenland reflect a very varied uplift in different parts of the mountain belt. At some places the rocks uncovered come from only the upper 10–15 km of the crust, while rocks in other areas have risen to the surface from depths of 50–60 km and, in the extreme, may have come from depths of about 100 km down in the crust.

The speed of uplift and subsequent erosion of the rocks of the mountain belt resulted in the production of an enormous volume of sediment that was transported away and deposited in the younger basins. The Franklinian Basin, which extends across North Greenland into Canada, was filled in by sandstones deposited by turbidity currents that can be linked to erosion of the Caledonian mountain belt, at a time when the Caledonian thrusts were beginning the build up of the eastern parts of North-East Greenland (see p. 83). Testimony to the earlier existence of substantial overlying sequences removed through erosion is exemplified in Kronprins Christian Land. Here Ordovician and Silurian sediments contain microfossils (conodonts) that have changed colour due to pressure and temperature effects imposed by a former pile of overlying rocks. Analysis of the colour index of the conodonts shows that they were buried beneath a pile of thrust sheets up to 15 km thick that has since been eroded away.

POSITION OF GREENLAND 390 Ma AGO

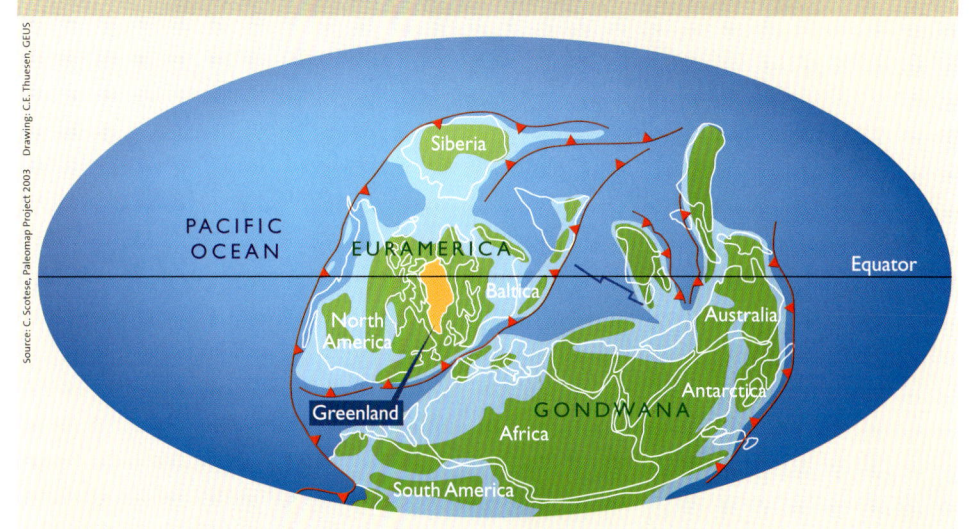

The distribution of the continents in the Middle Devonian, after the Caledonian orogeny had taken place, and the continents Laurentia and Baltica had joined together into a coherent block called Euramerica (after Europe and America). The supercontinent Gondwana that occupied a position in the southern hemisphere drifted northwards and later, together with Euramerica, formed the supercontinent Pangaea (see p. 123).

CHRONOLOGY OF THE CALEDONIAN FOLD BELT IN NORTH-EAST GREENLAND

The Caledonian fold belt in North-East Greenland incorporates a series of different sedimentary deposits and volcanic formations within the older basins, together with different units of the underlying basement. These reflect the long and complex geological history of the eastern margin of Laurentia. Their development spans more than 2600 Ma and within this period there is evidence for three Precambrian orogenies. The Neoproterozoic–Palaeozoic Caledonian cycle began with the opening of the Iapetus Ocean and ended about two hundred million years later with the collision between Laurentia and Baltica. A summary of the principal events in the chronology together with their approximate ages is given below.

Principle events	Million years
Collapse of mountain belt due to isostatic uplift, extension and erosion. Formation of the Devonian continental basins.	400–355
Continental collision and orogeny with folding, metamorphism, granite intrusion and thrusting.	465–400
Iapetus Ocean opening and closing with formation of sedimentary basins along the margins of the continent to the west (Tillite Group and Kong Oscar Fjord Group).	600–460
Neoproterozoic sedimentary basins in the eastern marginal zone of the Laurentian continent (Eleonore Bay Supergroup).	900–600
North-East Greenland branch of the North American–Scandinavian orogenic belt (Grenville/Sveconorwegian belt) with deposition and metamorphism of sedimentary formations (Krummedal sedimentary sequence) and granite intrusion.	1100–920
Development of an extensive volcanic province with plateau basalts (Zig-Zag Dal Basalt Formation) in eastern North Greenland.	1380–1350
Formation of a large continental sedimentary basin with associated plateau basalts in eastern North Greenland, part of which was later involved in the Caledonian fold belt (Independence Fjord Group; sandstones and basalts).	1750–1650
Development of a major Palaeoproterozoic orogenic belt with injection of juvenile granitic rocks. These rocks are widespread in the northern part of the Caledonian fold belt.	2000–1750
Metamorphic schists and gneisses of Archaean orogenic episodes form the oldest components in the Caledonian fold belt; occur mainly in the southern part.	3000–2700

FOLD BELTS IN NORTH AND NORTH-EAST GREENLAND

A mountain belt where only the upper part of the crust is uncovered

The Ellesmerian fold belt in North Greenland

The Palaeozoic Franklinian Basin in North Greenland formed due to rifting of an old continent, of which only the southern part is preserved. The northern part is unknown, but in the Lower Palaeozoic a land mass must have existed north of the present North Greenland and Arctic Canada. On the northern part of Ellesmere Island in Canada, immediately west of North Greenland, remnants of an old geological terrane are preserved that may be a part of this northern continent. The basement of the Franklinian Basin is only found at the southern margin of the basin, where it comprises Precambrian gneisses overlain by Proterozoic sedimentary and volcanic rocks. What lies hidden under the deeper parts of the basin is not known. It is accepted, however, that the basin was formed on the subsiding margin of a continent, as oceanic rocks have not been found in the sequence. Rifting of the old continent and the formation of the Franklinian Basin were followed by movements in the opposite direction causing closure of the basin and collision of the two bordering continents. The Ellesmerian deformation that formed the fold belt in North Greenland took place between the end of the Devonian and the earliest Carboniferous, about 365–345 Ma ago.

Banded sandstone with a cross-bedding like structure, but the bands in this special rock are secondarily formed and cut through the original bedding. The sample is about 30 cm across.

Fold intensity map showing increasing deformation from south to north across the North Greenland fold belt. In the most weakly deformed zone (shown in orange) the folds are open and overturned to the south. In the middle zone (lilac colour) the deformation increases in intensity northwards. In the most strongly deformed zone (purple), farthest north, there are three phases of tight folds overturned northwards.

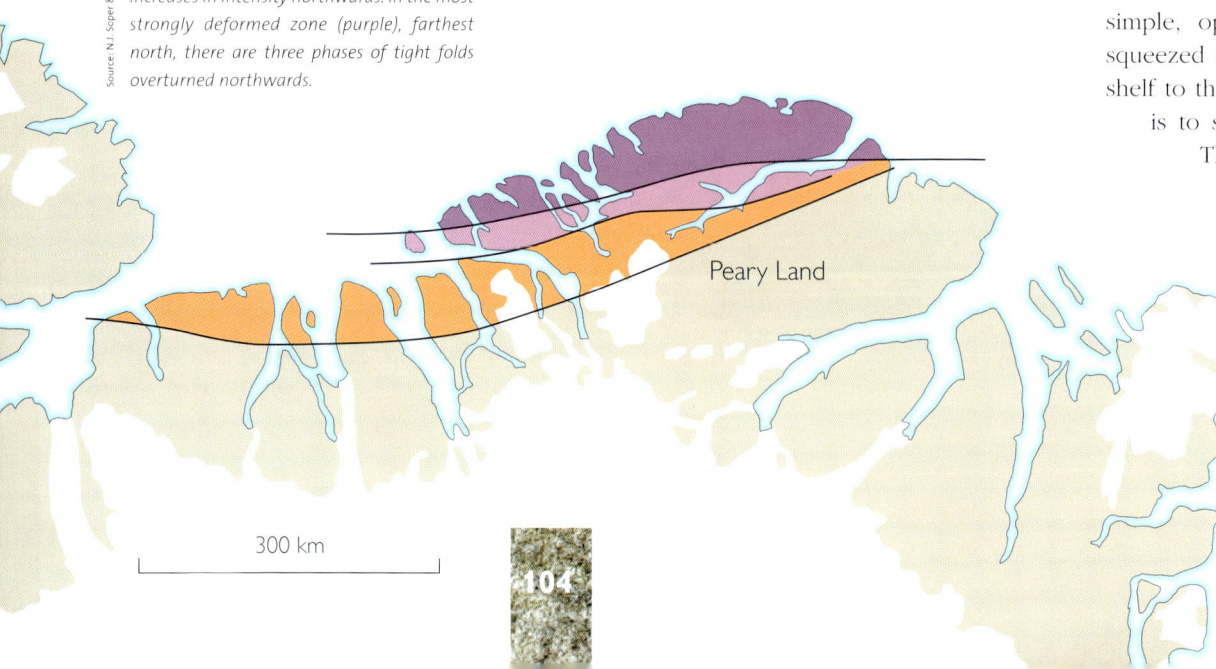

In contrast to the East Greenland Caledonian fold belt, which comprises a complex pile of thrusts and folded basement slices overlain by younger sedimentary basin deposits, the Ellesmerian fold belt is relatively simple. The involved rocks are exclusively composed of deep-water sediments from the Franklinian Basin and there are only few major thrust sheets. The total compression and shortening across the belt are limited. The structures and metamorphic facies reflect changes that have taken place in the upper parts of the crust, between 10 and 20 km deep, and granites or gneisses that are characteristic of deeper levels of mountain belts are not found. Since no crystalline basement rocks are involved in the Ellesmerian deformation in North Greenland, and thrust displacements are minor, the entire fold belt in North Greenland can be viewed as 'thin-skinned' (see p. 95).

The region of North Greenland that was affected by Ellesmerian deformation is called the North Greenland fold belt. This E–W-trending zone largely follows the distribution of the sediments in the Franklinian Basin, so that deformation is essentially confined to the deep-water sediments. Deformation is most intense in northern Peary Land where three phases of folding are distinguished. All three phases have the same orientation, with the younger sets refolding the slightly older sets, and are all accepted as of Ellesmerian age. Farthest to the north the most transformed rocks were recrystallised under amphibolite facies conditions. The intensity of the deformation decreases southwards and the folds gradually die out towards the boundary with the carbonate shelf. In North Greenland the fold belt can be followed almost 500 km from east to west, and at the widest it is a little over 100 km from north to south.

The three deformation zones in the fold belt

The North Greenland fold belt can be divided into three distinct zones that run parallel to the length of the belt. Each zone has its own characteristic deformation pattern with the most complex structures in the zone farthest to the north.

The southernmost zone comprises structurally simple, open folds and minor thrusts that were squeezed up against the edge of the shallow-water shelf to the south. The folds are south verging, that is to say the folds are overturned southwards. The minor thrusts have southwards displacements of generally a few kilometres each. The 3–5 km thick succession affected by folding consists of Cambrian–Silurian sandstones and shales, with Silurian deep-water, turbiditic sandstones dominant. This sedimentary package rests on an unexposed, rigid base of Proterozoic sandstones

Peary Land

300 km

and Precambrian gneisses which are probably unaffected by the Ellesmerian deformation.

The middle structural zone forms a transition between the southern and northern zones. In the south the folds are overturned southwards and in the northern part of the zone they verge northwards. The intensity of the folding in the middle zone is greater than in the south, and the folds are more steeply inclined. The deformed sedimentary rocks exposed in the steep mountainsides have undergone greater compression and have been subjected to higher temperatures than the similar rock units in the southern zone.

The northernmost structural zone was formed where the succession had its greatest thickness – up to 8 km. This zone is characterised by three phases of tight folding, strongly northwards verging and recrystallised under amphibolite facies conditions. The rock units affected by the deformation are Cambrian turbiditic sandstones that overlie carbonate rocks of probable early Cambrian age. The present southern boundary of this northern zone is a major fault that cuts and displaces the fold structures, and thus formed after the Ellesmerian deformation. However, there is evidence that major faults controlled the boundary between the shallow-water shelf and deep-water trough during the formation of the Franklinian Basin. To some extent, reactivation of these early faults has influenced the structural patterns that developed during Ellesmerian deformation.

The intensity of the metamorphism in the fold belt gradually increases from south to north and, as with the structural pattern, the belt can be divided into a series of parallel zones, each with its own characteristic mineral assemblages. Farthest to the south the sedimentary rocks are only weakly modified by a slight rise in the temperature of the crust. This is followed to the north by a zone with greenschist facies metamorphism (see p. 35). This again is followed by

Open zigzag folds in Silurian turbiditic sandstones in the southern fold zone. Northern end of Victoria Fjord in the central part of North Greenland. The fjord wall is about 350 m high.

In the intermediate deformation zone in the North Greenland fold belt, a local west-directed thrust system occurs, seen here on the 800 m high vertical fjord wall of Frederick E. Hyde Fjord in northern Peary Land. The structure forms a stack of slices which have a tiled pattern. The transport direction was from east to west (right to left on the photograph). The bedded sedimentary succession comprises Upper Cambrian and Ordovician mudstones and sandstones.

FOLD BELTS IN NORTH AND NORTH-EAST GREENLAND

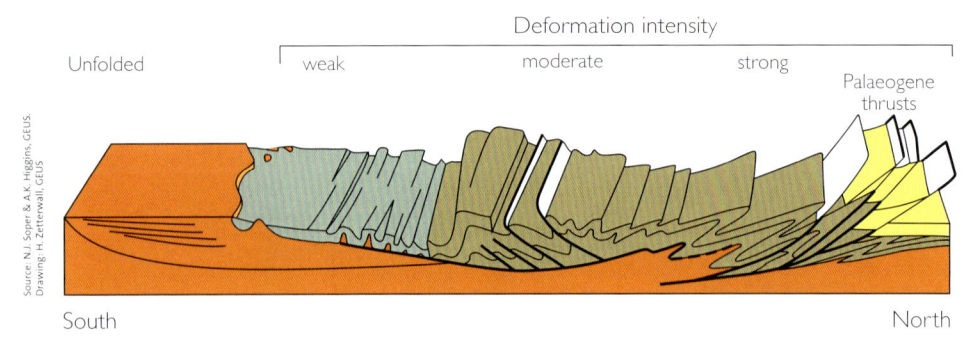

Schematic, three-dimensional profile across the North Greenland fold belt in western Peary Land. The profile shows the transition from unfolded, shallow-water shelf deposits (grey-green) in the south, into the Ellesmerian fold belt (brown) to the north. The intensity of the deformation increases from the south to the north. In northern Peary Land, both earlier Ellesmerian deformed Palaeozoic rocks and the later Wandel Sea Basin deposits (yellow, see p. 118) are deformed by Eurekan thrusts caused by Palaeogene compression in the crust.

a zone with somewhat higher-grade metamorphism (low amphibolite facies), and lastly, farthest north, all the rocks have been totally recrystallised under amphibolite facies conditions. Metamorphism has, however, not been so high that the sediments were transformed into migmatites and granites and, although overprinted by a strong schistosity, they have preserved some of their original sedimentary characters.

The rocks involved in the Ellesmerian deformation that produced the North Greenland fold belt represent only the southern part of the Franklinian Basin. The northern part of the basin, as well as the collision and subduction zones, that gave rise to the orogenic deformations, must originally have been positioned north of the present coast of North Greenland.

Palaeogene deformation and structures

The northernmost part of North Greenland was overprinted by renewed deformation in the Palaeogene (about 65–60 Ma ago). These younger structures

Tight folds that are overturned to the south (to the right) in a moderately folded part of the North Greenland fold belt. The sequence comprises dark, fine-grained slates and greyish sandstones from the Ordovician and Lower Silurian that were deposited in the Franklinian Basin. Locality is in northern Nares Land, North Greenland. The mountainside is about 350 m high.

belong to the Eurekan deformation episode, which deformed the sediments of the Mesozoic Sverdrup Basin in Arctic Canada. In the northern part of Peary Land, Eurekan deformation is characterised by strong N–S-directed compression, which led to displacement of older rocks from the Franklinian Basin being thrust northwards over a series of younger sediments from the Wandel Sea Basin (see p. 118). Elsewhere, the Eurekan deformation in North Greenland had only limited intensity, mostly seen in the formation of minor folds and schistosity associated with various fault structures.

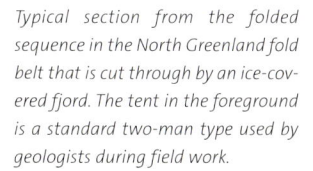

Typical section from the folded sequence in the North Greenland fold belt that is cut through by an ice-covered fjord. The tent in the foreground is a standard two-man type used by geologists during field work.

CONSTRUCTED PROFILES THROUGH DEFORMED SEDIMENTARY SUCCESSIONS

Deformation in mountain belts occurs mainly through horizontal compression so that the width of the units is shortened in the compression direction. During compression folds, thrusts and different faults may form. When unravelling the structural geological development of an area, the starting point is the observed structural pattern of the deformed sequence. Using mapping data and field measurements, the original form and extent of the sequence before deformation can be determined. Such an unravelling can, in some instances, be carried out through drawing of profiles that are constructed by a stepwise restoration of the deformation geometry of the original, layered sequence.

A series of good examples of such profile construction has been made across the central part of the Ellesmerian fold belt in North Greenland. Here a southern, thin-skinned tectonic zone with an original width of about 100 km has been compressed into about 60 km. Farther westwards in North Greenland the thin-skinned zone passes into a large-scale, 7 km high, one-sided fold (a monocline) that developed above the fault-bounded margin of the Greenland basement shield. The profile reconstruction shows that at an early stage of basin formation in the Lower Palaeozoic, a fault scarp formed in the underlying basement parallel to the length of the basin. Shallow-water deposits were laid down on the shelf side of this scarp, while deep-water sediments were deposited on the oceanic side of the scarp. The profile gives a glimpse into not merely the architecture of the hidden structures, but also information on the geometry of the Franklinian Basin during its development.

The profile is constructed in such a way that the length of an individual layer boundary is the same in both the deformed profile (as observed in the field) and in the reconstructed section depicting the succession prior to deformation. The construction is carried out stepwise, by removing the deformation and smoothing out the boundaries in the reverse order in which the structural elements were formed. Thus the youngest deformation is compensated for first, by moving the latest thrust slices back into their starting place in the profile. With this accomplished the next stage would be to straighten out the simple folds (provided they were next in the backwards sequence). The total development is illustrated by construction of a series of profiles, each of which represents a stage in the geometrical process. All steps must maintain the geometrical demands that all boundaries of the layers should keep their lengths unchanged.

Construction of profiles is an important geological method that allows the reconstruction of the original form and extent of a deformed sequence, so that it is possible to construct the unexposed structures in the substrata based on surface observations. The method is widely used in areas with thin-skinned tectonics and, in particular, in weakly deformed basinal areas, in connection with exploration for oil and gas.

1: Measured cross-section through the North Greenland fold belt. The numbers give the size of the displacement on the thrusts measured in kilometres.
2: Reconstruction of a section through a part of the original sedimentary sequence in the Franklinian Basin. The reconstruction is carried out as a profile that is constructed in stages by unravelling the deformation and structures one by one, and at the same time keeping the geometric dimensions constant. The constructed profile shows that at an early phase of basin formation there was a steep fault scarp, separating a shallow-water area in the south from deep water to the north. Deposition occurred through the Cambrian to Silurian and includes, from the bottom, sandstones (Skagen Group), calcareous deposits (Portfjeld Formation and Paradisfjeld Group), sandy deposits (Buen Formation and Polkorridoren Group), fine-grained mudstones (Frigg Fjord mudstone and Amundsen Land Group) and Silurian turbiditic sandstones (shown in yellow).

Succession of Jurassic sandstones (Charcot Bugt Formation) on Milne Land in East Greenland. The layering was not deposited horizontally but forms large-scale cross-bedding in a sand body about 50 m thick.

Photo: M. Larsen, GEUS

YOUNGER SEDIMENTARY BASINS

The basins contain flat-lying, unaltered, sedimentary strata

8

LARGE FOSSIL CONTENT

YOUNGER SEDIMENTARY BASINS

Following formation of the two Palaeozoic mountain ranges in North and North-East Greenland, sedimentary basins developed along what later became the margins of Greenland. Sediments several kilometres thick accumulated in these basins between the Devonian and the Neogene. With only rare exceptions the sediments are unaffected by later fold-belt deformation and are therefore found in the rather flat-lying 'layer-cake' attitudes in which they were deposited. The sedimentary rocks are consolidated, but are generally not altered into metamorphic rocks (metasediments), so their original composition, depositional structures and fossil content are preserved. It is therefore possible to study in great detail how they were formed and to interpret their original depositional environments.

Flat-lying sedimentary layers of Early–Middle Jurassic age in central Jameson Land in East Greenland. The lower layers consist of sandstone, but those higher up consist of finer grained, shaly sediments. The hillside in the picture is 400–500 m high.

The Earth's climate has undergone significant variation through geological time. The average global temperature has mostly been around 22°C for the last 500 Ma, but there have been times when it was about 10°C lower. Such warm and cold periods have alternated, and we are presently in a cold period with an average global temperature of about 12°C.

Huge volcanic provinces formed in both central East and West Greenland during the earliest Palaeogene, about 62–55 Ma ago, when thick series of lavas erupted (see p. 138). The volcanic rocks partly covered older depositions in the sedimentary basins, and neither the lavas nor the sediments have been affected by later folding.

The younger sedimentary rocks in North-East Greenland form a nearly unbroken succession that gives an excellent opportunity to study the geological development of the western half of the North Atlantic region. The sediments in East Greenland are important because they are an onshore equivalent to the sediments in the oil and gas provinces of the North Sea and offshore North-West Europe.

The supercontinent Pangaea – and its disintegration

The formation of the Caledonian mountains at the end of the Silurian assembled North America, Greenland, Baltica and the British Isles into a single, large continent called Euramerica, which straddled the equator at that time. Most of the other continents were assembled into another large, contiguous supercontinent called Gondwana, which was situated in the southern hemisphere. These two continent groups collided during the Carboniferous creating a giant supercontinent called Pangaea, which stretched from the South Pole to the middle of the northern hemisphere. New fold belts formed during the collision – the Appalachians in eastern North America, the Variscans in north-western Europe and the Mauritanides in western North Africa (see p. 96).

During the following millions of years Pangaea disintegrated slowly, and a number of younger sedimentary basins developed in Greenland over a period of about 380 Ma from the middle of the Devonian until the present day. Three hundred million years ago all the Earth's continents made up one land area, and Greenland was connected to the rest of the world, but shallow seas invaded Pangaea and it gradually became dissected by narrow straits. New oceans began to open and the outlines of our present-day continents began to appear. The disintegration of the supercontinent occurred because its lithosphere extended, which resulted in rifts – typically along lines of weakness that followed older continent margins. Continued extension led to plate-tectonic movements as the continents split, and they gradually moved to different parts of the globe. As they did so, each continent moved through new geographic and climatic zones, so the conditions under which sediments were deposited changed correspondingly.

This disintegration of Pangaea and reorganisation of the world's continents is the first part of the most recent Wilson cycle (see p. 93). The general disintegration was followed by collision of continental masses that started a new episode of earth-wide mountain building, the Alpine orogeny (see p. 53). This orogenic event occurred mainly between fifty and thirty million years ago and formed mountain belts, e.g. the Alps and the Himalayas, and along the western coasts of North and South America.

In Greenland the latest phase of this development took place from the beginning of the Palaeogene when Greenland first began to split from North America and then later from Europe by the creation of new oceanic crust between the continents. The start of this phase was dominated by the deposition of volcanic basalts many kilometres thick that are found in both East and West Greenland (see p. 138).

Greenland's drift through the climate zones

The younger basins formed from the Devonian until the present day. During this time continental drift moved Greenland fairly steadily from just south of the equator to its present position in the Arctic. As it moved, Greenland passed through different climate zones, from tropical to subtropical, through a region with desert conditions, to a warm and then a cold temperate zone, finally ending in its present cold, Arctic climate. At the same time as Greenland moved across the latitudes, the climate of the whole Earth varied between alternating warmer and colder periods (see p. 110). The sediments reflect these changes, with respect to both the environmental conditions that prevailed during their deposition and the remains of the different plant and animal life preserved within them.

POSITION OF GREENLAND 306 Ma AGO

The supercontinent Pangaea formed at the end of the Carboniferous when all the continents in the southern hemisphere – Gondwana – united with all the continents in the northern hemisphere – North America, Baltica and Siberia. Greenland found itself within a land area that stretched from the South Pole to half-way up the northern hemisphere.

The younger sedimentary basins in Greenland reflect these general global changes. The basins are found along the margins of the basement shield and the Palaeozoic mountain belts, especially in North-East Greenland and under Greenland's continental shelves.

The basins in Greenland

The younger sedimentary basins in Greenland define five discrete regions and periods of crustal subsidence that formed basins. The resulting depressions

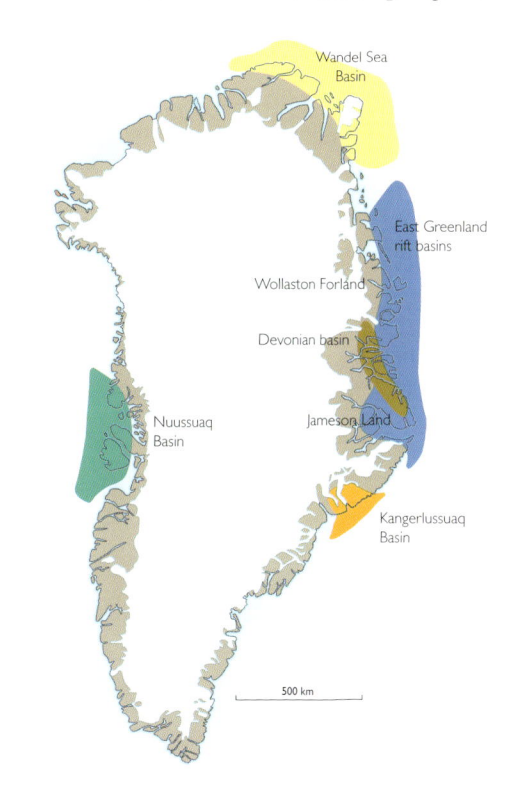

The younger sedimentary basins are found along the margins of the Greenland basement shield. They formed after the mountain-building events of the Palaeozoic. Most of the basins include depositions that today occur in the continental shelf areas. The map focuses particularly on the locations of the onshore portions of the sedimentary basins; their offshore extensions are considered later (see p. 156).

YOUNGER SEDIMENTARY BASINS

gradually filled with sediments derived from the adjacent uplands. The sediments are predominantly clastic: clays, sand and gravel whereas limestones and evaporites are found only locally. In this book the basins have been divided according to their location, age and depositional type:

1) The Devonian basin of North-East Greenland containing continental sediments
2) The Wandel Sea Basin in central and eastern North Greenland with Carboniferous–Palaeogene deposits
3) The rift basins in East Greenland that contain Upper Palaeozoic to Mesozoic sediments (the Jameson Land Basin and the Wollaston Forland Basin)
4) The Nuussuaq Basin in central West Greenland containing Cretaceous–Palaeogene sediments
5) The Kangerlussuaq Basin in South-East Greenland with its Cretaceous–Paleocene sediments.

Deposition in several of these basins extends into the present-day offshore areas (see p. 156).

PALAEOGEOGRAPHY – the changing positions of continents and oceans in the past

The distribution of land and sea we see today on our geographical maps is only a single frame in a moving picture of an ever-changing distribution of continents and oceans. Continental drift and the creation and destruction of oceans are the primary geological processes, collectively known as plate tectonics, that alter the Earth's crust and the constantly-changing geography on our maps.

The distance that continents move is very small during a human lifespan. Typically the crustal plates move only a few cm/year, although rates of up to 10–15 cm/year have been recorded. Even though such movements are very small, they accumulate to many thousands of kilometres during the long intervals of geological time. At a rate of 2–5 cm/year a continent will move 2000–5000 km over a period of 100 million years.

How do we know where the continents were in the past?

The past geographical locations of the continents and their relative positions are worked out using a whole suite of geological and geophysical measurements and observations. Primarily the palaeolongitude and -latitude of an area must be determined – this can be done by a number of methods. For older periods the most important method is palaeomagnetic measurements (see box on p. 113) that show how far north and south the continent lay at that time, and how it was oriented in relation to magnetic north–south. For developments during the last 150 million years the most important method is to reconstruct the palaeopositions of the continents by using the magnetic anomaly pattern in the oceans (see pp. 53 and 113).

The 'jigsaw method' is sometimes used to analyse the relative positions of continents in the past by seeing how their shapes fit together. The most well-known example is the close match between the west coast of Africa and the east coast of South America. These resemble two adjacent pieces of a jigsaw puzzle suggesting that they were once joined and have since separated. Other criteria used include the analysis of the distribution of fossil assemblages. An example is the distribution of some special 300 Ma old reptiles that originally must have lived in contiguous regions, but whose fossils are found today in regions of Africa and South America that were once adjacent.

Another method for determining how plates have moved is to use so-called 'hotspots', where rising bodies of high temperature material in the mantle (plumes) reach the surface and form volcanoes. If a hotspot stays stationary and the plate moves, a line of volcanoes forms, like pearls on a string, as the plate passes over the hot spot centre. The plate movement can be tracked in this way, back as far as the age of the volcanoes.

Maps showing the positions of the continents through time

Palaeogeographic maps can be drawn successively back through time from the present. Geologists have excellent data that show the development back to the Middle Jurassic (about 170 Ma ago), when the present oceans started to open. Generalised maps can be drawn with reasonable certainty for times before this, back to the beginning of the Cambrian, about 542 Ma ago. Uncertainty grows further back in time and it is common to see different interpretations of the global geography during the Proterozoic. One of the problems with such ancient reconstructions is that the sizes and shapes of the continents have altered. Every additional mountain-building episode added to the older continents, and both their sizes and shapes changed. Greenland can often be recognised on palaeogeographic maps, however, because its shield area has remained essentially unchanged during the last 1750 Ma.

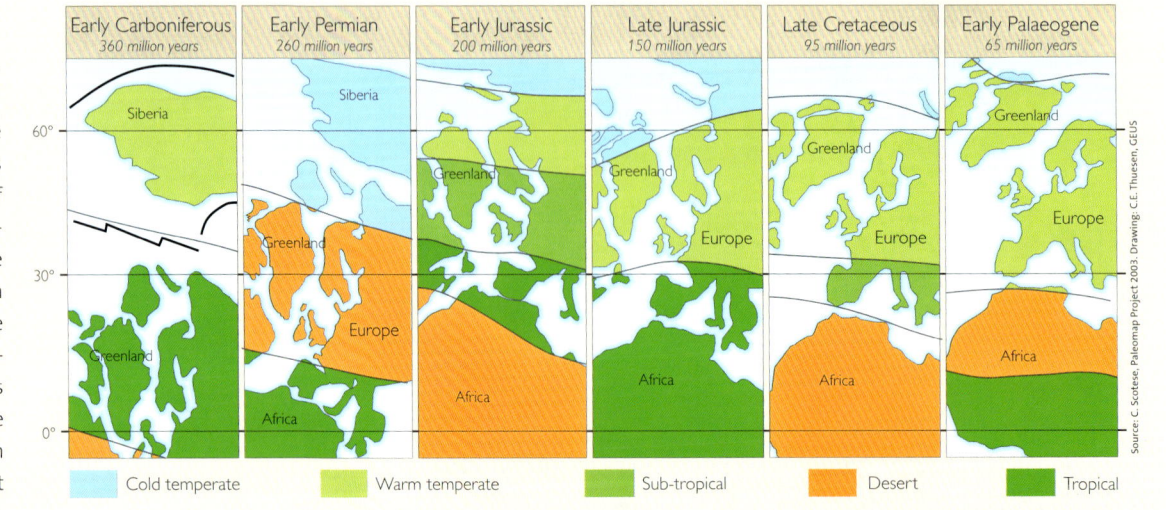

Greenland's northward drift over 300 Ma. Continental drift caused Greenland's latitude to change during this period from just north of the equator to the northernmost third of the northern hemisphere. This movement alone has caused Greenland's climate to change from warmer to colder conditions. Variations in the climate of the whole Earth must be added to these changes, so during this long period Greenland's climate has altered substantially. The figure demonstrates the climatic shifts that Greenland has experienced through this 300 Ma period.

PALAEOMAGNETISM – the Earth's former magnetism preserved in rocks

The Earth acts like a giant magnet with a N–S magnetic axis in close proximity to its rotation axis. The present-day North Magnetic Pole is about 8° (890 km) away from the geographic North Pole and lies north of the Canadian Arctic Islands. Although the direction of the Earth's magnetic field changes all the time, it does so very quickly compared to geological time. Its average position has coincided with the geographic pole through most of the Earth's history; so this direction can be used as a reference direction for plate-tectonic movements.

Rock magnetism

Many rocks contain minerals that are naturally magnetic, such as the iron oxides magnetite (Fe_3O_4) and hematite (Fe_2O_3). As a hot rock cools, individual grains of these minerals start to act like small compass needles and become oriented parallel to the prevailing magnetic field. As the rock cools through about 600–650°C (the Curie temperature) the orientation of the magnetic minerals becomes fixed in the direction of the Earth's field and the magnetism is 'frozen' into the rock.

Basaltic rocks are especially good at recording old magnetic orientations because they contain a large proportion of magnetic minerals. Some types of sedimentary rocks can do this too, because they contain small grains of magnetic minerals that become orientated parallel to the Earth's field as they are deposited. The orientation of the minerals is fixed when the loose sediments are cemented and consolidated. Generally the magmatic rocks contain much more magnetic material than the sedimentary rocks, so working with the latter requires more sensitive analytical procedures.

Measurement of the old 'frozen' magnetic directions in the rocks is done by cutting oriented samples in the field – commonly as small drill cores – and taking them home to a special laboratory where the magnitude and direction of their 'frozen' (remanent) magnetism can be measured using an instrument called a magnetometer.

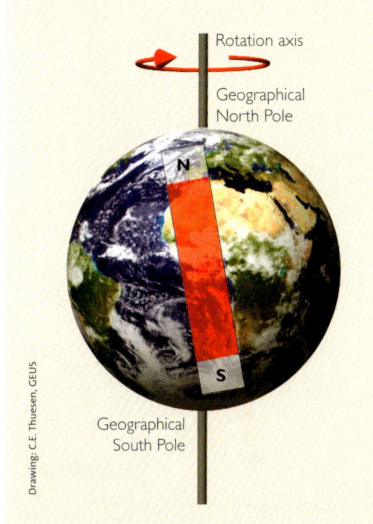

Switching of the magnetic poles

Measurements of thousands of samples, mostly of basaltic rocks and ocean-floor sediments from the last couple of hundred million years, has shown that the Earth's magnetic axis has changed polarisation in an irregular rhythm, causing the magnetic north and south poles to change place, (magnetic reversals). When the magnetic field points the way it does today, it is known as normally magnetised, and when it points in the opposite direction, it is called reversed. About 15 reversals have taken place during the last four million years. Further back in time the intervals have been variable. Long periods with the same orientation are known, such as one with reverse orientation in the Upper Palaeozoic.

The changes between normal and reverse orientation have been used to create a polarity timescale that is utilised as a stratigraphic reference over the whole Earth. Each epoch (Chron) is numbered backwards from today with the addition of an N or R for the normal or the reverse part respectively. The polarity timescale is often seen in the geological literature combined with a chronostratigraphical scale.

Magnetic stripes on the ocean floor

As the volcanic rocks produced by sea-floor spreading move away from the spreading axis, they cool down and become magnetised parallel to the Earth's normal or reverse magnetic field. The changing polarity can be registered as stripes parallel to the spreading centre that show anomalously high or low magnetic strength. The Earth's field is slightly strengthened above rocks that are normally magnetised, as today, and slightly weakened above those with reverse polarity. The stripes are called anomalies and are labelled with the same number as used to designate the chrons. The anomaly found today along the spreading axis is numbered anomaly 0, and rocks produced during earlier chrons are numbered sequentially. The first anomaly off Greenland's east coast is anomaly 24, because it formed during Chron 24 in the early Eocene about 54 Ma ago, when the ocean between Greenland and north-west Europe started to open (see p. 170).

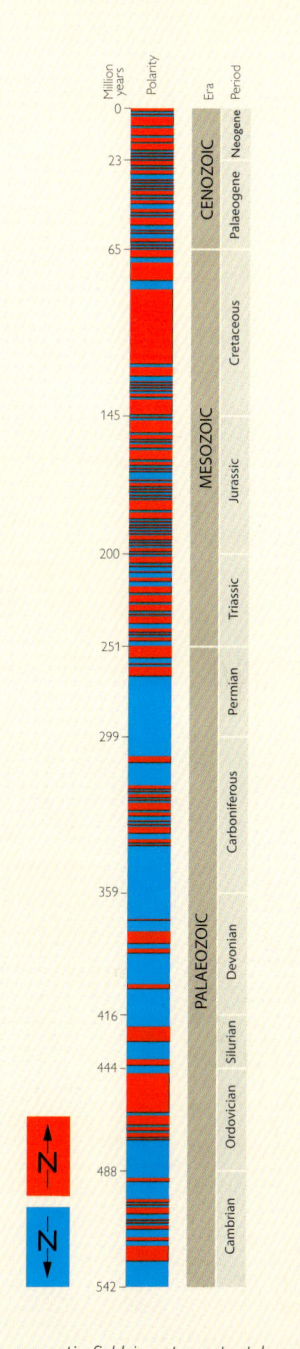

The Earth's magnetic field is shaped almost like that of a bar magnet. Today's magnetic north pole is not at the geographic pole, but about 890 km away at around 82°N, north of Canada's Arctic Islands. As a result the direction of the magnetic north shown by a compass differs from the geographic north by an amount that varies from place to place, and also varies a little with time. Far from the magnetic poles the difference is least but is very substantial close to the poles. The horizontal angle that the field makes with north is called the declination, and the angle it makes with the vertical is called the inclination. Palaeomagnetic studies determine both ancient declination and inclination from the magnetic field 'frozen' into rocks.

The orientation of the Earth's magnetic field is not constant because north and south change place at variable intervals. Studies of the orientation during the last few millions of years have shown that the changes take place irregularly at intervals between 0.25 and 0.5 Ma. Further back in the Earth's history there have been periods of more than 50 million years when the field was in the same direction. When the field is in the same direction as today it is known as 'normally magnetised', and when it is in the opposite direction it is called 'reversely magnetised'. The diagram shows how the Earth's magnetism has changed during the last 542 million years. Red shows normal magnetisation and blue shows reverse magnetisation. The changing orientation is particularly valuable in studies of the oceans to determine the age of the rocks in the oceanic crust by counting the stripes, much as one counts the growth rings of a tree.

YOUNGER SEDIMENTARY BASINS

Flood plains with lakes amid mountain ranges

The Devonian basin in North-East Greenland

After the formation of the Caledonian mountain range by the collision of Baltica (Scandinavia) with Laurentia (North America including Greenland), the stresses within the Earth changed. The compressional stresses that had forced the continents together relaxed and gave way to extension in the opposite direction. This in turn led to slow uplift of the roots of the mountain range by isostatic forces and the onset of erosion. The area broke up into linear blocks divided by steep faults; some blocks subsided into the upper crust and others were uplifted. Through a relatively short period of 30 million years during the Devonian, these movements created a landscape in Greenland comparable to the modern Basin and Range in western USA; sediments deposited in floodplains and lakes accumulated on subsiding blocks forming sediment thicknesses of about 8 km.

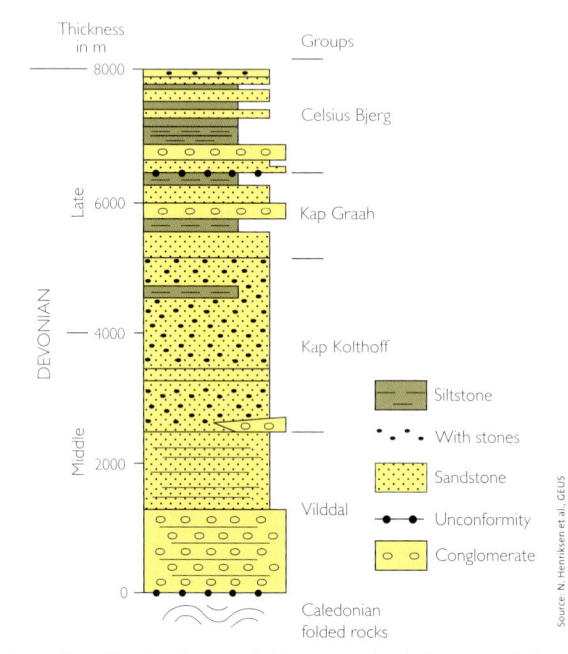

Composite sedimentary log compiled from many detailed logs through the 8 km thick Devonian continental succession in North-East Greenland. The sediments consist of conglomerates and sandstones with occasional siltstone beds. They were deposited in rivers and lakes and contain many fossils of vertebrates that illustrate the early stages of evolution from fish into amphibians.

The steep fjordside in North-East Greenland shows a 950 m high section of Devonian strata consisting of continental sandstones. Lowermost in the picture are light, medium- to coarse-grained sandstones of the Kap Kolthoff Group, overlain by reddish, cross-bedded, fine- to medium-grained sandstones of the Kap Graah Group.

During the Devonian, North-East Greenland lay near the equator where the climate was warm and desert-like. This led to the deposition of characteristically reddish continental sediments, often called 'Old Red Sandstone'. Much younger sediments that formed in a similar way in inland basins in the Alps are known as 'the Molasse', a designation that has also been used for the Devonian sediments in North-East Greenland.

The continental Devonian sandstones are found in the central fjord zone of North-East Greenland. Their present-day exposure extends more than 300 km from north to south, but seismic surveys over the Jameson Land Basin farther south have revealed the presence of Devonian sediments below the younger cover. The original length of the Devonian basin in East Greenland must therefore have been at least 500 km.

Deposition of the sediments was controlled by active faulting, in both the areas bordering the subsiding blocks and within the depositional area itself. Basin formation began with development of a major north–south-trending fault that forms the western boundary of the present-day outcrop. A later fault zone parallel to the western bounding fault forms the eastern margin of the outcrop. A series of N–S-elongated fault blocks (see block diagram p. 100) formed by extension of the crust within the basin.

The lowermost sediments in the succession consist of conglomerates and sandstones lying directly on Caledonian folded rocks from the Ordovician, showing that deposition started after most of the Caledonian movements had finished. The Ordovician–Devonian hiatus spans about 70 Ma.

The Devonian sediments in North-East Greenland consist mainly of sandstones with intercalations of conglomerates and occasional finer grained sediments. The entire sequence has been divided into four groups; each group separated from the next by local tectonic events that produced minor folding, overthrusts and faults.

Climate was a major factor controlling the type of sediment that was deposited. It varied in irregular cycles between wet and dry conditions. Most of the sediments were deposited in freshwater on large floodplains, but sediments in the middle and upper part of the succession were largely deposited by wind (eolian sediments), showing that desert-like conditions occurred at times.

The depositional patterns indicate that the rivers flowed mostly from the north but sediments were also shed into the basin from both east and west, demonstrating that the basins were surrounded by uplifted blocks of Caledonian rocks, so-called 'intramontane' basins. Various volcanic rocks, consisting of thin lava flows, ash layers and intrusions of flat-lying sills and cross-cutting, steep dykes are preserved in the lower part of the sedimentary succession.

Vertebrate fossils in the Devonian sediments

North-East Greenland's Devonian sediments contain a suite of important fossils of early vertebrates including skeletons of fish and the earliest transitional forms between fish and amphibians, popularly known as 'four-legged fish' (tetrapods; tetra = four, poda = feet). The fossils are found mostly in a few localities in the upper part of the Devonian succession around Kejser Franz Joseph Fjord. The first fossil fish were found in 1899, but investigations and research intensified when fossils of some of the earliest transitional forms between fish and amphibians were found in the early 1930s. Since then, around 11 000 individual pieces of vertebrate skeletons have been found, of which more than 10 000 are from fish. Preparation and examination of the large quantities of material have taken place mostly in Stockholm by a research group that brought home some of the early discoveries of these groups of animals. In more recent years scientists in Copenhagen and Cambridge have collected and worked on the Devonian material. Skeletal remains of about 40 different vertebrate species are known, together with a large number of plant fossils, indicating occurrence of swamps around the rivers and lakes in which the vertebrates lived. It is probable that the climate resembled those places on today's Earth that experience a monsoonal climate, with alternating dry and wet seasons.

The finds were made in the 360-million-year-old river deposits of the Celsius Bjerg Group of the Upper Devonian. The discoveries of 'four-legged fish' in North-East Greenland have undoubtedly been crucial for our understanding of the early phases of the evolution of four-legged amphibians (tetrapods) from fish. They show that the earliest vertebrates were exclusively fish, some of which later developed

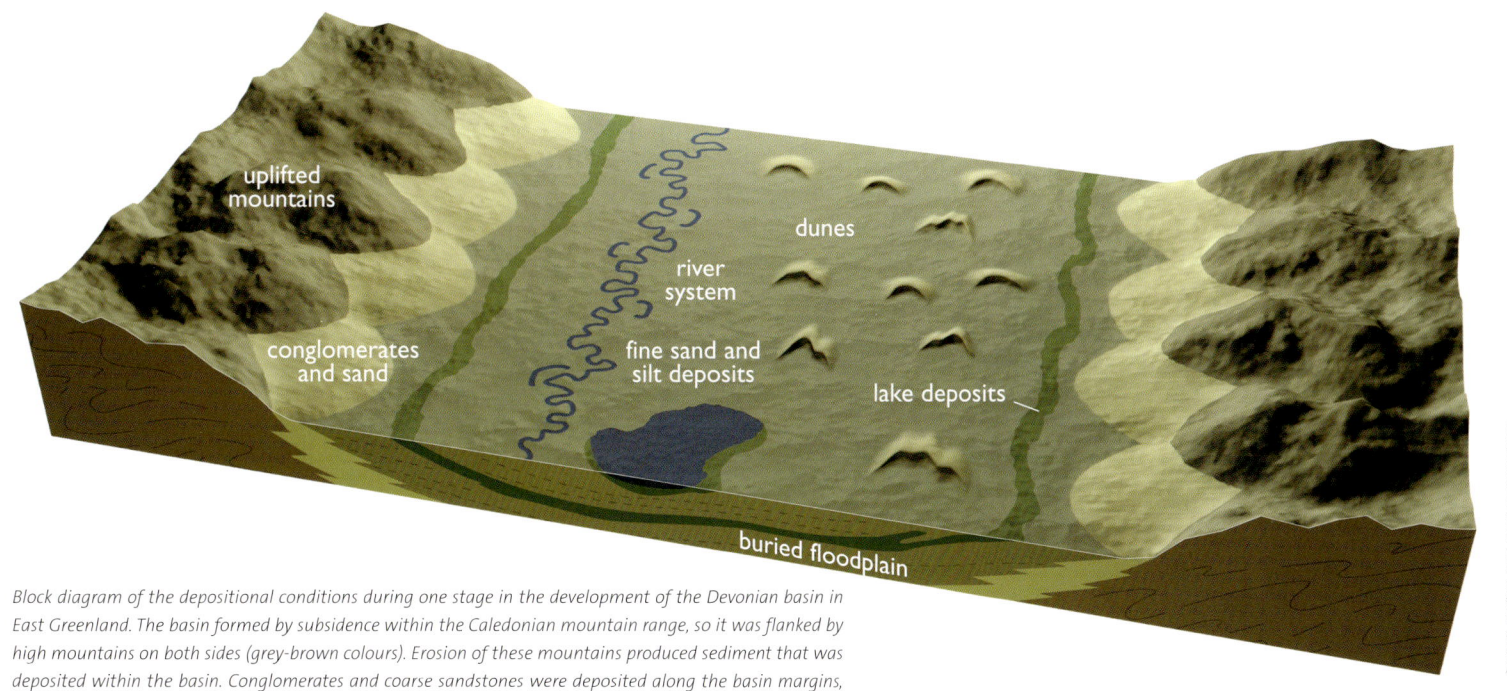

An armoured shark (Bothriolepis) whose skeletal remains have been found in the Devonian sediments. This now extinct animal is a close relation of today's sharks and rays. Its head, the front of the body and its pectoral fins were covered in bony armour. Several species have been found, each characterising a different stage within the Devonian, so the fossils can be used to classify the sediments stratigraphically. The shark was about 50 cm long.

Block diagram of the depositional conditions during one stage in the development of the Devonian basin in East Greenland. The basin formed by subsidence within the Caledonian mountain range, so it was flanked by high mountains on both sides (grey-brown colours). Erosion of these mountains produced sediment that was deposited within the basin. Conglomerates and coarse sandstones were deposited along the basin margins, close to the mountains, while finer sandstones and silts were dispersed farther into the basin.

YOUNGER SEDIMENTARY BASINS

ICHTHYOSTEGA – 'THE FOUR-LEGGED FISH'

Reconstruction of the 'four-legged fish' Ichthyostega (about 60 cm long).

The theory of how animal and plant life on Earth has developed gradually from a few simple forms to a diverse suite of complex organisms is called the theory of evolution. The theory, first advanced in its modern form by Charles Darwin in 1859, has survived many tests and is today regarded as a well-founded scientific theory.

The animal kingdom can be divided broadly into vertebrates, those with backbones, and invertebrates. The vertebrates branched into two evolutionary lines early in their development. One of those branches began with fish, then developed into amphibians followed by reptiles. A side branch of the reptiles developed first into dinosaurs and then into birds while another side branch developed into mammals. The evolution from fish to amphibians required substantial changes because the animals moved from living exclusively in water to being able to live partly on land and breathe through lungs. The early beginnings of this especially important and significant change have been elucidated and documented by studies of fossilised skeletons from the Upper Devonian of North-East Greenland. The fossils are found in river and lake sediments with an age of about 360 Ma.

The fossilised skeletal parts discovered here consist of both skulls and bones, including limbs. After lengthy and meticulous preparation work, it has been possible to assemble and reconstruct the animals' skeletons from these fragments and then to use these skeletons to reconstruct the animals' appearance and interpret how they lived. The skeletons are between half to just under a metre in length and their jaws bear a set of sharp teeth showing that the animals were predators. Several different species of these tetrapods ('four-legged fish') existed, divided into two genera that are named Ichthyostega (which means: having a fish-like skull) and Acanthostega (which means: having a skull with spikes). They are classified as primitive amphibians, but are really transitional forms that retain anatomical features that they share with their fishy ancestors but that are not found in later amphibians. The 'four-legged fish' from North-East Greenland lived in freshwater and breathed primarily with gills. They were also able to stay on dry land for short periods, when they breathed partly through the skin and partly through their gills as long as they could be kept moist.

The most marked development of the tetrapod skeleton was the formation of front and hind legs with their characteristic patterns of bones. The development can be traced back to a basic pattern among some fish ancestors and the tails of the primitive amphibians continued to resemble those of fish. The fish-like tail made it possible for them to swim freely in water. Ichthyostega's front legs were the stronger, while its hind legs were more like flippers. All four of Acanthostega's legs were paddle-like. From these observations it is thought that the animals would not have been particularly mobile on land. Ichthyostega may have been able to use its forelegs to drag the rest of its body, whereas Acanthostega may have been able to move a little as present-day salamanders. It is thought that the animals lived mainly in overgrown lakes and waterways where their limbs were used to crawl around between the plants in an environment where they could hide and find protection from the large predatory fish that were common at that time.

From an evolutionary point of view, the development of the limbs of these animals was most important. The structure of their limbs formed the basis for one of the most important features in the development of land vertebrates, the ability to grasp. This, of course, eventually led to the human hand, a crucial factor in the adaptive success of the human species.

Plants diversified and colonised widely at the beginning of the Devonian and their photosynthesis caused the content of oxygen in the atmosphere to increase. This created an environment that made animal life on land possible. Many water-living creatures evolved into forms adapted to the land, breathing with lungs and moving around on legs.

Reconstruction of the skeleton of the 'four-legged fish' Acanthostega gunnari. The skeleton shows clearly the four limbs and the fish-like tail. The limbs are paddle-like, especially the rear ones. The skeleton is about 65 cm long from snout to the tip of the tail.

Photograph of a fossilised skull of the 'four-legged fish' Ichthyostega – from the Devonian sediments in East Greenland. The head is about 20 cm long.

The Devonian basin in North-East Greenland

into primitive amphibians with limbs that made it possible for them to leave the water and move on land. The East Greenland finds are still exceptional because only very few comparable tetrapods have been discovered in other parts of the world.

Old Red Sandstone basins

Sandstone basins of similar type and age to the Devonian basins in East Greenland are found all round the North Atlantic. They all contain thick series of freshwater sandy, silty and conglomeratic sediments, and are all of about the same age. As their sediments are commonly red-coloured, they are all known informally as Old Red Sandstone. Their ages range from Silurian to Carboniferous, but most are Devonian. The basins occur on both sides of the northern Atlantic Ocean from 80°N in Svalbard to 40°N in Georgia in eastern USA – a distance of 4500 km.

Most of the basins appeared after formation of the Caledonian mountains around the North Atlantic. After formation of the fold belts, the crust was stretched in a direction at right-angles to the axes of the mountain chains. This stretching formed areas of subsidence within them, creating subaerial basins surrounded by areas of high ground. The marked relief between the subsiding areas and the adjacent mountains led to active, vigorous erosion so that the basins were filled with very thick sequences of sediment in a very short time. In East Greenland, the amount of sedimentation was hundreds of metres per million years, substantially higher than normal sedimentation rates in the sea.

By moving the continents back together from both sides of the Atlantic to their relative placing just after the Caledonian mountains formed and so compensating for continental drift, the Old Red Sandstone basins can be seen to have formed within a contiguous landmass. It consisted of the continents of Laurentia in the west, Baltica in the east and Gondwana (an assemblage of southern continents including Africa) in the south. The palaeogeographic position of the land areas was very different from today, as continental drift has moved the northern continents much farther north. At that time, the Greenland basins were in the tropics, in a monsoon area where rain fell in the summer. Many of the Old Red Sandstone basins, however, formed under desert-like conditions, with periods of severe drought. It is important to recognise these contrasting climatic conditions when interpreting the depositional environment of the Old Red Sandstone. The evolution of animals and plants was also highly influenced by the climate and many studies have been carried out on the evolution from freshwater fish into transitional forms of amphibians that lived on land.

Characteristic reddish sandstone layers of the Devonian Old Red Sandstone basin in East Greenland. The picture shows the north coast of Geographical Society Ø along the south side of Sofia Sund. The mountains in the foreground are about 1600 m high and the distance to the mountainside in the far background is about 50 km. The exposed section belongs to the Vilddal, Kap Kolthoff and Kap Graah Groups.

A globe showing the locations of the continents during the Middle Devonian about 390 Ma ago, after the formation of the Caledonian fold belts. The map shows how the various basins (in yellow) containing the Old Red Sandstone developed in a continuous land area at the margins of the continents of Laurentia and Baltica (see p. 103).

YOUNGER SEDIMENTARY BASINS

Break-up along Greenland's plate margin in the north-east

The Wandel Sea Basin in North Greenland

During the Carboniferous, most of the Earth's continents were assembled in the supercontinent Pangaea. Greenland was situated on the northern limit of the supercontinent with Scandinavia (Baltica) to its east and North America to its west. There was a large ocean immediately to the north of North Greenland, with a large flat continental shelf along this continent–ocean boundary, north of the landmasses. Sediment deposited on that shelf is found today on the Barents Shelf, around Svalbard, in parts of northern Greenland and in the Canadian Arctic Archipelago (the Sverdrup Basin). A widespread series of sediments was deposited on this shelf during the approximately 300 million year long interval from the Carboniferous to the early Palaeogene, with shallow-water sediments near the continent and deep-water sediments farther out.

The Wandel Sea Basin and its related, coeval, basinal areas, seen in a polar projection. The three contemporary basins – the Wandel Sea Basin, the Sverdrup Basin and the Barents Platform – were originally connected but were later separated by opening of the Arctic Ocean and the North Atlantic. The Svalbard archipelago lay north of Greenland before opening of the ocean between Greenland and the Barents Platform. Sea-floor spreading and strike-slip faulting of about 500 km have moved Svalbard to its present-day location about 600 km east of eastern North Greenland.

Greenland became separated from Svalbard and the Barents Platform 30–40 Ma ago when the northern part of the North Atlantic finally split up and the eastern part of North Greenland became an independent area known as the Wandel Sea Basin. The basin developed on the border of the stable Greenland continent, where the Caledonian and Ellesmerian fold belts meet (see p. 92). The Wandel Sea Basin was strongly influenced by this position between the stable continental area to its south-west and the tectonically active margin along the outer coast. It underwent faulting and other deformation, during the early phases of continental break-up and the beginning of sea-floor spreading in the Palaeogene. As a result, the basin does not represent a coherent, uniformly subsiding depositional setting (depocentre) but rather a series of separate, tectonically-bounded basins that, particularly in the north, were subjected to later deformation.

The development and division of the Wandel Sea Basin

The Wandel Sea Basin developed in two phases after the end of folding and the onset of collapse of the Palaeozoic fold belts in North and North-East Greenland. During the first phase, from the Early Carboniferous to the Early Triassic, a number of fault-bounded, subsiding sub-basins (grabens and half-grabens) developed, within which mostly shallow-water marine limestones were deposited. During the second phase, through the Mesozoic to the early Palaeogene, the basins were dissected by strike-slip faults and a number of small isolated basins were formed. These basins developed where the strike-slip movements caused extension and blocks of crust were pulled apart (pull-apart basins). The structures are separated from the stable part of continental Greenland by two major fault systems – the Harder Fjord Fault Zone and the Trolle Land Fault Zone. They both formed during the generation of the plate boundary that had North Greenland to its south and Svalbard to its north, leading to the opening of the ocean connection between the northern North Atlantic and the Arctic Ocean.

The southern part of the Trolle Land Fault Zone divides the basin sediments into northern and southern parts, each of which had its characteristic stratigraphic and structural development. The sediments in the north were affected by folding and other deformation during the latest Cretaceous, while the sediments south of the fault zone are flat-lying and unfolded. In northern Peary Land, in the most north-westerly corner of the basin, a suite of volcanic rocks more than 5 km thick was erupted during the Late Cretaceous, dominated by basalts intruded by steep dykes and flat-lying sills. This volcanic sequence and all of the sediments in the northern part of the Wandel Sea Basin were affected by overthrusting and other deformation in the Eurekan Fold Belt (see p. 107).

Carboniferous to Triassic sediments

The first sediments in the Wandel Sea Basin are from the Early Carboniferous, when a more than 1000 metres thick series of continental mudstones and sandstones accumulated. This reflects the start of rifting as the continent began to break up. The sediments were mostly deposited by rivers on floodplains, but they also include some lake deposits. In the Late Carboniferous, the depositional regime changed substantially. The sea invaded from the north and flooded the entire basin, i.e. the shelf region north of Greenland and the Barents Sea. During that period this region was at about 30° north of the equator. A widespread succession accumulated here, characterised by a warm-water fauna (now preserved as abundant fossils) dominated by limestones, with local, shallow-water, coral reefs. The sediments also include beds of rock salt, gypsum and anhydrite, testifying to the presence of bays and inlets with high evaporation rates and limited water circulation. Deposition of warm-water limestones with a rich fauna continued on this shallow, marine shelf until the Early Permian.

In mid-Permian times, an arm of the sea extended from the northern shelf south along the coast of East Greenland and established a connection with the land-locked sea that at that time covered the North

Carboniferous sediments on the east coast of Holm Land. The profile is about 400 m high. Its lower half shows alternating beds of marine sandstone and limestone that form gentle, partially scree-covered slopes whereas the steep wall above consists solely of more resistant limestone.

The strata in the Wandel Sea Basin in North Greenland have ages that range from Carboniferous to Paleocene. Apart from the Kap Washington volcanics, they all consist of sediments deposited in shallow water near the continental margin. The preserved outcrops are widely dispersed, often with limited lateral extent. The area has been extensively disrupted by faulting and folding associated with plate-tectonic movements between North-East Greenland and Svalbard.

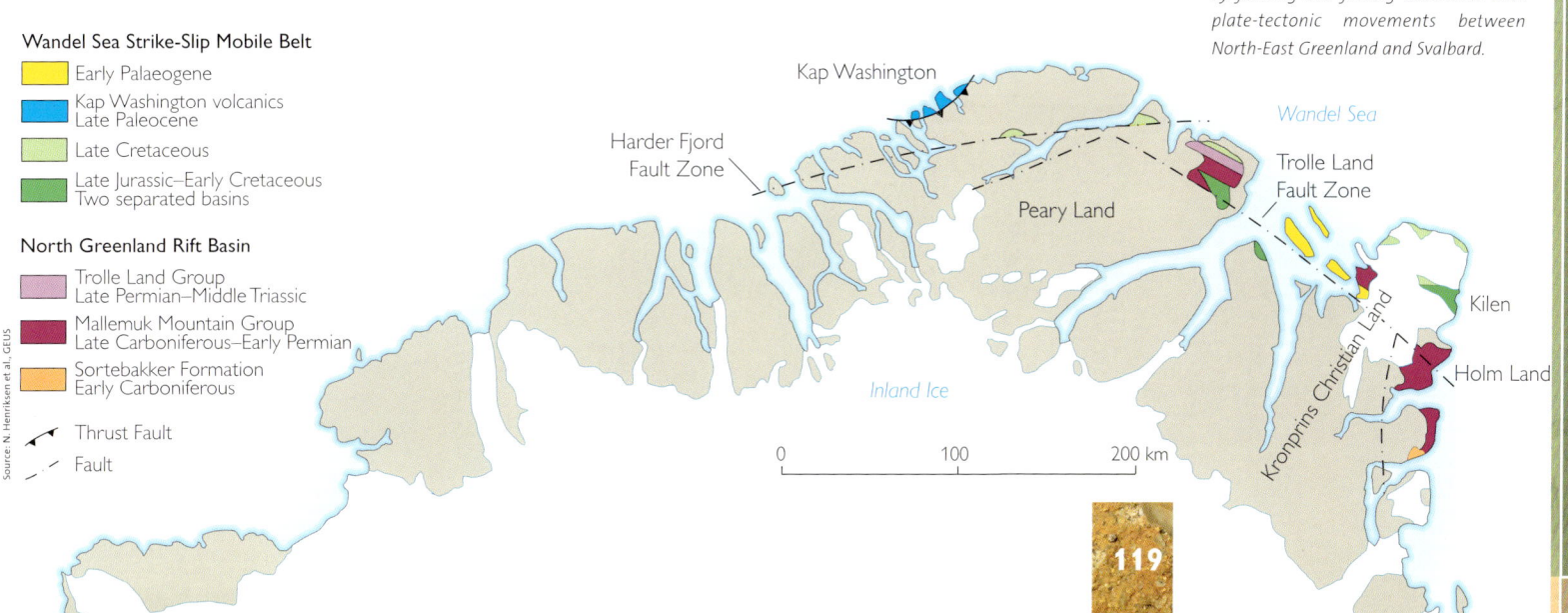

YOUNGER SEDIMENTARY BASINS

Fossilised impression of a leaf (ginkophyt) from the Upper Permian sedimentary series of northern Peary Land. A flora similar to the one here is known from contemporary strata in the Ural Mountains, Mongolia and north-eastern China, showing that these areas must have been connected at that time.

Sea, Denmark and northern Germany. This southern so-called 'Zechstein Sea' is characterised by extensive evaporitic salt deposits. In the Wandel Sea Basin area, which was much closer to the open shelf sea, deposition continued mostly as carbonates, with local shale and sand layers derived from erosion of the surrounding land areas.

Conditions on the shelf area north of Greenland and the Barents Sea changed at the beginning of the Triassic due to a change in global climate. Sedimentation then became dominated by deposition of sand and silt derived from erosion of the adjacent land. Large deltas formed in many places. Deposits from this period are preserved only locally in the Wandel Sea Basin area, with a thickness of up to 1000 m. It is assumed that sediments from this part of the early Mesozoic were originally up to 2000 m thick, but much of this succession has been removed by later erosion.

Active tectonics in the Jurassic and Cretaceous

There is a break (hiatus) in the Wandel Sea Basin deposits, from the middle of the Triassic to the Late Jurassic, corresponding to a 75 Ma interval. At some time during this interval, the structural regime changed as plate movement started between Greenland and Svalbard. The whole Wandel Sea Basin area was affected dramatically by fault zones and deformation belts that cut through the Earth's crust. These movements created a zone of NW–SE-directed deformation in eastern North Greenland that is known as the 'Wandel Sea Strike-Slip Mobile Belt'. A number of local, fault-bounded, subsidence areas developed, into which sandy and shaly sediments were deposited. The sediments consist of both marine and terrestrial deposits and they vary considerably in character and thickness, reaching to between one and two thousand metres. Sedimentation rates were often high, so the basins filled quickly

'PULL-APART BASINS'

Some of the Mesozoic sediments in the Wandel Sea Basin are found in small, restricted areas between NW–SE-trending fault zones. The strata in these local basins are unusually thick relative to their area and subsidence along the bounding faults was clearly significant. These depositional conditions are indicative of a particular style of basin development, associated with large, lateral earth movements, that contrasts with that of ordinary basins in which the subsidence rate is much less and the depositional areas are much larger.

Local basins of a limited area and thick sediment fills are commonly associated with steeply-dipping fault zones showing horizontal, sideways movements (strike-slip faults). Such zones exhibiting many parallel faults often form near plate boundaries, where one plate is sliding past the other and where the faults are generated parallel to this movement. The local basins are called 'pull-apart basins' because they form where a limited area of crust is pulled apart between two strike-slip faults, forming a 'hole' in which sediments can accumulate. Normally, the crust around a strike-slip fault system will be neither extended nor compressed, but where a strike-slip fault steps sideways, or the fault planes are not entirely parallel to one another, they can generate strains in the Earth's crust that lead to local extension and the formation of a 'pull-apart basin'.

The best-known 'pull-apart' basins are found along the San Andreas Fault in southern California. This fault system is the active plate boundary where the Pacific Plate is moving north-west relative to the North American Plate and the movement of the plates frequently causes earthquakes. There are many examples of recently and even currently active 'pull-apart' basins in this strongly faulted zone. One of the basins near Los Angeles contains a sediment pile about 13 km thick that was deposited in less than 8 million years within an area only 10–15 km across. The sediments are mainly coarse-grained conglomerates and sandstones derived from erosion of the immediately surrounding terrain and deposited by rivers and in lakes. Crushed rocks (breccias) are found along the faults, showing that fault movement and subsidence happened at the same time as deposition.

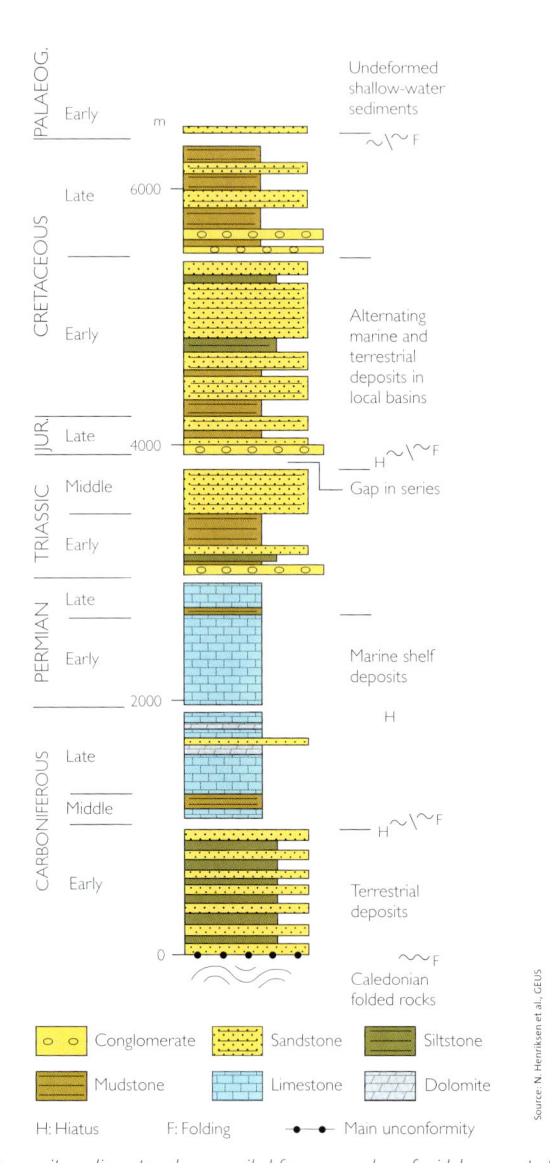

Composite sedimentary log compiled from a number of widely-separated profiles in the Wandel Sea Basin. Due to the complex stratigraphic history and segmented nature of the basin, individual sediment packages are representative only of particular sub-basins and cannot be traced over the entire basin complex. The sedimentary log is therefore idealised and the full succession, as depicted, cannot be found in any single locality.

The Wandel Sea Basin in North Greenland

with immature clastic sediments originating from nearby erosion.

From the Middle Jurassic to the earliest Paleocene, a period of about 100 Ma, four phases of tectonism took place involving block-faulting and fault-related folding. The most intense deformation took place late in the Cretaceous when the whole sedimentary sequence was folded during the so-called 'Kronprins Christian Land Orogeny'. The most recent deformation involved extensional movements during the Paleocene. The uppermost part of the sedimentary sequence in the Wandel Sea Basin consists of sandstones of Paleocene–Eocene age that contain coal, plant remains and microfossils. These shallow-water sediments are unaffected by the deformation recorded in the underlying sediments. From a regional perspective, the tectonic development ended with ocean-floor spreading and the formation of oceanic crust between Greenland and Svalbard. It was accompanied by a right-hand directed (dextral), strike-slip movement of about 500 km along the NW–SE-striking fault zone between Greenland and Svalbard (the Spitzbergen Fracture Zone).

The Kap Washington volcanics

In the most north-westerly corner of the Wandel Sea Basin in northern Peary Land, a layered series of volcanic rocks over five kilometres thick crops out. It includes basaltic lavas and other types of volcanic rock. The rocks have been dated as 64 Ma old, close to the boundary between the Cretaceous and the Paleocene. Below the volcanic rocks are older Wandel Sea Basin sediments of Carboniferous and Permian age. The most northerly part of Peary Land was affected by earth movements during the early Palaeogene. A large overthrust formed – the Kap Cannon thrust – that brought old, folded rocks from the North Greenland fold belt over and into contact with the younger volcanic rocks. This late deformation took place during the Eurekan Orogeny that overprinted the structures formed by the older Carboniferous deformation (Ellesmerian Orogeny) of the North Greenland fold belt (see p. 104).

The Cape Washington Group comprises a range of volcanic rocks that indicate that the volcanic province formed on continental crust. The group contains basalts, volcanic sediments and breccias formed by erosion of the newly-formed lavas, tuffs, ash flows with welded tuffs (ignimbrites) and acid lavas (rhyolites). A north–south-oriented dyke swarm that extends over much of northern Peary Land records additional volcanic activity in the region. The dykes were intruded at the same time as the volcanic rocks of the Cape Washington Group were extruded.

The formation of these volcanic rocks in the most northern part of Greenland – close to the boundary between the continent and the oceanic crust to the north – is interpreted as being related to formation of oceanic crust in the Arctic Ocean. The volcanic province in North Greenland probably formed where an extension of the N–S-directed spreading axis met continental crust and added melted material from the mantle beneath the oceanic part of the plate. The dyke swarm striking N–S shows that the area underwent substantial E–W extension – up to 50% in some areas. This extension was also in the same direction as the ocean-floor spreading that took place north of the continent. This indicates that the contemporary volcanic activities in the ocean area and on land are different expressions of the same large-scale geotectonic development.

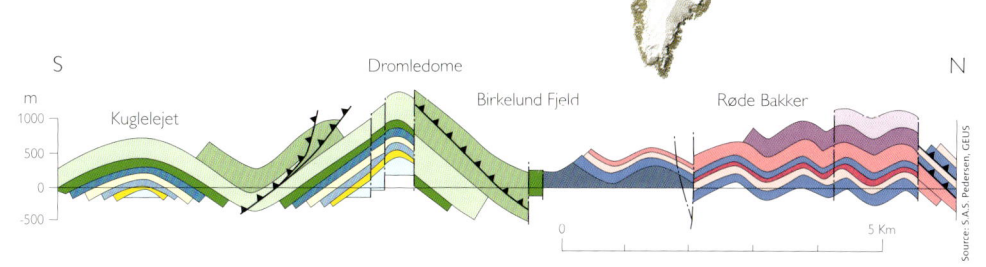

North–south profile through the folded strata of the Wandel Sea Basin on the northern part of Kilen, Kronprins Christian Land. The southern part of the profile consisting of folded sediments of Early Cretaceous age is shown in green colours while the strata shown in blue, red and violet are Upper Cretaceous.

Part of the volcanic succession at Kap Washington comprising lavas and volcanic sediments. They crop out on a peninsula extending into the frozen Arctic Ocean, west of Greenland's northernmost point, Kap Morris Jesup. The volcanic strata are found below a large overthrust fault that has pushed older rocks of the Ellesmerian fold belt northwards over the Paleocene volcanic rocks.

Agglomerate consisting of ash and blocks of volcanic rocks deposited in water. From the base of the succession in the Kap Washington volcanics.

YOUNGER SEDIMENTARY BASINS

Greenland and Norway were originally joined but later separated by sea

Rift basins in East Greenland

A large sedimentary basin complex developed along the north-eastern margin of the Greenland shield during the latest Palaeozoic and the whole of the Mesozoic. The basin complex formed as a result of substantial, regional E–W extension across the border region between the two continents of North America (Laurentia) and Scandinavia (Baltica) that had collided in the Late Silurian to form the Caledonian fold belt. The basin formed by a combination of subsidence and stretching (rifting) of the crust and developed gradually over a period of about 300 Ma. A total thickness of 6–7 km of sediment was deposited from the Carboniferous to the Cretaceous. Initial deposition in the basin was predominantly of non-marine sediments from rivers and lakes, but later the sea invaded the rift zone and marine sediments were deposited. The basin complex was a precursor to the formation of oceanic crust in the North Atlantic that started about 57 Ma ago.

The basin complex in East Greenland formed partly during the assembly of the supercontinent of Pangaea and partly during its subsequent disintegration. Subsidence at one stage of its development led to the formation of a marine connection between the oceanic area to the north of Greenland and the great Tethys Ocean between Europe and Africa. Greenland drifted slowly northwards as Pangaea broke up, moving the East Greenland basin from around 20°N to nearly 60°N. This drift through the different climate zones and the variations in climate and sea level that affected all of the Earth are reflected in the sediments. They contain tropical and subtropical marine sediments with coral reef limestones, desert sandstones and evaporitic sediments containing salt.

Reddish continental sandstones of Triassic age with impressions of dinosaur footprints on the surface in the foreground. The impressions are from small predatory dinosaurs about 1.5 m long. North-East Jameson Land with Carlsberg Fjord in the background.

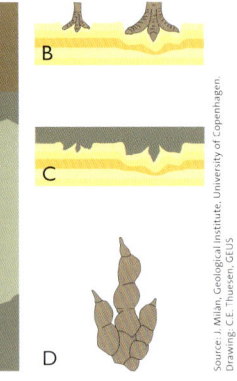

A) Drawing of the bedding surface showing the distribution of the footprints.
B) Vertical section showing how the prints formed. The larger foot pressed deeper into the layers than the smaller foot. This allows the ratio of the animals' weights to be calculated.
C) Section through a layer with fossil impressions as they are found today.
D) A dinosaur foot reconstructed from the shape of the impressions. The tracks show that the animals had three toes, long sharp claws and pads – like modern birds.

However, the sediment type that predominated throughout the whole development of the basin consists of sands derived from the erosion and breakdown of the nearby continent of Greenland to the west.

The total thickness of sediment in the East Greenland basin complex is very large, but the older sediments are largely concealed under younger sediments. However, since the sedimentary deposits have been subjected to block faulting and subsequently uplifted by up to 3 km, much of the sequence can now be studied at the surface. It is therefore possible for geologists to interpret depositional models that chart the step-by-step evolution of the basin.

For 250 Ma prior to the opening of the North Atlantic between East Greenland and Norway, the

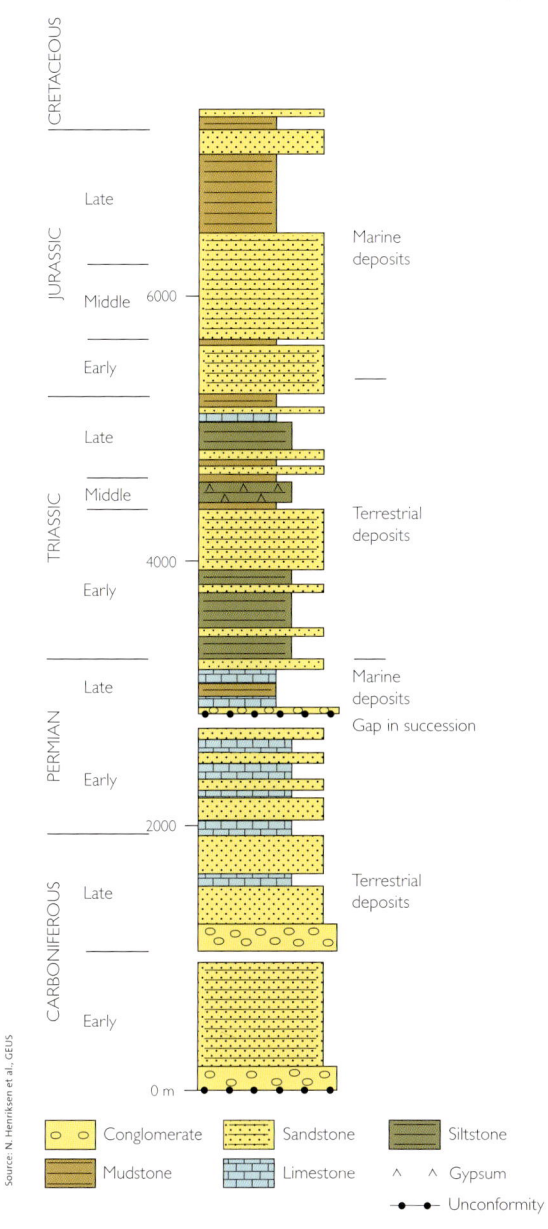

POSITION OF GREENLAND 237 Ma AGO

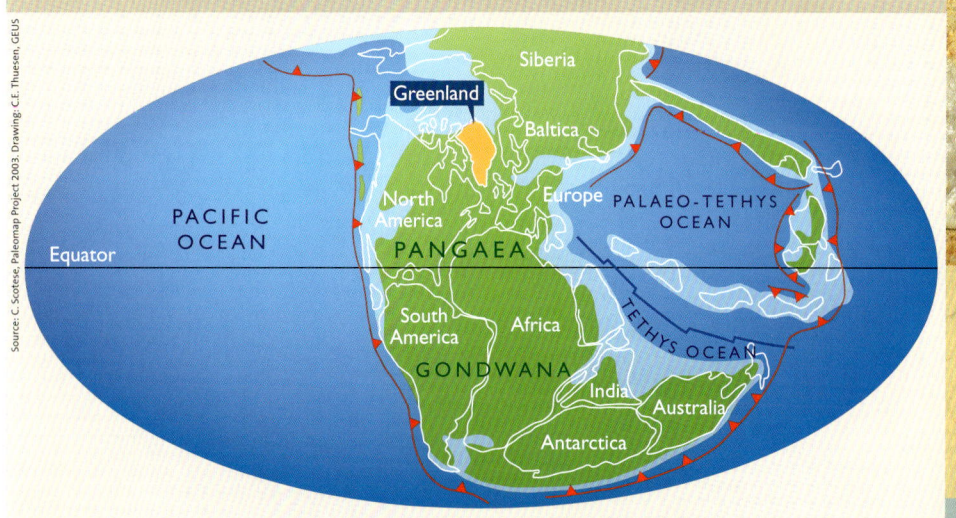

Palaeogeographic map showing the location of the continents during the Middle Triassic, 237 Ma ago. At that time, North-East Greenland had a desert climate and was connected to most of the supercontinent Pangaea.

POSITION OF GREENLAND 255 Ma AGO

Palaeogeographic map showing the location of the continents during the Late Permian, 255 Ma ago. The continents were assembled into the supercontinent Pangaea and a narrow strait between Greenland and Scandinavia connected the marine basins in East Greenland with those in the North Sea.

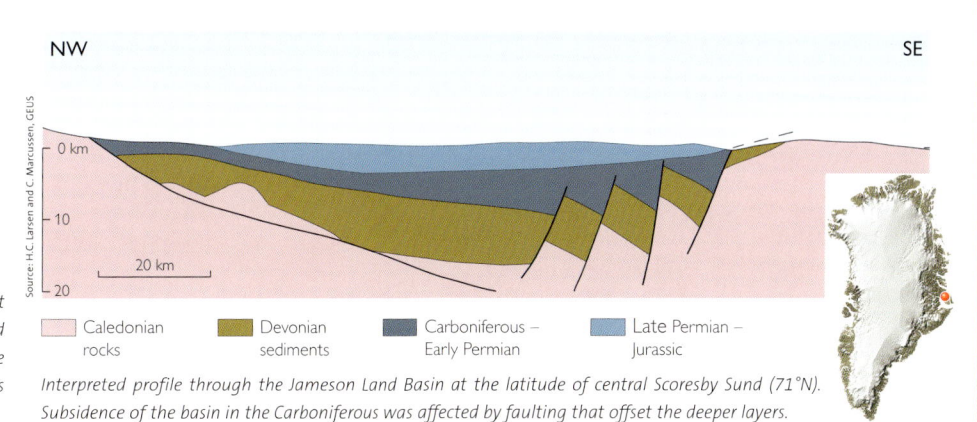

Composite sedimentary log constructed from profiles from the East Greenland rift basin. Measurements of the strata of Carboniferous and Early Permian ages in the lower part were made in the central part of the basin while the upper part shows sediments of Late Permian to Cretaceous age representing developments in the Jameson Land Basin farther south.

Interpreted profile through the Jameson Land Basin at the latitude of central Scoresby Sund (71°N). Subsidence of the basin in the Carboniferous was affected by faulting that offset the deeper layers.

Sandstone strata of Carboniferous age in northern Jameson Land south of Kong Oscar Fjord. The sandstones are cut by two steeply dipping intrusive basalt dykes of Palaeogene age. The hillside is about 400 m high.

SUTURE LINES

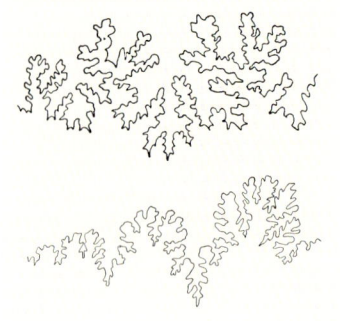

Examples of structure lines from ammonite shells similar to the ones from East Greenland. The elaborately fluted lines can be seen on the inner surface of the shell and mark where the walls of the internal chambers met the outer wall. The detailed pattern varies significantly and is critical in defining and identifying the species.

present-day continental shelves between these two land areas were subjected to E–W extension. It has been calculated that the part of the Norwegian shelf that lay opposite central East Greenland was stretched by a factor of two between the Permian and the start of sea-floor spreading. We can assume that the amount of extension in East Greenland was about the same, which explains the thinning of the crust and the subsidence of the basins, creating space for their thick fill of sediments.

The sediments are found in a belt about 100 km broad in onshore East Greenland, but they also extend offshore for about 100 km off the southern part and more than 300 km off the northern part of North-East Greenland. Studies have shown that the present-day shelf areas west of Norway and east of Greenland formed parts of the same inland basin system during the Carboniferous and Early Permian.

Basin development and continental sedimentation in Carboniferous – Early Permian time

Rifting began in East Greenland during the Carboniferous with the creation of a N–S-oriented area of subsidence into which was deposited a thick series of continental sandstones with local occurrences of more fine-grained rocks. The sediments were distributed by rivers or laid down in lakes in a warm and humid climate. The area of subsidence was affected and partly bounded by large steep faults. These movements created a number of 'half-grabens' with tilted blocks, a structural development that also pre-

vailed during a later period of extension in the Mesozoic in the northern part of the area.

The extension during the Carboniferous between Norway and Greenland took place at the same time as North Greenland was undergoing N–S-directed compression (the Ellesmerian Orogeny) from a collision between Greenland and an unknown continent to its north (see p. 104). Continental collision and mountain belt formation also occurred in other parts of the world at this time, for instance the formation of the Variscan fold belt in Europe and the Urals in Russia. By the end of the Carboniferous, most of the Earth's continents were assembled into the supercontinent of Pangaea.

The first marine sedimentation late in the Permian

Sedimentation under continental conditions in East Greenland continued throughout the Early Permian. Late in the Permian, however, pronounced subsidence of the area between Greenland and Norway caused the sea to encroach from the north. An arm of the sea extended southwards as a narrow N–S-trending strait all the way to the North Sea area in northern Europe. At that time, the basins in East Greenland lay at about 30°N and the climate was warm and dry. More than 900 metres of sediment were deposited in the basin, with a layer of sandy sediments at its base. The following series includes marine limestones containing reefs, black shales and various evaporitic sediments including gypsum and salt.

Desert conditions in the Triassic

An abrupt change in conditions happened early in the Triassic. Pangaea, including Greenland, had moved farther north and the basin now lay at around 45°N. The arm of the sea retreated towards the north and the climate in East Greenland became desert-like. An elongated inland basin existed in East Greenland throughout the Triassic, in which sediments were deposited in temporary lakes, in rivers with highly fluctuating amounts of water and by wind, just as in the Sahara today. Deposition in this basin amounted to around 2000 m of continental sandstones and finer sediments. The sediments around the margins of the basin in Jameson Land consist of coarse-grained conglomerates and sandstones while fine-grained lake sediments and dune sandstones with sporadic, thin layers of limestone and gypsum, were deposited in its inner parts. The climate became more humid towards the end of the period and a large lake formed over much of the basin into which was shed material eroded from all sides.

In recent years, a number of vertebrate fossils have been found in the Triassic sediments of East Greenland, including small early mammals and dinosaurs.

AMMONITES – SHELL-BEARING CEPHALOPODS

An extinct group of shell-bearing cephalopods

Ammonites were a form of mollusc and, together with other shell-bearing molluscs such as snails and bivalves, are very important for classifying sedimentary strata into their biostratigraphical divisions and thereby determining their geological ages. Most of the animals lived in the sea and their shells consisted of calcium carbonate that is preserved within sediments as a fossil, while the soft parts of the animal decomposed after its death. The only modern animal that resembles the ammonites in shape is the nautilus that lives in tropical waters.

The class of animals called cephalopods has existed for almost 500 Ma, since the Early Palaeozoic, and has developed significantly during that time. The class was particularly common and widespread during the Mesozoic. Fossils of these animals play an important role for geologists as an aid in the stratigraphic classification of sediments and in mapping the distribution of the ancient landmasses and shelf seas. These divisions are based on variations in the composition of the fauna with time (faunal assemblages) found in different geographic regions (faunal provinces). One of the main principles of biostratigraphy is that rocks that contain the same fossil assemblages are of the same age. The contrary is, however, not the case; not all rocks of the same age contain the same fossil assemblages. Whether or not a species can live in a particular place depends on the environment, so contemporaneous sediments from different climatic zones normally contain different faunal assemblages.

Ammonites are a highly specialised group of animals characterised by a spiral shell made of aragonite (a form of calcium carbonate). The inner part of the shell is divided into chambers whose dividing walls merge with the outer shell. The join between the walls of the internal chambers and the outer wall creates elaborately-fluted lines (suture lines) that are visible on the inner surface of the outer shell. The living animal was in the outermost chamber. As it grew, a new segment was added to the outermost chamber and the animal gradually moved outwards following the extension.

The three-dimensional forms of ammonites, their external shell sculpture and especially their suture lines, are so varied that many thousands have been described and can be distinguished by specialists. They can be identified in the field with no other tools than a magnifying glass and they can be collected using a hammer to free them from their enclosing rock. Each form is a species and represents one development stage and a style of life belonging to a particular biotope, such as warm or cold water. The ammonite larvae were not attached to the bottom, but could drift over long distances in open water, spreading over huge areas. Ammonites were therefore very common and widely dispersed and their fossils are generally well-preserved so they have become an important aid in geological dating all over the world. They are more suitable than most other animal groups as index fossils for stratigraphic subdivision. They evolved more quickly than most other marine animals and new species quickly replaced those that became extinct. Ammonites reacted quickly to changing environments by evolving new species that coped better with the new conditions, so they could live in deep or shallow water or in a warm, temperate or cold climate.

The Jurassic strata in Jameson Land are rich in ammonites, which means that up to 50 different stratigraphic units can be recognised within a typical sedimentary succession about 200 m thick. Each unit contains its characteristic assemblage of fossils including ammonites. Some horizons are paved with well-preserved ammonites that sometimes lie loose on the surface having being freed from their host rock by weathering.

The Jurassic succession in Jameson Land is nearly complete and represents uninterrupted sedimentation over about 55 Ma. The presence and distribution of ammonites within these sediments has been studied in great detail and, as a result, the sediments have now been divided into about 60 ammonite zones. Jameson Land has therefore become a commonly used key reference area for studies of Jurassic sediments over the whole of northern Europe and around the North Atlantic.

Reconstruction of an ammonite. The living animal (a cephalopod) sat in the outermost chamber of its shell with its arms extended outwards. As the animal grew, a new chamber formed and the animal moved into it, so it always lived in the outermost chamber. The old chambers were sealed off by walls that formed a characteristic pattern on the inner surface of the outer shell wall.

Ammonite (Pavlonia sp.) from the Upper Jurassic of the Jameson Land Basin. The shell is about 12 cm across. Ammonite shells are well-preserved and common fossils in sand- and siltstones from which it is often easy to free them. The photograph of the shell was taken in a laboratory after collection.

YOUNGER SEDIMENTARY BASINS

Cliff section through sandstones of latest Late Jurassic age (about 148 Ma old) showing large scale, low angle cross-stratification. The sandstones are on Milne Land in the south-west corner of the Jameson Land Basin. The people in the left foreground give the scale.

A 200–300 m thick series of sandstones with coal horizons that contains many plant fossils from two different flora types was deposited in southern Jameson Land around the Triassic–Jurassic boundary. The depositional environment was mostly swamps into which the sands were transported by rivers. Detailed studies of the plant fossils show that they are comparable to contemporary floral assemblages found in southern Sweden and Germany.

The marine sediments of the Jurassic and Cretaceous

The sea re-invaded the strait between Greenland and Norway early in the Jurassic and spread over most of the rest of the basin. A shallow shelf-sea with a water temperature like that of the North Sea today, covered the area of subsidence in East Greenland. A nearly complete Jurassic sequence was deposited in the Jameson Land Basin, whose lower part is dominated by shallow-water marine sandstones, but whose upper part contains fine-grained mudstones deposited in

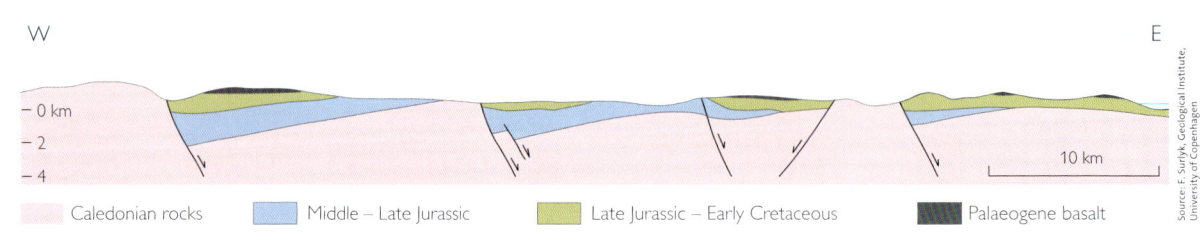

Interpreted profile through the Wollaston Forland area (74°30'N). The section shows that the basin formed as a result of fault movements that caused subsidence along the western parts of the sub-basins. In this way a number of 'half-grabens', fault-blocks with dipping strata, were created.

Caledonian rocks — Middle – Late Jurassic — Late Jurassic – Early Cretaceous — Palaeogene basalt

deeper water. The sediments were transported from the north by large river systems that brought erosional products from the higher-lying crystalline hinterland. The northern and southern parts of the East Greenland basin complex developed rather differently at this time. While slow uniform subsidence characterised the southern part of Jameson Land, the more northerly areas around Wollaston Forland were affected by faulting that resulted in tilted fault-blocks (half-grabens). Sediment was deposited into the resulting depressions at the same time as the fault movements.

The uniform subsidence in the Jameson Land Basin resulted in the preservation of an almost unbroken series of Jurassic sediments which therefore makes it a key area for studies of animal evolution through the period. In particular, it has been possible to produce a complete picture of the development of the ammonites, an extinct cephalopod, which is a sub-class of shell-bearing mollusc related to the present-day squid and octopus. These fossils make it possible to resolve the stratigraphy of the Jurassic, which lasted about 55 Ma, into intervals of about 1 Ma (see p. 125).

In global terms, Pangaea started to break up during the Jurassic with the onset of sea-floor spreading in the central Atlantic between America and Africa. The narrow arm of the sea between Greenland and Norway connected the oceanic area north of Greenland with the Tethyan Ocean to the south and made it possible for marine animal life to migrate between the two areas. The faunas in East Greenland thus contain species that are representative of both warm- and cold-water types, making East Greenland a key area for correlation between the different faunal provinces.

The basins continued to develop into the Cretaceous. The gentle subsidence in the Jameson Land Basin resulted in an uninterrupted sedimentary series into the lowermost Cretaceous. At the same time, in the Wollaston Forland area farther north, Cretaceous sedimentation was influenced by continuing tectonic movements. They resulted in tilted fault blocks (half-grabens) and sedimentation of coarse conglomerates and sandstones in the deeper parts of the half-grabens. Finer sands and clayey sediments higher in the sequence show that the tectonic movements decreased over time.

The Cretaceous was also the time when the continents around the North Atlantic started to break up, when global sea-level is thought to have been 100–200 m higher than today and the climate much warmer. In the East Greenland basins, shallow-water sediments were deposited in a belt west of the present-day coast, but deep-water sediments were probably deposited farther to the east under the present continental shelf. Sedimentation ceased in the earliest Cretaceous in the Jameson Land Basin, but continued into the mid-Cretaceous around Wollaston Forland. Sediments of Late Cretaceous age are, however, present in the offshore basins to the east showing that the land areas were uplifted and eroded and the products transported into the offshore area.

POSITION OF GREENLAND 152 Ma AGO

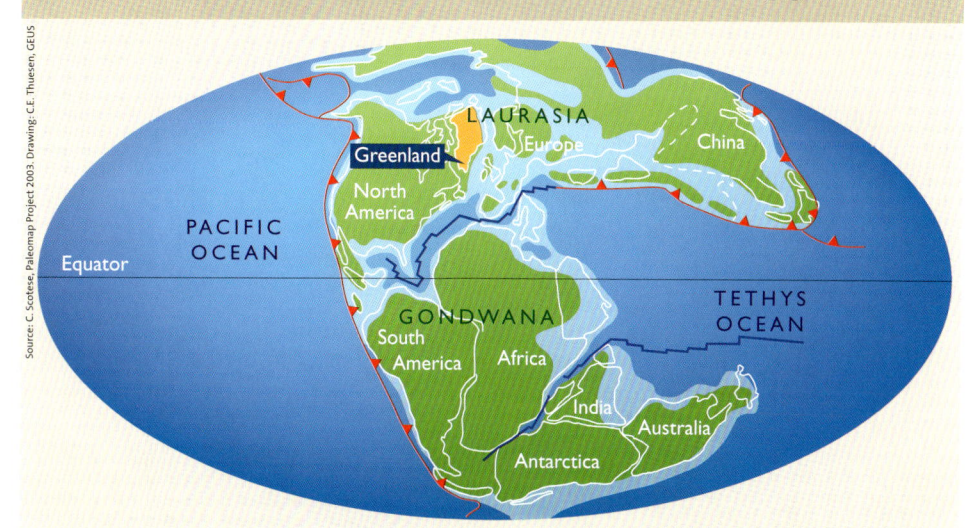

Palaeogeographic map showing the location of the continents during the Late Jurassic, 152 Ma ago. As Pangaea started to disintegrate, Africa and North America started to split and move away from one another, forming the earliest part of the Atlantic Ocean between them. A shallow shelf covered western Europe and eastern Greenland where various shaly sediments were deposited. These deposits were so rich in organic material in many places that they later formed one of the most important source rocks for oil and gas in the North Sea area.

POSITION OF GREENLAND 195 Ma AGO

During the Early Jurassic, the supercontinent Pangaea was still contiguous but started to disintegrate soon after. The map shows that at this time there was a shallow sea between Greenland and North America on one side and western Europe on the other.

YOUNGER SEDIMENTARY BASINS

From a forested delta to deep-sea sedimentation

The Nuussuaq Basin in West Greenland

Until the start of the Cretaceous, Greenland was attached to the North American continent. The oldest sediments within the Nuussuaq Basin, on the western margin of the present-day Greenland shield in central West Greenland, are the first known signs of extension. They were deposited about 100 Ma ago in the Cretaceous. The basin continued to develop for a period of 50–60 Ma during the Late Cretaceous to earliest Paleocene. The location of its western boundary is not well-established but was probably close to the eastern margin of the Canadian basement shield. The sediments in the basin range from continental lake and delta deposits closest to the eastern margin, to fine-grained, marine sediments deposited far offshore in a deep marine setting.

As ocean-floor spreading continued in the Atlantic Ocean throughout the Cretaceous, extension and subsidence of continental crust between Greenland and Canada caused an arm of sea to extend into what became the Davis Strait between Greenland and Canada. The strait formed a marine connection between the Nuussuaq Basin in the north and the Atlantic Ocean. Central West Greenland became the scene of major volcanism about 60 Ma ago in the mid-Paleocene. A thick sequence of basaltic lavas was erupted and this covered much of the present-day coastal region and the shelf area to its west (see p. 142). This volcanic activity changed conditions in the Nuussuaq Basin radically. Most of the basin was covered by plateau basalts at the same time as sedimentation continued in other areas that are now offshore of West Greenland (see p. 174).

The total thickness of sediment in the Nuussuaq Basin is more than 8 km, of which only the uppermost 3–4 km are exposed onshore while underlying sediments, interpreted as of Early Cretaceous age, have been discovered using geophysical methods. The eastern boundary of the Nuussuaq Basin is now a NNW-trending fault system along the edge of the Precambrian basement shield, but in the Late Cretaceous, sediments may have been deposited farther east, covering the basement. Sediment was carried into the basin from the east by a major river system that probably drained much of the interior of Greenland, forming a major delta in the Nuussuaq Basin with open sea to the west. The basin became unstable at the end of the Cretaceous with episodes of both rapid uplift and subsidence, segmentation of

Section through part of the Nuussuaq Basin showing Cretaceous–Paleocene sediments on the south-west side of Nuussuaq. The lower half of the exposed sedimentary section consists of deltaic sediments – light-coloured sandstones alternating with darker mudstones and coals. This is incised by a submarine canyon infilled with turbiditic marine sediments. The rocks forming the skyline consist of volcanic hyaloclastites (see p. 145). The well-exposed steep slope is about 250 m high.

the basin by faulting, and the erosion of deep canyons into the older sediments. Volcanism during the Paleocene changed conditions dramatically. The volcanic rocks dammed the basin, initially creating an inlet of the sea that developed into a lake (see p. 144) and finally the volcanic rocks spread to cover the whole basin.

Fossil-rich sediments

The oldest sediments exposed in the Nuussuaq Basin are about 100–110 Ma old. They are found in the south-eastern part of the basin and consist of river and delta deposits of sandstone and mudstone. Towards the north-west, the delta deposits are replaced by marine mudstones and turbidites (see p. 130). These fine-grained sediments in the most northern and western parts of the basin on Nuussuaq and on Svartenhuk Halvø were deposited on the sea floor in deep water and at distances up to 250 km north of the front of the delta. In many places, the sandy delta deposits contain a well-documented series of plant fossils showing that the area was covered by forest in the Late Cretaceous. The rich plant life has resulted in the formation of a widespread series of coal beds found on both Disko and Nuussuaq on both sides of Vaigat.

The marine sediments contain fossils of invertebrates such as bivalves, snails, belemnites, ammonites, corals, sponges, brachiopods and bryozoans as well as many forms of microfossils. The fauna is well documented and has been the subject of intensive studies since before the Second World War, when early expeditions started systematic, geological investigations of the Nuussuaq Basin region. Among the most interesting fossil discoveries is the world's largest bivalve, an *Inoceramus* species that is more than 2 m long. This fossil can be seen today, along with many other examples of fossils from the area, in the Geological Museum in Copenhagen.

The fossils have been described scientifically and assigned to their appropriate place in the biological classification system consisting of phyla, classes, orders, families, genera and species. The stratigraphic distribution of the animals and their development has also been studied. After completion of the studies, the fossil material was deposited in the Danish national geological archive in the Geological Museum in Copenhagen, where the entire collection is available for future studies.

The final development of sedimentation in the Nuussuaq Basin, with the formation of a lake limited to the west by active volcanism, is described together with the West Greenland volcanic province (see p. 142). Those parts of the Basin that lie offshore are described together with the rest of the offshore area (see p. 174).

POSITION OF GREENLAND 94 Ma AGO

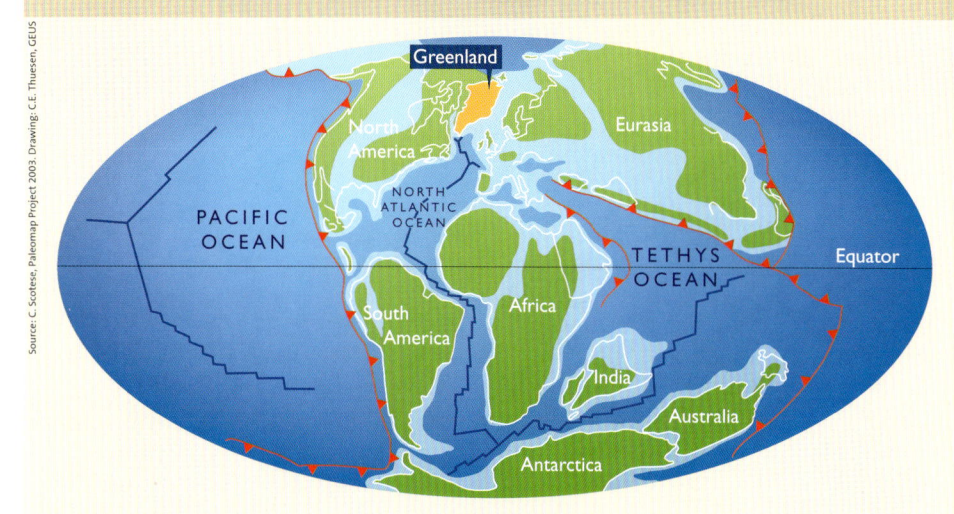

Opening of much of the Atlantic Ocean was underway by the middle of the Cretaceous. In the north, however, sea-floor spreading had not yet started along Greenland's margins. Tension within the crust was causing them to stretch and subside, forming basins between Greenland and Scandinavia on one side and Greenland and Canada on the other. Sea-floor spreading on both sides of Greenland began between 65 and 57 Ma ago, shortly after the beginning of the Palaeogene.

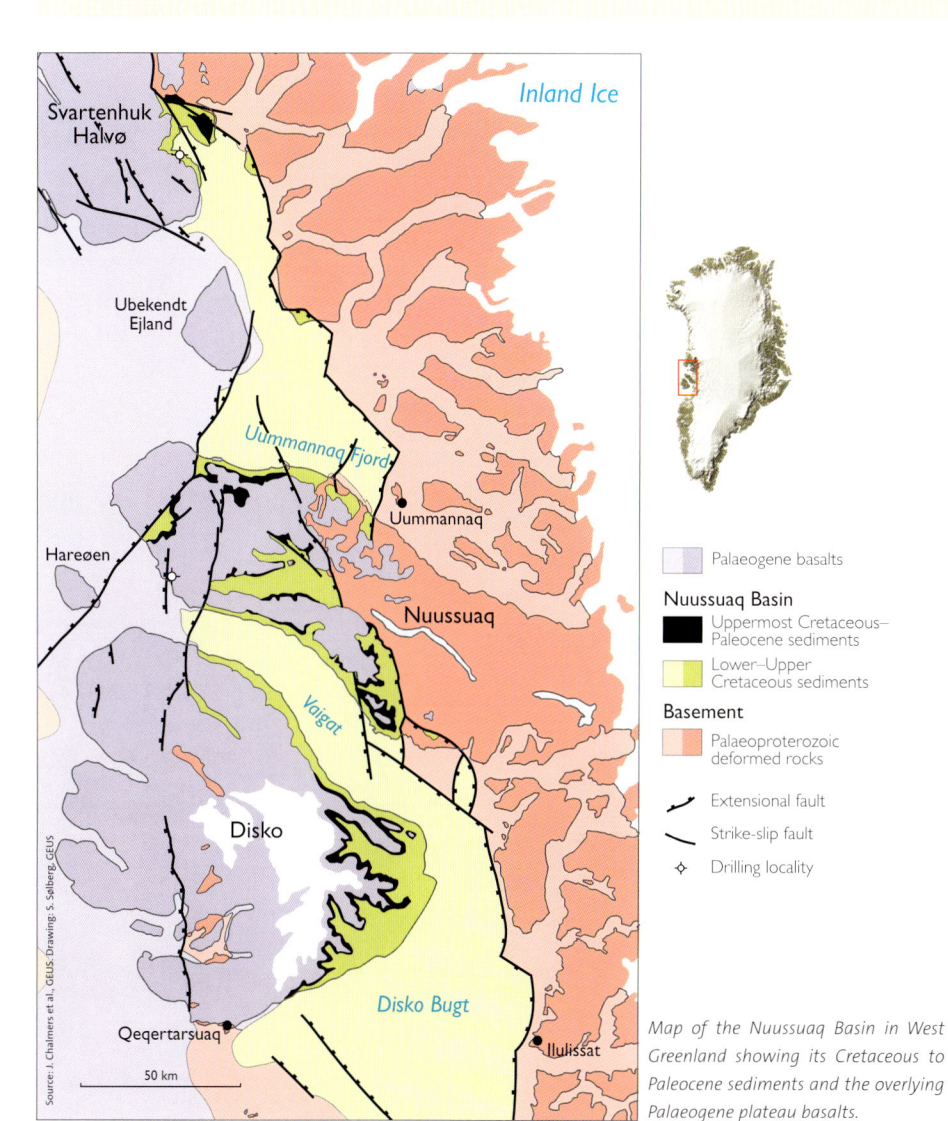

Map of the Nuussuaq Basin in West Greenland showing its Cretaceous to Paleocene sediments and the overlying Palaeogene plateau basalts.

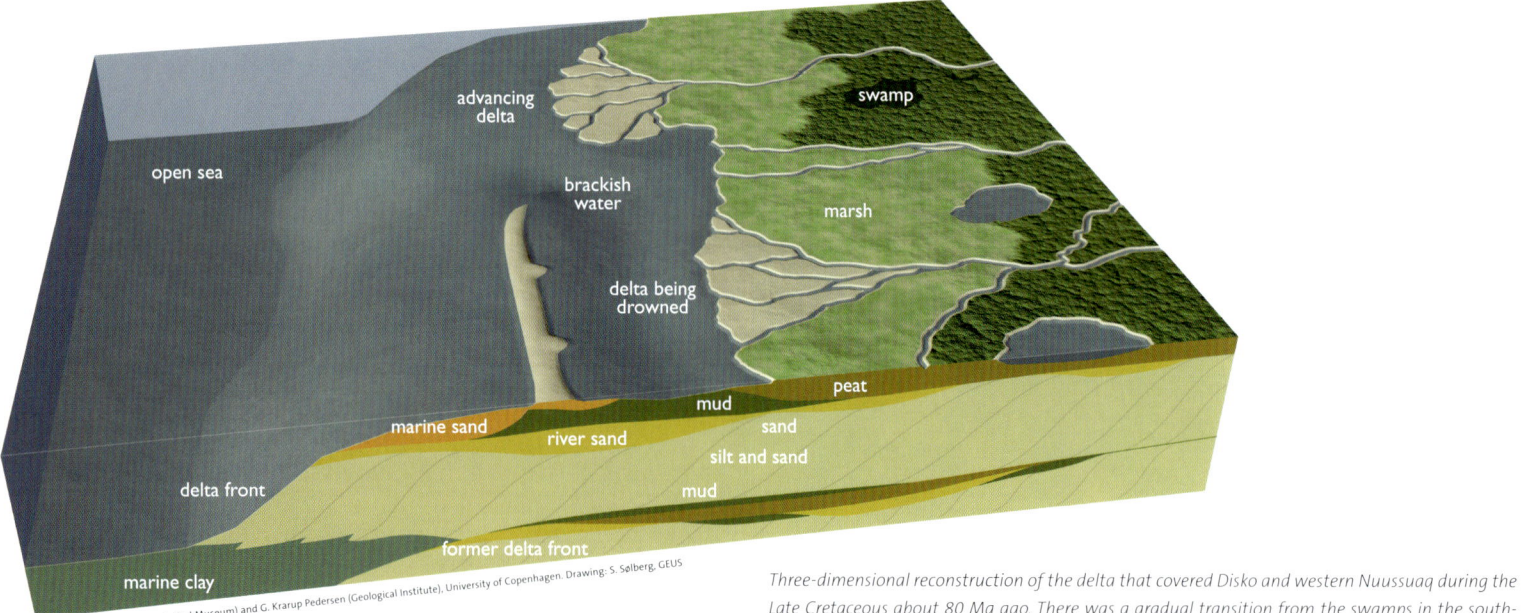

YOUNGER SEDIMENTARY BASINS

Three-dimensional reconstruction of the delta that covered Disko and western Nuussuaq during the Late Cretaceous about 80 Ma ago. There was a gradual transition from the swamps in the south-east to a brackish lagoon behind a sand-bar along the edge of open sea. The resulting sedimentary types are shown on the vertical cross section of the block diagram. The location of the coast was not fixed but moved cyclically towards and away from the land in the south-east. The resultant changes in deltaic sedimentation are illustrated on the vertical section.

SEQUENCE STRATIGRAPHY

Sequence stratigraphy is the analysis of how sediments within sedimentary successions change as sea-level changes, which often takes place in repeated cycles. This is an important modern, analytical tool that makes it possible to predict the development and lateral extent of rock types within a sedimentary succession by applying a model of how sediments are deposited as sea level rises and falls.

At any one time, different types of sediment are deposited in different places in a sedimentary basin. Coarse sandy sediments are deposited in shallow water near land, fine-grained sands and silts farther offshore and clays in deeper water. Abyssal oozes and turbidites are deposited in very deep water. When sea level rises or falls, the water depth changes at any given place and the transition zone from shallow to deep water shifts landwards or seawards. As relative sea level rises, the sea moves landwards to cover low-lying coastal regions – a change called a transgression. When sea level falls, the sea withdraws to reveal new land, a change called regression.

Global sea-level changes take place in a broadly rhythmic fashion cycling from low to high to low again (see p. 66). Sediments deposited during a single cycle – a sedimentary sequence – vary following a certain pattern (see figure). A cycle starts when sea level is at its lowest, moves through a transgression as sea level rises to its highest level (maximum flooding) after which it again falls back to a new lowstand level. Erosion of areas exposed during such lowstands forms a break in the succession (an unconformity) onto which the next sequence is deposited when the water levels are higher again. When sea level once more reaches a low level, a new unconformity is eroded, so the sequence is bracketed between two episodes of erosion that created unconformities.

Sequence stratigraphy provides a framework for understanding the distribution of different sediment types in time and space. It thus explains the relationships between sandstones, siltstones and clays and their distribution in the basin. Sequences can often be mapped on seismic profiles and sequence stratigraphic analysis is thus an important tool in hydrocarbon exploration to predict where oil and gas may be generated and trapped.

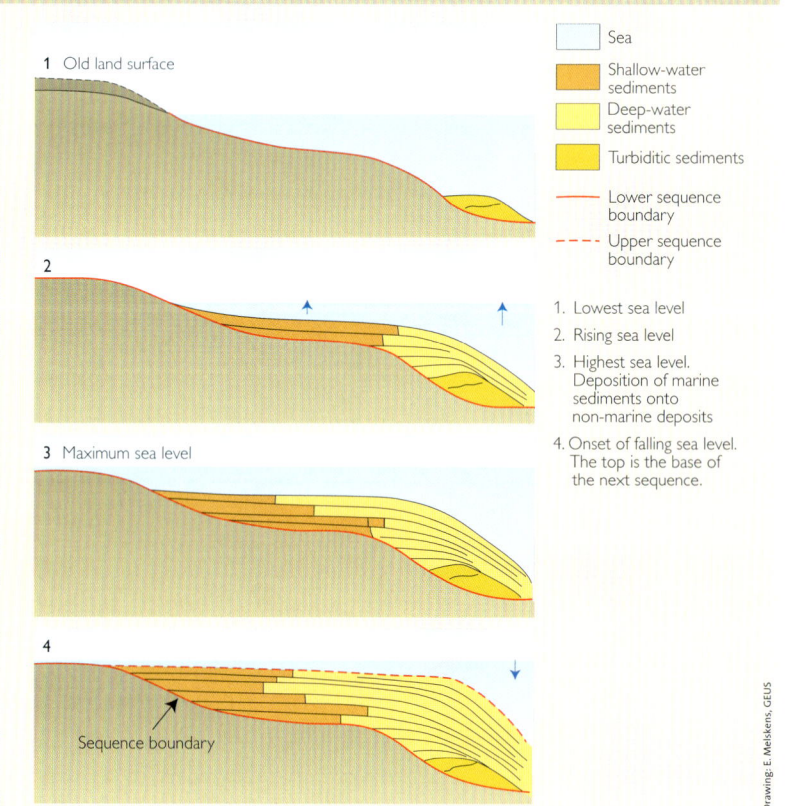

- Sea
- Shallow-water sediments
- Deep-water sediments
- Turbiditic sediments
- Lower sequence boundary
- Upper sequence boundary

1. Lowest sea level
2. Rising sea level
3. Highest sea level. Deposition of marine sediments onto non-marine deposits
4. Onset of falling sea level. The top is the base of the next sequence.

Sections showing the principles of deposition of a sequence. The four profiles show different stages in a sea-level cycle and the accumulation of a sequence, beginning with a sea-level lowstand (1) followed by sea-level rise to its maximum (2, 3) and the subsequent turnaround and beginning sea-level fall (4). The profiles are drawn with vertical exaggeration.

The Nuussuaq Basin in West Greenland

FOSSILS FROM THE NUUSSUAQ BASIN

The sediments of the Nuussuaq Basin, deposited under both freshwater and in marine conditions, are very rich in fossils. The freshwater forms are dominated by plant remains, while the marine fossils are mostly small invertebrates such as bivalves, snails and ammonites. Plant fossils from the deltaic sediments are shown upper left and middle. Below them are examples of invertebrate animals that lived in the sea outside the delta area. On the upper right is a photograph of a giant mussel.

'BEECH LEAF'

DAWN REDWOOD (Metasequoia)

Part of a twig from a Dawn Redwood preserved as an impression in a Lower Paleocene sandstone found in central Nuussuaq. A species of Dawn Redwood tree survives today in China where it grows to a height of 30–35 m. It is now common in European gardens.

Impression of a leaf from Lower Paleocene sediments in central Nuussuaq. The leaf is from an extinct species of beech. This discovery shows that there were deciduous trees in this part of the world 63 Ma ago.

The world's largest mussel 'Inoceramus (Sphenoceramus) steenstrupi' from Nuussuaq. Most mussels of this type (Inoceramus) were much smaller and only a few species were as big as the example shown. These types of mussels are important as stratigraphic indicators. The Upper Cretaceous example shown is about 83 Ma old. The picture shows two shells stuck together, both with their points upwards. The shells are 178 cm long.

AMMONITE (Scaphites)

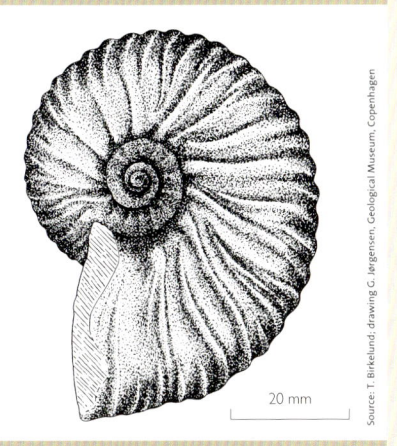

Upper Cretaceous ammonite shell from inner Nuussuaq. About 30 different species of ammonites have been found in this area and they have been used to divide the sediments from 90 to 65 Ma into 12 biozones, each with its characteristic assemblage of fossils.

FORAMINIFER (Anomalinoides)

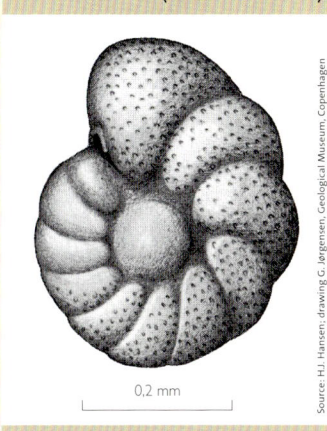

Example of a foraminifer shell, a microfossil from a group of single-celled animals belonging to the order Foraminifera in the phylum Protozoa. The shell is made of chambers that were added as the animal grew. From the Danian stage (65–62 Ma years ago). Inner Nuussuaq.

SNAIL SHELL (Buvignieria)

Fossil snail shell from the Paleocene (65–56 Ma ago) of inner Nuussuaq. More than 260 different named snail species have been found from this epoch in this area, most of which come from a single 50 m thick sandstone bed.

'LIMPET SNAIL' (Semperia)

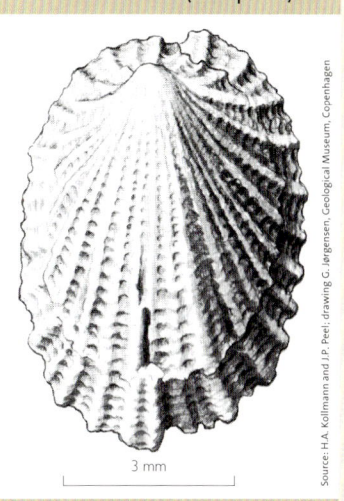

Fossil snail shell with a flat hood-shape that is not coiled like most snail shells. From Paleocene sediments about 60 Ma old, in inner Nuussuaq.

YOUNGER SEDIMENTARY BASINS

First sea, then land – before the whole basin was buried under lava flows

The Kangerlussuaq Basin in South-East Greenland

Southern East Greenland is dominated by basement rocks with a cap of Palaeogene basalts to the north. Only in a small area north of the great Kangerlussuaq fjord is there an isolated series of undefomed sediments, which form the Kangerlussuaq Basin. It lies on the Precambrian basement and is itself covered by a thick sequence of Palaeogene lavas. The sediments onshore are 600–700 m thick and their ages range from Early Cretaceous to mid-Paleocene (from about 115 to 60 Ma old) and the exposed sequence is the most westerly part of the basin. The Kangerlussuaq Basin formed much later than the other younger sedimentary basins along the coast of East Greenland and is interpreted as an independent development. The basin formed along the western margin of a narrow seaway that later developed into the North Atlantic Ocean that has separated Greenland from Western Europe since the Palaeogene.

The area between Greenland and western Europe was affected by extension and rifting during the Cretaceous and a seaway stretched from north to south between the two land areas (see p. 129). This sea connected an ocean north of Greenland with the newly-formed central Atlantic Ocean and a further connection to the Tethys Ocean between Europe and Africa.

Cretaceous marine sediments

The oldest sediments in the Kangerlussuaq Basin are from a period known as the Early Cretaceous when global sea levels were rising and the sea flooded progressively over the nearby land areas (transgression). The sedimentary succession in South-East Greenland begins with coarse-grained and cross-bedded sandstones deposited by rivers, but the transgression caused these sediments to be succeeded by finer-grained marine mudstones separated by beds of sandstone. The marine sediments contain fossils including ammonites and belemnites and traces of bivalves and crustaceans.

The sediments from the latest Cretaceous consist of a series of sandstone beds separated by shaly deep-water sediments. The sandstone beds are interpreted

Fossil mussel shells from the marine Lower Cretaceous sediments (Aptian – about 120 Ma old) of the Kangerlussuaq Basin. The shells are grouped on a surface in the same way that many present-day shells often cluster on a sandy surface. Most of the fossil mussels are only impressions and the shell material is preserved only in a few specimens.

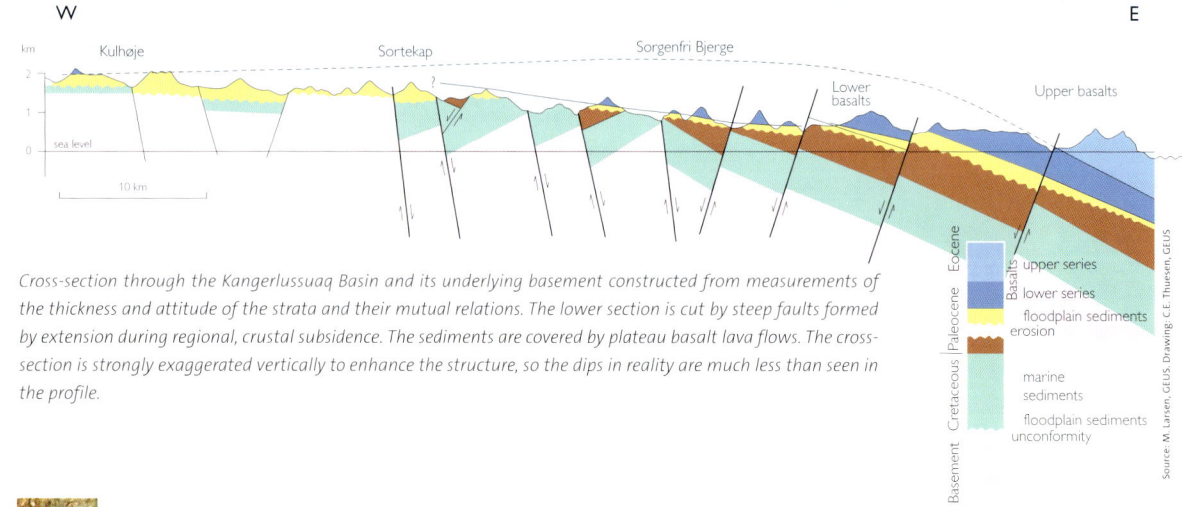

Cross-section through the Kangerlussuaq Basin and its underlying basement constructed from measurements of the thickness and attitude of the strata and their mutual relations. The lower section is cut by steep faults formed by extension during regional, crustal subsidence. The sediments are covered by plateau basalt lava flows. The cross-section is strongly exaggerated vertically to enhance the structure, so the dips in reality are much less than seen in the profile.

Strata from the Kangerlussuaq area comprising light-coloured sandstone below and brown volcanic ash above. The sandstone unit in the foreground is about 7 m thick and consists of sediments deposited in a channel carved by a river. Such coarse-grained sandstones were deposited as a result of the substantial uplift of the land that took place in the mid-Paleocene about 60 Ma ago, just before the start of volcanism. The dark volcanic beds at the top of the picture consist of hyaloclastites (see p. 145).

as the deposits of submarine avalanches (turbidity currents) that removed sandy sediments from shallow-water shelf areas and transported them out into the deep-water areas, where normally only silt and clay were deposited. These turbiditic sandstones contain impressions of fragile leaves and twigs that could not have been transported intact over long distances. Their presence therefore shows that the land area to the west of the basin could not have been far away.

Land area in the Paleocene

About 60 Ma ago, in the middle of the Paleocene, a complete change took place in the geological conditions in the area. The land areas surrounding the narrow seaway between Greenland and Western Europe had been part of a gently subsiding stable Precambrian shield. Now they were lifted upwards as a huge plume of magma (a mantle plume – see p. 139) rose up from the mantle and spread out under the lithosphere (see p. 20), leading to thinning and uplift of the crust. The resulting greater topographic relief around the basin led to rapid erosion of its hinterland. The formerly marine Kangerlussuaq Basin was uplifted so that it became an area of land where large river systems deposited gravel and sand. These sediments in the Kangerlussuaq Basin frequently contain remains of leaves, cones and small pieces of wood, revealing that vegetation covered the land areas in the Paleocene. The large rivers stretched far out onto the shelf and fed sediments into submarine canyons incised into the earlier finer-grained sediments. These processes resulted in the deposition of a series of sandy sediments along the basin margin and tongues of sand being transported out over the formerly fine-grained sediments in the deeper water areas.

A giant volcanic province formed in central East Greenland as the North Atlantic started to open (see p. 146). Volcanic eruptions gave rise to flows of basaltic lava and ash falls that accumulated in the low-lying land areas and covered the sediments that had previously been laid down in the Kangerlussuaq Basin.

Composite sedimentary log compiled from numerous detailed sections of the Kangerlussuaq Basin sediments. The strata consist of sediments of Cretaceous and earliest Paleocene ages, after which time they were buried under plateau basalt lavas. The log shows only the lowest part of the basalt succession. The depositional environments are listed to the right and the sketches show the distribution of different fossil types.

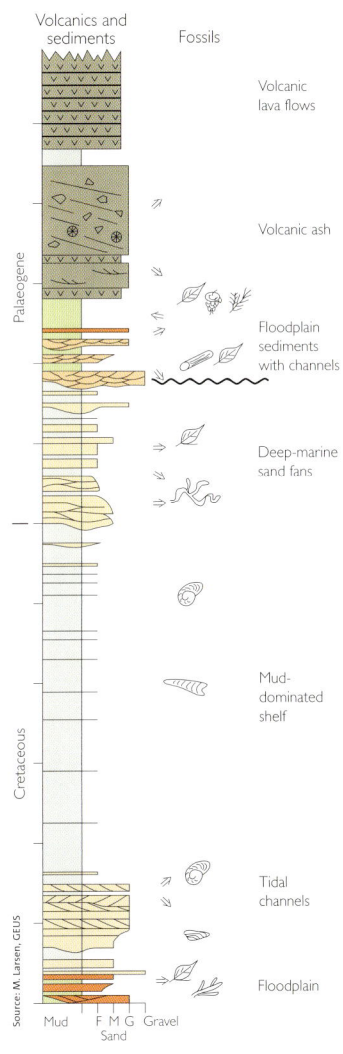

YOUNGER SEDIMENTARY BASINS

Oil exploration around the Faroes aided by results from Greenland

Exploration for oil and gas offshore of the Faroes began in the 1990s and the first wells were drilled in 2001. Interpretation of the deeper sediments under the sea around the Faroes is more difficult than normal because they are covered by a sequence of basalt lava flows. Before drilling deep and expensive wells, it was important to understand what sedimentary sequences might be expected below the lavas and studies of nearby areas were one of the methods used. Prior to opening of the North Atlantic, the present-day Kangerlussuaq Basin was only 100–150 km away from the Faroes and the geological history of the two areas was expected to be similar. Mesozoic and Paleocene sediments are found under the lava flows in both areas (see p. 148).

Exploration on the United Kingdom shelf near the Faroes, but outside the area covered by basalt, had shown that thick sandy units of Paleocene age were reservoirs for oil. Comparable sandy sediments were expected to occur around the Faroes, but would be difficult to study because of the overlying cover of volcanics. Measurements of the flow directions in the Paleocene river channels in Greenland showed that the sediments originating there were transported towards the Faroes. Sands in the Faroese area can therefore be expected to thicken north-westwards, in contrast to comparable sandy units on the United Kingdom shelf that originated from Scotland to its south-east. The studies from Greenland therefore indicated that additional potential reservoir rocks could exist in the sediments under the lavas around the Faroes.

The first wells drilled on the Faroese shelf in 2001 found traces of oil in reservoirs of Paleocene age but insufficient to begin production. Special studies of the source of the sandstones (provenance studies, see p. 76) showed that a large proportion of the sand probably originated in Greenland. Geological models based on studies of the sedimentary basins in Greenland continue to play an important role in oil exploration around the Faroes.

Large-scale cross-bedding in sandstone; each of the cross-bedded units is many metres thick. Many of the sandstone units have dark mudstone beds above and below them. Cretaceous sediments from the lower strata in the Kangerlussuaq Basin. The section shown is about 25 m high.

Marine sandstones of Paleocene age showing large-scale cross-bedding, illustrated by the dipping nature of the sandstones to the right of the waterfall compared to the more flat-lying, darker sandstone beds above. The section shown is about 6–8 m thick. Younger basalt lava flows are seen in the background.

The Kangerlussuaq Basin in South-East Greenland

TRACES OF PLANT LIFE OF THE PAST

The types of plants found in Greenland today reflect its present Arctic climate and the country has no real forests. A total of only about 500 plant species live in the whole of this large area today and most of the present-day flora came to Greenland during the 10 000 years since the end of the latest ice age – the Weichselian (see p. 186). The sparse nature of the vegetation is due to the land's present high latitude, but Greenland has not always been so barren. Many geological strata contain impressions and remains of plants showing that previously there has been rich vegetation at various periods and that the country has been periodically covered by forests. The reason for this is, of course, that Greenland has moved through various climate zones during its geological history (see p. 112) and that at times the climate of the whole Earth has been warmer, so conditions for plant growth were more favourable.

The plant remains that are found as fossils occur almost exclusively in terrestrial deposits such as sandstone and finer grained sediments. Where plant remains are found, it is nearly always an indication that the sediments were deposited in lakes or rivers or in the sea very close to land.

Through the history of the Earth, plants have evolved just as much as animals in response to evolutionary pressure. Thus, as conditions changed, new species adapted better to the changing environment and replaced older species, resulting in continuous renewal of the plant communities. Plant fossils can be used just as well as animal fossils, to subdivide the strata in which they are found into stratigraphic units. The study of fossil plants and their evolution is known as palaeobotany or phytopalaeontology.

The land plants developed from primitive forms that grew in water in the earliest Palaeozoic. Terrestrial forms established themselves first in wet areas on land during the Silurian and Devonian. A major evolutionary step took place during the Carboniferous and Permian, when woody plants resembling trees emerged. Other significant developments occurred during the Mesozoic, with the appearance of conifers followed by flowering plants.

The oldest plant fossils found in Greenland are in continental sandstones of Devonian and Carboniferous–Permian ages in North-East Greenland, but the diversity of the flora from these rocks is limited and the preservation of the fossils is poor. One of the best-preserved groups of plant fossils in Greenland is much younger. It is found in sandstones deposited during the transition from the Triassic to the Jurassic (Rhaetic–Liassic, about 204–195 Ma ago) along Hurry Inlet in the Scoresby Sund region of East Greenland (see p. 126). The large collection of more than 200 species of plant fossils that has been sampled from this area has led to a number of classic palaeobotanical investigations and scientific descriptions.

Another very large occurrence of plant fossils is found in the Disko Bugt area of West Greenland. Terrestrial sandstones of Cretaceous and Paleocene ages (about 115–60 Ma ago) in the Nuussuaq Basin here contain well-preserved remains of a diverse group of around 500 species of plants. This flora grew at the same time as the plants found in the continental sediments of the Kangerlussuaq Basin in South-East Greenland. The examples of plant fossils shown in the box are from the Kangerlussuaq Basin. They are of Paleocene age between 65 and 56 Ma old.

A much younger plant assemblage with fossils from a warm interval at the beginning of the Ice Age is found in Neogene (Pliocene) deposits near Kap København in easternmost North Greenland (see p. 185).

ELLIPTIC LEAVES

A well-preserved leaf showing its serrated rim and structure of petiole and ribs. The enclosing rock is a fine-grained Paleocene sandstone. Length of leaf 7 cm.

LANCEOLATE LEAF

The good preservation of this fossil leaf shows that it settled gently through calm water and was then quickly covered by sediment, so the structure of the leaf is preserved intact. The leaf is from a woody angiosperm – probably a tree. The rock is a fine-grained Paleocene sandstone. Length of leaf 14 cm.

Areas with strata containing abundant plant fossils.

EARLY PALEOCENE PLANT REMAINS

Plant remains, including pieces of leaf and branch, in a medium-grained sandstone. The fossils are preserved in a submarine, turbidite sand that was deposited in a marine environment, but the plants originated from a terrestrial site not far to the west. Leaf impressions about 3–8 cm long.

Palaeogene volcanic rocks from the Kangerlussuaq area (East Greenland). In the foreground are conglomerate deposits above which is a basaltic lava flow with columnar jointing.

Foto: L. Melchior Larsen, GEUS

PALAEOGENE VOLCANISM

When Greenland and northern Europe drifted apart widespread volcanism occurred in the North Atlantic region

9

LAVA IN KILOMETRE-THICK SEQUENCES

PALAEOGENE VOLCANISM

The whole of the North Atlantic region is marked by widespread volcanism that began in the Paleocene about 62 Ma ago and continues to the present day. Today there are active volcanoes on Iceland, and along the mid-Atlantic ridge active submarine volcanism is continually forming new oceanic crust. The entire sea floor beneath the Atlantic Ocean comprises volcanic rocks that were originally formed at the central spreading ridge. The oldest parts now occur in the marginal zones of the ocean, close to the continental edge, while the youngest rocks are found at the spreading ridge. Thus on both sides of the North Atlantic, the oldest volcanic rocks occur closest to the old continents; in the east they are found on the Faroe Islands and in Scotland, and to the west in Greenland. At all three places an enormous lava pile was built up between 62 and 54 Ma ago, but as the Atlantic Ocean widened the activity moved away from land, following the spreading ridge in the newly formed ocean.

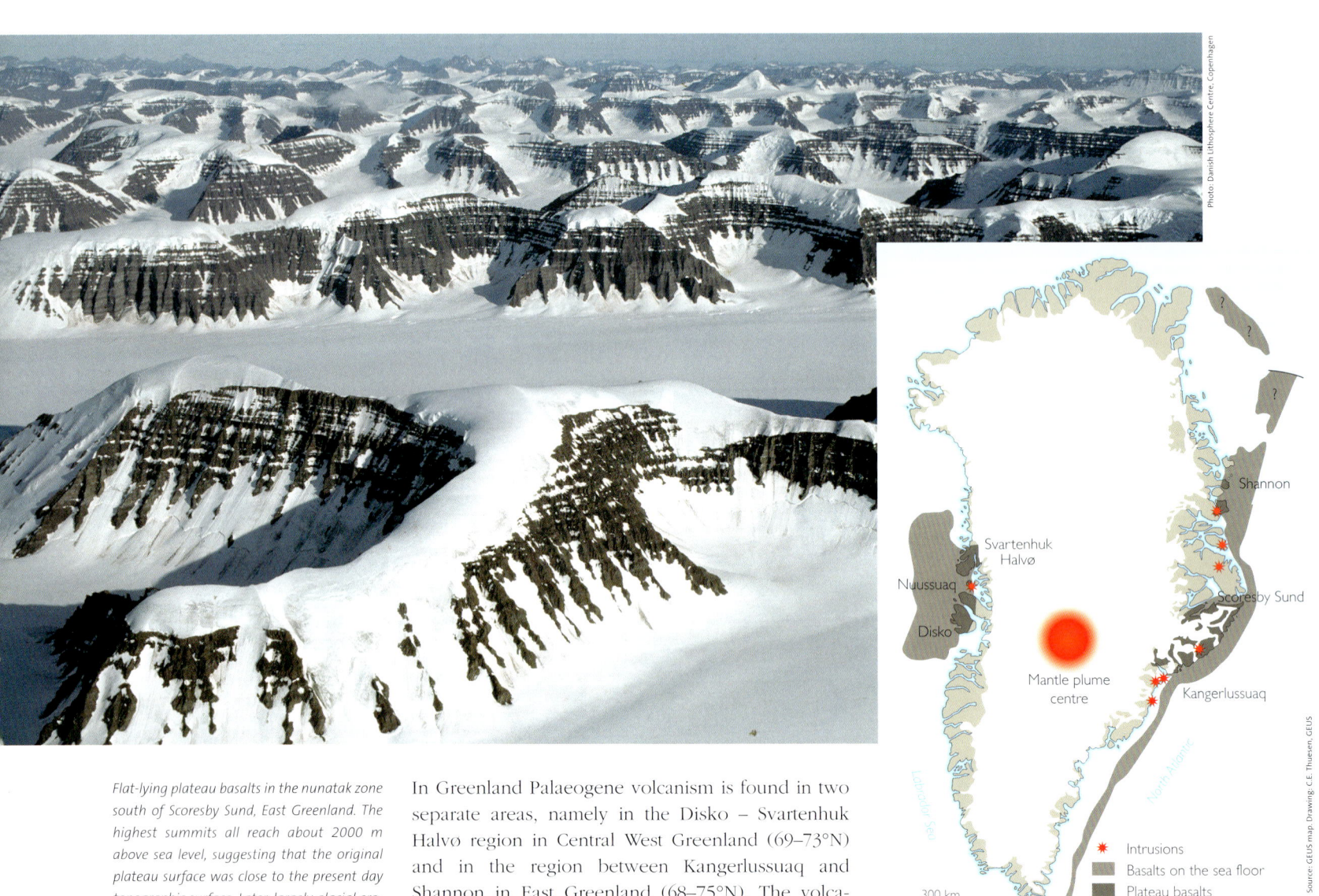

Flat-lying plateau basalts in the nunatak zone south of Scoresby Sund, East Greenland. The highest summits all reach about 2000 m above sea level, suggesting that the original plateau surface was close to the present day topographic surface. Later, largely glacial erosion has remodelled the original plateau landscape to produce cliff exposures 1000–2000 m high. In the photograph the height difference between the mountain summits and the glacier surface is about 1000 m. The lavas in the photograph are a part of the uppermost basalt sequence.

In Greenland Palaeogene volcanism is found in two separate areas, namely in the Disko – Svartenhuk Halvø region in Central West Greenland (69–73°N) and in the region between Kangerlussuaq and Shannon in East Greenland (68–75°N). The volcanism in both regions is related to the onset of Paleocene sea-floor spreading. It began in West Greenland, with the opening of the Labrador Sea between Greenland and Canada about 62 Ma ago, and continued some few million years later by the beginning of the opening of the North Atlantic between Greenland and northern Europe. Both in West and East

Map showing the occurrences and extent of the Palaeogene volcanism and associated intrusions into the volcanic provinces in West and East Greenland. Volcanism in West Greenland began a little earlier than in East Greenland, but in both regions took place in the Paleocene and Early Eocene about 60–54 Ma ago. The map shows the postulated position of the 'hot spot' (the mantle plume, see p. 139), which reached the base of the crust and generated the magmas leading to the volcanic activity.

Greenland a flat-lying succession of plateau basalts up to 10 km thick was erupted that covers many thousands of square kilometres onshore and extends out onto the shelf, where a correspondingly large area of the sea floor was covered by basalts. In East Greenland the surface volcanism was followed by a series of later intrusions, including important dyke swarms and scattered intrusive centres that cut through the basalts and the underlying sedimentary and basement rocks.

The opening of the Atlantic Ocean began a little north of the equator in the Middle Jurassic (180–170 Ma ago), when the supercontinent Pangaea started to break up, and the future Africa and America began to move apart. The central spreading ridge associated with this opening slowly extended northwards so that 100 Ma later it had reached a position between the future North America and Europe. The spreading axis continued to extend northwards, initially following a split between Greenland and Canada; however, after a while, this gradually stopped developing and in the Palaeogene (about 59–54 Ma ago), the spreading axis took another direction and continued northwards between Greenland and northern Europe. Since then, continued spreading has opened the North Atlantic so today the distance between the formerly connected areas of South-East Greenland and the Faroe Islands is about 1000 km.

The volcanic material

The enormous quantity of volcanic material that was poured out onto the Earth's surface comprises predominantly basaltic melts that formed through partial melting of the upper mantle. Melting was caused by the supply of hot material from the deepest parts of the lower mantle at the boundary to the outer core at depths of 2900 km. From here the lighter, hot, buoyant material flowed upwards into the upper part

POSITION OF GREENLAND 14 Ma AGO

Palaeogeographical map of the Middle Miocene, about 14 Ma ago. The mid-Atlantic spreading ridge has forced its way up between Greenland and North-West Europe and formed new ocean floor between Greenland and Norway. In the Alps and Himalayas large-scale mountain belt complexes were formed following continental collision. Along the Pacific coast of the Americas the Earth's longest mountain chain, the Cordillera–Andean orogen, developed at a plate boundary between the American continental plates and the oceanic plate to the west.

POSITION OF GREENLAND 65 Ma AGO

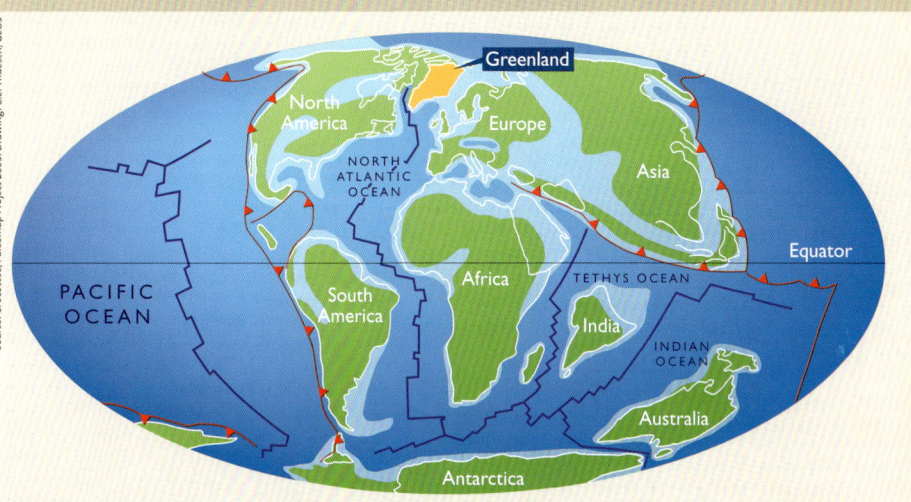

Palaeogeographical map showing the position of the continents at the Cretaceous–Paleocene boundary, 65 Ma ago. The opening of the Atlantic Ocean initially formed a spreading zone between Greenland and Canada. At this point there was no ocean formation in the North Atlantic region between Greenland and North-West Europe. India was moving rapidly (at 13 cm per year) north-eastwards towards southern Asia.

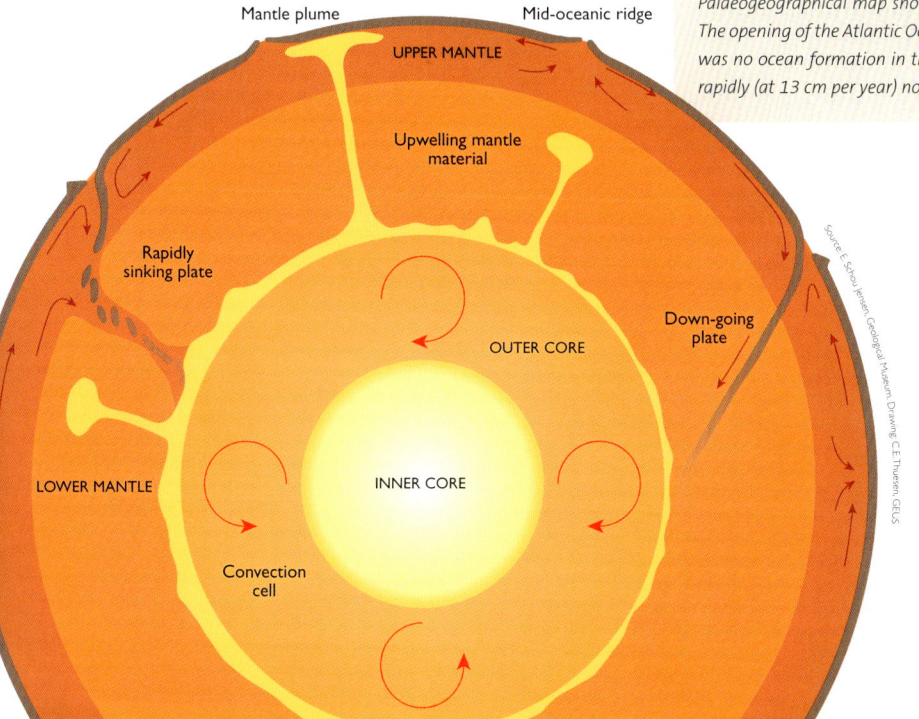

A cross-section of the Earth showing its internal structure and dynamic character. Convection cells (red arrows) in the outer core resulted in transference of heat energy upwards. At the boundary between the outer core and the lower mantle, bulges of hotter, more mobile material (yellow) formed. In some cases these grew and moved upwards to develop into 'mantle plumes'; ultimately these reached the upper mantle at the base of the crust (brown) forming mushroom-shaped structures. Such mantle plumes transferred vast amounts of heat energy into the overlying mantle, causing parts of the upper mantle to melt resulting in volcanism at the surface.

Cliff section showing lava flows from the upper lava series on Disko, West Greenland. The photograph shows a section through eight flows with similar construction. The flows are 20–40 m thick, and each has a lower, columnar-jointed zone, passing upwards into a zone with irregular small fractures (jointing). The uppermost cindery tops of the flows often have a red colour due to the high iron content of the lava, which oxidises by reaction with the air (i.e. forms rust) during the several hundred years between each major eruption.

of the mantle as a cylindrical up-welling body called a 'mantle plume'. When the plume nears the uppermost part of the mantle, it tends to broaden into a mushroom shape, bringing deep-sourced hot material into contact with the cooler and more rigid ultrabasic rocks of the upper mantle. The mushroom shape of the plume head can reach a diameter of about 1000 km, and a temperature up to 250°C higher than the surrounding rocks. The pressure release induced by the rise causes the mantle in the upper parts of the plume head to melt, and some of the basaltic magmas that form can pass up to the surface. The amount of heat energy introduced may be so large that enormous volumes of plateau basalts can be formed in a very short period of time. However, in some instances a long-lived active volcanic centre may develop over the central part of the plume, producing melts that rise to the surface creating local volcanoes. Such persistent volcanic centres are called 'hot spots' (see p. 112) and Iceland is a modern example of such a volcanic region.

When the ultrabasic mantle rocks in contact with the top of the mushroom-shaped plume melt, and the magma subsequently crystallises with falling temperatures, a number of different rocks develops that constitute a reaction series. Partial melting in the upper mantle initially produces a magnesium-rich, olivine basaltic melt that migrates away and can force its way into higher levels in the crust. As this magma begins to cool, olivine basalts of different composition separate out, with olivine crystallising first, thus being more concentrated in the early magma types. The remaining, olivine-depleted magma now comprises principally basaltic liquids that typically erupt onto the surface and solidify when temperatures have fallen to 1200–1000°C. The process whereby magmas derived from melted ultrabasic rocks modify their composition by mineral separation during cooling is called magmatic differentiation (see p. 152).

Eruption sizes and frequencies

The volumes of magmatic material that are produced over a mantle plume within a very short time period can be enormous. Estimates based on data from East Greenland and the Faroe Islands show that in this region Palaeogene basalt covered an area of about 220 000 km^2, and the volume of the volcanic sequence, extruded within a few million years, was approximately 370 000 km^3. Estimates of the frequency of the major eruptions in connection with the volcanism in South–East Greenland show that at any one place there were approximately 3000 years between each major lava flow unit. In the Scoresby Sund area a similar calculation indicates there were about 8000 years between each major flow unit. The highest known frequency of eruption is found in West Greenland, where the intervals between flows in some periods may have been as short as 300 years.

The basaltic formations in both West and East Greenland comprise a flat-lying succession many kilometres thick. Each layer represents an individual eruption around a fissure from which the rapidly flowing lava could spread out on either side. The thickness of a flow ranges from a few metres to several tens of metres. A lava flow can often be followed over several tens of kilometres, but typically will wedge out and be superseded by another flow from a different eruption centre. The lava flows usually flooded out over a low-lying, comparatively flat land area or, in some cases, were erupted under water and formed pillow lavas (see p. 46). When the eruption took place on land the uppermost parts of the flow became modified due to interaction with the atmosphere. These upper parts are full of bubbles formed by gasses escaping from the magma and their surfaces are brecciated and rubbly due to rapid cooling. Between the different lava flows there may be a layer of volcanic ash, which commonly has a more reddish colour and easily disintegrates. Individual flows can often be distinguished from each other, and their occurrence can been mapped in great detail over a wide area. During fieldwork samples are collected up through the lava succession, and in subsequent laboratory studies the mineral and geochemical compositions of each sample are determined, showing possible systematic variations in composition. On the basis of these variations and the stratified successions of the basalts, the lava sequences can be divided into stratigraphic units, and classified into formations and groups. In both West and East Greenland very detailed and accurate mapping has been carried out so that it is possible to follow the extent of a single flow, or a complex of flows, over hundreds of kilometres.

In East Greenland intrusive activity continued long after the last lava flow in the sequence had erupted over the surface about 54 Ma ago. During the following part of the Palaeogene, the crust was cut by a series of small intrusions that made their way up through the overlying rocks and solidified in the upper crust before they reached the surface. These intrusions are exposed today over a distance of about 1000 km along the coastal zone between 67–74°N. They were formed in a period when the main part of the volcanic activity in the North Atlantic region had moved out to the mid-ocean spreading ridge. The many intrusions show that even though the mantle plume had moved away, it left behind a series of areas in the underlying lithosphere that were hotter, more enriched in volatiles and capable of melting to form local magmas that produced the intrusions. The latest volcanic activity found in the inland areas resulted in the eruption of lavas 14 Ma ago.

MAGMA – A MELTED ROCK MASS

Magmas occur through partial melting of existing solid rocks, either from part of the mantle or from the crust. When rocks change from a solid to a liquid form they become mobile and move away from their place of origin, and solidify again higher up in the crust or at the surface. When liquids solidify within the crust they are called plutonic, while when they erupt and solidify on the surface, they are called volcanic rocks. Movement from the depths to the surface is primarily driven by a response to pressure differences, because typically magmas are less dense than the surrounding solid rocks and therefore move upwards.

Melting is dependent upon temperature and pressure, which means that rocks deep down in the crust and the mantle need higher temperatures to melt than rocks nearer to the surface. Melting temperature also varies with the rock composition. Acid rocks, with high contents of alkalis and silica, melt at lower temperatures than more basic types. Thus ultrabasic magmas have a temperature of 1500°C, while the more common basaltic liquids have temperatures of 1200–1000°C. The lowest magma temperatures are found in water-rich, granitic liquids that are only about 700°C. The content of volatile components, such as water (steam), carbon dioxide and other gases in the magma, also influences melting point and viscosity (rigidity). The higher the water and gas content in a magma, the lower the melting temperature at a given pressure.

The mobility of a magma depends on its fluidity, which is controlled by both composition and temperature. Basic magmas with high temperatures, e.g. basalts, have a low viscosity and are comparatively fluid which, when erupted at the surface, produce calm lava streams that can flow out over an extensive area. In contrast, acid granitic magmas with lower temperatures have a higher viscosity, and usually crystallise in the crust as granitic plutons. Acid magmas only rarely reach the surface, and when they do, erupt explosively, producing limited quantities of lava, but often large quantities of ash together with volcanic bombs of solidified magma.

Gabbroic rock viewed under the microscope in transmitted, cross-polarised light that shows the different minerals in 'false' colours. The microscope section is prepared from a thin slice of rock glued to a glass slide and polished until it has a thickness of 1/30th of a millimetre. This section becomes translucent, making it possible to study its components under a microscope. The minerals seen in the photomicrograph are: feldspar (plagioclase), grey and white colours; olivine, yellow-green; pyroxene, dusty, greyish yellow and reddish colours. The section shown is approximately 30 mm across.

Type	Volcanic/ intrusive rocks	SiO_2 %	Melting temp. °C	Minerals – main components in %						
				Olivine	Pyroxene	Garnet	Other dark minerals	Plagioclase	Alkali feldspar	Quartz
Acid	Rhyolite/ Granite	70	700–900				10	35	20	35
Acid	Andesite/ Quartz diorite	57	900–1100		30			50	10	10
Basic	Tholeiite/ Quartz gabbro	52	950–1150		48			50		2
Basic	Basalt/ Gabbro	50	1000–1200		50			50		
Basic	Olivine basalt/ Olivine gabbro	48	1150–1250	2	48			50		
Ultrabasic	Pyroxenite	44	1200–1500	20	80					
Ultrabasic	Peridotite	40	1300–1600	70	30					
Ultrabasic	Mantle, parent rock	40	1300–1600	58	29	13				

Schematic representation of common volcanic and plutonic rock types that occur in the Palaeogene volcanic regions in Greenland. All proportions are approximate and there are large variations. The melt intervals vary, as they are dependent on composition, content of volatile components (liquids and gases) and the pressure.

PALAEOGENE VOLCANISM

Flat-lying lava flows that were poured out into a deep lake, producing sedimentary-like volcanic deposits with inclined bedding similar to a delta front

Plateau basalts in the Nuussuaq Basin

Between Disko and Svartenhuk Halvø in Central West Greenland, Cretaceous sedimentary rocks of the Nuussuaq Basin are overlain by a 4–10 km thick sequence of plateau basalts. The smallest thickness occurs in the southern part of the area on Disko, while the greatest thickness is found in the middle, on Ubekendt Ejland. The sequence has been mapped on the land areas, but the lavas also extend onto the shelf where they cover a large part of the sea floor. The known eruption centres were mainly in the middle of the basin in a zone between Disko and Svartenhuk Halvø. From here the lavas spread out to all sides, both towards the boundary of the basin with the basement in the east and out onto the sea floor to the west.

The main part of the volcanic sequence on land was formed in the Paleocene, during a period of around 1 Ma 60–61 Ma ago. A smaller, 2 km thick sequence of basalts at the top of the volcanic pile comes from the Eocene and has an age of 55 Ma.

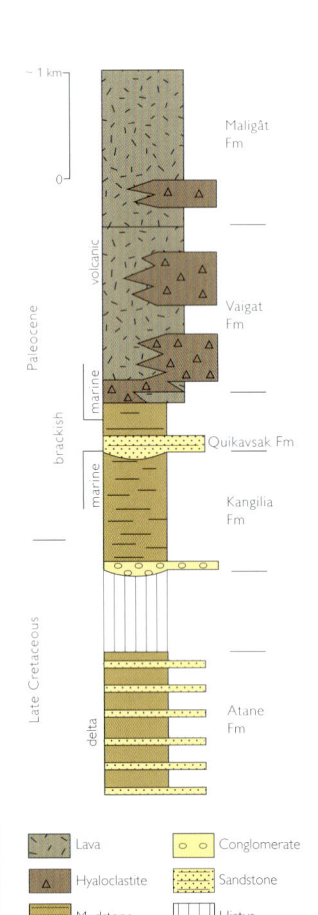

Composite log measured through the succession in the central part of the Nuussuaq Basin showing the sedimentary rocks and overlying plateau basalts. Only the part of the stratigraphic column measured on land is shown. During basin formation depositionary environments changed several times at any one place, so that the diagram is a schematic and simplified representation of the conditions.

Volcanism began with submarine activity, where the first eruptions formed lava flows with pillow lava structures (see p. 46) together with units of brecciated, hydrated basaltic rocks (hyaloclastites). As the layered complex accumulated, the eruption centres rose above sea level and the rest of the sequence was erupted on land.

During the course of the volcanic events much of the lava sequence was extruded eastwards from the eruption centres towards the continent, and the uppermost lava flows conceal the entire land area to the east.

Basaltic dyke (4 m wide) that stands out on the mountainside due to erosion of the surrounding softer sediments. The dyke represents the remains of a feeder channel from an underlying magma chamber through the sediments to the overlying plateau basalts.

The lower lava sequence (Vaigat Formation)

The lower, 1–4 km thick part of the lava sequence is dominated by magnesium-rich, olivine-bearing basalts that were separated from mantle material at high temperatures. The magmas moved rapidly up to the surface and on the way they sometimes incorporated material from the rocks that they passed through. These early magmas represent a more primitive, original lava type than the subsequent, more normal, plateau basalts. The lower, olivine-rich sequence was formed in three eruption cycles that comprise slightly different magma types. In each cycle there is a progressive modification of the magma from an early olivine-rich type to a later type with less olivine. Estimates suggest that each cycle probably lasted less than 70 000 years and that all three cycles occurred within a period of about half a million years. It was mainly this eruption sequence that formed the island Ubekendt Ejland, although lavas of this period are also present on Disko, Nuussuaq and Svartenhuk Halvø.

The lower lavas formed both submarine hyaloclastites (see box p. 145) and terrestrial flows. On land they have 'pahoehoe' surfaces, and form tabular sheets with a smooth, gently undulating or wrinkly surface that shows the lavas were formed from low viscosity magma solidifying under quiet conditions. As the volcanic pile accumulated there was a slow subsidence of the crust. This meant that the lavas were erupted onto a surface that remained close to sea level. Subsidence and build-up of the sequence were thus evidently closely related.

The upper lava sequence (Maligât Formation)

The overlying and most widespread of the lava sequences has a thickness of 2–6 km, and comprises normal plateau basalts with a composition termed tholeiitic. In contrast to the underlying types, tholeiitic basalts contain only a little olivine and often have phenocrysts (large crystals) of plagioclase feldspar. The surfaces of the individual flows are covered with

Plateau basalts of the upper lava sequence on the southern part of Disko. The section shown is the 900 m high Skarvefjeld, east of Qeqertarsuaq (Godhavn). The lowest submarine lava flows show several different structures formed in water, e.g. pillow lava breccias. The terrestrial flows were poured out onto a low-lying land surface, and due to their exposure to weathering they have a weakly reddish colour at the top and a crumbly surface. The contact between the submarine and terrestrial lava flows is concealed by the base of the modern talus cones in the middle of the photograph.

loose, angular blocks of different shapes and sizes – a surface structure called 'blocky lava'. This form shows that the magma was rich in gasses that, during eruption, migrated upwards to the top of the flow. The gasses escaped explosively as the magma solidified, and this, combined with movement in the still liquid magma underneath, broke up the surface into rubble that is full of bubbles and holes giving a coke-like appearance to the blocks.

The upper tholeiitic part of the plateau basalts also accumulated quickly – probably in less than half a million years. This sequence covered the whole of the underlying sequence and gradually spread out eastwards, finally covering the whole area of the Nuussuaq sedimentary basin; some flows extended over the basement, notably in the area east of Svartenhuk Halvø.

Native iron in the basalts

In the lava sequences on Disko a series of volcanic units occurs with a markedly different chemical composition, reflected in their rusty brown or yellow colours. Seen from a distance they can often be

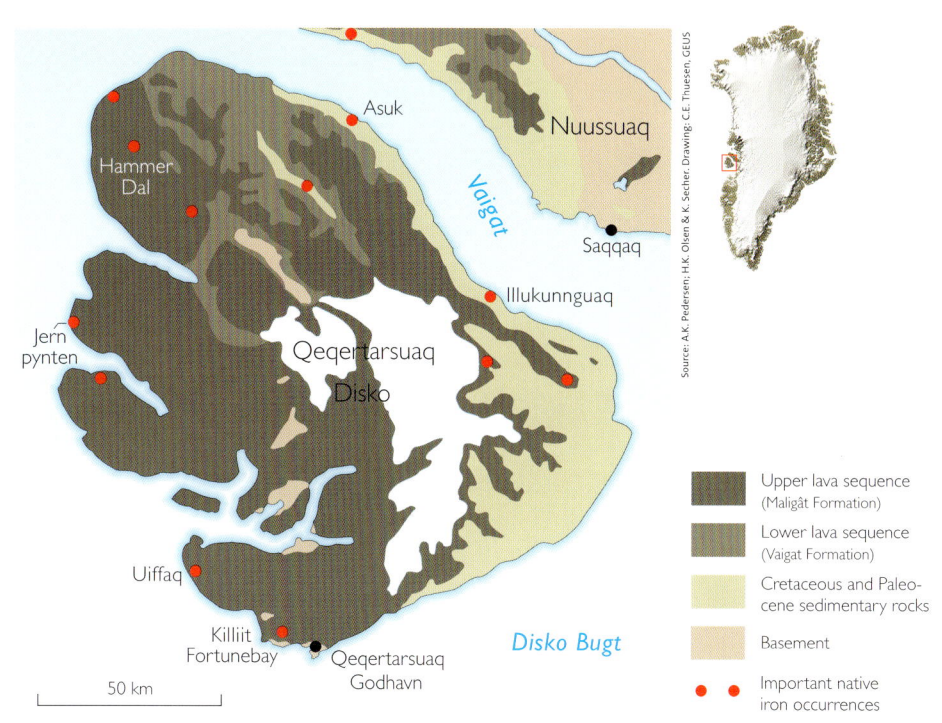

Sketch map of the Disko area in West Greenland, depicting the most important localities at which iron occurs as a native metal.

PALAEOGENE VOLCANISM

distinguished from the other types by their special colour, caused by iron compounds in the volcanic rocks which oxidise and form rust. Their yellowish colour is due to inclusions of sandstones and mudstones. These units with the different chemical composition were formed by natural contamination of the primary magmas reacting with material derived from the country rocks, during their passage from the magma chamber to the surface. The sedimentary successions found underneath the basalts are Cretaceous and Lower Palaeogene sandstones and mudstones, some of which contain coal. When the magmas incorporate and melt such material their chemical composition becomes modified, and when the carbon reacts with the basaltic magma, iron oxides in the magma are reduced and precipitated as native iron, similar to processes that take place in a blast furnace.

The rocks that contain native iron make up only a very minor part of the whole sequence in the entire volcanic province, but nevertheless the occurrences throughout the area of metallic iron have attracted considerable attention; large blocks of up to 5 tonnes in weight are represented in the collections of several geological museums. When first found in the early 1800s, they were interpreted as iron meteorites, but about 1875 it was realised that the iron was of a terrestrial origin and subsequently has been described as telluric (earthly) iron.

The Naajaat Lake

The formation of volcanic centres in the western part of the Nuussuaq Basin led to a build-up of a land area about 100 km west of the high-lying Precambrian basement. Between this new land area and the Greenland continent, a long N–S-trending bay was formed that, during the following volcanic activity, was gradually filled up with volcanic products. It was

Mountainside exposing a 1000 m thick section through the plateau lavas in the northern part of Disko. Uppermost is the brown-coloured, flat-lying sequence of the upper lavas. Below this comes a series of thin, flat-lying, yellowish lava flows of the lower lavas. The contrast between these two sequences and an underlying sequence of greyish volcanic rocks with inclined layering is seen in the middle of the photograph. This greyish sequence comprises hyaloclastites formed from lava flows that ran from the land into an arm of the sea. On contact with the water the hot lava solidified and broke into small pieces that rolled down a slope into deeper water. The sloping layers dip from left to right in the photograph. At the bottom of the section there are again flat-lying lava flows that belong to the lower basalts.

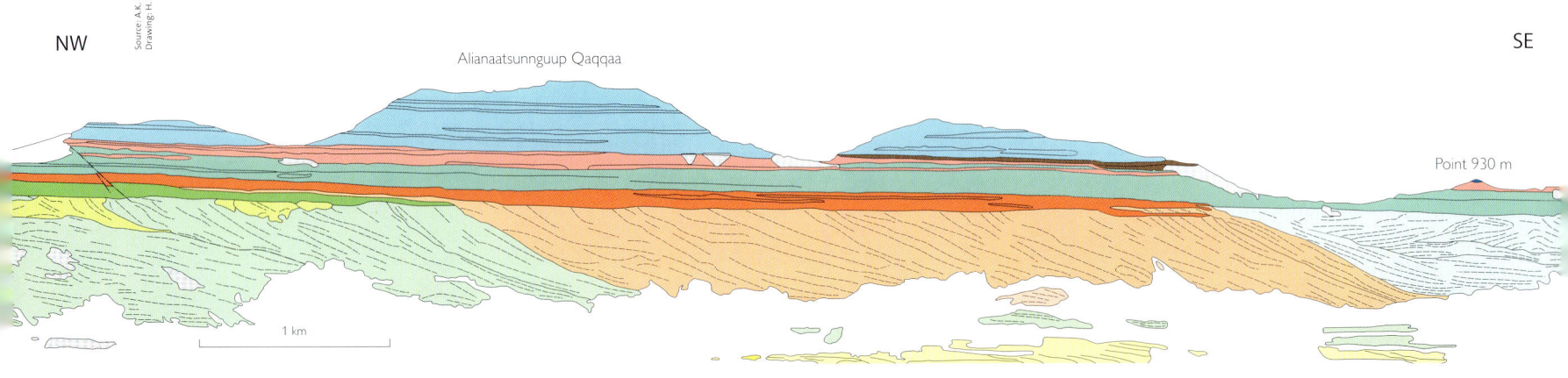

Part of a 75 km long profile along the south-west coast of Nuussuaq showing the architecture of the volcanic sequence in the central part of the Nuussuaq Basin. The base of the sequence here comprises hyaloclastite units (see p. 145) that formed when magma flowed from the land into water and the fragments rolled down the sloping side of the basin to depths of up to 700 m. The upper, flat-lying sequence of volcanic layers was laid down after the basin had filled up, so that the younger lava flows ran out across a land surface and covered the earlier lake-deposited lavas. The colours show different parts (members) of the sequence, with the pale colour tones representing the inclined hyaloclastites, and the darker tones the flat-lying lavas.

Plateau basalts in the Nuussuaq Basin

the lavas from the lower sequence that formed a barrier in the bay that gradually narrowed as the lavas continued to spread eastwards. The northern outlet from the bay was blocked early on, so that the bay was converted into a freshwater lake termed 'Naajaat Lake'. Subsequently the lake was infilled in stages by material from the continuing volcanism and with sedimentary material derived from the land area to the east. It is estimated that the lake was completely filled in less than half a million years.

The deposits in the lake comprise volcanic material and different types of normal freshwater sediments that amongst other things contain plant remains and different types of pollen and spores. The volcanic deposits comprise mainly hyaloclastites and were built out into the lake from the west, forming a giant delta with oblique bedding planes. The dipping layers (foresets) can be up to 450 m high and provide a minimum figure for the water depth in the lake. The lake probably existed no longer than 1/2 million years. It developed in five stages with first accumulation of hyaloclastite breccias and a rise in water level of the lake. The last stage reflects a stop in the volcanic activity and deposition of stable clastic lake sediments. The lake deposits reach a thickness of up to approximately 300 metres.

At its largest the lake is estimated to have had an area of around 2 500 km². Studies of the form and extent of the lake have shown that it originally covered the inner part of Nuussuaq and the north-western corner of Disko. As it gradually filled with volcanic and sedimentary deposits from the west, the lake was reduced in size and restricted to the northern part of the Disko Bugt area.

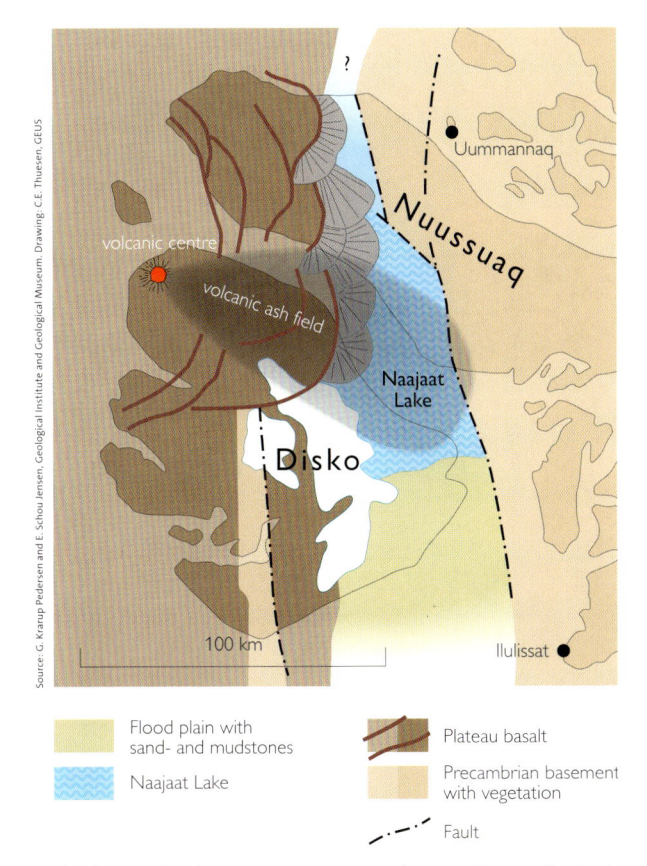

Sketch map showing the Nuussuaq Basin about 60 Ma ago. The basin formed west of the margin of the basement with a delta area and freshwater deposits in the south, and marine deposits to the north-west (see p. 130). About 60 Ma ago, violent volcanic activity occurred in the sea area 100–200 km west of the coast, which created a series of volcanic islands that formed a barrier and left an arm of the sea between the coast and the islands. Continued volcanism subsequently closed the mouth of this arm creating a lake – Naajaat Lake.

HYALOCLASTITES

Some volcanic rocks are built up in the same way as sedimentary rocks, in that they comprise water-lain beds of fragments and grains of eruption products. One type of volcanic rock formed in this way is called hyaloclastite. The name is derived from the Greek words hyalinos, glass and klastos, broken.

Hyaloclastites form under water or, where a lava flow comes into contact with the underside of an ice sheet. Due to rapid cooling (quenching) and solidification of the magma, it does not crystallise but instead forms a solid mass of glass that immediately shatters into small fragments and forms a granular mixture of rock components. Hyaloclastites are poorly sorted breccias comprised of small, angular single grains. Within a single bed, these rocks are usually not stratified and there is a complete mixture of all the components. The rocks are for the most part basaltic, but hyaloclastites may also form from lavas with other compositions.

In the Nuussuaq Basin a thick sequence of inclined hyaloclastites rocks occurs that formed as a result of volcanism about 60 Ma ago. Flood basalts erupted from fissures and poured out over the surface, spreading eastwards as thin lava flows over a flattish landscape, until they reached the shore of an up to 700 m deep bay that later became a lake (Naajaat Lake). When the incandescent magma flowed into the water it was instantaneously quenched so that it solidified as glass and shattered into small fragments. The hyaloclastite fragments accumulated as sediment-like deposits on the inclined side of the sea-arm or lake, typically with dips of 10–30°. The individual hyaloclastite layers are from 1 to 10 m thick. Through repeated eruptions the lake was gradually filled and the sequence of inclined hyaloclastites was eventually overlain by flat-lying plateau basalts that solidified on the surface of the land.

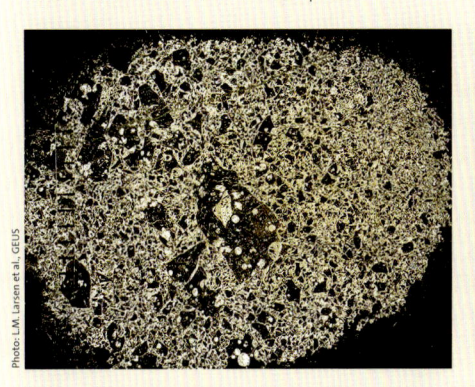

A microscope photograph of an hyaloclastite rock showing the individual, angular glass fragments (black) that have a highly variable grain size with a structure that resembles glass shards from a broken window. View is 20 mm across.

Close-up photograph of a typical hyaloclastite rock. In the left centre an entire pillow is preserved, but most pillows have been broken into pieces and make up the larger and smaller fragments in the rock. The pale brown groundmass comprises basaltic glass shards. The knife is 23 cm long.

The East Greenland volcanic province

Similar sequences of basalts, occurring on both sides of the North Atlantic, show that Greenland and the Faroes, formed a coherent landmass when the basalts were erupted

The Palaeogene volcanic rocks in East Greenland extend for more than 1000 km in the coastal region between Kangerlussuaq (68°N) and Shannon (75°N), covering an area of about 65 000 km². Kilometre thick successions of flat-lying plateau basalts dominate the volcanic rocks. The province is divided into a southern region between Kangerlussuaq and Scoresby Sund (69–70°N), and a northern region between Hold with Hope and Shannon (73–75°N). Between these two regions the volcanic rocks occur in the form of flat-lying intrusions (sills) and locally, as dykes intruded into the underlying Mesozoic sedimentary rocks. The volcanic activity concluded with emplacement of a series of plutonic intrusions.

In the inner parts of the Scoresby Sund area (70°N) the lowest units of the plateau basalts rest directly on gneisses from the Caledonian fold belt. After formation of the fold belt there was a period of about 350 Ma during which the mountain belt was eroded, resulting in a fairly flat and smooth surface. When volcanism began, this surface allowed the liquid basalt magma to flow unhindered over long distances; shallow valleys and low hills in the gneiss surface were buried by the earliest flows. At right, the photograph shows the old gneiss landscape with a valley filled in by basalt flows. The basalt succession shown is about 800 m thick.

The earliest basaltic lavas were erupted onto a gently undulating surface of either earlier sedimentary rocks or eroded crystalline basement rocks of the Caledonian fold belt. In some areas the gneissic basement rocks had an irregular, topographic surface with hills and valleys, and the first lava flows filled in the depressions leaving a relatively flat surface on which succeeding flows were laid down.

After the buildup of the thick basalt sequence, volcanic activity continued in East Greenland for a period of about 30 Ma with emplacement of more than 100 small and large intrusions. These bodies of magma forced their way up into the crust, and generally solidified as plutonic complexes before they reached the land surface.

The many different individual intrusions have varied compositions indicative of strong magmatic differentiation in the underlying magma chambers that fed the intrusions. This variation is in marked contrast to the rather uniform composition of the basalts. Most of the intrusions form compact, cylindrical-shaped bodies with steep, cross-cutting contacts against the surrounding country rocks, but some occur as steep, sheet-like, basaltic dykes or flat-lying sills. The intrusions occur in the coastal region from about 66° to 74°N, with an areal extent similar to that of the plateau basalts, but is displaced about 200 km towards the south.

The East Greenland volcanic province was formed during the break-up and separation of the North American and European continents and the opening of the North Atlantic Ocean. In East Greenland the volcanic sequence comprises a lower part formed immediately prior to the first formation of oceanic crust, and an upper part that was formed during the first spreading and break-up between the continents.

The later, mainly intrusive rocks in East Greenland were emplaced into the crust after the mid-oceanic spreading ridge had moved away from the continental margin. These intrusions did not have their source in any of the magma chambers beneath the spreading ridge.

Plateau basalts in the southern region

The thickest and most widespread plateau basalts occur in the region between Kangerlussuaq and Scoresby Sund (68–70°N), where today the basalts crop out over the whole of the coastal region and the nunatak areas along the margin of the Inland Ice. The basalt sequence extends from sea level to the top of the highest summit in Greenland, Gunnbjørn Fjeld (3693 m), north of Kangerlussuaq. The basalts dip gently towards the coast, so that the lowest parts of the succession occur inland, and the upper levels in the coastal areas. The maximum total thickness is more than 6 km, but the sequence is generally thinner to the west and north-west, and exposures wedge out over the basement composed of Caledonian rocks.

The basalt sequence is divided into five formations that have slightly different characters and chemical compositions. The lowest, and oldest, formation rests on either crystalline basement or on Paleocene sedimentary rocks. It comprises basaltic and olivine basaltic material, mainly formed by eruptions on land, but also contains units of hyaloclastite (see p. 145), ash layers and sedimentary rocks. This lowest formation is only exposed in the southern part of the region, whereas the remaining four formations can be followed over the whole of the region. These four formations comprise 5–50 m thick flows of tholeiitic basalts, with or without feldspar and olivine phenocrysts. Some lava flows can be traced over very long distances, and an individual flow in the Scoresby Sund area has been followed for 170 km, indicating that the lava had a very low viscosity and that the underlying landscape was both even and flat. The largest individual eruption has flows with an estimated volume of about 300 km^3; the total volume of the basalts in the whole of the southern region is about 160 000 km^3 (equivalent to a body measuring 220 × 145 × 5 km).

Scoresby Sund (see p. 149) marks the northern boundary of the outcrop of the plateau basalts. However, it is likely that the basalts originally extended

Dipping sequence of lavas along the outer coastal region of the Blosseville Kyst, south of Scoresby Sund. The basalts originally flowed out over a flat landscape as horizontal flows and were subsequently cut by a series of coast-parallel faults that divided the region into a number of fault blocks that were tilted so the lavas now dip eastwards towards the sea. The heights of the mountains are around 1000 m.

PALAEOGENE VOLCANISM

A block from the surface of a lava flow. The surface solidifies first, and movement of the still liquid lava beneath produces the wrinkles and ribs; it is known as 'ropy lava'. The reddish colour is due to weathering and oxidation.

Unweathered block of the inner part of a lava flow. The rock has a fine-grained, dark groundmass and contains small phenocrysts of feldspar and olivine.

over part of the Jameson Land area, but were later removed by erosion. Analysis of the metamorphic grade of some of the sediments at the present-day land surface indicates that they have been subjected to temperatures equivalent to an overburden of several kilometres. This suggests that the sediments probably have had a thick sequence of basalts on top of them.

Detailed age determinations have shown that the lava sequence in East Greenland comprises two phases of eruptions; the first phase between 60 and 56 Ma ago and the second phase between 55 and 54.5 Ma ago. Eruption activity was greatest in the second phase. In all, the basalts in the whole volcanic region covered an area of 220 000 km² and had a total volume of about 370 000 km³.

Correlation between basalts in South-East Greenland and the Faroes

If the two opposing sides of the North Atlantic are brought back together, removing the effect of ocean-floor spreading, East Greenland and north-western Europe including their shelf areas would be in direct contact. This means the Faroe Islands, with their shelf area to the north-west, would be situated to the south-east of Kangerlussuaq in an almost perfect fit. Geologically the Faroes comprise a basaltic sequence about 6 km thick that shows remarkable similarity to the southern part of the basalt sequence in East Greenland. Detailed chemical and lithological comparisons between the basalts in these two areas have shown that there is indeed a direct correlation of the development of the two areas.

Measurement of the succession of strata and analyses of the chemical compositions of the various minerals and rocks, together with radiometric dating, have shown that in both East Greenland and the Faroes the lower parts of the lava sequences can be divided into six nearly identical units. In Greenland that part of the succession has a total thickness of about 2700 m, while on the Faroes it is over 3000 m. The lower formation at both places comprises homogeneous basalts without any large crystals (phenocrysts) and beds of ash and tuff between flows at both localities have characteristic weathering horizons. On the Faroes, an approximately 10 m thick coal seam occurs above this formation that is interpreted to correspond to a bed of sedimentary rocks in Greenland. This lowest part of the lavas was erupted about 59–56 Ma ago, when Greenland and the north-western part of Europe were still together, but at a time when the crust between them was undergoing stretching and thinning. At this time the distance between the Faroes and South-East Greenland was roughly 100 km, but, 20 Ma earlier, at the end of the Cretaceous, they were even closer, about 50 km apart.

The succeeding five units in both East Greenland and the Faroes can also be correlated, and are composed of lavas that evolved from olivine-rich types with phenocrysts of olivine, through phenocryst-free basalts to types that contain feldspar (plagioclase) phenocrysts. The systematic change in chemical compositions shows a development from lavas that were erupted through a thicker part of the crust to magma types generated below a mid-oceanic spreading rift, where the lithosphere was only 6–8 km thick. These five formations were formed at the same time as the first development of the mid-Atlantic spreading ridge, commencing about 55 Ma ago when the distance between the Faroes and East Greenland had increased to about 120 km; today the distance between these two areas is about 1000 km across the North Atlantic.

Plateau basalts in the northern region

The volcanic sequence in the northern region of East Greenland cannot be correlated directly with the sequence in the southern region. In the north the measured thickness is only 800 m. Radiometric dating and measurement of the magnetic properties of the basalts show that these lavas were erupted about 59–57 Ma ago. The lowest lava flows were erupted under water, probably in lakes, but the higher levels of the volcanic pile comprise entirely lavas that were erupted on land. In the lowest levels, thin beds of sandstone and ash also occur.

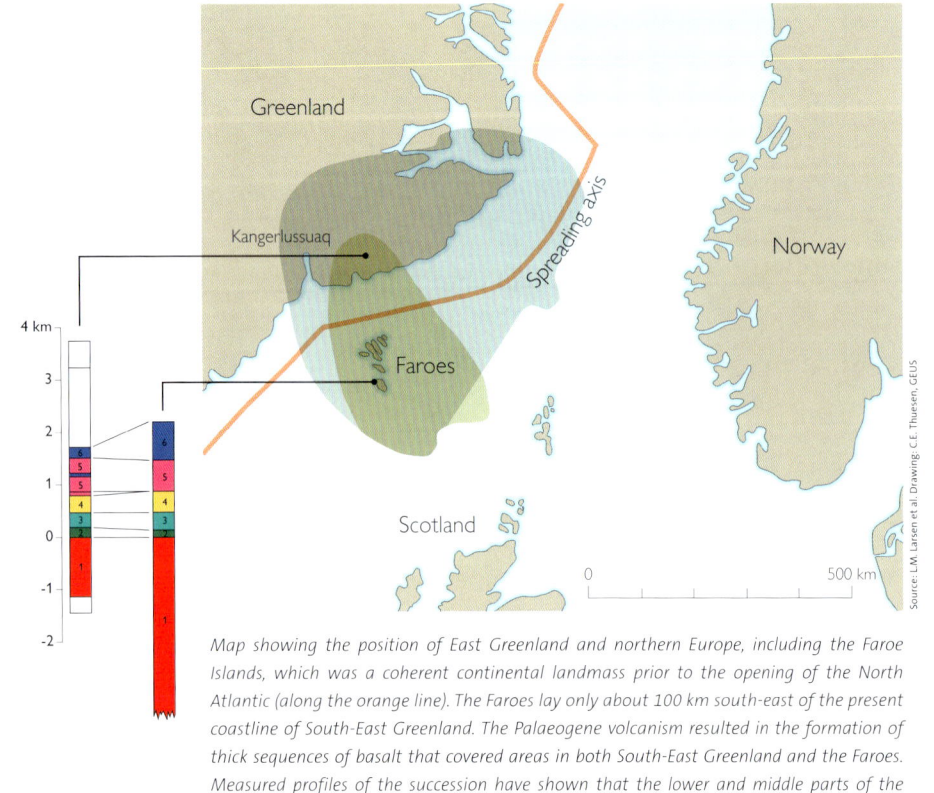

Map showing the position of East Greenland and northern Europe, including the Faroe Islands, which was a coherent continental landmass prior to the opening of the North Atlantic (along the orange line). The Faroes lay only about 100 km south-east of the present coastline of South-East Greenland. The Palaeogene volcanism resulted in the formation of thick sequences of basalt that covered areas in both South-East Greenland and the Faroes. Measured profiles of the succession have shown that the lower and middle parts of the basalts in both areas are nearly identical and that it is possible to correlate a series of units. Today they are separated by the Atlantic Ocean and lie more than 1000 km apart.

The lower succession comprises lava flows up to 50 m thick of homogeneous basalt without phenocrysts. At the base the flows are massive, but higher up the individual flows often have a weathered and partly eroded surface, which indicates a long time gap between each eruption. The upper about 350 m thick lava sequence comprises olivine-bearing rocks with frequent feldspar phenocrysts. On average the individual flows here are only about 4 m thick, and have a massive appearance. The massive basalts from this part of the succession are not easily weathered. Throughout the whole of the lower and upper series, the successive flows rest on top of each other without any marked discordance. This indicates that the lavas were very fluid and that they were erupted during a tectonically quiet period with no crustal movements. Subsequently, the sequence was disrupted by vertical faults that have moved individual blocks up and down relative to each other.

Palaeogene intrusions in East Greenland

The volcanic activity that produced the huge volume of plateau basalts in East Greenland moved gradually eastwards following the relative movement of the spreading rift under the mid-Atlantic ridge. However, magmatic activity continued along the East Greenland continental margin on a smaller scale for about 30 Ma. During this period about 100 intrusive centres and a series of dyke-like intrusions were emplaced over a distance of about 1000 km. While the earlier plateau basalts are generally uniform in composition, the intrusive centres are extremely varied. This is due to magmatic differentiation in the deep-seated magma chambers supplying the intrusions and to the primary magmas having melted and incorporated country rock material as they rose through the crust. Most of the magmas have crystallised high in the crust, forming intrusions at depths of just a few kilometres below the surface. Some may have had connections with local surface volcanism, while others have crystallised in closed magma chambers. The intrusions are divided into two main groups, one comprising pale, granitic and syenitic rocks while the other is dominated by dark gabbros and related rocks. In addition to these, the many dyke-like intrusions that occur are principally composed of rocks similar to the plateau basalts. The oldest intrusions occur in the southern region and have ages between 55 and 35 Ma. The intrusions in the northern region are all younger, emplaced between 48 and 28 Ma ago.

The Skaergaard intrusion

The most celebrated and best-investigated Palaeogene intrusion is the Skaergaard intrusion, a layered magmatic complex about 56 Ma old located on the north side of Kangerlussuaq fjord (68°N). The complex has

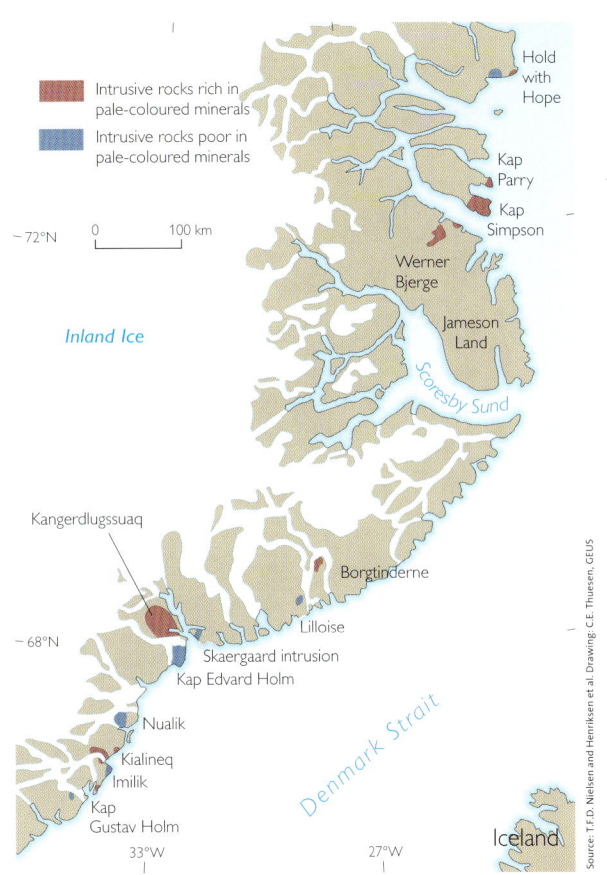

The intrusive centres in the Palaeogene East Greenland volcanic province. Only the 17 largest of the about 100 centres can be shown at this scale. The intrusions include two groups of rocks; one that comprises pale, feldspar-rich rocks such as granites and syenites, while the second comprises dark, more mafic rocks such as diorites and gabbros. The latter group is closely related to the plateau basalts, while the first represents different rocks that have formed through magmatic differentiation of basic parent material from a magma chamber deep in the crust (see p. 152).

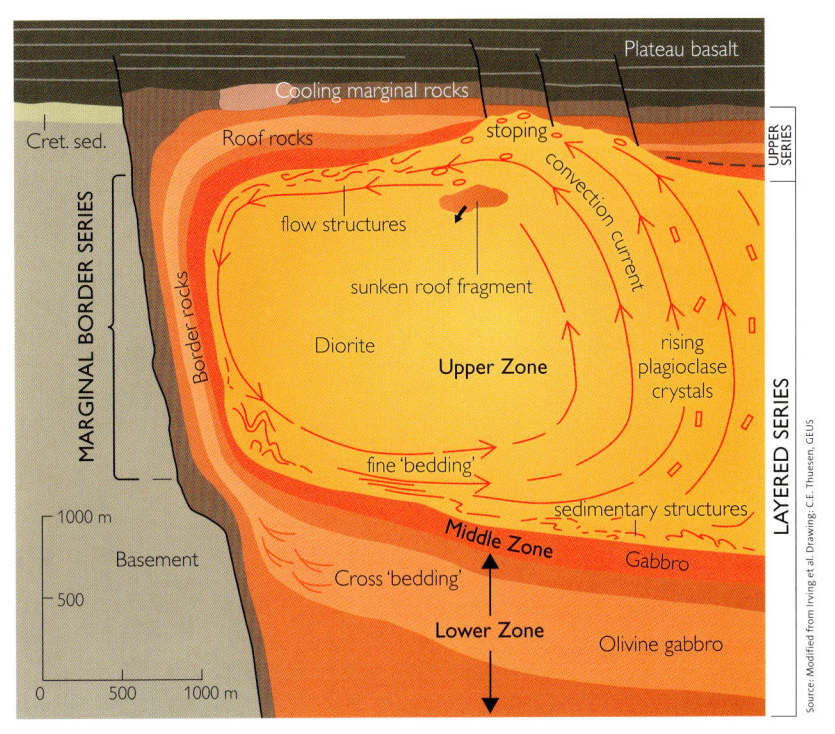

A vertical section model through the western part of the Skaergaard intrusion, showing how the magma chamber was built up at a late stage during its solidification. The model represents a time when the rocks at the top, sides and bottom had solidified, while the material in the central part, the Upper Zone, was still fluid magma. As is shown in the drawing, magmatic differentiation produced different rock types (see p. 152) related to cooling and solidification. In the still fluid part circulating convection currents carried the crystallised, less dense minerals upwards while the denser minerals sank to the bottom and were deposited. This process led to a layering (magmatic layering), similar to that produced during the deposition of sediments.

PALAEOGENE VOLCANISM

Magmatic layering in gabbros from the lower part of the Upper Zone of the Skaergaard intrusion magma chamber. The different bands in these magmatic rocks have variations in mineralogical and chemical composition and are similar in structure to the alternating bedding in sedimentary rocks. The darker parts have a higher content of dark minerals such as pyroxene and olivine, while the paler parts are more plagioclase-rich. The banding is caused by repeated, stepwise deposition of minerals gradually settling from the liquids in the circulating currents within the magma chamber.

been studied in detail since 1939 by many different geological expeditions, resulting in a very large number of international, scientific publications about the intrusion, which has become a classic geological locality. The ice-polished and unvegetated surfaces of the intrusion provide an ideal section through this magma chamber, some 10 km across and 3–4 km deep. It is possible to study the rocks and the processes that lie at the heart of understanding the special internal architecture of the original magma chamber. It is a so-called layered intrusion where the magma crystallised gradually over a long period. The different rock types separated sequentially, influenced by their different crystallisation temperatures, the convection currents and the precipitation of the denser components. The magma originated from the mantle and forced its way up into the crust to form a high-level magma chamber between the basement and the overlying plateau basalts at a depth of around 5 km beneath the then surface. Today, due to uplift of the crust and erosion of the overlying rocks, it is exposed at the surface. The primary magma had a basaltic composition that is equivalent to a plutonic gabbro. The main process in the formation of a layered body was magmatic differentiation (see p. 152). This differentiation occurred in the magma chamber by deposition of crystallised minerals (cumulus) that separated from the remaining, still fluid magma (intercumulus). At the same time, circulating currents formed in the magma chamber carried the lighter, less dense parts of the magma upwards while the denser parts sank down. These convection currents allowed crystallised minerals and other components to be deposited in the magma chamber, in much the same way as sediments are deposited by water currents. The layered magmatic deposits in the magma chamber thus resemble in many respects the layered structures in sedimentary rocks.

The detailed studies of the Skaergaard intrusion have shown that the magma chamber cooled from all sides at the same time. The magma crystallised from the top downwards and formed the 'Upper Border Series', at the same time as crystallising inwards from the side to form the 'Marginal Border Series'. The main part of the intrusion crystallised from the bottom upwards to form the 'Layered Series' that comprises four different zones (Hidden Zone, Lower Zone, Middle Zone and Upper Zone). These zones contain variations of gabbroic rocks ranging from mafic, olivine gabbro in the lowest parts of the Lower Zone to olivine-dioritic rocks in the upper part of the Upper Zone (diorite is a pale gabbroic rock with a greater SiO_2 content and fewer dark minerals than gabbro).

Dyke swarms

Another geologically well-known part of the East Greenland Palaeogene intrusions is a very dense swarm of basaltic dykes that cut through the basement, the plateau basalts and some of the early intrusions. The dyke swarms occur in a 10–20 km wide zone along the outer coast from 67°N, north of Tasiilaq (Ammassalik) to Scoresby Sund (70°N) – a distance of more than 500 km. The dykes follow the main coastline along a trend varying from NNE–SSW to roughly E–W near Kangerlussuaq. Inland there are few dykes, but the number increases towards the coast and at the outer coast where the swarm is densest there can be more than 50% dykes. Where there are few dykes, a single intrusion may be 4–8 m across, but their width decreases to 3–4 m where the swarm is densest. The individual dykes were not intruded at the same time, and frequently there was a sufficiently long time between them for a dyke to crystallise and solidify before the next fracture opened and magma was emplaced along it to form a new dyke. The intersecting system of dykes that cut through one another allows a detailed, relative chronology to be constructed.

The dykes were formed by the emplacement of basaltic magma into fractures that developed due to stretching (tension) of the crust in the early phase of continental break-up and opening of the North Atlantic that took place about 55 Ma ago. The dykes were formed a few million years prior to opening and until seven million years afterwards. Although the dykes were emplaced into vertical fractures, flexuring of the continental margin since their emplacement means they are now inclined up to 30–45° from the vertical at the coast. The flexuring occurred after the formation of the plateau basalts by subsidence of the inner part of the shelf area off the coast at the same time as the land was uplifted. This resulted in the original flat-lying strata of plateau basalts becoming tilted towards the coast with dips of up to 45–60°. The existing dykes followed the flexure, and their now tilted planes of intrusion are still more or less perpendicular to the inclined layers of the plateau basalts. Like the basalts, the underlying basement rocks in the south are also influenced by the flexure. After the flexuring, additional dyke emplacement occurred resulting in a new vertical dyke swarm that cuts through the earlier, now inclined swarm. The early dykes were formed prior to initiation of continental break-up, and the magma forming these came from magma chambers at the transition between the mantle and the crust. On the other hand, the younger dykes were derived from larger magma chambers higher in the crust and from these, dykes were emplaced both upwards and sideways.

Basaltic dykes also occur elsewhere in the East Greenland volcanic province, and cut through the plateau basalts and many of the smaller intrusive centres. In the sedimentary basin deposits that underlie the plateau basalts, the basaltic intrusives appear both as dykes and as flat-lying bodies (sills).

Many scientists interpret the dense dyke swarm in southern East Greenland to have formed in connection with the passage of the 'hot spot' that today comprises the mantle plume under Iceland. The East Greenland dyke swarm and plateau basalts extend from the coast out onto the adjacent shelf region, as far as 40–50 km away where the oceanic crust begins. Throughout this shelf zone the dyke intrusions occur with a high frequency, and the overall structure is very similar to that formed by dyke swarms that develop immediately above an oceanic spreading ridge (see p. 169).

Intrusive dykes of dark basaltic rocks cutting through pale-coloured gneissic basement. The dykes were intruded in several phases, and the younger dykes cut through the older. The intrusions were originally emplaced into a vertical fracture system that, due to later flexuring, is now tilted, so they now dip towards the sea (to the right in the photograph). The locality is on Langø, about 150 km south of Kangerlussuaq and the cliffs in the foreground are about 80 m high.

MAGMATIC DIFFERENTIATION

Rocks are aggregates of minerals that have different physical properties, e.g. melting temperature. When a rock melts forming a magma, the different minerals melt or dissolve at different temperatures and similarly, when the magma solidifies, the individual minerals crystallise at different temperatures and times depending upon their composition. Minerals crystallise in a certain order related to falling temperatures during cooling, according to their unique crystallisation temperatures. Silica (SiO_2)-poor minerals crystallise first and the water (H_2O)-bearing phases crystallise last. During crystallisation mineral grains form in the magma and between the crystals a fluid phase of the magma remains. On the basis of density differences between the mixtures of solid and liquid material, a series of processes occur in the magma chamber whereby the less dense material floats upwards and the denser material sinks. The chemical composition of the crystallising minerals is different from that of the magma from which they form. Therefore, as the minerals crystallise, the composition of the magma changes in a predictable way. Typically there is a development from basic (SiO_2-poor) to more acid (SiO_2-rich) magma compositions in step with falling temperature and increasing solidification. These processes are very complex and are called magmatic differentiation.

As crystallisation proceeds the composition of the remaining melt gradually changes, and normally there is no chemical equilibrium between the minerals and the remaining melt. If the separated minerals remain in the melt they may partially or completely dissolve back into the liquid and a new mineral forms in place of the first. Such earlier separated and related, later-formed minerals are called reaction pairs. Several reaction pairs after each other are called a discontinuous reaction series (e.g. olivine – pyroxene – amphibole – biotite).

The chemistry of some minerals varies within a so-called isomorphous (with the same crystal form) series. For example, the feldspar mineral plagioclase varies from a Ca-rich type to a Na-rich type and there is a complete mixture with a smooth transition (solid solution) between the two end members. When a magma solidifies the minerals that make up an isomorphous, solid solution series crystallise with changing compositions depending on temperatures and pressures. At the highest temperature Ca-rich plagioclase (anorthite-rich) separates first, and the plagioclase becomes progressively more Na-rich with falling temperature and ends with Na-rich (albite-rich) plagioclase providing that the chemical composition allows it. Separation of isomorphous minerals with varying chemical compositions contributes significantly to magmatic differentiation.

Many volcanic rocks are formed through partial melting of ultrabasic rocks in the mantle. For example, a Mg-rich, parental magma may separate an olivine basalt phase, which then rises and begins to crystallise. During this process the composition of the magma changes through magmatic differentiation. The first liquids to separate are different olivine-rich types, by which the magma comes closer to a more normal basaltic composition. Through magmatic differentiation in this basaltic magma, crystallised minerals will separate from the remaining liquid; the denser minerals will sink while the lighter minerals float upwards in the magma. This process is called fractional crystallisation. It controls the way the magma modifies its composition from the primary Mg-rich, olivine basalt melt via a magma with basaltic composition to more acid residual liquids with andesitic to rhyolitic/granitic compositions. Different rocks from the whole of the fractionation series can be represented in a single, large, layered intrusion as, for example, the Skaergaard intrusion.

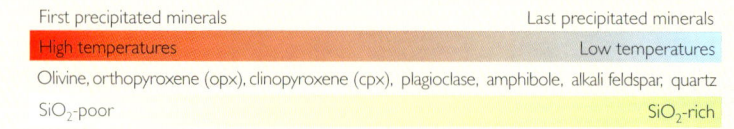

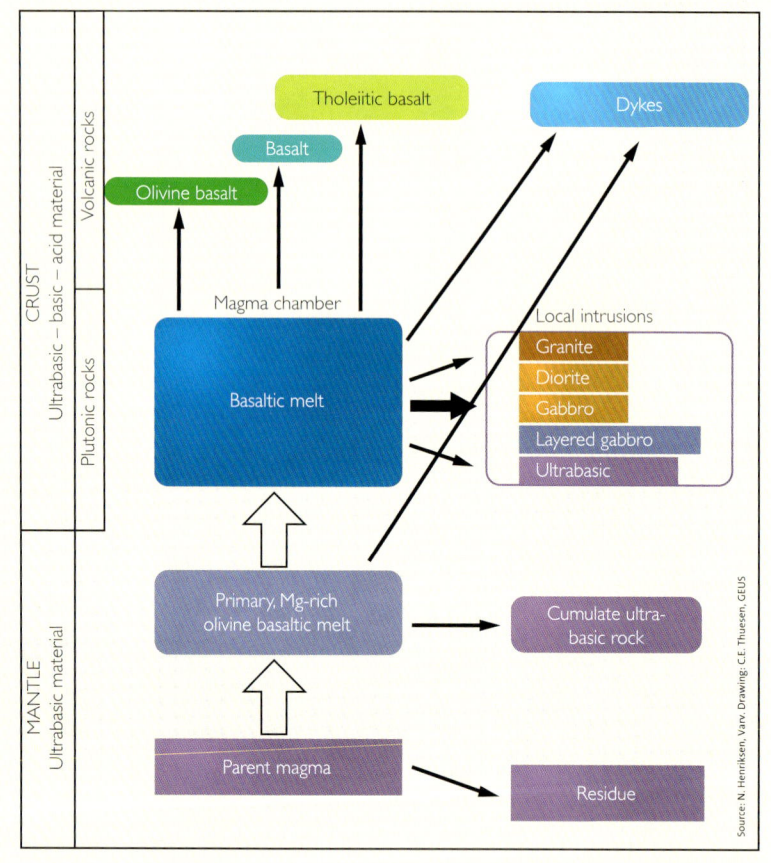

Schematic diagram that shows how rocks form and magmas develop within a magmatic province, such as the Palaeogene province in East Greenland.

Ultrabasic rock
peridotite

Basic rock
gabbro

Intermediate rock
diorite

Acid rock
granite

Cliff showing a 150 m high section through part of the basalt sequence in the Scoresby Sund area. Three of the lava flows comprise basalt with columnar jointing that is nearly vertical. Each flow represents a volcanic eruption. The columns were formed in the solidifying magma perpendicular to the top and bottom surfaces. As cooling progressed shrinkage caused the columns to contract and separate from each other.

An example of columnar basalt from one of the lava flows in the Palaeogene basalts of the Scoresby Sund area in East Greenland. The columnar structure formed during cooling of the solidifying magma and the development of regular, vertical fissures (joints) was due to contraction of the rock mass. The joints form a polygonal pattern (often hexagonal) and the columns are mostly perpendicular to the cooling surfaces. One theory is that columnar structures frequently form when basaltic lava has flowed onto a very shallow body of water such as a lake or a marsh area. Columnar structures may occur in only a few individual flows in a sequence, while the remainder may be entirely without columnar structures. Similar columnar structures can also be found locally in intrusions such as dykes and sills, where the columns form perpendicularly to the margins that are the cooling surfaces.

Survey ship 'Dana' carrying out a seismic survey in the Disko Bugt area. The ship is 78 m long, 15 m wide and displaces 2483 tonnes.

Photo: C. Marcussen, GEUS

GEOLOGY OFFSHORE

Exploration using geophysical methods and drilling

10

GEOLOGY UNDER THE SEA

10 GEOLOGY OFFSHORE

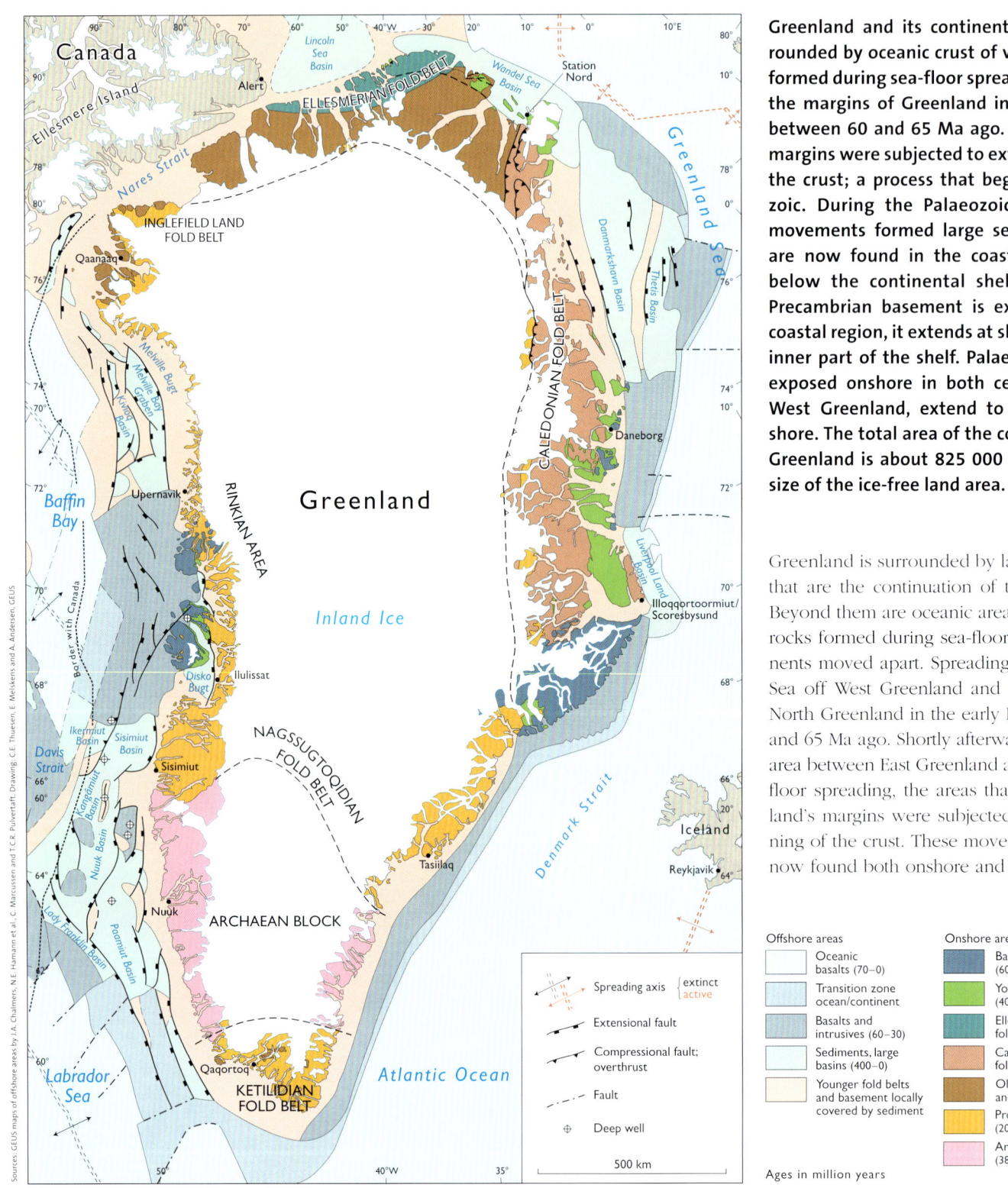

Greenland and its continental shelf is almost surrounded by oceanic crust of volcanic origin that was formed during sea-floor spreading that started along the margins of Greenland in the early Palaeogene, between 60 and 65 Ma ago. Prior to spreading, the margins were subjected to extension and thinning of the crust; a process that began in the Late Palaeozoic. During the Palaeozoic and Mesozoic these movements formed large sedimentary basins that are now found in the coastal areas onshore and below the continental shelves. Where extensive Precambrian basement is exposed in the onshore coastal region, it extends at shallow depth below the inner part of the shelf. Palaeogene plateau basalts, exposed onshore in both central East and central West Greenland, extend to cover large areas offshore. The total area of the continental shelf around Greenland is about 825 000 km², about double the size of the ice-free land area.

Greenland is surrounded by large continental shelves that are the continuation of the land under the sea. Beyond them are oceanic areas composed of volcanic rocks formed during sea-floor spreading as the continents moved apart. Spreading started in the Labrador Sea off West Greenland and in the Arctic Ocean off North Greenland in the early Palaeogene, between 60 and 65 Ma ago. Shortly afterwards, it also began in the area between East Greenland and Europe. Prior to sea-floor spreading, the areas that now comprise Greenland's margins were subjected to extension and thinning of the crust. These movements formed rift basins now found both onshore and offshore (see map).

Simplified map showing the geology of the areas offshore Greenland. Closest to the land is the continental shelf whose width varies considerably in different parts of Greenland, from 25 km off South-East Greenland to more than 350 km off North-East Greenland. The continental shelf can be considered to be the continuation of the land under the sea and it consists of similar rocks to those found in the ice-free land areas. A slope leads from the continental shelf into deep water where the basement consists of oceanic crust, whose volcanic rocks formed during sea-floor spreading. They are covered by a thin veneer of deep-water sediment.

Because of its inaccessibility under the seabed, the study of geology offshore requires rather different techniques from the traditional studies of onshore outcrops. There is, of course, no difference in the rocks themselves, just in the methods used to study them. In general, the term 'marine geology' is used for the study of the unconsolidated sediments at and just beneath the seabed, whereas the term 'offshore geology' is used for the study of consolidated sediments and other rocks down to depths of many kilometres. This chapter is about 'offshore geology'.

The primary methods used to study the offshore areas are indirect, using geophysics, particularly seismic, magnetic and gravity studies. However, none of these methods on their own, or even in combination, lead to an unambiguous interpretation of the offshore geology and it is necessary to calibrate the geophysical methods with information from drill holes or samples from the seabed.

The most fundamental geological boundary around Greenland is that between the continental and oceanic crust. Between them is a transition zone that varies greatly in width, from as little as 10–15 km to well over 100 km. The continental shelves are really just extensions of the land areas and consist of the same wide spectrum of rocks, from Archaean basement with an age greater than 3000 Ma to present day sediments. The width of the continental shelf around Greenland varies from about 20 km to more than 350 km. Water depths above the continental shelves are mostly less than about 500–1000 m, but can be as great as 2500 m.

The oceanic crust consists of basaltic material, formed by volcanism during sea-floor spreading commonly buried below a layer of younger sediments. The oldest oceanic volcanic rocks around Greenland were erupted during the Paleocene, about 62 Ma ago, and the youngest volcanic rocks offshore Greenland are only 10 Ma old.

The geological and political limits of the seas around Greenland

The 'offshore areas' around Greenland are the seas and their subsurface up to several hundred kilometres from the coast. Most of the areas consist geologically of continental crust that is the natural extension of the land. Some of the oceanic areas are equivalent to exposures onshore particularly that off southern and northern East Greenland, where basalts of the oceanic crust are contiguous with the Palaeogene volcanic province exposed on land. The transition between oceanic and continental crust in this area is sharp and can be mapped quite precisely, whereas the transition zone off southern West Greenland is wide.

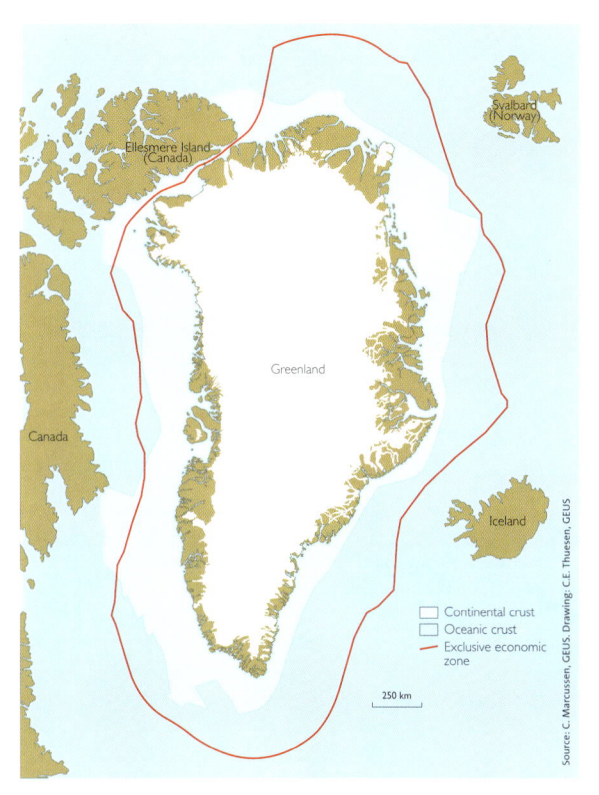

Greenland is surrounded by a continental shelf underlain by continental crust. Beyond the shelf is the deep ocean where the basement is volcanic oceanic crust with the exception of an area off southern West Greenland whose basement consists of subsided continental crust. According to international agreements, exclusive economic zones extend 200 nautical miles (370 km) from the coast or to an agreed boundary with a neighbouring state. The large sedimentary basins on the shelf where oil and gas could be found are the areas of most economic interest. The figure shows the boundary between the economically interesting continental shelf and the volcanic oceanic crust that has very limited economic potential. The limit of Greenland's exclusive economic zone is shown with a red line.

Even though there are natural geological boundaries in many offshore areas, the political boundaries of Greenland's economic territory are determined by international agreements with neighbouring states and by United Nations Convention on the Law of the Sea. The latter determines the limits of the so-called exclusive economic zone where the territory of a neighbouring state is more than 200 nautical miles (about 370 km) away from the outer limits of the coast. Where the coast of a neighbouring country lies less than 400 nautical miles from the coast, the boundary between the two exclusive economic zones is determined by bipartisan agreement, usually the median line between the two coasts. Such boundaries have been determined between Canada and West Greenland and between East Greenland and both Iceland and the Norwegian island of Jan Mayen. The exclusive economic zones beyond 200 nautical miles are still undetermined in the Arctic Ocean and off South Greenland, so joint Danish/Greenland surveys in recent years, some in partnership with Canada, have been concentrated in these areas.

Investigating the geology of the offshore areas

Our understanding of the offshore areas is based principally on interpreting geophysical, primarily seismic, data calibrated where possible by information from deep boreholes. Where there are no boreholes, the interpretations are guided by analogy with the geology of nearby land areas, but the resulting inter-

GEOLOGY OFFSHORE

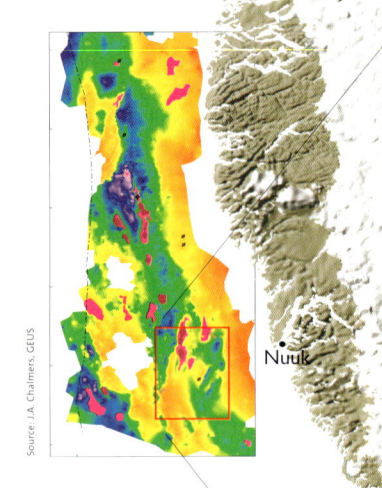

Map produced from interpretation of seismic data. This so-called isopach map, shows the thickness of sediments below the seabed in an area about 150 km from the coast south-west of Nuuk in West Greenland. Each contour shows where the sediments are equally thick and the thickness intervals are colour coded, with red showing sediments between 0 and 1500 m thick, yellow 1500 to 3000 m, green 3000 to 4000 m and blue 4000 to over 5000 m. The sediments vary in thickness in the area shown from nearly nothing to over 5000 m.

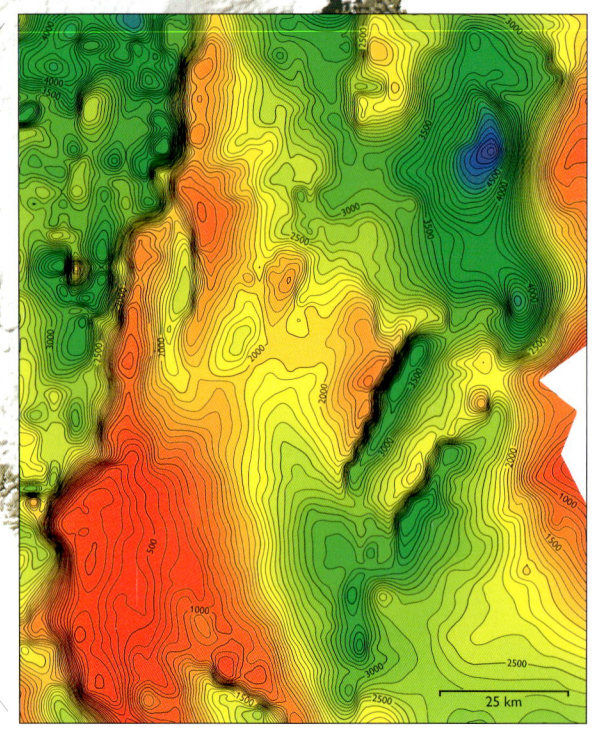

Example of a survey grid showing the lines along which geophysical data were acquired in an area off West Greenland. The ship or plane acquiring the geophysical data normally travels in a systematic pattern making the lines in one direction closer together than those in the other. The latter lines are then used to validate and correlate between the primary lines. To draw a map from such a grid requires interpolating the data or interpretation of the areas between the lines.

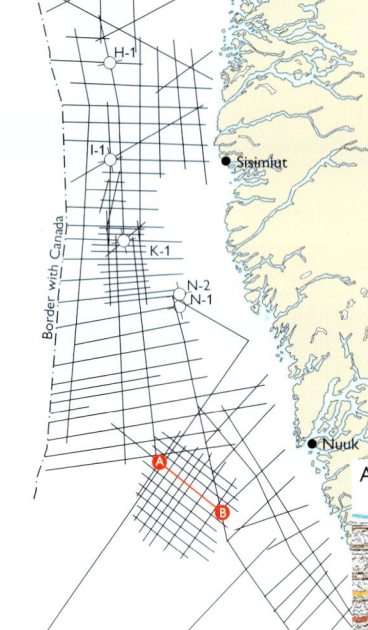

Seismic profile (location shown on the map, A–B) through sediments below the seabed. The top of the profile is at the sea surface and the first strong reflection shows the depths to the seabed. The upper sediments are broadly flat-lying and overlie faulted older sediments. Various horizons within the sediments have been interpreted as shown by the coloured lines. The steep black lines indicate faults where the sediments are offset.

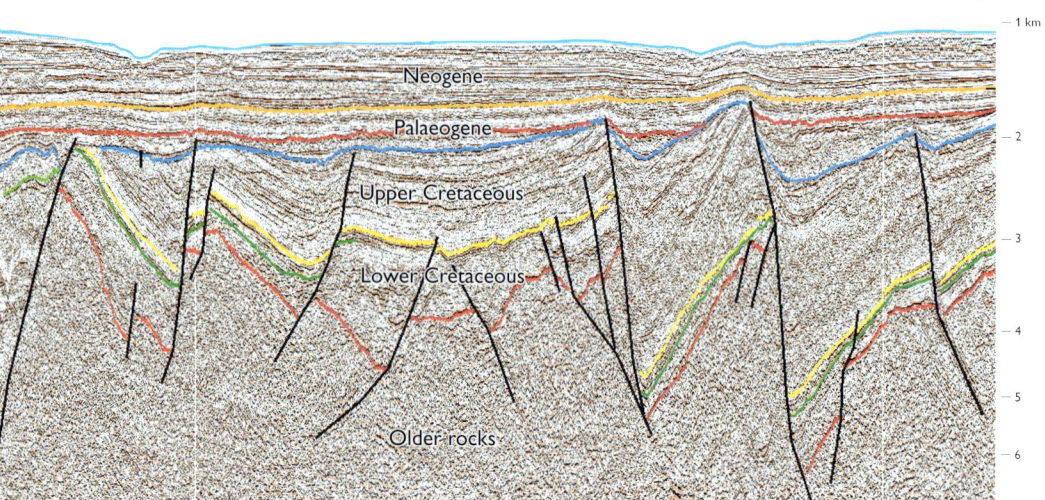

pretation is less reliable. The geophysical data can readily determine the presence of such fundamentally different rocks as basement, sediments and basalts, but the ages of the rocks cannot be determined from the seismic data alone. This is especially true of crystalline basement rocks where geophysical data cannot distinguish, for example, between Archaean and Palaeoproterozoic rocks.

Seismic methods are particularly suitable for mapping sedimentary rocks and are commonly used to derive a 3-D model of the extent, structure and thickness of a sedimentary basin. Maps showing depths to a particular stratigraphic 'horizon' and maps depicting the thickness of sedimentary units, so-called isopach maps, are commonly used to present such models. The top of folded and metamorphosed basement under the sedimentary series is an important horizon, commonly visible on the seismic data, but this surface may be difficult to trace where the sediments and the basement have similar seismic properties.

The number of sedimentary units of different ages that are interpreted on seismic sections varies considerably from project to project, depending on both the nature of the data and the aim of the study. A regional reconnaissance interpretation aimed at identifying large prospects for oil and gas exploration may focus on only very few horizons, for example, the seabed, the base of the Cenozoic succession and a marker horizon in the mid-Cretaceous sediments in the basins off southern West Greenland (this page and p. 237). Investigations aimed at understanding the structural history of an area may involve interpretation of 5–20 horizons. A more detailed study of the upper Paleocene – lower Eocene sediments of offshore southern West Greenland involved interpretation of 29 horizons in this very restricted stratigraphic interval.

Geophysical methods used to investigate the subsurface

The geophysical methods used to study the subsurface, both onshore and offshore, are all based on measuring various physical properties of rocks. The Earth's gravity and magnetic fields are regionally uniform, but small local variations caused by variations in the density and magnetic minerals in rocks can be used to map these properties. Seismic surveys use sound waves sent down into the subsurface that are either reflected or refracted back to the surface to detect the layering and thicknesses of rocks – a process not unlike a ship's echo sounder that measures the depth of water under it.

The various offshore geophysical measurements are normally carried out using specialised equipment carried either by ship, aircraft or sometimes satellite. After acquisition of the data they are normally subjected to a complicated processing by computer before being interpreted; in the case of seismic surveys, the amount of processing requires large, advanced and fast computers. The results of the gravity and magnetic surveying are normally presented as maps and those of seismic surveys are normally presented as profiles. They are the starting points for interpretation of the data to determine what geological conditions and structures have given rise to the geophysical parameters measured.

The geophysical data are normally acquired along lines travelled by the ship or aircraft. The separation between the lines depends on how detailed a survey is required. During reconnaissance phases, the lines may be between 10 and 100 km apart, but the separation may be reduced to as little as 1 km during investigations of oil and gas prospects. Nowadays, it is common to acquire so-called 3-D seismic data from ships towing several receiver cables simultaneously, so the resulting line separation may be as little as 25 m, although no data of this type has yet been acquired around Greenland.

Magnetic surveys

The Earth's magnetic field is like that of a bar magnet with a north and south pole. If the total magnetic field across an area is measured and the Earth's regional field is subtracted from the measurements, the remaining anomalies can then be used to infer the magnetic properties of the subsurface. Rocks with a large content of magnetic minerals, such as basalt, give large magnetic anomalies, whereas rocks with a lower content of magnetic minerals, like most sediments, produce little or no anomaly. The size of the anomaly also depends on how deep the magnetised rock is buried. From knowledge of the magnetic properties of rocks, it is possible to deduce what kind of rocks occur and at what depth they are found.

The strength of the magnetic anomalies is measured in a unit called a nanoTesla (nT). The strength of the total field measured at a locality depends on its position on the Earth and the time of measurement, because there are small daily variations. The Earth's total field can also be affected by large-magnitude, irregular changes, known as magnetic storms, whose strength can be 50–200 nT, which is of the same order as the anomalies. They are caused by the emission of charged particles from the sun. To compensate for the daily variations, the Earth's magnetic field is monitored by fixed magnetic stations and small variations are subtracted from the survey measurements. During large magnetic storms, magnetic surveying is abandoned. The date and year when the measurements were recorded must also be known, because the strength of the Earth's field and the positions of its poles change slowly with time. Once all these corrections are made, they can be subtracted from the measured field and the result shown on a magnetic anomaly map like the one on p. 45.

The anomaly map produced from all these measurements shows differences between the local total field and the background field and modelling of these differences is used to interpret both the type of rocks and the depth of the rocks causing them. The magnetic anomalies produced by rocks are generally less than about 1/1000th of the strength of the Earth's

Aircraft used for magnetic surveys in Greenland. The plane is a small single-engine Cessna Grand Caravan that flies with a speed of about 250 km/h while acquiring data. The tail boom protruding from the plane contains sensitive instruments that measure the magnetic field and the cabin of the plane is fitted out as a laboratory with recording and navigation equipment.

GEOLOGY OFFSHORE

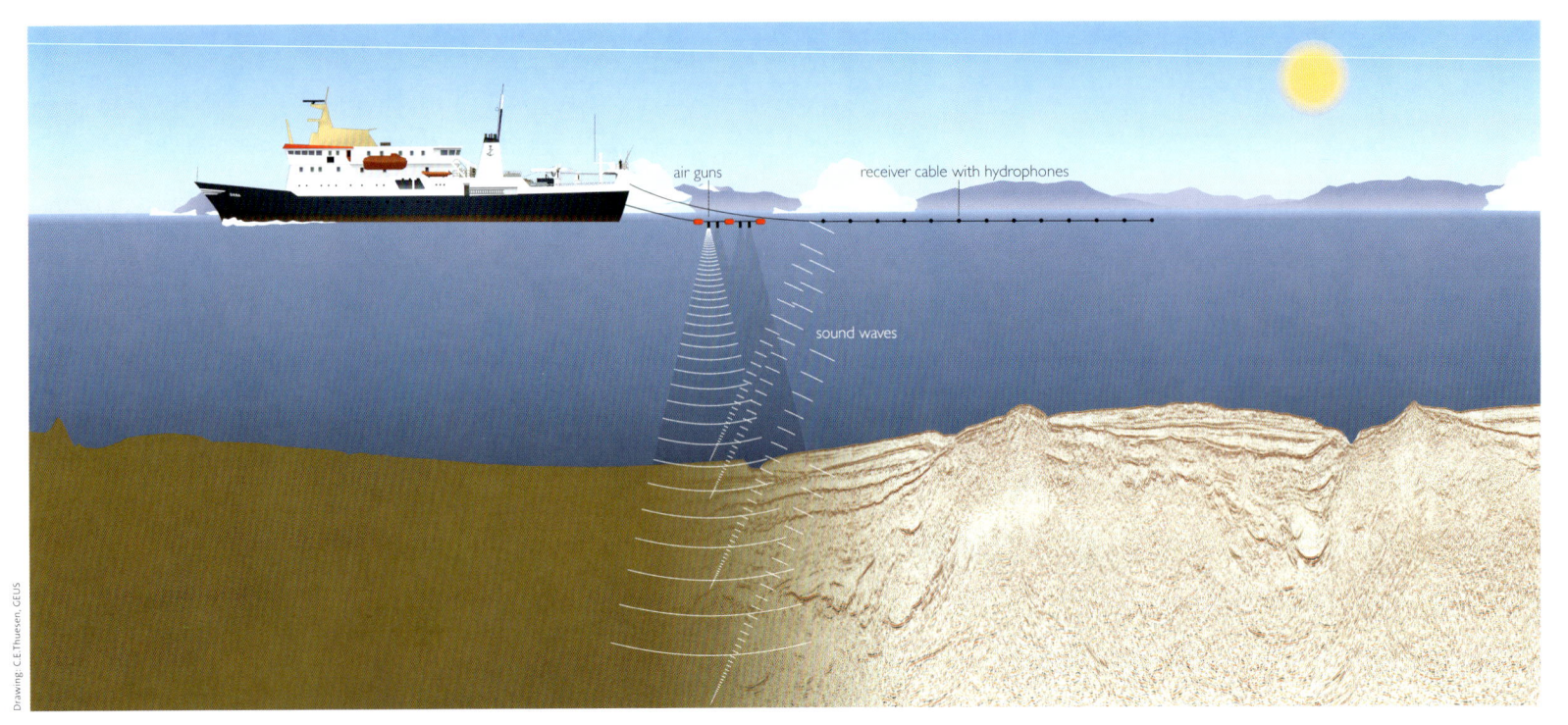

Seismic surveys at sea are carried out from a ship that emits sound waves at regular intervals from high pressure airguns that are pulled behind it. The sound waves travel down through the water and into the underlying layered sediments, from which they are reflected and received by many water-tight microphones (hydrophones) along the length of a 3–6 km-long cable that is also dragged behind the ship. The sound waves are received by groups of hydrophones typically 25 m long and spaced evenly at 25 m intervals along the cable. The signal is then sent to the recording instruments on the ship. The recorded data is normally sent ashore to be processed using powerful computers and the end result is presented as a cross-section through the upper part (or even the whole) of the Earth's crust (a seismic profile). One ship can acquire data continuously along a track several hundred kilometres long. The drawing shows a seismic profile inset under the seabed on the right to illustrate the type of data that results from these operations. The lower photograph shows red buoys that support airguns being pulled just behind the stern of a ship during a seismic survey.

total field and to provide detail within them, measurements and corrections of the total magnetic field have to be done to an accuracy of about 1% of their size. So the measurements and corrections of the total field that need to be made have to be very accurate indeed, to one part in 100 000.

Seismic surveys

Seismic waves are sound waves that travel through the Earth. They do so with speeds that depend on the material through which they travel, so measurements of the speed of the waves indicate the distribution of the type of rocks through which they have passed. Seismic waves are generated naturally by earthquakes but can also be generated artificially by explosions or by pressure waves from special generators. The most common artificial generators at sea are either airguns, equipment that uses bubbles of high-pressure air to generate waves, or slow-burning explosives. Onshore, the most common sources of seismic waves are explosives or mechanical vibrators.

Seismic waves spread into the subsurface where they encounter boundaries between the rocks within which they travel at different speeds. They are reflected at some boundaries and refracted at others, depending on the angle at which they strike the boundary. Recording reflected and refracted waves requires different techniques and results in different types of information. Using reflection techniques, the seismic waves are sent downwards almost vertically, reflected from the boundaries between different layers of rock and the returning waves are detected by receivers within a few kilometres of the source, similar to a ship's echo sounder. The receivers are sensitive devices like microphones, called geophones if they are on land and hydrophones if they are in water. Using refraction seismic techniques, the sound waves are sent down at a shallower angle and are refracted at the various boundaries. The returning sound waves can be detected by receivers, such as seismometers, often a long distance from the source – up to several hundreds of kilometres in some surveys.

In both types of survey, the time taken for the wave to travel down and return to the surface is measured. In reflection surveys, this time is a good indication of the depth to the reflecting surface, although the actual depth has to be calculated from knowledge of the speed of sound in the rocks. An indication of the speed of sound in rocks can be obtained from reflected waves, but is measured more accurately by refraction surveys, where the important measurement is the difference in the time of arrival at the different receivers.

The data received by the arrays have to be processed using high-speed computers to produce either 2-D or, in the case of reflection surveys, 3-D displays that can be interpreted geologically. Refraction surveys are particularly good at determining the types of rock in the subsurface, whereas reflection surveys are better at detecting structures and delineating details within sediments.

Gravity measurements

Everything on Earth is attracted by its gravity, by an amount that varies from place to place. The largest variation is related to latitude, because the Earth is slightly flattened at the poles, so the distance from the surface to the centre there is less than at the equator. The Earth's spin also causes a centrifugal force that points away from its axis, thus acting in the opposite direction to gravity. The centrifugal effect is greatest at the equator and zero at the poles. The combination of these effects means that the acceleration due to gravity is 9.83 m/s^2 (9830 gals) at the poles and 9.76 m/s^2 (9760 gals) at the equator, a difference of about 0.75%.

In addition, there are much smaller variations that are caused by different distributions of mass in the Earth's crust and upper mantle, because rocks with a low density attract mass less than rocks with a higher density. Very sensitive instruments can measure these differences from the 'normal field' and the variation of anomalous gravity from place to place can be used to estimate what rocks of different density might be found underground. The slight changes in the force of gravity from place to place are known as 'anomalies' and are measured in a unit called a milligal. One milligal (mgal) is about one millionth of the total acceleration due to gravity on Earth. Careful measurements onshore can achieve accuracies of 0.1 mgal, but the practical limit of accuracy offshore is about 1 mgal. Large anomalies can be as great as 200–300 mgal and sedimentary basins give rise to negative anomalies of tens of milligals. A good example lies under Melville Bugt off northern West Greenland where a 12 km thick sedimentary section gives rise to a negative gravity anomaly of about 50 mgal.

Deep drilling

The only direct method to sample subsurface rocks is to drill into them. Such drilling offshore requires special equipment and drilling projects can cost well

The drill-ship 'West Navion' that drilled off West Greenland in 2000. The well, named Qulleq-1, was drilled about 150 km south-west of Nuuk where the water depth was 1150 m. The well reached a depth of 2973 m under the drilling platform. The tall tower in the middle of the ship is the drilling tower from where the drilling is carried out and where the many drill pipes are stacked. The ship is kept on position while drilling by a series of propellers that can move the ship in all directions.

Schematic illustration of a drilling tower. In principle the drilling technique used on a floating drilling platform or a ship is the same as used onshore. A motor and gear box at the base of the tower rotate the drilling pipe. Lengths of drill pipe are stacked at the side of the tower to be lifted one at a time by a crane at the top of the tower and screwed onto the end of the pipe already below ground. A tank at the side of the tower is used to store drilling mud that is pumped down into the well.

GEOLOGY OFFSHORE

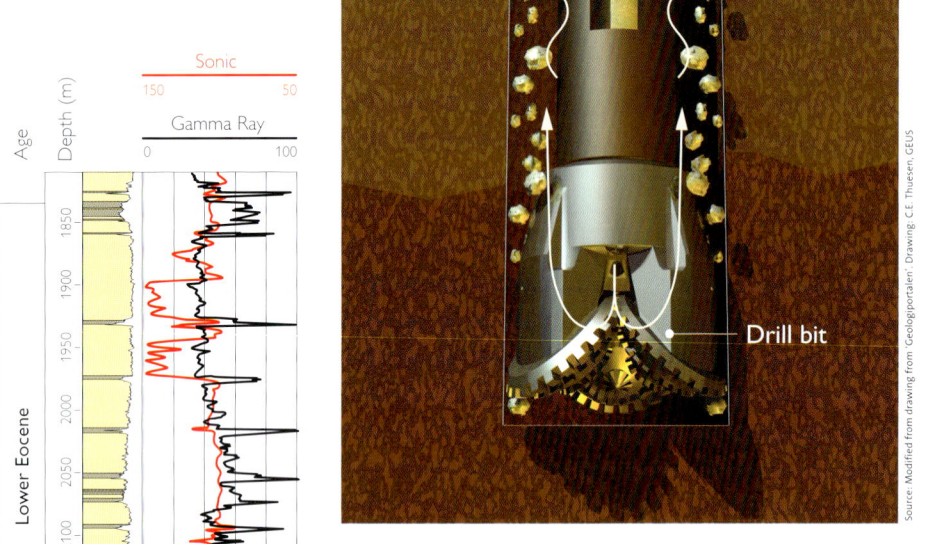

Pieces of rock (cuttings) from the bottom of a deep well. The cuttings are chipped off the rock by the drill bit and carried to the surface in the drilling mud to be studied by geologists and analysed both chemically and for their content of microfossils.

Schematic drawing showing a drill bit at the bottom of a well. The drill pipe rotates and the bit chips pieces of rock (cuttings) from the bottom of the hole. Drilling mud flowing down through the centre of the pipe washes the cuttings back to the surface in the annulus around the drill pipe.

Data from a section of Lower Palaeogene sediments and volcanic rocks drilled at the Nukik-2 well about 80 km west of Maniitsoq (about 65°40´N). The section shown is between depths of 1850 and 2550 m and comprises upper Paleocene and lower Eocene sandstones and mudstones with lower Paleocene volcanic rocks at the base. The coloured column shows the type of rock, and the geophysical data measured from a sonde sent down into the hole are shown on the right. The speed of sound in the rocks (the sonic velocity) is depicted by the red curve and the black curve shows natural gamma radiation.

over a million dollars for every well. Such drilling is normally undertaken only by the oil industry in search of hydrocarbons or by large international scientific projects such as the Ocean Drilling Program and its predecessors and successors. In consequence, there are only a small number of deep wells around Greenland, some of which have been drilled by the oil industry and others for scientific research. The wells are drilled not just to discover what rocks and fluids are present, but also to measure their physical properties to calibrate the much cheaper and more extensive geophysical surveys so that their results can be interpreted geologically. The wells therefore give information not just about the rocks in their immediate vicinity but, when combined with geophysical, especially seismic, data, their results can be extended over wide areas.

Wells offshore are drilled either from specially constructed ships or drilling platforms. Platforms can stand on shallow seabed or be floating. A floating drilling rig or ship is kept in position either by an anchor system or by a dynamic positioning system that uses either satellites or sonar emitters on the seabed to measure its position and special propellers to keep it at the desired location. Only floating drill platforms and drillships have been used in Greenland waters because they have to be moved at short notice from the location if threatened by collision with an iceberg.

Oil exploration wells are carried out in sedimentary basins where they have to be drilled to depths of 3000–4000 m below the seabed; present-day equipment can drill such a well in water depths up to 2000 m. Scientific wells can be drilled in much deeper water, up to 3–4 km, but only in places where there is no possibility of encountering hydrocarbons. With this different technique, holes can be drilled to less than a kilometre below the seabed.

A drilling rig consists of a high tower into which the drill pipes can be hoisted vertically. The tower is positioned over the locality to be drilled and winches lift and lower the drill bit and pipes. The pipes pass through a bushing at the base of the tower that can grip them and be rotated, so rotating the entire drill stem. The drill pipe consists of steel sections that can be screwed together and the pipe is extended as drilling proceeds to greater depths or shortened as it is pulled up. A drill bit is mounted at the bottom end of the drill pipe. It can consist either of a set of hard metal teeth mounted around a downward-pointing cone with which the rock is crushed or a cylindrical hollow tube with diamond teeth around its periphery, used to cut cores from the rock as it proceeds.

Cutting cores takes much more time than just crushing the rock and washing the crushed material (called 'cuttings') away, so the latter technique is normally used in oil exploration wells. The cuttings give valuable information about the rocks through which the well has been drilled. Cuttings are normally brought to the surface by pumping drilling fluid down the centre of the drill pipe. At the bottom of the hole, the drilling fluid flows through holes in the drill bit and is then diverted back to the surface in the annulus around the drill pipe, either between it and the sides of the hole or through any casing that has been set around the hole. The drilling fluid consists of a suspension of clays in either diesel oil or (normally) water (which it why it is normally called drilling mud) and its density is carefully calculated so that its weight offsets the pressure from any hydrocarbons that may be encountered, thereby preventing them from flowing to the surface and 'blowing out'. The flowing mud also cools the drill bit. Because the mud is usually fairly thick and viscous, powerful pumps are required to circulate it and in order to return it from the seabed to the drilling platform, a double pipe called a 'riser' has to be set between the platform and the seabed. The weight of the riser is supported by the rig's buoyancy, so this limits the length of riser that any drilling rig can carry and therefore in what depth of water it can drill. Scientific drilling often takes place without a riser, but then there is no guard against a blow-out, so great efforts are made to ensure that no hydrocarbons are encountered. Because cuttings cannot be returned to the surface, cores are normally cut continuously during scientific drilling.

The uppermost layers often consist of poorly-consolidated material that is drilled with a large-diameter bit and then lined with a steel liner, called casing, so the sides of the hole do not collapse. Such liners are inserted at intervals as drilling progresses, typically every kilometre. In addition to protecting the well from collapse, the liners also ensure that cuttings from deep levels are not contaminated with material from shallower levels in the hole. As each casing is set, the deeper well is drilled with a smaller diameter bit, so the diameter of the well decreases gradually downwards.

When coring, the drill bit has to be lifted to the surface regularly to recover the core. To do so, the entire drill-string is raised to the surface then re-lowered to the bottom of the hole. This is a time-consuming and expensive process when the bottom of the hole is at depths of 1–4 km, so cores are taken in hydrocarbon exploration wells only for special purposes.

Equipment called the 'well head' is placed on the seabed at the top of the hole that couples the riser together with the lined hole. The blow-out preventer, which is also placed here, allows the hole to be sealed rapidly, thus avoiding extensive pollution if hydrocarbons should flow uncontrolled to the surface – a blow-out. The well head also makes it possible to re-enter the hole should it be necessary to leave the drilling location for a time, for example if there is danger of collision with an iceberg. The well head used for scientific drilling is less sophisticated and consists of a large funnel plugged into the hole into which the bit is lowered after the drill-string has been taken to the surface to recover a core.

Geophysical measurements within boreholes

Once a well is drilled, a wide variety of different instruments is lowered into it to measure various properties of the rocks through which the hole has been drilled. This procedure is normally carried out several times with different arrays of instruments on a sonde. It is lowered to the bottom of the hole and then pulled up steadily and its location measured accurately. Different geophysical instruments give various types of information about the rocks and the fluids within them.

The determinations are made electrically and transmitted to the surface, where they are recorded by computer for later processing. Measurements are commonly made of:
1) electrical resistance, which gives information about porosity and the amount and type of fluid in a rock – saline waters or hydrocarbons
2) electromagnetic properties, which also determine fluid content
3) natural electrical potential in the borehole sediments, a method particularly directed at identifying clays
4) radioactivity, either natural radioactivity or radioactivity induced by bombarding the sides of the well with neutrons. These methods give information about the clay minerals in the surrounding rocks, which contain higher amounts of radioactive elements (particularly potassium) and information about porosity
5) speed of sound in the rock, which varies from rock to rock and thus provides information about the constituents of the sediments. It is also used to calculate the time taken for seismic waves to travel to a given depth and back again, and is used to calibrate seismic sections
6) angle of dip or orientation of the sedimentary beds by registering the different depths on either side of the borehole at which the same parameters are measured.

Geological studies of material from boreholes

Examination of material recovered from boreholes gives information about the lithology of the rocks drilled and often about their ages. Its mineral composition is determined by the same methods as for

Electron microscope photographs of microfossils in Palaeogene sediments from a well drilled in the Davis Strait, West Greenland.
1–3: Foraminifera of late Paleocene and early Eocene age
4–5: Diatoms of late Paleocene age
6: Ostracod of late Paleocene age.

GEOLOGY OFFSHORE

Cretaceous dinoflagellate microfossils (about 50 microns in diameter) from West Greenland photographed through a microscope. (see also p. 165).

rocks sampled at outcrop, i.e. by chemical analysis and microscopy. The age of sedimentary material is determined principally from its content of microfossils.

Micropalaeontological studies are carried out using normal stratigraphic principles (see p. 69). The fossils can therefore be used to determine the age of a given rock when a reference is available that links fossil assemblages and time. When the material available consists of cuttings, not cores, the individual grains are usually too small for pieces of macrofossil to be identified (fossils large enough to be identified with the naked eye). Luckily, enormous numbers of smaller organisms, plant spores and pollen have accumulated in the sediments. These fossils are mostly in the size-range 20–2000 µm (1 µm = 1 micron = 1/1000 mm), but some of them can be up to a few centimetres across. Many of these organisms had hard shells or contained hard 'skeletons' that have been preserved as fossils after death. Many of these relics from the fauna and flora of the past are therefore preserved in sedimentary layers as microfossils. Their size means that they can be preserved in the material recovered from wells and they are so common that that they are often found in thousands.

Micropalaeontological studies begin by washing the content of fossils from the rocks before examining them under the microscope. When the fossils consist of non-calcareous material such as silicon or chitin, they are commonly recovered by dissolving the rock in concentrated acid and separating the insoluble remains of the organic skeletal parts. This

DEEP-SEA DRILLING

The discovery that oceanic crust was formed by spreading around the mid-oceanic ridges was made in the early 1960s. This led to the development of the theory of plate tectonics that explains how developments in the oceans cause many of the processes operating on the continents, particularly the formation of mountain ranges (see p. 53). Prior to the 1960s, geologists had focused almost exclusively on the land areas, but the new theory showed that there was a great need for geological studies in the areas covered by sea. Many projects started during the following decades, including the drilling of many deep-water boreholes under the auspices of large international collaborative projects. The first project was called the Deep Sea Drilling Project, DSDP. Techniques were developed and a drilling ship, 'Glomar Challenger', was built to drill into the seabed under very deep water and penetrate through the sediment cover and into the upper part of the crust, recovering cores from the entire section. It started operations in 1968 and during the following years drilled more than 600 boreholes in all the world's oceans. A new ship, 'JOIDES Resolution', was converted from a conventional oil-drilling rig and continued the work of 'Glomar Challenger' after 1985 under a new international agreement, the Ocean Drilling Program, ODP. This has now entered a third phase, the Integrated Ocean Drilling Program, IODP. Deep-sea drill ships have operated in Greenland's waters several times. 'Glomar Challenger' drilled in the Labrador Sea in 1970 and 'JOIDES Resolution' off South-East Greenland twice, in 1993 and 1995, on both occasions because of Danish initiatives and partly under Danish leadership. The programme in the Labrador Sea was to prove that it consisted of basaltic crust and to date the crust. The programme for the work off South-East Greenland was to study the early development of oceanic crust in the North Atlantic during the Palaeogene and to study the transition from continental to oceanic crust by collecting samples from the transition zone and correlating them with samples taken onshore.

Drill ships and drilling techniques

'JOIDES Resolution' is 143 m long and 21 m wide and displaces nearly 3000 tons. It is a floating drilling platform with a 60 m high drilling tower built onto it. While drilling, the ship is kept stationary over the location of the borehole by a so-called dynamic positioning system that uses 10 sets of propellers mounted along and across the ship, to counter the effects of wind, current and waves. The ship is equipped to drill in water depths up to about 6000 m and can penetrate up to 2 km below the seabed and recover long,

The deep-sea drill ship 'JOIDES Resolution'.

Sketch of the drill ship 'JOIDES Resolution' showing the ship's positioning equipment, drill tower and the funnel at the entry to the hole in the seabed.

treatment can be used on microfossils from many different animal groups as well as organisms from the plant kingdom such as algae, pollen and spores.

The occurrences and distribution of microfossils from different periods of the Earth's history are just as varied as those of macrofossils. The most common single-celled animals are foraminifera and radiolaria (animals with a covering of silica). Additional groups that are not found today can also be utilised, such as conodonts and chitinozoans. Small multicellular animals such as ostracods are also found as microfossils and are used in the stratigraphic analysis of drilled material. The most important groups among plants that form microfossils are the coccolithophoridae (cells that possess 'skeletons' of calcite – the so-called coccoliths that form most of the rock we call chalk), dinoflagellates (cells armoured with cellulose), diatoms and a long list of various algae, spores and pollen. Thousands of species are found in each group and their identification requires specialist knowledge. In fact, specialists in micropalaeontology tend to work only with one group of organisms and final detemination of the stratigraphic and palaeoenvironmental information from drilled material depends on the collaboration of groups of specialists.

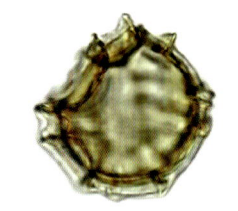

Dinoflagellate. Studies of microfossils from the 1500 m thick sedimentary series in the Nuussuaq Basin in West Greenland have led to the definition of 10 biostratigraphic intervals within a time interval of 20 million years in the Late Cretaceous.

DEEP-SEA DRILLING (continued)

continuous cores. The drill is driven by a motor connected to the drill pipe by a rotatable clamp and a compensator allows the ship to move vertically by up to 5 m without the drill pipe in the hole being moved. Sections of drill pipe 90 feet (approximately 30 m) long are attached to and removed from the drill string by means of a crane mounted high in the tower. A large funnel is mounted on the seabed at the top of the borehole to guide the drill pipe during re-entry of the hole after it has been pulled entirely onto the ship. An acoustic sender mounted nearby is used to guide the end of the drill pipe into the re-entry cone.

More than 100 people can be accommodated on board and the ship can operate at sea for up to three months without visiting a harbour. It is equipped with the latest navigation and communication equipment and with onboard laboratories that are quite comparable with those found in research institutions ashore. The rock drill cores arriving on deck are described in great detail and studies of the chemical and petrological composition and content of microfossils of the rocks recovered in the cores can be carried out on board together with such geophysical studies as palaeomagnetic measurements. Up to twentyfive scientists, each with their own speciality, are on board during a cruise so the results can be analysed continually and decisions made about future courses of action.

Studies off South-East Greenland

Deep-water boreholes have been drilled off South-East Greenland along two lines at right angles to the coast, traversing the inner shelf, the continental margin and oceanic crust as far as the spreading ridge. Nine holes in all have been drilled, those farthest out by 'Glomar Challenger' and the later ones by 'JOIDES Resolution'. The early holes on oceanic crust were drilled to study the oceanic crust and to confirm the relationship between the linear magnetic anomalies observed and the age of the oceanic basalts producing them. These boreholes also confirmed the nature of the oceanic crust – a thin veneer of sediment capping on an igneous complex of pillow lavas, dyke swarms, gabbros and at the base, pyroxene- and olivine-rich ultrabasic rocks (see p. 169).

The aim of drilling closer to the coast was to study and date the transition zone between continental and oceanic crust formed during the early rifting of the North Atlantic. A detailed seismic survey of the whole shelf margin was carried out before drilling started, with the aim of studying the sediments above the basaltic lavas and the uppermost volcanic rocks themselves. The seismic data showed the presence of continental crust below the inner shelf that was hardly affected by rifting. Farther offshore are so-called 'Seaward-Dipping Reflector Sequences (SDRS)', a series of layered basaltic lavas that dip towards the ocean and create a characteristic pattern on the seismic sections. When the innermost well was drilled, it found steeply dipping sediments of unknown age below the basalts, but these sediments were not visible on the seismic data so their extent could not be mapped. Younger sediments lie above the dipping basalts, with thick successions of Neogene sediments along the margin of the shelf and oceanic sediments under the deep water. The transition zone is about 200 km wide at 63°N, south of Ammassalik. The interpretation of the seismic data was confirmed by the drilling and the age of the earliest basalts was dated as 61 million years. These lavas are among the earliest erupted as the rising mantle plume that now underlies Iceland (see p. 171) impacted the base of the lithosphere. Sea-floor spreading proper off South-East Greenland first started 55 million years ago. The data from the boreholes also confirmed that the entire SDRS were produced above sea level (sub-aerially) in much the same way as we see on Iceland today. As it cooled, the lava series sank below the surface of the sea long after its formation, at a time when sea-floor spreading was happening farther offshore.

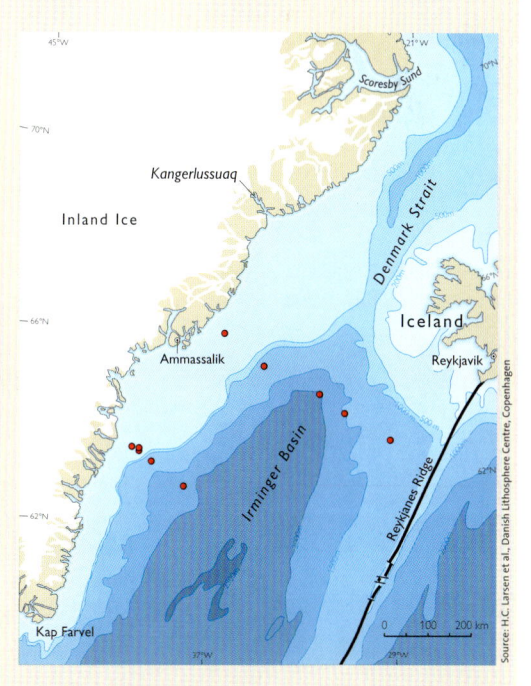

Map showing the locations of the boreholes drilled by 'JOIDES Resolution' off South-East Greenland.

GEOLOGY OFFSHORE

Sediments and basalts under the seabed

Geology offshore North and East Greenland

Seismic surveys have shown the presence of large sedimentary basins under the continental shelves off East Greenland between 67–72°N and 75–80°N. The two areas are separated by a large area underlain by Palaeogene basaltic lavas that may in turn conceal basins containing Mesozoic and older sediments similar to those farther north. Basement and basalts covered by Cenozoic sediments underlie the shelf off South-East Greenland, south of 67°N. The area north of Greenland is poorly known.

The German ice-breaker RV 'Polarstern' is specially equipped for marine studies in the ice-filled waters of polar areas and for relief of research stations in Antarctica. The ship is 118 m long and displaces 12 000 tons. It can carry up to 50 scientists and technicians and has a crew of 40. The ship can sail at a constant speed through ice up to 1.5 m thick and can break through even thicker masses of ice.

The most important geological boundary around Greenland is that between the continental crust and the volcanic oceanic crust. This boundary first developed off East Greenland between 60° and 76°N early in the Palaeogene about 55 Ma ago along a line of weakness that started as a fracture (rift) within the supercontinent of Pangaea (see p. 122). There is a narrow transition zone between the continental and newly-formed oceanic crust where outermost continental rocks were riddled by a swarm of dykes consisting of the same type of volcanic material as the ocean bed. The continental crust west of the transition zone consists of crystalline basement overlain by younger sedimentary basins and regions covered by plateau basalts (see map p. 156). Oceanic crust was formed at a spreading axis east of the transition zone.

The boundary between oceanic and continental crust north of 78°N first started to form at the beginning of the Oligocene about 34 Ma ago. Prior to that a microcontinent with the island of Jan Mayen had been part of the Greenland plate. Plate movement between Svalbard and eastern North Greenland took place along a large NW–SE-trending fault. The transition between continental and oceanic crust in the region north of North Greenland is still poorly known because only studies at reconnaissance level have been carried out, but studies planned for the next few years should improve this situation greatly.

Offshore North Greenland

The area north of Greenland's north coast is poorly known because it is covered by sea ice all year round, so geophysical surveys are difficult to carry out. Aeromagnetic and gravity data do, however, exist together with: observations from drifting ice islands, a few profiles from icebreakers, as well as information from nuclear submarines, the latter not being available to the public. Considerable interest in this area has arisen recently so a substantial Danish/Greenlandic/Canadian effort is planned for the next few years. Its goal is to investigate the extent and nature of the areas north of Greenland and Canada and to chart the transition between continental and oceanic crust. These activities will be carried out with the aim of defining the prolongation of the national exclusive economic zones as defined by Article 76 of the United Nations Convention on the Law of the Sea (see pp. 157–158) and to provide data that can be used to determine the boundaries between neighbouring states around the Arctic Ocean.

The continent–ocean boundary in the eastern part of the area is thought to lie somewhere around the base of the continental rise; the fairly abrupt change in angle where the continental slope meets the deep ocean floor (abyssal plain). In the most easterly area, off Kronprins Christian Land, the shelf is only about 30 km broad, but its width increases westwards, up to about 250 km north of western North Greenland and Ellesmere Island (Canada). A plateau called the Morris Jesup Plateau (or Morris Jesup Rise), where water depths are less than 1 km, exists north of Peary Land. It is about 200 km wide and extends NNE for a distance of 300 km before water depths increase abruptly to around 4 km.

Two large sedimentary basins are separated by the Morris Jesup Plateau. The plateau itself appears to consist of volcanic rocks underlain by either continental or oceanic crust. An area with a similar geological structure – the Yermak Plateau – occurs north of Svalbard. The two areas were originally contiguous but have been separated by sea-floor spreading around the Gakkel Ridge, the spreading axis in the eastern Arctic Ocean.

To the east of the Morris Jesup Plateau is an elongated narrow basin that appears to be a part of the Wandel Sea Basin which contains a succession of Carboniferous to Palaeogene sediments (see p. 118). The succession is more than 8 kilometres thick and the dimensions of this basin are about 150 × 350 km. To the west of the Morris Jesup Plateau is another sedimentary basin that forms part of the shelf north of western North Greenland and is called the Lincoln Sea Basin. It extends into Canadian waters and is about 600 × 600 km. The sediments in the basin are more than 8 km thick. It is thought that the oldest

In 1993, a German oceanographic expedition on the ice-breaker RV 'Polarstern' discovered some tiny islands 70 km east of the snout of the Nioghalvfjerdsfjorden glacier in North-East Greenland. Since then the islands have been visited and their position measured. They lie in the middle of an area of shallow shelf and could be used in the future as a platform to study the petroleum geology under the shelf. The islands consist of six groups of islets that hardly rise above the sea ice. The largest group, shown on the picture, covers an area of about 75 m × 25 m. During a visit in 1999 it was found to consist of tiny islets and melt-water lakes surrounded by sea ice. The islands have been named 'Tuppiap Qeqertai' (Tobias' islands).

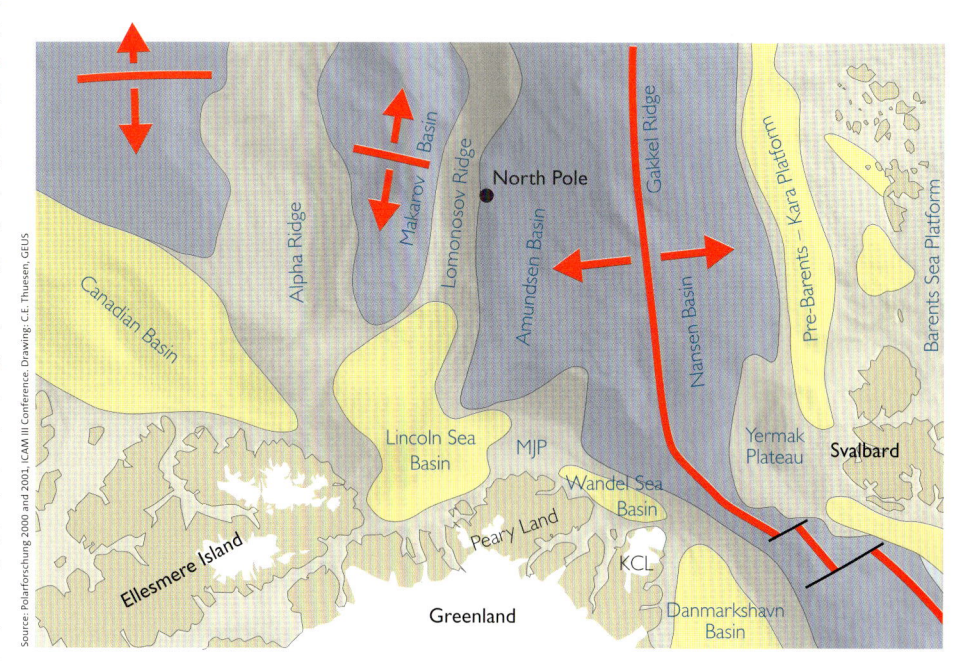

Map showing the general geological structure of the area north of Greenland emphasising the division between oceanic and continental crust. In the sector north of Scandinavia and western Russia, sea-floor spreading takes place around the Gakkel Ridge, an extension of the spreading axis in the Mid-Atlantic. The sector north of North America developed differently, involving older sea-floor spreading in an entirely different direction. The areas of oceanic crust in these two sectors are separated by a narrow ridge, the Lomonosov Ridge, that is thought to be a sliver of continental crust that either was contiguous with the Barents Sea Platform, east of Svalbard, or an original part of the North American shelf.

- Oceanic crust
- Large sediment basin
- Continental crust and other areas of shallow water
- Land

MJP Morris Jesup Plateau
KLC Kronprins Christian Land

GEOLOGY OFFSHORE

Oceanic crust
- Younger sedimentary basins (55–0 million years old)
- Oceanic basalts (55–8 million years old)
- Transition zone ocean/continent

Continental crust
- Palaeogene basalts and intrusions onshore/offshore
- Platform sediment Palaeozoic – Mesozoic onshore/offshore
- Basin: Cretaceous – Cenozoic offshore
- Wandel Sea Basin sediments onshore/offshore
- Basin: Upper Palaeozoic – Mesozoic onshore/offshore
- Basin with salt diapirs

Fold belts and foreland
- Caledonian and older rocks onshore/offshore

- Magnetic anomaly, numbered
- Fracture zone
- Fault
- Thrust fault

ILLB: Inner Liverpool Land Basin
OLLB: Outer Liverpool Land Basin

Map showing the sedimentary basins under the shelf off North-East Greenland, between Scoresby Sund (SS, 70°N) and Kronprins Christian Land (KCL, 82°N).

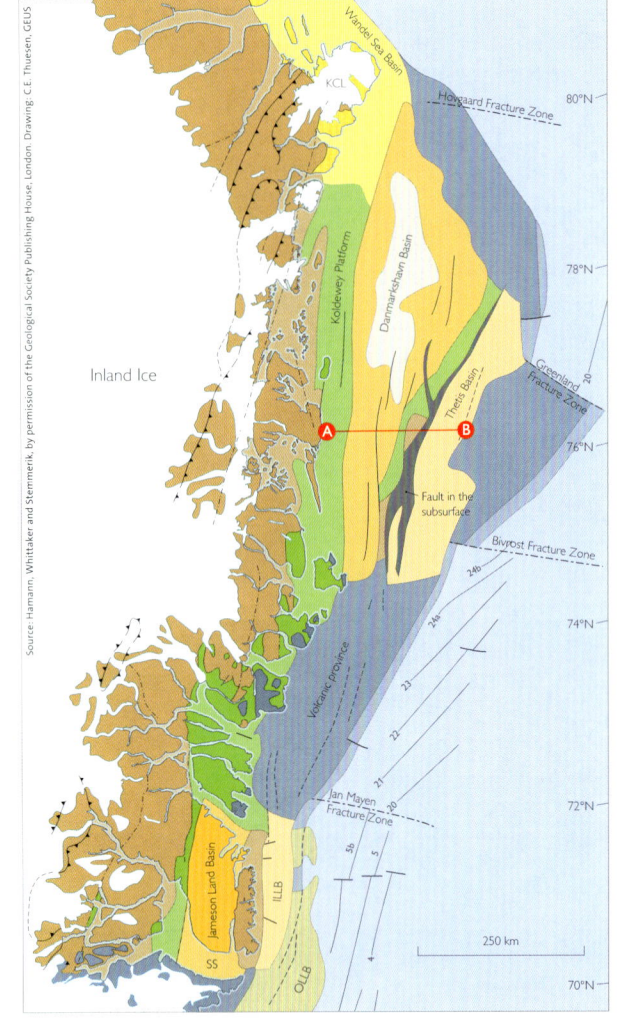

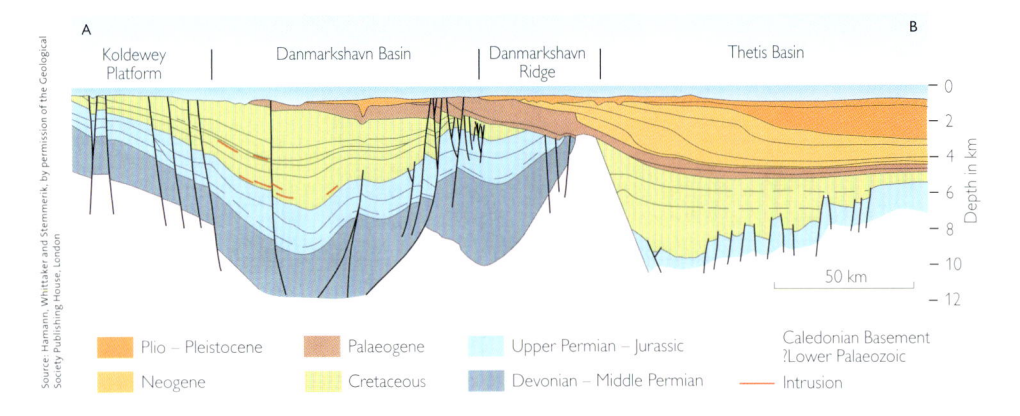

Cross-section through the widest part of the shelf at about 78°N (A–B on the map above) showing the structure of the sedimentary basins extending down to a depth of 13 km. The Palaeozoic–Mesozoic basins are divided by a ridge of crystalline basement whereas the Cenozoic sediments blanket much of the area.

that the Eurekan fold belt formed when these two plates collided.

To the north of the Lincoln Sea Basin is a submerged ridge – the Lomonosov Ridge. The nature of the crust under this ridge is important in determining the extent of the exclusive economic zones of Greenland, Canada and Russia. Evidence that the ridge is continental was provided by drilling by IODP in 2004. The ridge is interpreted as a detached sliver of a continental margin from one of the shelf regions now bordering the oceanic North Pole basin. The sliver represents either a part of the Barents Sea Platform to the east or may be an original part of the North American shelf. The Lomonosov ridge forms a bathymetric high with water depths of around 500 m over its top. The southern part of the ridge is overlain by a succession of more than 1 km of sediments, interpreted as of Upper Jurassic to Pleistocene age. There is a deep-sea passage between the continental crust of the Lomonosov Ridge and the Lincoln Sea continental margin to its south. The nature of the subsurface structure of this deep-sea region is important in determining if the Lomonosov Ridge is a natural prolongation of the continental shelf of Canada and Greenland.

New data, as yet not released, were obtained by a joint Danish/Canadian refraction seismic survey across the Lincoln Sea Basin and onto the Lomonosov Ridge in 2006 and an international expedition using ice-breakers that acquired geophysical data and seabed cores in 2007.

The shelves off North-East Greenland (72–80°N)

The shelf between 75°N and 80°N off North-East Greenland is Greenland's widest. Water depths between 200–500 m extend 300–400 km from the coast. The sea is very shallow locally, for example around the recently discovered Tuppiap Qeqertai (Tobias' islands) about 50 km offshore at about 79°N. Extensive sedimentary basins underlie much of the shelf. Their sedimentary fill is comparable to that of the Wandel Sea Basin in eastern North Greenland (see p. 118), to basins under the Barents Sea off north Norway and on Svalbard, and to younger basins onshore East and North-East Greenland (see pp. 122–127). Investigations of the area have been carried out using aeromagnetic, gravity and seismic methods. Interpretation of the data has suggested the occurrence of a sedimentary section from ?Devonian to recently deposited sediments, corresponding to an interval of 360–380 Ma. The total thickness of sediment is up to about 13 km. The later development of the basins included extension (rifting) as part of the separation of Greenland from Scandinavia. The opening is thought to have begun in the Middle Jurassic

may be of Carboniferous age, but that most are from the Mesozoic and Cenozoic.

South of the Lincoln Sea is the narrow Nares Strait between Greenland and Ellesmere Island. The northern part of this strait is underlain by a marked strike-slip fault with a substantial displacement between Greenland and Ellesmere Island. Narrow pull-apart basins are associated with the fault zone. Recent investigations on both sides of the Nares Strait suggest that this region forms a plate boundary between northern Greenland and north-eastern Canada and

and reached maximum intensity at about the Jurassic–Cretaceous boundary. New basins containing deepwater deposits developed under the outer shelf regions during the Cretaceous.

Basins underlie most of the shelf between about 75° and 81°N and cover an area of about 700 × 250 km. The rift basins make up a series of N–S-trending tectonic elements each with its characteristic geological development and bounded by faulted margins. The Danmarkshavn Ridge consists of an upfaulted (horst) block of crystalline basement that separates two basins. To its west is the Danmarkshavn Basin whose sediments are up to 13 km thick and appear to have been deposited as an almost complete succession between the Devonian and Palaeogene. Salt structures are visible in the centre of the basin. East of the Danmarkshavn Ridge lies the Thetis Basin whose sediments are up to 10 km thick and appear to consist predominantly of Cretaceous to present-day sediments, suggesting that this basin formed rather later than that farther west.

The seismic data over central Danmarkshavn Basin show structures interpreted as salt diapirs. This kind of structure forms when low density, mobile salt is overlain by heavier sediments. When the salt layers occur at varying depths, the shallower parts are at a lower pressure than the deeper parts. This creates a pressure gradient that makes the mobile salt flow sideways into the low pressure areas where it accumulates and forms dome-shaped structures. The more salt that flows into the domes, the shallower they become and the greater the pressure gradient. The result is cylindrical columns of salt (diapirs) that force themselves upwards through the surrounding sediments. The interpretation that the diapirs are salt is confirmed by gravity data showing that the diapirs are lighter than their surroundings. The diapirs of the Danmarkshavn Basin resemble those that have been drilled in the Barents Sea where the salt was deposited during the Carboniferous and Permian in subtropical conditions when the area was about 25–30° north of the equator.

Epochs in the Cenozoic

Period	Epoch	Age in Ma
Neogene	Holocene	0.0117–0
	Pleistocene	1.8–0.0117
	Pliocene	5.3–1.8
	Miocene	23.0–5.3
Palaeogene	Oligocene	33.9–23.0
	Eocene	55.8–33.9
	Paleocene	65.5–55.8

FORMATION AND DEVELOPMENT OF OCEANIC CRUST

Oceanic crust forms at a spreading axis where two tectonic plates are moving apart. The gap created is filled progressively by molten material from the mantle. The mantle consists of ultrabasic rocks that melt as they move upwards into lower pressure without losing heat. The melt differentiates (see p. 152) and a primary magma consisting of magnesium-rich olivine basalt moves into magma chambers under the spreading axis. The magma differentiates further in these magma chambers and the component with basaltic composition flows upwards through fractures to form dykes and to be erupted at the seabed, while the remaining denser magma solidifies onto the base and walls of the magma chambers and remains at depth. This process leads to the formation of a layered crust of igneous rocks whose chemistry is nearly uniform from layer to layer, but whose structure is very different in the different layers. From bottom up, the igneous layers consist of:
1) layered gabbros with scattered ultrabasic rocks
2) unlayered gabbros with scattered bodies of less basic rocks
3) a layer consisting of contiguous dyke intrusions
4) a layer of basaltic pillow lavas.

A thin layer of sediment may lie above this sequence. The magma chambers that are the source of the volcanic rocks lie at shallow depth below the spreading axis. The chambers are tens of kilometres in length along the axis, but are only some kilometres wide across it. Material from the mantle replenishes them periodically and they drain upwards by injection of dykes and eruption of magma to form pillow lavas (see p. 46) on the seabed.

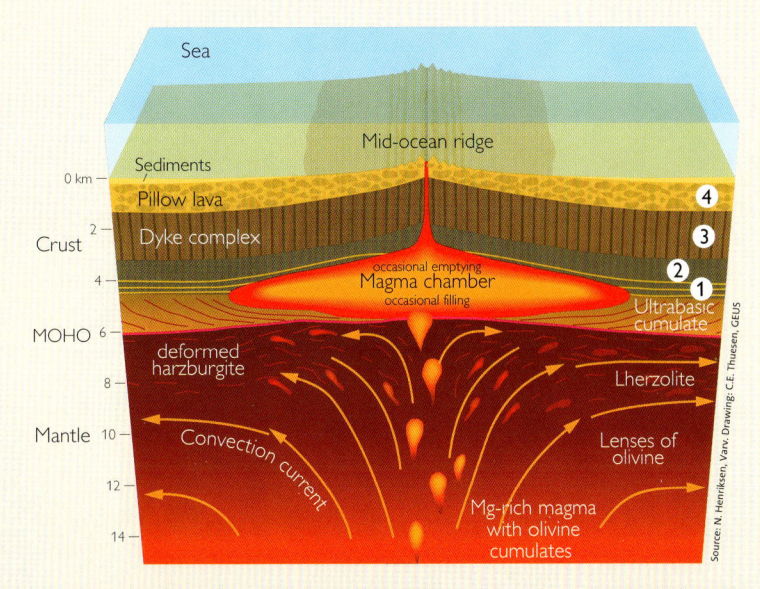

Block diagram showing the development of the upper mantle and crust at a spreading axis. Oceanic crust forms where melt from the mantle collects into magma chambers from where basalt is injected into dykes and erupted at the seabed forming pillow lavas.

The Earth's crust is only about 6 km thick around a spreading axis and the transition to the mantle is a marked increase in the velocity of seismic waves (the seismic Moho). Because oceanic crust is thin, the individual layers are never more than a couple of kilometres thick. The top of the asthenosphere at the spreading axis is only about 4–8 km below the seabed, but the new material of the oceanic crust is added to the lithosphere and becomes part of a tectonic plate. As the newly formed part of the plate moves away from the spreading axis and becomes older, it cools slowly and the lithosphere/asthenosphere boundary moves farther and farther downwards into the mantle. The lithosphere thus increases in thickness from 4–8 km at the spreading axis to 60–80 km along the ocean margins. The thicker areas are also the densest, so isostatic compensation (see p. 27) causes their surface to become deeper and deeper below sea level. The oldest parts of the oceans are thus also the deepest with water depths of around 5 km.

GEOLOGY OFFSHORE

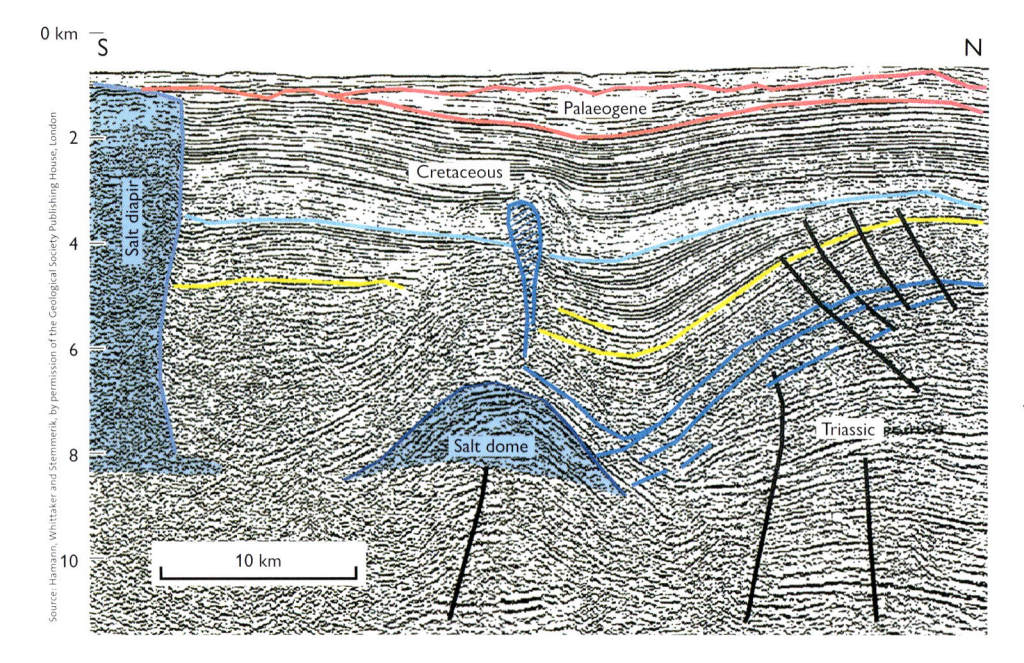

Seismic section from the northern part of the Danmarkshavn Basin showing how salt from deep in the sedimentary series has pierced the overlying sediments and risen upwards as a diapir. The salt is thought to be of Carboniferous–Permian age, deposited 310–280 million years ago. Some diapirs have risen through most of the sedimentary section until their tops lie less than 1 km below the seabed. Note that the horizontal and vertical scales are different.

The Mesozoic offshore succession is thought to be related to the sequences known from the onshore Jameson Land Basin and its correlatives in North East Greenland (see pp. 122–127). It shows similarities to the development known under the Norwegian shelf and it is considered highly likely that the offshore succession in this region contains the same Upper Jurasssic source rock for oil and gas that generates the major oil fields in the North Atlantic region. If so, the Mesozoic sediments were deposited when there was a marine connection through the strait between Greenland and Scandinavia. Water depths in the outer parts of the basins were fairly large at times, so it is thought that the sediments there consist of sandy and shaly units with occasional limestones.

The Liverpool Land Basin (70–72°N) off central East Greenland

A set of N–S-trending normal faults cuts across the mouth of Scoresby Sund and along the outer coast of Liverpool Land. They form the western margin of a narrow, coast-parallel basin containing Mesozoic and Cenozoic sediments with a total thickness of more than 8 km. Most of the basin is on continental crust, but the youngest sediments form a large fan that extends out over oceanic crust (see p. 24). These younger sediments, of Miocene, Pliocene and Quaternary ages, are more than 6 km thick and originate from erosion of the rocks that once formed the uppermost sedimentary and basaltic layers of the Jameson Land Basin (see p. 123). The particularly thick series of sediments fanning out from the mouth of the large fjord system behind Scoresby Sund indicates an enormous amount of erosion of the surrounding land areas. The large fjord and valley systems that cut through the Palaeogene basalts to a depth of 3–4 km also indicate the enormous magnitude of the erosion.

The Blosseville Kyst Basin (67–70°N) off East Greenland

A narrow, coast-parallel basin underlies the outer part of the continental shelf and the transition to oceanic crust. Its sedimentary fill is more than 4 km thick and accumulated between the Eocene and the Quaternary. The sediments rest on subsided Palaeogene basalts, some of which may in turn lie on Mesozoic sediments, but seismic surveys have not been able to determine the geology beneath the Palaeogene basalts because the seismic energy cannot penetrate the basalts and image what is below them.

The area of basalts between 72°N and 75°N off North-East Greenland

Palaeogene plateau basalt lava flows, similar to those found in the adjacent onshore area (see p. 148), underlie much of the shelf in this area. The basalts crop out at the seabed near the coast, but are covered by a layer of younger sediments further offshore. The

Cross-section through the Liverpool Land Basin just north of the mouth of Scoresby Sund. The boundary between oceanic and continental crust is shown in the deeper part of the figure. Cenozoic sediments have buried the boundary and have built out onto oceanic crust. Note that the horizontal and vertical scales are different.

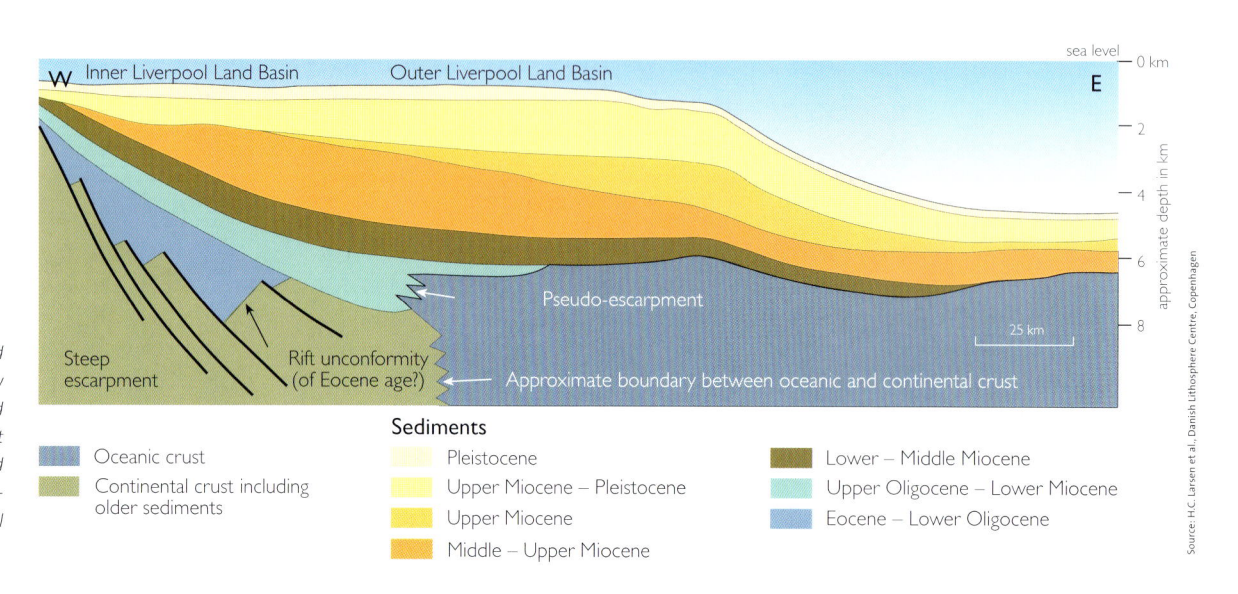

Sediments
- Oceanic crust
- Continental crust including older sediments
- Pleistocene
- Upper Miocene – Pleistocene
- Upper Miocene
- Middle – Upper Miocene
- Lower – Middle Miocene
- Upper Oligocene – Lower Miocene
- Eocene – Lower Oligocene

Geology offshore North and East Greenland

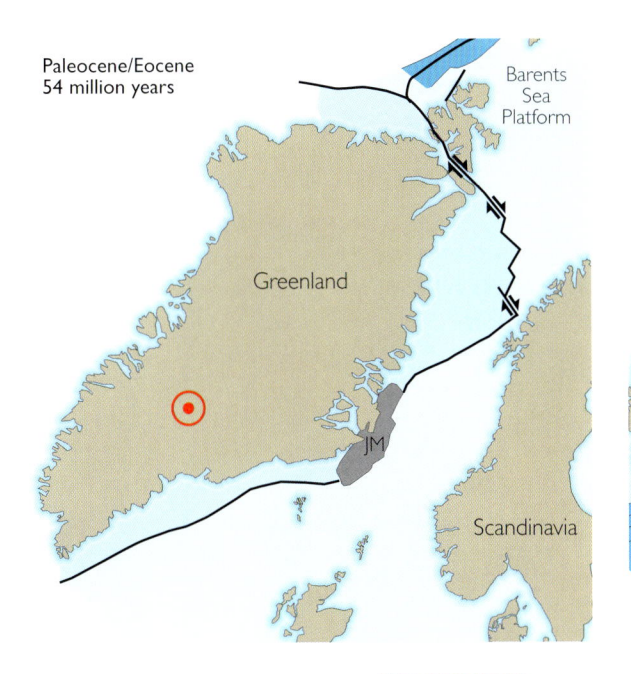

MJP	Morris Jesup Plateau
YP	Yermak Plateau
JM	Jan Mayen microcontinent
GFZ	Greenland fracture zone
SFZ	Senja fracture zone
JMFZ	Jan Mayen Fracture Zone

⊙ Plume centre

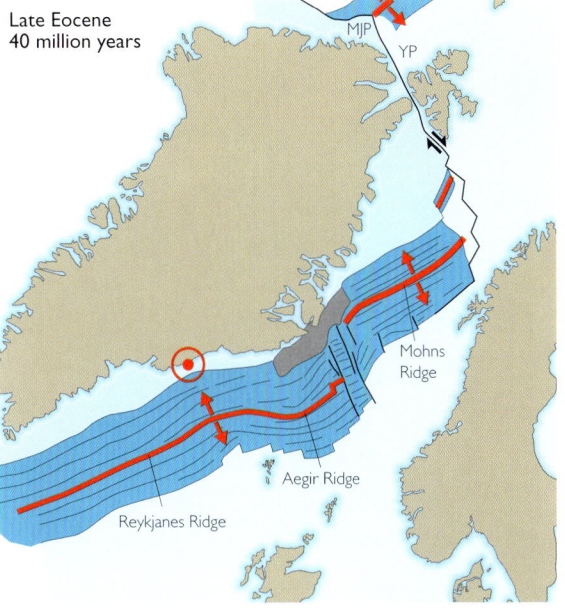

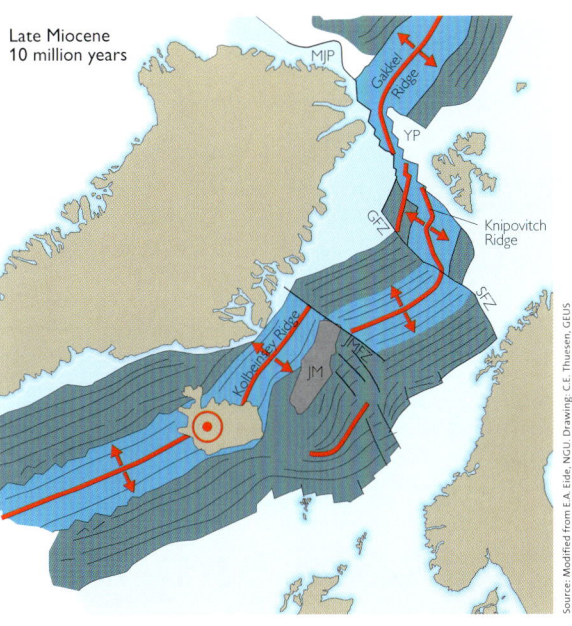

Development of oceanic crust in the North Atlantic between Greenland and Europe at four different times from the Paleocene/Eocene boundary to the Miocene. The figures show how the spreading axis has changed location with time and how spreading in the North Atlantic has become connected to spreading in the Polar Basin north of Greenland. Note that the volcanic island Jan Mayen lies on a fragment of continental crust that was attached to the Greenland shelf in the Palaeogene. The figure also shows how the centre of the Iceland plume has moved out from under Greenland to lie today under the spreading axis, so forming Iceland, whose oldest rocks are about 14 Ma old.

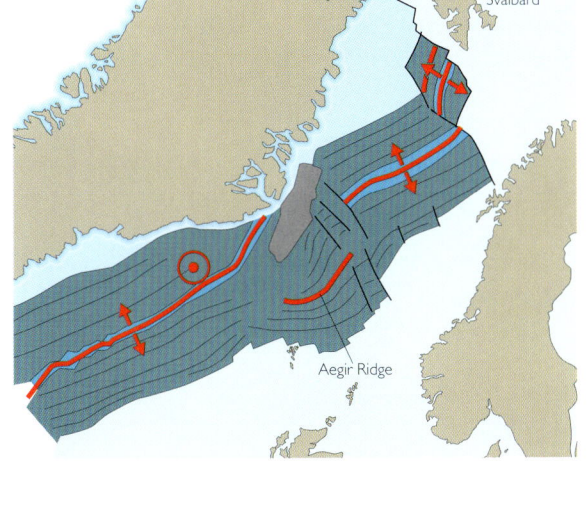

basalts appear to have been erupted sub-aerially, but later subsided below the sea. The basalts between 72°N and 75°N probably cover a thick succession of Upper Palaeozoic and Mesozoic sediments, equivalent to those exposed in the adjacent coastal areas onshore.

A number of roughly circular areas showing large positive magnetic anomalies are found near 72°N. They are interpreted as indicating the presence of vertical cylindrical structures containing magnetised rocks below the surface. Such structures are typical of magmatic intrusions that have been forced upwards through sediments and in places through plateau basalts. Intrusions of this type are known in the coastal area along the entire Palaeogene volcanic province in East Greenland (see p. 149).

Formation of oceanic crust off East Greenland

The oldest oceanic crust in the North Atlantic between Greenland and Europe formed about 55 Ma ago and continental drift and formation of new oceanic crust continue today. The distance between Greenland and Europe continues to increase at a rate of about 4 cm/year. Oceanic crust consists of volcanic rocks that are erupted symmetrically around a central axis. As time proceeds, oceanic crust of the same age is thus found in belts parallel to and arranged symmetrically on either side of the spreading axis with the youngest rocks closest to the axis.

Europe and Africa began to separate from North and South America during the Middle Jurassic, about 170 Ma ago. The earliest separation was between north-western Africa and eastern USA, creating the

GEOLOGY OFFSHORE

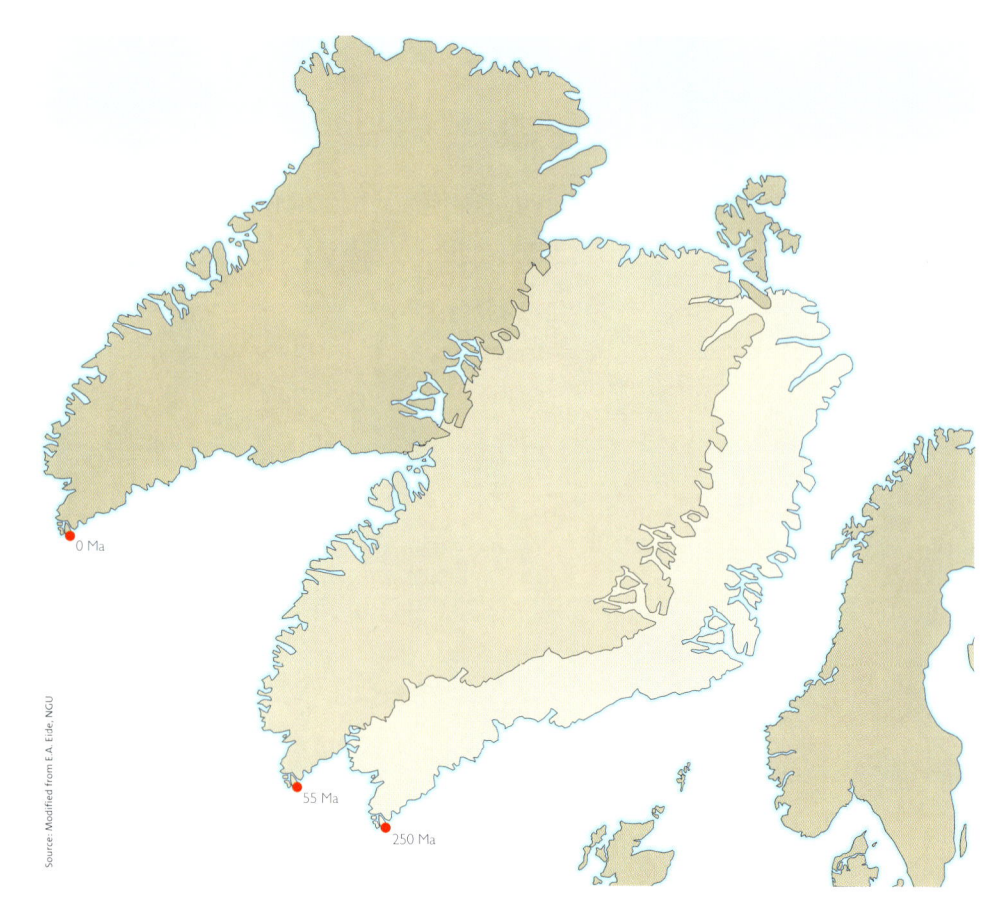

Changes in the position of Greenland relative to Europe as a result of extension of continental crust and sea-floor spreading in the North Atlantic. The continental crust of the two areas was contiguous 250 million years ago, but extension of the continental crust gradually moved them apart, giving rise to the large sedimentary basins under the continental shelves on both sides of the North Atlantic ocean. Sea-floor spreading started about 55 million years ago resulting in much more rapid movement of Greenland away from Europe.

central Atlantic Ocean and the Gulf of Mexico. Spreading then gradually extended to the south and the north, to produce new spreading axes between Africa and South America and between Europe and North America. Spreading began between Canada and Greenland around 62 Ma ago. The direction of movement changed about 55 Ma ago when a new spreading axis developed between Greenland and Europe, after which the speed of movement between Canada and Greenland gradually declined, stopping altogether about 33 Ma ago. The new spreading axis east of Greenland was the beginning of the North Atlantic Ocean. Its spreading axis at that time was divided into segments known as the Reykjanes Ridge, the Aegir Ridge and the Mohns Ridge.

At the same time as sea-floor spreading began between Canada and Greenland, 62 Ma ago, a mantle plume – a diapir of very hot rock (see p. 112) – impacted the base of the lithosphere after rising up through the asthenosphere. The impact gave rise to very widespread volcanism in a region that now occurs on both sides of the North Atlantic, from eastern Baffin Island in Canada, through both East and West Greenland to the Faroes, western Scotland, northern Ireland and Lundy in the Bristol Channel between England and Wales. The most extensive series of products of this early event in Greenland are the volcanic rocks of the Disko–Nuussuaq area of West Greenland (see pp. 142–145). Very hot material from the diapir was still present under East Greenland seven million years later when the North Atlantic started to open, so the thinning of the continental crust as sea-floor spreading started in the North Atlantic gave rise to a second major episode of volcanism. As Greenland moved away from Europe, it also moved north-west across the head of the diapir that gradually became narrower. Today the plume stem is found under Iceland where active volcanism still takes place.

Early spreading in the North Atlantic started somewhat later than that west of Greenland. The first oceanic crust formed off South-East Greenland about 55 Ma ago. Spreading farther north first started a couple of million years later east of what is now the microplate containing the island of Jan Mayen, at that time part of Greenland. As spreading proceeded, the Greenland plate moved gradually across the head of the mantle plume and by 25–30 Ma, the plume stem emerged out from under Greenland and moved to lie under the spreading ridge where it is today and gives rise to the active volcanism on Iceland. This caused a shift in the location of the North Atlantic spreading axis north of Iceland. Spreading ceased on the Aegir Ridge east of Jan Mayen and a new spreading axis formed, the Kolbeinsey Ridge, that cut between Jan Mayen and Greenland and continued northwards until it was displaced eastwards by the NW–SE-trending Jan Mayen fracture zone.

Spreading in the Arctic Ocean began in the Paleocene, at the same time as that west of Greenland. The spreading axis here is known as the Gakkel Ridge. Svalbard lay north of Greenland and the opening in the Arctic Ocean moved it gradually eastwards along a major strike-slip fault. This movement persisted until between 15 and 25 Ma ago when the continental crust of the Barents Sea Platform lost contact with that of Greenland and a spreading axis formed that connected the Kolbeinsey and Gakkel Ridges, separating Greenland from Svalbard.

Because of these complex movements, the age of the oceanic crust in immediate contact with the continental crust of East and North Greenland varies considerably. The oldest crust is 55 Ma old and is found off South-East Greenland. Farther north, the oldest crust is a couple of million years younger. The oceanic crust nearest Greenland in the Scoresby Sund region is only about 20 Ma old, much younger than that to its south or north where it is once again more than 50 Ma old. The oldest crust in the Fram Strait between Greenland and Svalbard is less than 20 Ma old and the crust in the Arctic Ocean is of all ages between 62 Ma and today.

The continental shelf off South-East Greenland

The physiographic continental shelf consists of an area of relatively shallow water flanking the continental landmass. Its outer edge, usually at water depths of between 200 and 500 metres, is relatively sharp, the shelf break, beyond which is the continental slope that descends towards oceanic depths. This configuration developed off South-East Greenland a considerable time after sea-floor spreading started about 55 Ma ago. Below the sea floor of the shelf, a series of sediments up to 2000 m thick lies directly on oceanic basalts and consists of material that has been eroded from Greenland to the west. The uppermost few hundred metres consist of sediments deposited during the Ice Age and the effects of extension of the ice cap can be traced all the way to the shelf break. A significant proportion of the sediments under the continental slope consist of sediments that have been bulldozed by the ice over the edge of the shelf.

The Greenland continental shelf is very broad (up to 300 km) where it faces Iceland, double the width of most other areas around South-East Greenland, and is separated from the Iceland shelf by a narrow trench in the Denmark Strait where water depths are only 600–700 m. Under this area of shallow water are volcanic rocks that were erupted above sea level forming a 'proto-Iceland' area that is now submerged. This thick volcanic pile accumulated as oceanic crust much thicker than normal because it developed above the hot Iceland mantle plume. This 'hotspot' moved relatively from below Greenland to the present position of Iceland over a period of about 50 Ma years.

Just as on land, the continental shelf area has been affected by ice during the Quaternary. Sea level was much lower than today during the Ice Age, and the ice sculptured a landscape that is below the sea.

Deep erosional, fjord-like valleys cut across the sediments and can be traced out to the shelf break. These valleys have formed transport routes for very large amounts of sediment that have been washed off the continent and into the deep sea. Moraines that originally formed on land are today found on the flat, plateau-like banks between the valleys. Some of the youngest traces of ice seen on the shelf are the many plough marks that formed when icebergs scraped across the seabed, ploughing sediment up in front of them and leaving a trench behind them. The trenches can be up to 25 m deep and 250 m wide and are found in water depths down to about 350 m.

The continental slope dips at an angle of 1–5° down into the deep ocean. The slope is built of sediment that is transported across the shelf and shed across its margin which moves progressively seaward.

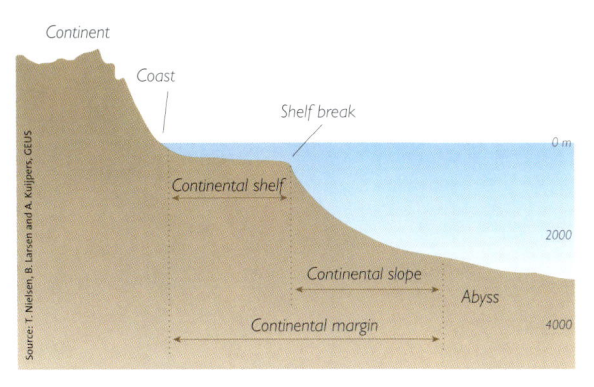

Schematic cross-section of the transition from continent shelf to oceanic depths off South-East Greenland. The section shows how the seabed topography varies from the shelf, down the continental slope to the deep sea. The profile is exaggerated vertically.

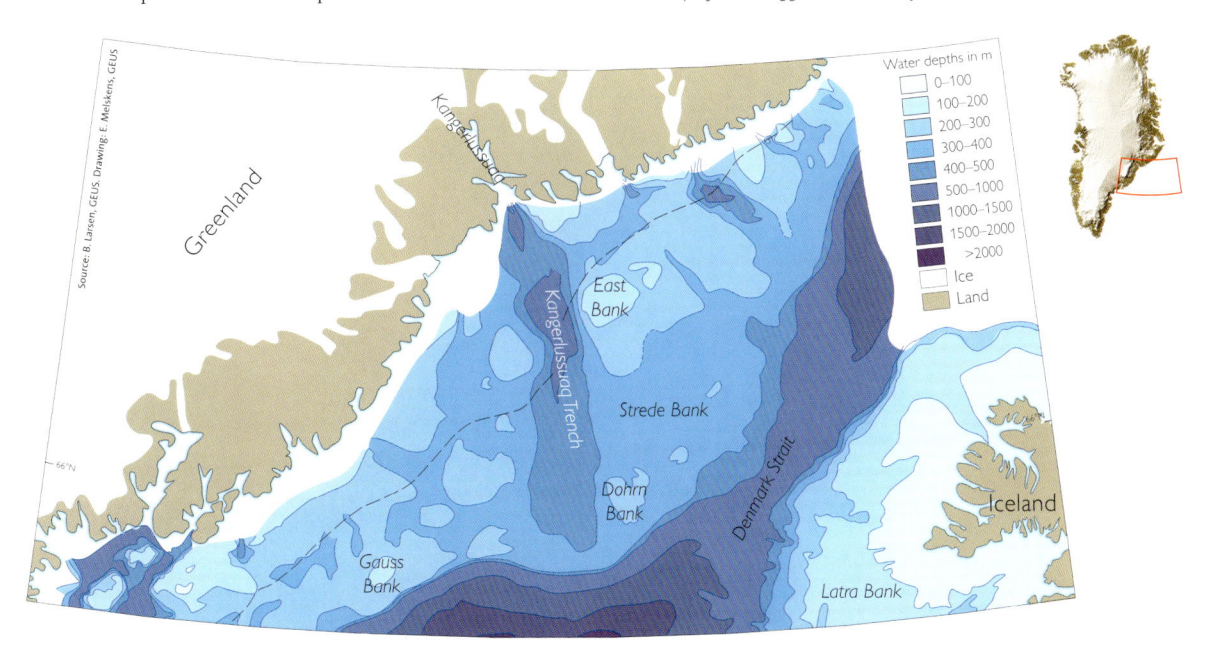

Seabed topography in the area off South-East Greenland and between Iceland and Greenland. The shelves around the two land areas are almost contiguous and separated only by the narrow channel of the Denmark Strait. A deeply eroded submarine valley extends across the flat shelf from the mouth of the Kangerlussuaq fjord. The dashed line shows the boundary between the inner shelf, where basement or basaltic lavas crop out at the seabed, and the outer shelf, where Cenozoic sediments up to 1–2 km thick have been deposited onto oceanic crust.

GEOLOGY OFFSHORE

Greenland was joined to Canada until about 130 million years ago

Geology offshore West Greenland

Continental crust extends under the whole of the extensive shelves west of Greenland and may continue all the way across the Davis Strait to Baffin Island in Canada. Farther south, block-faulted continental crust also extends under deeper water in the Labrador Sea, beyond which is a broad transition zone to oceanic crust formed by sea-floor spreading during the Palaeogene. Continental crust also extends well offshore in Baffin Bay, again across a poorly-known transition into Palaeogene oceanic crust covered by thick Cenozoic sediments. Deep sedimentary basins, containing both Mesozoic and Cenozoic deposits, occur under the shelves and in the deep-water zones landward of the transition zones. Extensive areas of Palaeogene plateau basalt lavas, underlain by inferred Mesozoic sediments and crystalline basement and mostly covered by Cenozoic sediments occur off central West Greenland.

The area offshore West Greenland formed by processes that rifted Greenland away from Canada. The two landmasses formed a contiguous area during the Palaeozoic in the most north-easterly corner of the North American continent, Laurentia. The crust between the two areas began to stretch slowly at some time in the middle of the Mesozoic forming sedimentary basins. In one area, the extension was so great that the continental crust broke open and unmelted mantle material reached the seabed. Sea-floor spreading in the central Atlantic extended during the early Palaeogene into both the Labrador Sea off southern West Greenland and Baffin Bay off northern West Greenland. The Davis Strait off central West Greenland was cut by major strike-slip faults as a result of these movements. Sea-floor spreading ceased at the end of the Palaeogene about 33 Ma ago after which the sedimentary basins mainly continued to subside although some areas were uplifted during the Neogene.

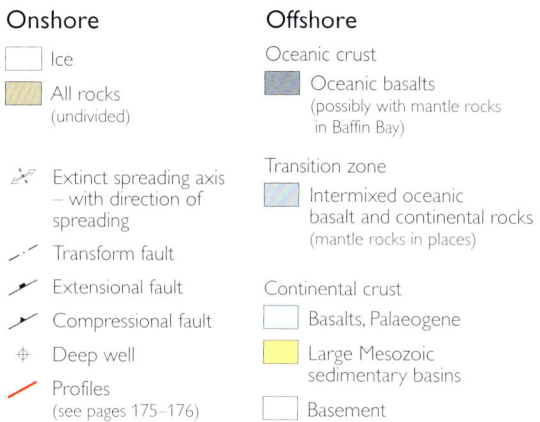

The area west of Greenland is divided into three geographical sectors: the Labrador Sea (60–63°N), Davis Strait (63–69°N) and Baffin Bay (69–77°N). The geological development was different in each of these regions, though there are many similarities and overlaps between them. Water depths are also different; oceanic depths of 2000–3000 metres occur in the Labrador Sea and Baffin Bay, but water depths in the Davis Strait are moderate, only 500–1000 metres. The reason for these differences is that both the Labrador Sea and Baffin Bay are underlain by oceanic crust while the crust under the Davis Strait is continental.

West Greenland's offshore areas are the most intensively studied around Greenland. Since the 1970s there have been many geophysical studies in search of hydrocarbons, particularly on the shelf and areas of moderate water depth. Six deep wells have been drilled on the Greenland side of the Davis Strait that have calibrated the uppermost few kilometres of sediment visible on the seismic records. The basins on the Canadian side of the Labrador Sea resemble those on the Greenland side and more intense drilling has taken place there, the results of which have been essential in interpreting the geophysical data from Greenland, particularly the part of the sedimentary section below that penetrated by the Greenland wells.

The Baffin Bay shelf of northern West Greenland (69–77°N)

The shelf in this area is up to 200 km wide and very large 'half-grabens' are found under it, formed by extension of the crystalline continental basement. Several thick sedimentary series have been deposited into and above the grabens. A sediment thickness of up to 12 km has been measured in the central part of the area in the largest basin, the so-called Melville Bay Graben. The thicknesses of sediments in the other basins farther north are somewhat less and there are large variations from basin to basin. These

Onshore
- Ice
- All rocks (undivided)
- Extinct spreading axis – with direction of spreading
- Transform fault
- Extensional fault
- Compressional fault
- Deep well
- Profiles (see pages 175–176)

Offshore
- Oceanic crust
- Oceanic basalts (possibly with mantle rocks in Baffin Bay)
- Transition zone
- Intermixed oceanic basalt and continental rocks (mantle rocks in places)
- Continental crust
- Basalts, Palaeogene
- Large Mesozoic sedimentary basins
- Basement

Schematic map showing the structure of the area off West Greenland. The map shows the sedimentary basins and areas of volcanic rocks on continental crust, the eastern limit of oceanic crust in the Labrador Sea and the area of transition between continental and oceanic crust off southern West Greenland.

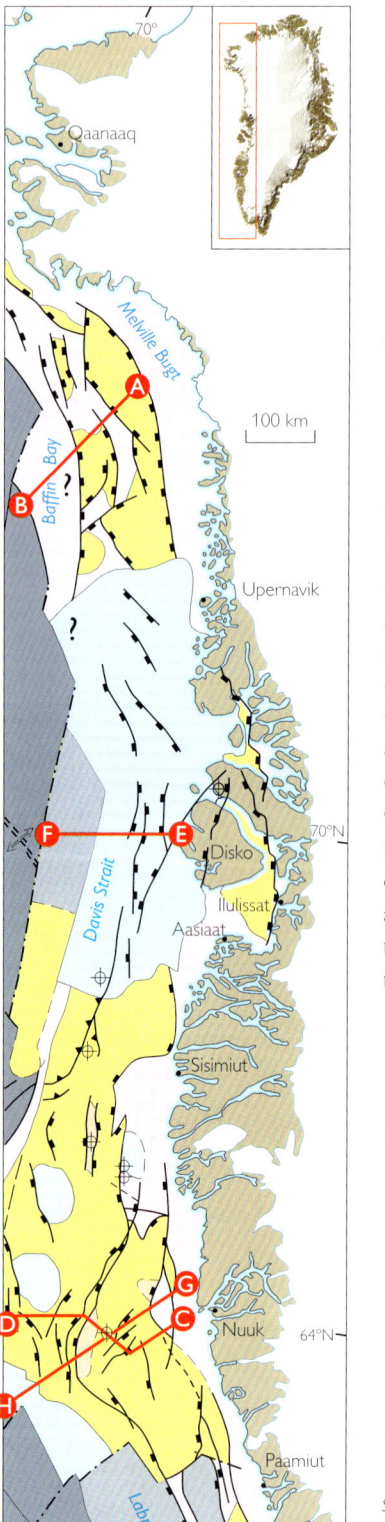

Source: J.A. Chalmers and T.C.R. Pulvertaft, GEUS. Drawing: E. Mersløv, GEUS

174

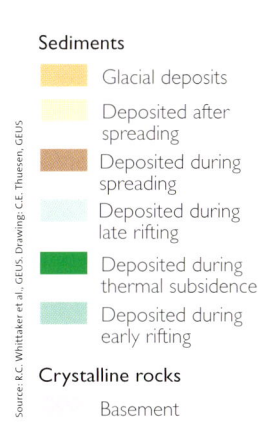

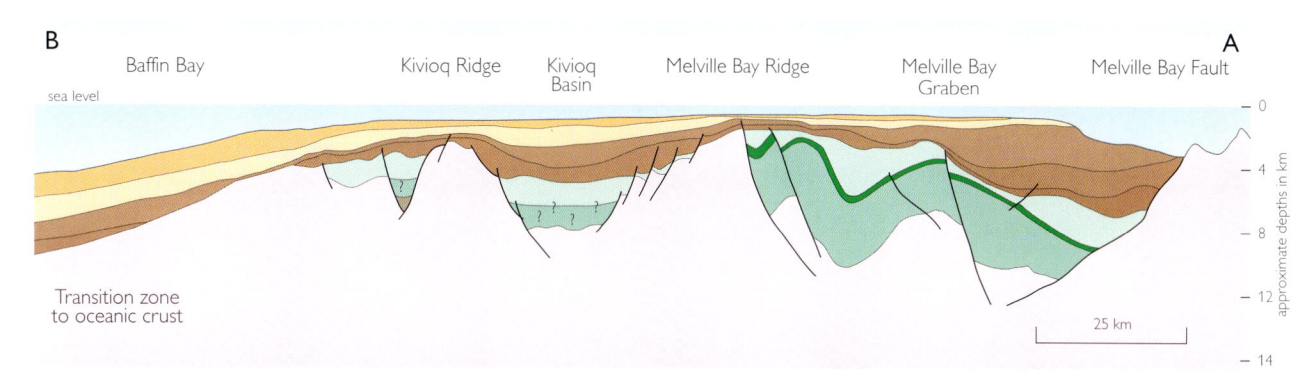

Cross-section through the sedimentary basins offshore northern West Greenland at about 75°N in Melville Bugt. The basins are on continental crust and are contained in a number of fault blocks that have been folded during later tectonic activity, probably during the Eocene. The youngest sediments were deposited after the cessation of tectonic activity. The vertical scale is exaggerated by a factor of about 2.5 compared to the horizontal to make the structures more clearly visible. The location of the section A–B is shown on the map on the opposite page.

northern basins have also been disturbed tectonically so that the older sediments in them are folded. South of 73°N, the older sediments are buried below Palaeogene basalt lavas, the offshore part of the basalt province exposed onshore in the Nuussuaq Basin (see below). There are no wells in the area, so the composition and ages of the sediments are interpreted entirely from seismic data and by analogy with basin development farther south and on the Canadian side of the Labrador Sea.

The basement under the sediments lies at widely different depths, because the basins occur in a number of grabens and half-grabens (see p. 177) that are separated from one another by coast-parallel blocks. There appear to have been two main phases of extension creating these rifts, one probably in the Cretaceous about 100 Ma ago and the other probably around the end of the Cretaceous – early Palaeogene, around 65 Ma ago, just prior to the volcanism and start of sea-floor spreading in central Baffin Bay. This has resulted in the sediments being divided into a number of packages, two of which were deposited during active extension and two others deposited during regional subsidence after each of the phases of block-faulting. The lowest part of the uppermost series and all the underlying series have been folded in the northernmost basins in the area, probably during the Eocene as Greenland moved northwards and collided with Ellesmere Island. Under the deep-water areas in central Baffin Bay to the west of the shelf are very thick sediments that probably lie on oceanic crust around the margins and altered mantle material in the centre.

Sedimentary basins off southern West Greenland (60–69°N)

A number of sedimentary basins extend NW–SE under the eastern shelf of the Davis Strait (see map on p. 237). The basins are limited to the east by crystalline basement exposed on the inner shelf and to the west by a basement high in the central Davis Strait. Their northern extensions are under Palaeogene basalts extending west from the Nuussuaq Basin. The southernmost basins merge into the transition zone to oceanic crust in the Labrador Sea.

The sediments in the basins are up to 10 km thick but there are large variations in thickness mainly because the basins are, like the basins in eastern Baffin Bay, divided into fault blocks within which sedimentation kept pace with subsidence. Eight main sequences of sediments are visible on the seismic sections, of which only the uppermost five have been sampled by drilling. The lithology and age of the lower three therefore remain uncertain. The stratigraphic packages and evolutionary stages are listed below in chronological order:

1) possibly Palaeozoic (Ordovician ?) sediments or Jurassic – Lower Cretaceous sediments, possibly interbedded with volcanic rocks (undrilled)

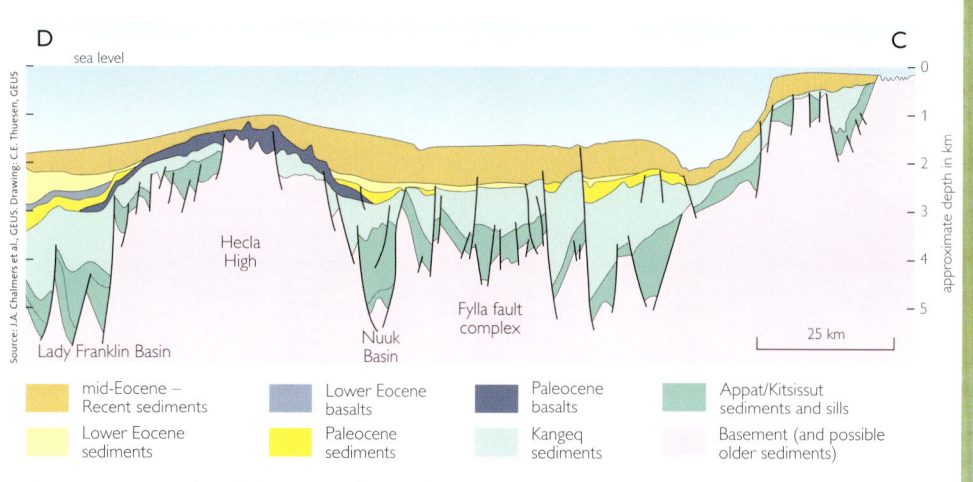

Zig-zag cross-section through the area south-west of Nuuk at about 64°30'N. The sedimentary basins formed on continental crust dissected into a number of blocks bounded by faults. From west to east, the profile shows the Lady Franklin Basin, the combined basement and volcanic high called the Hecla High, the narrow Nuuk Basin, the Fylla fault complex and the shallow crystalline basement under the narrow shelf off southern West Greenland. These faulted basins are buried under younger Cenozoic sediments. The non-linear vertical scale is exaggerated compared with the horizontal to make the structures more clearly visible. The location of the section C–D is shown on the map on the opposite page.

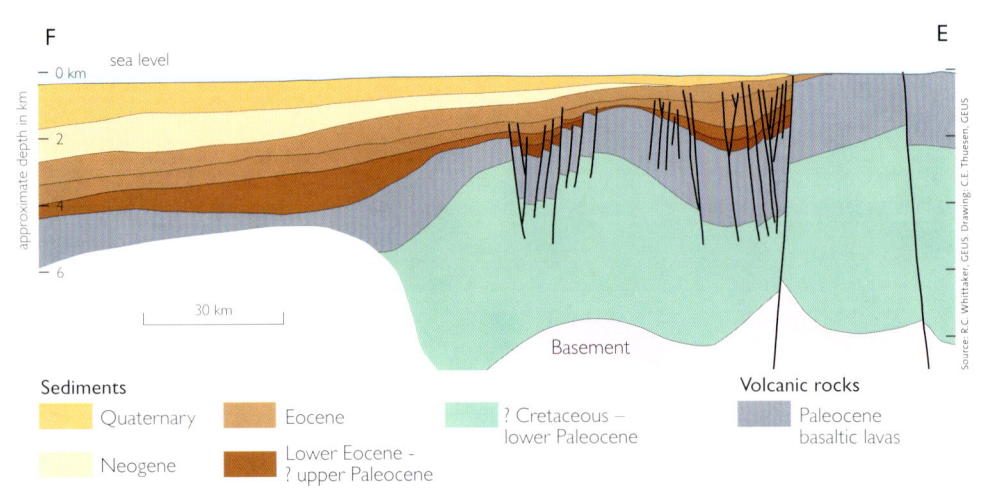

GEOLOGY OFFSHORE

Sediments
- Quaternary
- Neogene
- Eocene
- Lower Eocene – ? upper Paleocene
- ? Cretaceous – lower Paleocene

Volcanic rocks
- Paleocene basaltic lavas

Cross-section through the area west of Disko at about 69°30′N. The cross-section shows that the basaltic lavas exposed onshore on Disko and Nuussuaq continue offshore where they are offset by younger faults. A thick series of sediments probably exists under the basaltic lavas in the east, overlying continental basement, and may extend for at least 100–120 km west of the west coast of Disko. Cenozoic sediments, deposited later than the basaltic lavas, have been eroded away at the eastern end of the profile during uplift movements in the Neogene. The vertical scale is greatly exaggerated (by a factor of about 7) compared with the horizontal to make the structures more clearly visible. The location of the section E–F is shown on the map on p. 174.

2) possibly Lower Cretaceous sediments about 130–115 Ma old, deposited during crustal stretching (undrilled)

3) possibly mid-Cretaceous sediments deposited during active rifting and block faulting about 115–100 Ma ago (undrilled)

4) Upper Cretaceous sediments (drilled) consisting of sandstones in the lower part and mudstones in the upper part deposited 100–80 Ma ago (and probably until 63 Ma ago). Their deposition was followed by a period of renewed rifting and thereafter by uplift that removed the upper parts of this sequence by erosion

5) the period of uplift and erosion continued until the middle of the Paleocene when it ceased at the same time as volcanic activity started, 62 Ma ago. Large areas of plateau basalt lavas covered the area between 68°N and 73°N and also the western part of the area south of 66°N. A second, less-widespread, volcanic episode followed about 55 Ma ago. Subsidence of the basin started again during the first episode of volcanism and deltaic sediments, comprising alternating sandstones and mudstones, were deposited north of about 66°N and marine mudstones to the south

6) this sedimentation pattern continued during the Eocene (55–35 Ma ago). In the western parts of the basins, between 66°N and 68°N, the deposits were folded during strike-slip movements through the Davis Strait. Uplift starting at 35 Ma led to more erosion and formation of a major unconformity and no sediments of Oligocene or early Miocene age appear to be present in the basin

7) renewed subsidence during the Miocene led to deposition of sandy marine sediments over the whole area of the basins (23–5 Ma)

8) uplift of the coastal areas of Greenland by 1–2 km during the Pliocene and Pleistocene led to major erosion and deposition of large amounts of sediment across the shelf, first transported by rivers, later by ice.

The area of basalts between 68°N and 73°N off West Greenland

The Paleocene–Eocene basalt lavas found in the Disko – Svartenhuk Halvø region onshore (see p. 142) continue offshore to the west forming a large contiguous area at and under the seabed. Near the coast, the lavas are exposed at the seabed, but further west they are covered by younger sediments whose total thickness increases westwards. Similar volcanic rocks are found under the seabed on the Canadian side of the strait. What lies under the basalts is poorly known, because seismic energy is scattered by the basalts so the underlying rocks are imaged poorly. Interpretation of gravity data, however, suggests that the presence of Mesozoic sediments overlying continental basement is likely.

The basalt lavas consist of a series several kilometres thick that developed initially as submarine eruptions in the area on western Disko and Nuussuaq and offshore just to the west. The lava pile built up quickly until the lavas were being erupted on land. The development offshore appears to resemble that described from the Disko – Svartenhuk Halvø

Sediments
- Palaeogene – Neogene
- Mesozoic

Oceanic crust and transition zone
- Basaltic lavas
- Altered mantle

Continental crust
- Mainly basement

Mantle rocks
- Ultrabasics

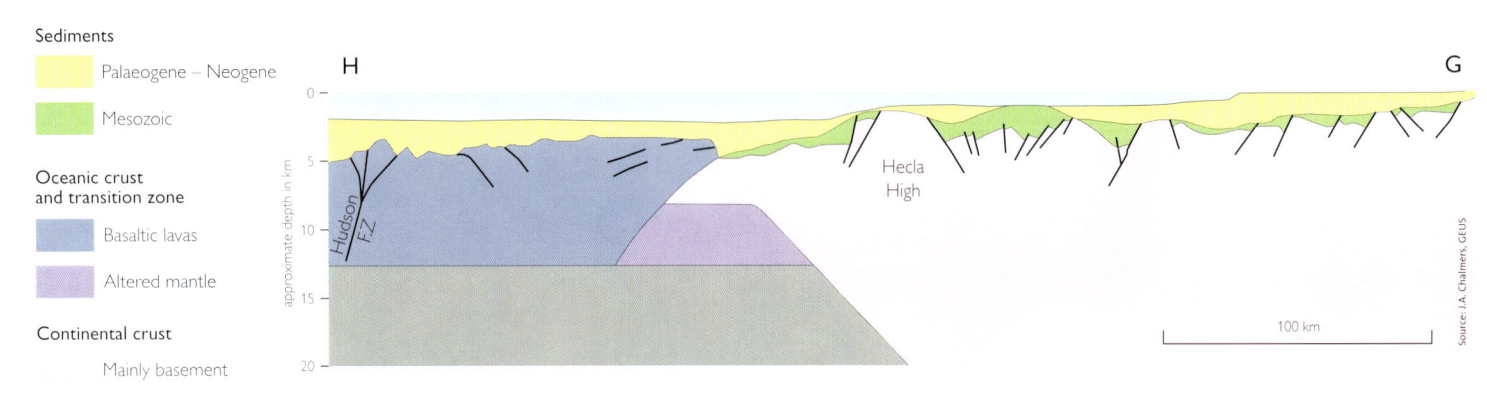

Schematic cross-section from SW to NE across the continental margin off southern West Greenland. The basement under the north-eastern half of the section consists of continental crust that is overlain by Mesozoic and Cenozoic sediments. The south-western end of the profile shows the oceanic crust of the Labrador Sea formed by movement of Greenland away from Canada in the Palaeogene. The oceanic crust in this area is thicker than normal, indicating that the ultrabasic rocks forming it were hotter than normal when emplaced. The Cenozoic sediments observed in the north-east continue south-westwards over the oceanic crust. The vertical exaggeration is 5 times. The location of the section G–H is shown on the map on p. 174.

Geology offshore West Greenland

STYLES OF FAULTING IN OFFSHORE AREAS

Many sedimentary basins are formed when the crust breaks along faults and subsides on one side relative to the other. Such faults are commonly recognised, both onshore in areas where the sedimentary basins are exposed and offshore under the sea. On the other hand, certain types of faults form in direct response to movements at lithospheric plate boundaries. They exist mostly on oceanic crust and are only rarely seen onshore.

Normal faults

A fault is a crack or break in the Earth's crust, where the rocks on one side have moved relative to those on the other. Three types of faults are recognised: 1) strike-slip faults with a nearly vertical fault plane where the rocks on either side have moved sideways relative to one another, 2) normal or extensional faults with a dipping fault plane where one side has moved down relative to the other and 3) reverse or thrust faults with a low-angle or even horizontal fault plane where one side has moved upwards or across the other. In normal faults the fault plane usually dips at angles of about 45°–60° and the block above the fault plane has moved downwards relative to the block below it (see p. 94).

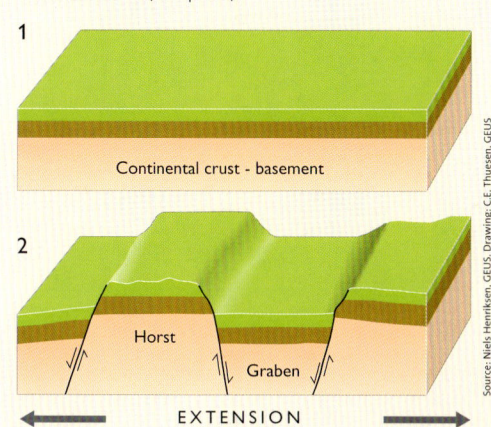

Block diagram showing extension and vertical movement on normal faults:
1) deposition of sediments (brown and green signatures) before the start of faulting
2) movement on faults forming 'horsts' and 'grabens'

Sedimentary basins with faults

Sedimentary basins form in areas where the Earth's crust is subsiding, most commonly because it is also extending. The areas of subsidence are commonly bounded by fractures in the crust called faults and faulting within the basins commonly divides them into 'fault blocks' that are relatively long compared to their width. The movements tend to be larger vertically than horizontally, by a ratio of about 2:1.

Sedimentation in a basin normally continues during a period of faulting, so it is common to differentiate sediments into those that have been deposited before (pretectonic), during (syntectonic) or after (posttectonic) a faulting episode.

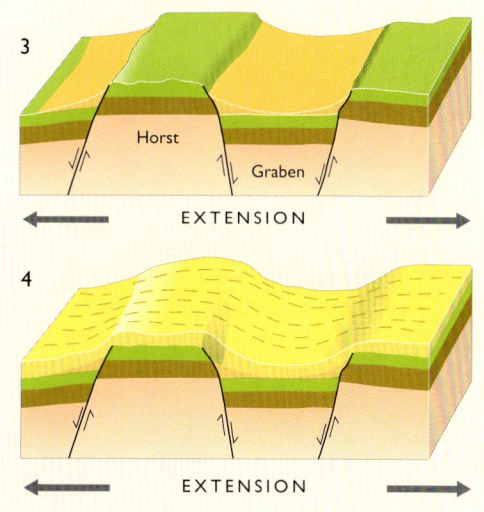

3) Block diagram showing sedimentary deposits (yellow brown) accumulated during normal faulting (syntectonic deposits)
4) Block diagram showing the development of a sedimentary basin with sedimentation after cessation of faulting (posttectonic)

Half grabens

Not all fault blocks subside by the same amount and many of them subside more on one side than the other, so that they rotate, causing previously horizontal layers of sediment to tip. Blocks that have subsided with little rotation are called 'grabens' and those that subside more on one side than the other are called 'half-grabens'.

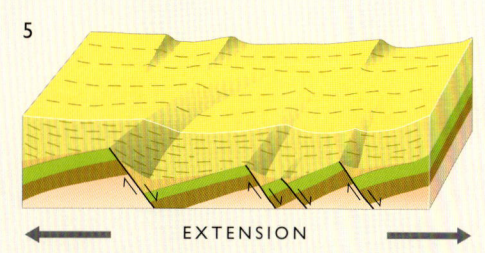

5) Sedimentation similar to 4) but the movement on the faults has formed several 'half-grabens' instead of 'horsts' and 'grabens' as in 2).

The structure shown in 5) develops where the faults, formed in response to extension, all dip in the same direction. The sediments covering the tops of these blocks are often deposited later than those deposited in the subsided areas, so there is often an unconformity on the top of a fault-block.

Faulting in areas with oceanic crust

The spreading axes in oceanic crust are often divided into sections that are offset relative to one another. The offsets are caused by a special class of faults with steeply dipping or vertical fault planes at high angles or normal to the spreading axis. The faults are known as 'transform faults'. To some extent they resemble strike-slip faults in that the movement is horizontal and parallel to the fault strike, but the movements along the transform faults are in the opposite direction to the offsets of the spreading axis and the displacements occur only between the offset spreading axes (see the figure).

Offsets can be several hundred kilometres long, but the fault zones are rarely more than 1 km wide. Great differences in structure and topography can develop on each side of such a large-scale fault zone which is called a 'fracture zone'.

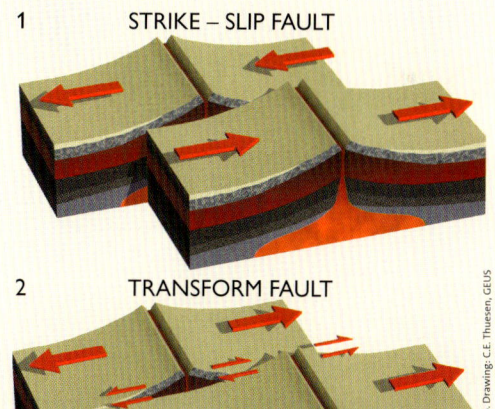

Block diagrams showing the patterns of movement along two types of fault through oceanic crust, a 'strike-slip fault' and a 'transform fault'. In both cases the fault plane is vertical and the movements are sideways along the fault zone. The arrows show the direction of movement.

1) An ordinary strike-slip fault where relative movement takes place along its entire length. This type of fault is also common in continental areas.

2) A transform fault connecting two sea-floor spreading axes. The two axes move steadily away from one another as oceanic crust is formed. Movement between the two sides of the transform fault takes place only in the segment between the two spreading axes and is in the opposite direction in which the axes are separating. There is no horizontal movement between the sides of the transform fault outside the segment between the two axes, although a topographic difference remains because the oceanic crust on each side of the fault is of different age.

GEOLOGY OFFSHORE

The Russian survey ship 'Professor Logachev' that was used in August 2003 to acquire seabed samples off West Greenland. The samples shown are from south-west of Nuuk where they were dredged up from depths of between 1950 and 2145 m. The dredged samples consist primarily of loose blocks of rock.

Blocks dredged up from the seabed lying on the deck of the ship after being cleaned of mud and sand.

Selection of blocks dredged up from a submarine canyon. The blocks are up to 30 cm across and consist mostly of granitic and gneissic basement.

area onshore with subaerial lavas building lava deltas into the sea away from the local palaeo-coasts. Offshore, the deltas appear to have migrated westwards whereas the onshore lava deltas built out eastwards. Seismic and magnetic data suggest that there are several separate units of volcanic rocks – several older units from about 62 Ma and a younger one from about 55 Ma. The older volcanism happened as the Iceland mantle plume impacted the base of the lithosphere (see p. 171) and sea-floor spreading started in the Labrador Sea and Baffin Bay whereas the later phase happened when the direction of sea-floor spreading in the region changed in response to the onset of opening of the North Atlantic (see pp. 170–172).

After volcanism ceased, the entire area subsided below the sea and was covered by sediments. Uplift during the Neogene tilted the basin, exposing the present-day Nuussuaq basin onshore, removing its cover of sediments, and eroding the Cenozoic sediments and basalts offshore.

Oceanic crust west of Greenland

Oceanic crust is found in two areas. The southern is in the central Labrador Sea between Greenland and Labrador in Canada, south-east of Baffin Island, and the northern is in Baffin Bay between northern West Greenland and the north-east coast of Baffin Island.

Sediments with a total thickness of up to 12 km lie under the seabed in central Baffin Bay. Below that is some kind of crust that has been difficult to interpret and, until recently, the neutral term 'Baffin Bay crust' has been used to designate it. Gravity data show two linear anomalies in the centre resembling that from the extinct spreading centre in the Labrador Sea, but there are few magnetic stripes and they are difficult to understand. A recent plate refit suggests that the crust of central Baffin Bay consists of altered (serpentinised) mantle material of Eocene age symmetrically arrayed around the extinct spreading centre. This Eocene crust is surrounded by more typical basaltic oceanic crust of Paleocene age, just seaward of the offshore Paleocene basalts.

The northernmost oceanic crust in the Labrador Sea is just south of the latitude of Nuuk (64°N). The margins of the oceanic crust in the Labrador Sea are complex and consist of a broad transition zone in which development can be interpreted in different ways. The uncertainty is caused mainly by lack of borehole data to calibrate the geophysical data. The magnetic data, however, clearly show sea-floor spreading anomalies, the oldest of which is known as anomaly 27, indicating that magmatic sea-floor spreading was underway by the middle Paleocene, 62 Ma ago. The first spreading axis ran NNW–SSE but that changed after about 7 Ma to spreading around an axis in a more north-westerly direction. Magmatic spreading continued until about 43 Ma ago after which it slowed and finally stopped about 35 Ma ago.

The oceanic crust of the Labrador Sea connects directly with that of the North Atlantic. Until recently it was thought that the Baffin Bay crust was entirely isolated with no oceanic crust connection to any other oceanic area, but a new interpretation suggests that the deep basins on the Canadian side of the continental crust region of the Davis Strait may be floored by Paleocene oceanic crust. During the Eocene, the movements in the Labrador Sea and Baffin Bay were connected by major N–S transform faults through the Davis Strait. Movements on these faults caused substantial deformation of the continental crust and sediments in the area and of any Paleocene oceanic crust that may exist under the western Davis Strait.

Neogene sediments and uplift

Marine sediments of Paleocene age are found intercalated with volcanic rocks at a height of 1200 m in western Nuussuaq, whereas volcanic rocks of the same age are found at a depth of 2–3 km west of Disko. The volcanic rocks and many of the overlying sediments dip steadily westwards from the west coasts of Nuussuaq and Disko showing that the area has been tilted and the rocks now exposed onshore have been lifted to their present heights above sea level after the sediments were deposited.

The Cenozoic sediments offshore southern West Greenland have been studied using a dense network of seismic lines with a total length of over 40 000 km. The Cenozoic sediments have been divided into five sequences by calibrating the seismic data using data from the six wells in the area. The uppermost sequence contains sediments deposited during the Pliocene and Pleistocene (5 Ma ago and younger). Below it is a unit of Miocene sediments and three sequences of Upper Eocene (45–35 Ma), Lower Eocene (55–45 Ma) and Paleocene age (62–55 Ma). The Miocene and Upper Eocene sediments are separated by an unconformity, so that deposits of all ages between 35 and about 13 Ma are missing. The Palaeogene sediments and basalts are in places almost 3 km thick and lie on either basement or Mesozoic sediments. In many places, this entire series is tilted and dips to the west and is cut either by the seabed or by an angular unconformity at the base of the Plio-Pleistocene sediments, showing that sediments older than Pliocene have been tilted and eroded. At the same time, the basement areas to the east were uplifted and erosion during the subsequent time, either by rivers or by ice, formed the source for the thick Pliocene and Quaternary sediments deposited across much of the shelf west of Greenland.

The lower Paleocene and Cretaceous sediments under the basaltic lavas on Nuussuaq have been drilled to a depth of about 3 km. These sediments contain grains of the mineral apatite that contain small amounts of uranium. The radioactive decay of the uranium atoms leaves trails where the crystalline structure of the apatite is disturbed (fission tracks) and by studying how these dislocations become annealed, it is possible to reconstruct the temperature history of the sediments. This has been used to show that the area has been uplifted and eroded three times since the end of the Eocene, 35 Ma ago. Prior to that, the Nuussuaq Basin (and probably many of the present-day mountains of West Greenland) was below sea level. Uplift that started about 35 Ma ago lifted the area above sea level and erosion removed both the Eocene sediments that had been deposited on top of the basalts and perhaps the uppermost 500 m of the basalts themselves. By 10 Ma, the area had been reduced to near sea level again when a second episode of uplift started, lifting the onshore areas by about 1 km. A third uplift episode that started between 7 and 2 Ma ago tilted the basalts and sediments offshore and lifted the rocks close to their present heights. The combined result of the last two uplift episodes has been to elevate what was a coastal plain 10 Ma to heights up to 2100 m. Rivers and glaciers have cut deep valleys into the terrain after it was uplifted, but much of the basalts and their underlying sediments remain, because the basalts have been resistant to erosion.

Equivalent uplift during the Neogene has also taken place in the basement areas. There are no young sediments or basalts in these areas that can be used to calibrate the movements, but analysis of the landscapes high in the mountains shows that, starting 35 Ma ago, they were uplifted and eroded by rivers in the same way as the sediments and basalts in the Nuussuaq Basin, and a flat, low-lying coastal plain, a so-called peneplain, had formed by 10 Ma. Any Neogene sediments that may have been deposited onto that peneplain have subsequently been removed. These areas were then lifted in the same two episodes as the Nuussuaq Basin so that today the peneplain is found in many places as extensive flat-lying or gently dipping plains at heights of up to 2000 m (see p. 184). The relief formed by the uplift resulted in powerful erosion that formed valleys and the eroded material forms the thick Plio-Pleistocene sediment wedges found over much of the shelf off West Greenland.

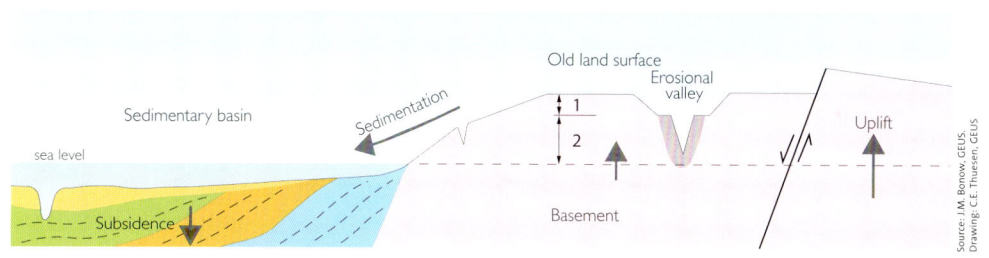

Schematic profile through West Greenland after uplift during the Neogene. The uplift took place in stages and after each stage, rivers eroded the landscape either until it was flat (a peneplain) or incompletely to leave river valleys. The bottom of the valleys and the peneplain surfaces were close to sea level prior to the next phase of uplift. The uplift has also caused older sediments in the area just offshore to rotate and they now dip seawards.

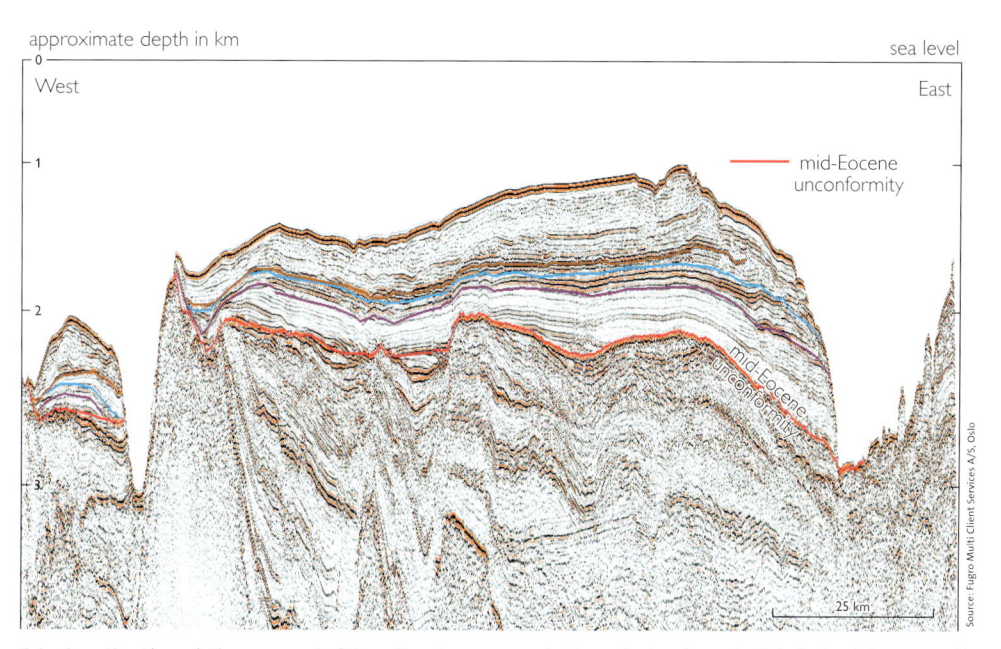

Seismic section through the upper part of the sedimentary cover under the seabed south-west of Nuuk. An obvious unconformity on the section is interpreted to be of mid-Eocene age (about 45 million years old). The sediments under the unconformity are mostly Mesozoic with a very thin succession of overlying Palaeogene deposits. The sediments above the unconformity were deposited during the Neogene uplift of the land areas to the east. The sediments are cut by two deep submarine canyons that were probably formed from erosion by sediment-laden submarine flows derived from rivers during melting of the ice that formerly covered the coastal region.

In many places glaciers descend from the Inland Ice through the mountainous terrain all the way to the fjords, where they calve to form icebergs. The head of Danell Fjord, South-East Greenland.

Photo: B. Thomassen, GEUS

THE ICE AGE 11

The patterns of glacial erosion and deposition are imprinted on all the older geological formations

GLACIATION FOR MORE THAN 2 MILLION YEARS

11 THE ICE AGE

The Earth's climate has undergone great variations during its geological history, changing from periods when the Earth's average temperature was about 22°C to periods when its temperature was about 10°C lower (see p. 110). During the cool periods, the polar regions were covered by large ice sheets and local ice caps formed in high mountain areas in other parts of the world. During the last 5–7 Ma the average temperature of the Earth has been low enough for ice sheets to build thicknesses of up to several kilometres on land areas near the poles and at times the ice sheets have extended as much as 3000–5000 km from the poles. Temperatures alternated during the ice ages between slightly colder and slightly warmer conditions, causing the ice to advance and retreat. The colder intervals are called glacials and the slightly warmer periods interglacials. Each glacial lasted for about 100 000 years and the interglacials were typically 10 000 years long. There have been extensive ice sheets in both the northern and southern polar regions throughout the last 1.8 Ma, the period of Earth's history known as the Quaternary.

A glacier tongue and its moraine. The moraine is a rampart of sand, gravel and boulders that the glacier ploughed up in front of itself and left stranded after it melted back. The picture shows a glacier that flows down a steep hillside near Kap Ammen in North Greenland. The glacier tongue is about 2 km wide.

Greenland's Inland Ice is the remainder of a much larger Quaternary ice sheet that covered nearly the whole of Canada, Greenland, Northern Europe, parts of Siberia and parts of the Arctic Ocean, and there was a similar ice sheet over Antarctica. At its maximum extent, the Inland Ice covered the whole of Greenland and its continental shelves. Since then, the ice has retreated to its present limits with a few readvances. Viewed in the perspective of geological time, our present climate is an interglacial and extrapolation from the climate swings that have occurred previously leads us to expect the start of a new glacial epoch in about 5000–10 000 years.

During glacial periods, large amounts of water from the oceans are bound up in the great ice sheets causing sea level to fall, so that some areas formerly covered by shallow seas become dry land. An opposing effect is produced by the weight of the kilometre-thick ice sheets pressing the land downwards until isostatic equilibrium with the underlying mantle is reached (see p. 27). As the climate warms and the ice melts, the resulting sea-level rise is countered by the slow uplift of the land as the weight on it decreases due to thinning of the ice sheet. As the land and sea rise and fall, the location and height of the coastline varies relative to the land surface. By studying glacial/interglacial sediments and their ages, geologists can unravel this complex interplay between sea-level change and the vertical movements of the land surface due to the varying isostatic loads of the ice sheets in glaciated regions.

The Quaternary period is divided in two main intervals. The older is the Pleistocene, from 1.8 Ma to 11 700 years ago, and the younger is the Holocene, or post-glacial time, from 11 700 years ago to the present. The Pleistocene is the period during which much of the polar regions was glaciated, while the Holocene comprises the interglacial since the last retreat of the ice. Ice advanced and retreated several times during the Pleistocene, but in general only the most recent cycles can be studied from the glacial sediments preserved onshore. In contrast, the entire history of glacial advances and retreats can be studied in material recovered by drilling offshore, and temperature curves for the whole period of the Ice Age have been obtained by analysing oxygen isotope ratios in shell remains from the drill cores.

The oldest material affected by the ice sheet preserved on land in Greenland has been dated at 2.4 Ma, near the end of the Pliocene. Drill-core material from the shelf, however, suggests that there may have been glacial activity as long as 7 Ma ago. The Inland Ice began to develop late in the Pliocene, but it melted away completely several times until about 2 Ma ago, near the start of the Pleistocene, when the ice cover became much more stable. Since then it has existed continuously, although its size has varied considerably. Most of the glacial sediments found on land in Greenland are less than 400 000 years old and were deposited during the latest two glacials, which were separated by the Eemian interglacial, from 130 000 to 115 000 years ago. Most of the glacial sediments in Greenland are from the last glacial epoch, the Wisconsian or Weichselian, which lasted from about 115 000 to about 11 700 years ago. At its maximum extent, about 150 000 to 200 000 years ago, the Inland Ice reached the outer edge of the continental shelf, up to 200–300 km beyond the present-day outer coastline. At this time, the Inland Ice was so thick that it covered almost the entire country and probably only a few, small, isolated areas were ice-free.

Glacial erosion

Evidence of the complete ice cover during the Quaternary can be seen everywhere in Greenland and far out onto the surrounding continental shelf. That the ice previously covered a much greater area is testified particularly by the landforms in the present ice-free zones around the Inland Ice – features such as fjords and deeply incised, U-shaped valleys. Glacial deposits such as moraines and meltwater sediments are widespread on land, and submarine canyons and gigantic, kilometre-thick sediment fans (see pp. 24, 173 and 179) are found on the continental shelf in front of some of the major meltwater drainage systems. These fans are the final resting places of much of the material eroded from the adjacent land areas.

The erosive power of the ice was enormous, which is not difficult to appreciate when contemplating how the glaciers have sculptured the giant U-shaped valleys and fjords that have a relief of up to several kilometres. The rate of erosion by the ice in the valleys can be as much as 2–3 m per thousand years, which is up to 10 times faster than normal non-glacial erosion. This happens in four main ways: 1) by glaciers and large ice caps transporting rock debris that falls onto their surfaces from surrounding cliffs, 2) by glaciers plucking blocks from their sides, backwalls and bases, 3) by powerful erosion by sub-glacial rivers and 4) by the rock debris frozen into the base of the moving ice causing a sandpaper-like abrasion of the bedrock. The many meltwater rivers that drain the ice are also important agents of erosion. They transport material through the steeper valleys to deposit it in places where the water flows less rapidly; the material may be moved up to several hundred kilometres from its place of origin.

Quaternary deposits

Ice-related sediments consist of fragments of rocks and minerals. They can be deposited directly from melting ice or be material ploughed up by the ice itself. The latter can be till deposits (boulder clay) formed at the base of the ice, or it can be moraines within, in front of or along the margins of the ice (lateral moraines, end-moraines and marginal moraines). The moraines consist of an unsorted, heterogeneous mixture of clay, silt, sand, gravel and boulders. Meltwater sediments, in contrast, are better sorted and commonly consist of clay, silt, sand or gravel layers that often form flat terraces at the bottoms of the large river valleys. Both moraines and meltwater sediments are found far out on the present-day continental

ICE SHEETS DURING THE LAST GLACIAL

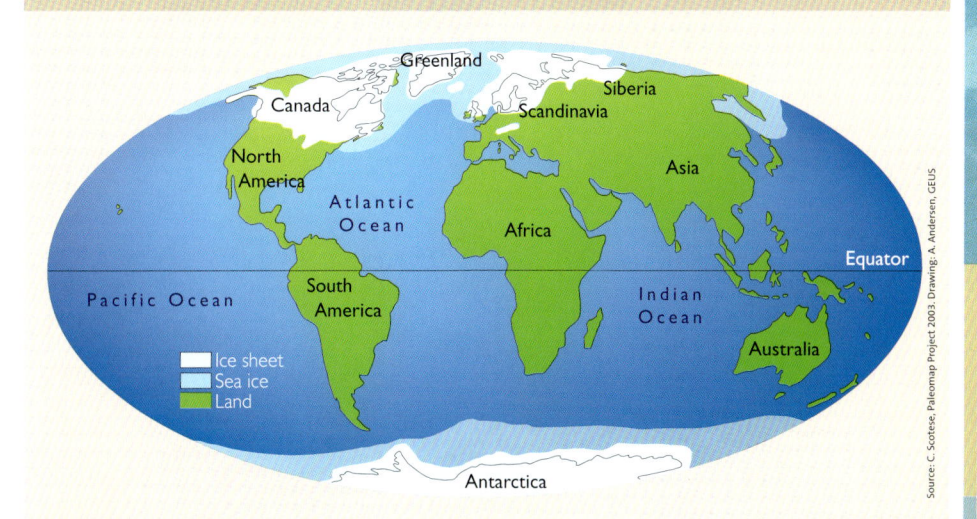

The large ice sheets during the last glacial about 18 000 years ago. Ice caps covered much of the land in both the north and south polar areas. The seas around them were partly covered by large contiguous areas of floating ice that probably resembled the ice shelves found today around Antarctica.

Time column showing glacial conditions in Greenland. Glaciation had already begun before the end of the Pliocene, before the Ice Age proper started in the Quaternary. Note that our postglacial time, called the Holocene by geologists, consists of only the last 11 700 years of the Ice Age that has lasted more than 2 Ma. Blue: Glacials. Red: Interglacials.

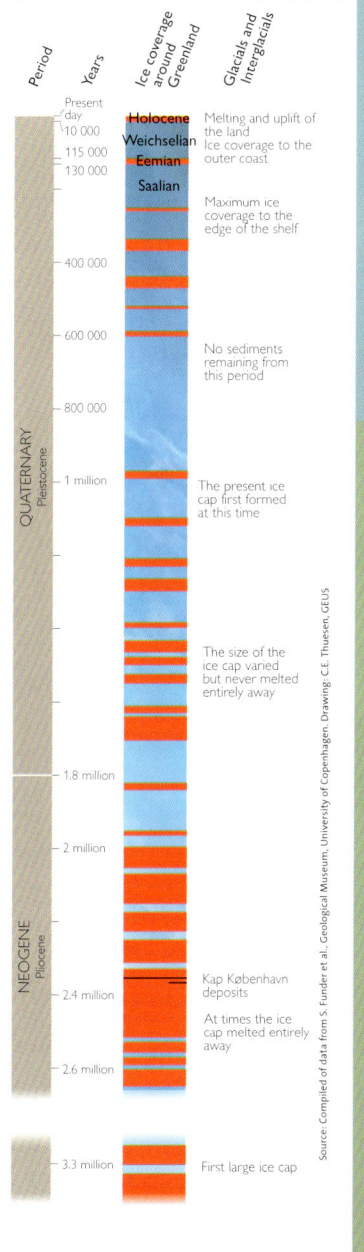

THE ICE AGE

CLIMATE CHANGES DURING THE ICE AGE

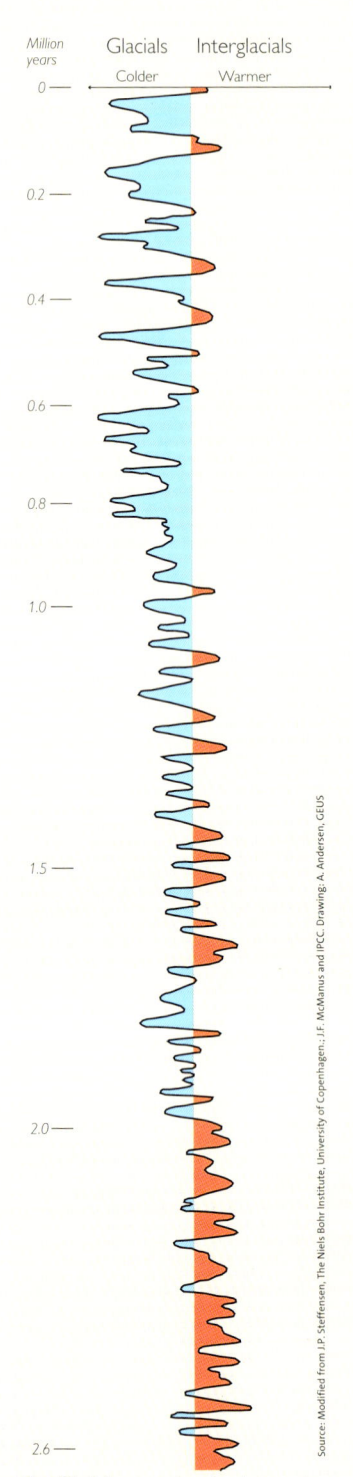

The glaciation of the northern polar regions began during the last part of the Cenozoic era, about five million years ago, whereas an ice sheet started to form over eastern Antarctica 20–30 million years earlier. Early glaciation caused a significant lowering of global sea level by about 50 m and the later expansion of the ice cover in the polar regions caused a further drop in sea level to about 130 m below its present level. The onset of glaciation is a clear sign of a fall in the average temperature of the Earth. Temperatures during the long period of glaciation have not been constant but have varied between colder and warmer intervals. This has caused the ice sheets to expand during cold periods (glacials) and retreat during the intervening warmer periods (interglacials). A total of about 20 oscillations between glacial and interglacial conditions has been recognised, during which the Earth's average temperature varied from about 10°C to about 20°C. Traces of the last four glacials are visible in the Alps, but in Greenland only limited evidence has been found of the very earliest glacials older than 2.4 Ma; much of the preserved record from Greenland concerns the most recent glacials and interglacials within the last 300–400 thousand years.

Our knowledge of the variations in the Earth's temperature during the last few million years comes primarily from drill cores from the deep oceans. Studies such as oxygen-isotope analysis of the shells of single-celled animals (foraminifera) show that the temperature of the oceans has oscillated during the last 5 million years in a fairly regular fashion. These variations in temperature must also have affected the land areas, but the sparse information has made it possible to find evidence of only a few of the swings. This is because more recent processes on land tend to erase traces of earlier events, especially in glaciated areas, where more recent ice erodes away the evidence of earlier glaciations.

The reasons for the variations in temperature are primarily astronomical. The Serbian astronomer Milutin Milankovich used detailed observations to show that the angle that the Earth's rotation axis makes with its orbit and the eccentricity of its orbit are not constant, but vary in three 'Milankovich cycles' that result in variations in the amount of sunlight falling on the Earth's polar regions. The three cycles are of different lengths. The main one is about 100 000 years long and determines the length of the glacials. The two other cycles are somewhat shorter: 41 000 years and 23 000 years.

These slow variations do not explain all of the observed changes in the glacial record. The frequent and sudden variations in temperature that have been documented from studies of the Greenland ice cores are not explained by these astronomical cycles, and scientists have recently suggested that the abrupt changes are caused by sudden shifts in the circulation systems in the oceans. Further research is needed to identify the reasons and mechanisms behind these abrupt changes in climate.

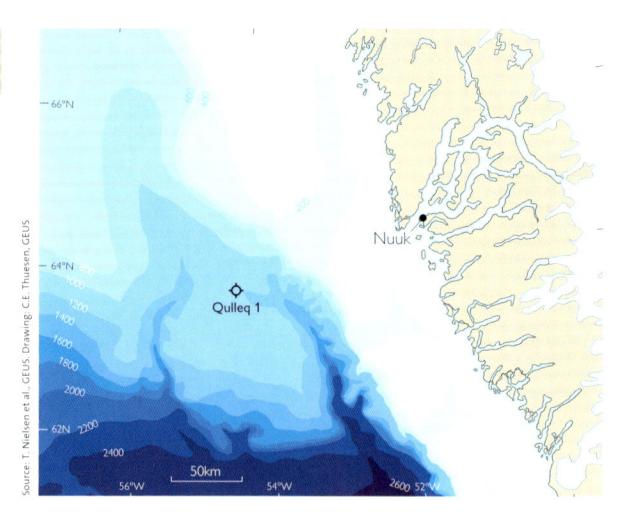

Simplified diagram showing temperature variations during the last 2.6 Ma, compiled from a variety of sources including ice cores and sediments recovered in deep-sea cores. Temperatures have varied rhythmically during that long interval of time creating 'glacial' and 'interglacial' conditions.

Bathymetric map of the area west of Nuuk in West Greenland, showing water depths contoured in metres. A steep escarpment separates areas of shallow water along the coast from water depths of more than 2600 m to the west. The shelf is dissected by a number of submarine canyons caused by erosion by meltwater streams from ice. Qulleq-1 is an oil exploration well.

shelves. Another special type of deposit, found both on the shelves and in deep-water areas farther out, consists of fine-grained sediments that contain isolated patches of anomalously coarse sediments – gravel and boulders. Such deposits are formed when the exotic (i.e. derived from a distant source), coarse-grained material is transported out to sea by floating glaciers or icebergs. When the ice melts, the exotic material falls to the seabed and is incorporated in the fine-grained sediments normally deposited in this environment. This type of heterogeneous sediment is clear evidence that ice existed at the time of its deposition. The exotic material is known as ice rafted debris.

Quaternary marine sediments are found principally on the shelves and as fine-grained clays and silts in the adjacent deep-water areas. Fine-grained sediments found closer to land are mostly of Holocene age and form terraces along fjords and around river mouths. These sediments commonly contain mussel and snail shells, animals that lived in the sea around Greenland when the sediment was deposited. Some of the animals lived in warmer waters than exist today around Greenland, so they are clear evidence of some of the climate changes that have happened. The marine terraces are found at different levels, from a few metres to more than 100 metres above present sea level. Since all of these silt layers were deposited at shallow depths in the sea, their present height above sea level shows roughly how much the land has been uplifted relative to the sea surface since they were deposited.

Sediment deposited in lakes has been particularly useful in elucidating the climatic and biological de-

THE ICE AGE

Large U-shaped valley 4 km wide with sides about 500 m high. The glacier terminates in a lake. Meltwater deposits of sand and gravel are visible in the foreground. The ice that formerly covered the whole area has modified the cross-section of the valley. The locality is in southern Peary Land in North Greenland.

Marginal moraine of sand, gravel and boulders left by an ice tongue that has now melted back and is not seen in the picture. Traces of the former extent of the ice are found as deposits along the former ice margins, commonly as rampart-like moraines like those shown in the picture. The moraine is 10–20 m high. The picture was taken in Hall Land in North Greenland.

velopment since the last retreat of the ice. Plant remains in drill cores taken from the sediment at the bottom of the lakes have been especially important because they show the progression of how plants colonised the land following glacial retreat and how the vegetation responded to variations in climate since the end of the last glacial.

The older glacials and interglacials

An important upper Pliocene sedimentary succession about 100 m thick has been found at Kap København in easternmost Peary Land in North Greenland. It contains bivalves, insects and other fossils showing clearly that the climate at that time was substantially warmer than now. The sediments have been dated to be 2.3 Ma old and moraine layers both below and above them show that they were deposited during an interglacial, the earliest interglacial known from Greenland. Comparison with the global Ice Age stratigraphy suggests that the underlying moraine was formed during a glacial about 2.5 Ma ago, when central parts of Greenland were covered by an ice cap.

Bivalve shells lying on the surface of a marine silt deposit of Holocene age. Marine deposits like this are commonly found as flat terraces at altitudes up to 140 m above sea level along the sides of fjords and the mouths of rivers. The shells often lie in small banks that are weathered out of the sediment, as shown in the picture. Nordpasset, Peary Land, North Greenland.

Large, flat, meltwater terraces about 10 m high consisting of layered sands and gravels that were deposited by rivers emerging from the Inland Ice. There are commonly several terraces at different levels, showing that the height of the water in the river and the uplift of the land have changed with time. Wulff Land, North Greenland.

THE ICE AGE

The ice melted away completely during the period when the Kap København beds were laid down, but the ice cap formed again about 1.9 Ma ago and, despite fluctuations in size, has persisted ever since.

Evidence for the end of a major period of glaciation about 3.3 Ma ago has been found in cores drilled from below the sea-bed offshore Greenland. It is therefore thought that a large contiguous ice cap must have existed prior to this time. It is likely that these early ice caps occasionally melted away completely. Only after the beginning of the Quaternary has the Inland Ice been present continually, even though it has varied substantially in size.

The colder climate after the start of the Quaternary 1.8 Ma ago caused ice to spread out from the mountains near the coast of Greenland to cover the centre and, on several occasions, to expand out onto the present-day continental shelf. There have been several glacials and interglacials between 1.8 Ma and 11 700 years ago, but deposits from the last two glacials dominate in Greenland. The last but one glacial, known as the Illinoian (in North America) or Saalian (in Europe), lasted from 400 000 to 130 000 years ago and the last interglacial, the Eemian, lasted from about 130 000 to 115 000 years ago. The last glacial, known as the Wisconsian (in North America) or Weichselian (in Europe), lasted from about 115 000 to 11 700 years ago and was followed by the so-called postglacial interval the Holocene, which really is an interglacial. The Weichselian and Holocene are described separately below.

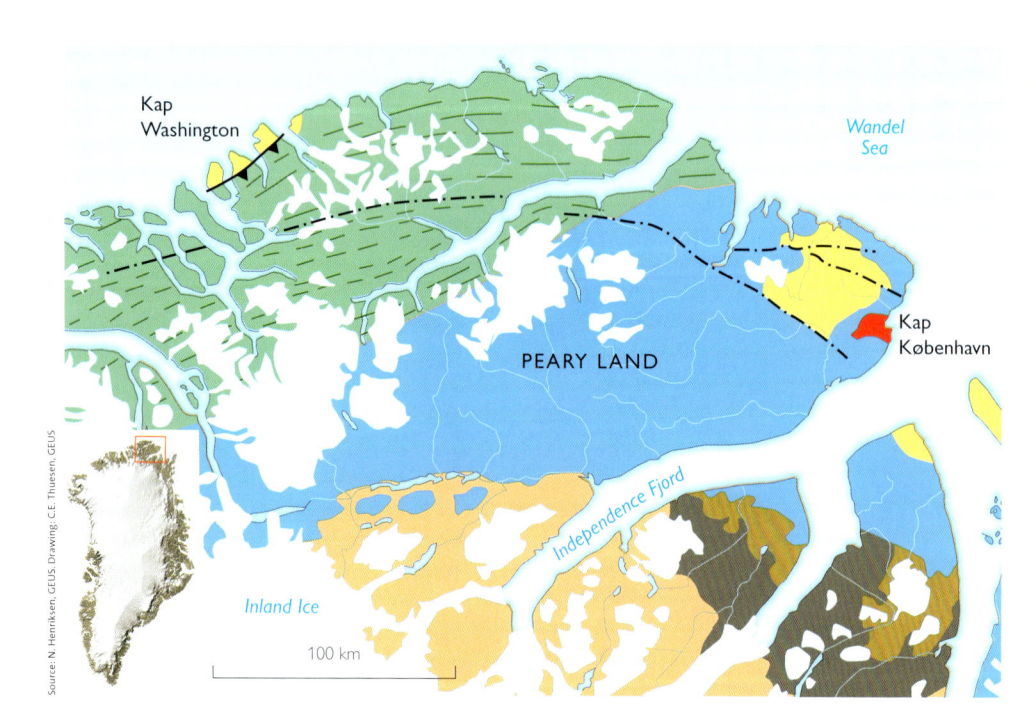

Traces of the last but one glacial are found widely in Greenland. The whole country was covered in a great thickness of ice and many deposits from this time are found isolated in the high mountains as deeply weathered tills and scattered meltwater sediments. It is difficult, however, to correlate these deposits from place to place and to measure their ages with any accuracy. Correlation with sediments offshore suggests that this glacial activity took place a few hundred thousand years ago. Its ice reached the outer edge of the shelf, 50–300 km from the present-day coast.

The average temperature during the last interglacial (the Eemian) was 4–5°C higher than that of the present day. Global sea level was also somewhat higher than today's because a large proportion of the world's ice sheets had melted. The ice cap was substantially smaller than today's Inland Ice, although large parts of interior Greenland were still ice-covered. Marine sediments from the Eemian are found in Greenland's outer coastal area as flat beds and terraces up to 50 m above present sea level. The sediments contain many plant and animal remains, such as bivalves and snail shells. A warmer climate than that of today is revealed by the presence of shells of the blue mussel (*Mytilus edulis*) in these sediments. This sub-Arctic species is not found in northern Greenland today, but is widespread on the coasts of northern Europe. Remains of many other animals that do not exist in present-day Greenland are also found in these sediments.

The last glacial – the Weichselian

The phrase 'the last ice age' is often used for the last glacial, the Wisconsian or Weichselian, that lasted from 115 000 to 11 700 years ago. Calling it 'the last ice age' is really a misnomer because the Holocene is, in fact, just an interglacial in the general Ice Age, that has lasted more than two million years, so we really ought to call it 'the latest glacial'.

This period in Greenland was charactised by advances and retreats of the ice cover, leaving various ice-margin deposits such as tills, terminal moraines and meltwater deposits. These glacial deposits have been used to reconstruct the different phases of ice movement, to document fluctuations in the degree of ice cover and to gain an oversight into the gradual retreat of the ice at the end of the period.

Greenland's annual average temperature was 25°C colder than now during the coldest phase of the latest glacial, but there were short periods during which temperatures were only 5°C lower than those of today. Boreholes have penetrated the 3000 m thick central zone of the Inland Ice, yielding ice cores that give a very precise picture of temperature variations all the way back to 120 000 years ago. In this way we have detailed temperature data from about the last 5000 years of the warmer Eemian interglacial, all the

Using a simple corer in a small lake to take samples of the uppermost sediment layers under its floor. The equipment consists of a single pipe that is pressed into the soft sediments at the bottom. The sediment core with its contents of animal and plant remains is analysed in a laboratory to show how animal and plant communities have varied with time as a result of climate change.

Simplified geological map showing the 2.3 Ma old Kap København deposits (red) in a small area of eastern Peary Land in North Greenland. The geology of the region is dominated by the much older Precambrian (orange and brown) and Lower Palaeozoic (blue) sedimentary basins, the North Greenland Fold Belt (green) and the Carboniferous–Palaeogene Wandel Sea Basin (bright yellow).

THE ICE AGE

way through the Weichselian glacial to the present-day.

Evidence of slightly colder and slightly warmer periods during the earliest part of the last glacial between about 115 000 to 70 000 years ago, when the ice advanced and retreated several times, has been found in the Scoresby Sund region in East Greenland. The ice cover was not nearly as thick as during the previous glacial, so large areas of high ground were ice-free.

During the middle of the last glacial, the East Greenland ice cover was very stable, without large movements for nearly 50 000 years. The large glaciers filling the East Greenland fjords were relatively thin and reached the present-day fjord mouths. Conditions may have been less stable in West Greenland, with more advances and retreats.

The Inland Ice reached its maximum extent towards the end of the last glacial (last glacial maximum), when the extent of the ice was very different in various parts of Greenland. The last and most extensive advance came 22 000 to 21 000 years ago when the highest mountains may have remained free of ice, but glacial ice covered all the low ground and filled the valleys and fjords. In central East Greenland, the ice reached only as far as the present-day fjord mouths, but off Kangerlussuaq in South-East Greenland, it reached as far as about 200 km out from the present-day outer coastline.

In southern West Greenland at this time, the ice stretched out as far as the shelf edge and the mountain tops were completely buried by ice. Several large valleys formed under the ice on the shelf that were drainage channels for the main glaciers. It is thought that the northern West Greenland ice was substantially more stable with an ice margin a little beyond the present-day coast, but data are insufficient for us to be sure.

During the glacial maximum in North Greenland, the ice margin was to the north of the present coast and the area between Greenland and Ellesmere Island (Canada) was under a continuous ice sheet that covered both land areas. A large ice shelf like those known today around Antarctica covered part of the polar basin north of Greenland, Canada and Russia.

The retreat of the ice began between 19 000 and 16 000 years ago, when the glaciers and ice caps that covered the outer shelf gradually disappeared and their melting contributed to a general rise in sea level. Melting reached the present-day onshore area about 11 700 years ago. Sea level continued to rise and marine sediments were deposited along the contemporaneous coast. These sediments are found today as terraces up to 140 m above present-day sea level, due to isostatic uplift of the land.

The retreat of the ice to its present margins

The retreat of the ice accelerated after the start of the Holocene. At the beginning of the period, the outer limits of the ice were close to the present-day outer

The first ice cap probably formed about 3 Ma ago and ice caps must have melted completely and reformed several times during the following million years. The ice cap was entirely missing at the time the Kap København deposits were laid down, about 2.3 Ma ago when the interior of Greenland was probably covered by coniferous forest (dark green) and deciduous trees were found only near the coast in South Greenland (light green). The mountainous areas at that time were covered by heath (brown) with local ice caps (white). The size of the ice cap has varied substantially since the start of the Pleistocene, 1.8 Ma ago, but has never melted away entirely since then.

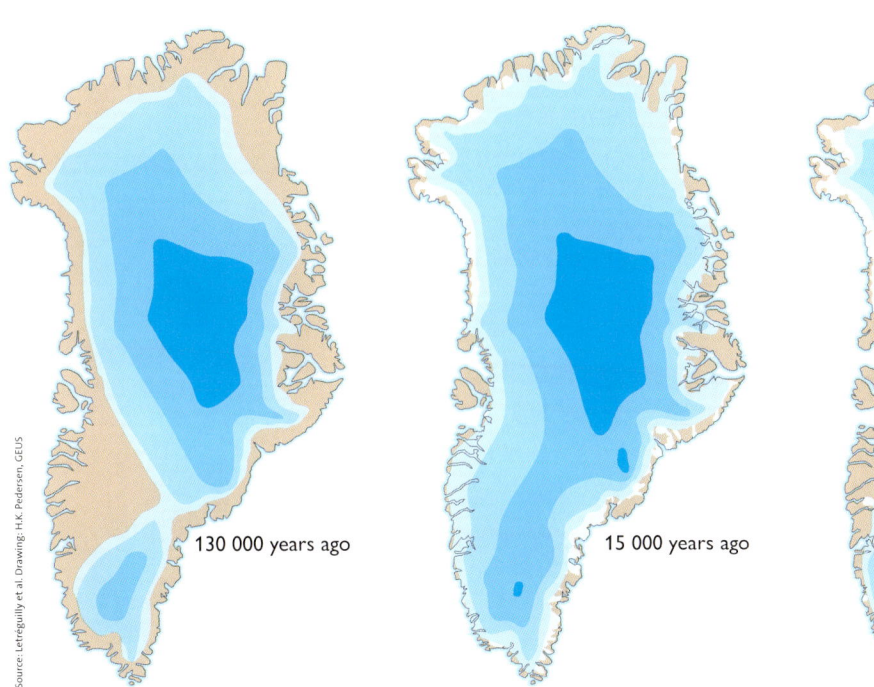

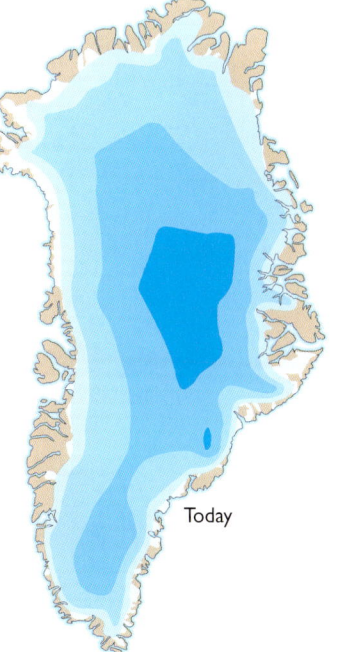

Calculations of the thickness and extent of the Inland Ice at three different times:
1) at the start of the Eemian interglacial about 130 000 years ago, when the temperature was 4–5°C higher than today,
2) near the end of the last glacial 15 000 years ago, when the temperature was 10–12°C lower than today, and
3) the present extent of the ice.

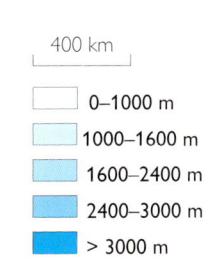

- 0–1000 m
- 1000–1600 m
- 1600–2400 m
- 2400–3000 m
- > 3000 m

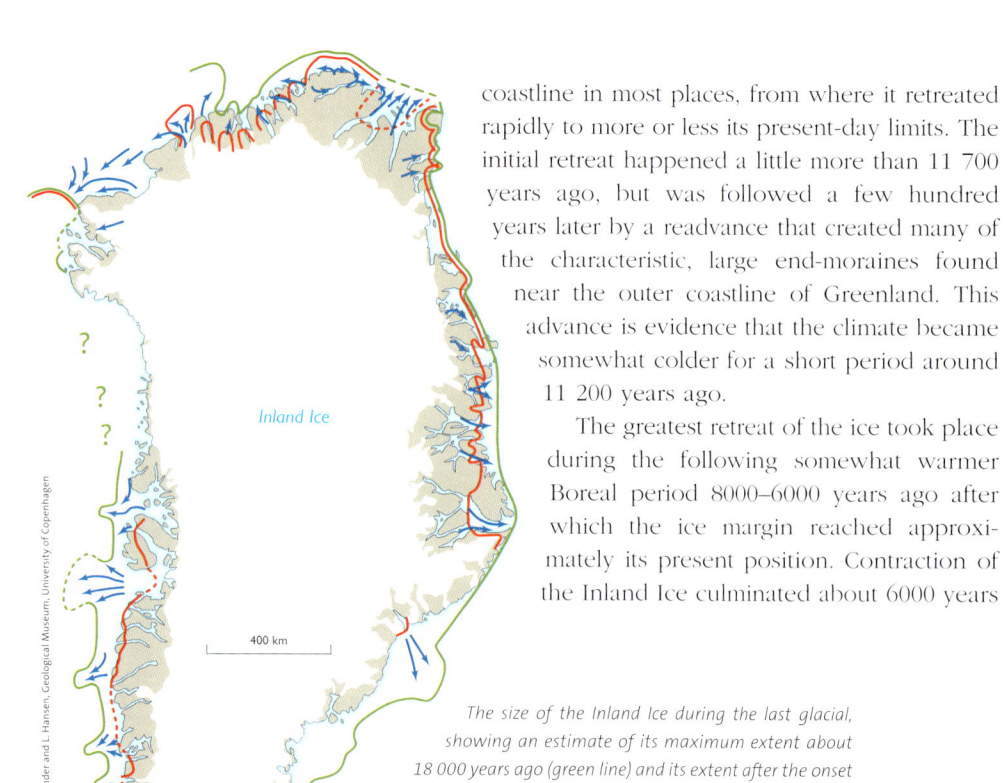

THE ICE AGE

coastline in most places, from where it retreated rapidly to more or less its present-day limits. The initial retreat happened a little more than 11 700 years ago, but was followed a few hundred years later by a readvance that created many of the characteristic, large end-moraines found near the outer coastline of Greenland. This advance is evidence that the climate became somewhat colder for a short period around 11 200 years ago.

The greatest retreat of the ice took place during the following somewhat warmer Boreal period 8000–6000 years ago after which the ice margin reached approximately its present position. Contraction of the Inland Ice culminated about 6000 years ago when in many places its margin was between 20 and 40 km farther back than now. Between 8000 and 4000 years ago, the average temperature was somewhat higher than that of today and the summers were particularly warm. Temperatures began to fall again about 3500 years ago and the ice margin advanced again to reach its present-day position.

Uplift of the land after the Ice Age

Flat-lying terraces consisting of uplifted Holocene marine sediments occur in many places along the coast and in the inner fjords and bays. The sediments accumulated in the sea near the contemporary coast and consist of fine-grained sands and silts deposited in shallow water. They contain the shells of various bivalves and snails whose ages can be determined by the radiocarbon-14 method and whose ecology indicates the temperature of the water in which they lived. The melting of ice on land began approximately 11 700 years ago and the earliest raised marine sediments and lake deposits are younger than that. The marine sediments were uplifted to their present altitude by the isostatic response of the land

The size of the Inland Ice during the last glacial, showing an estimate of its maximum extent about 18 000 years ago (green line) and its extent after the onset of melting about 10 000 years ago (red line). The blue arrows show where the largest glacier streams flowed out.

Characteristic view of the margin of the Inland Ice showing moraines and small lakes in the ice-free landscape beyond the ice. The distance from the foreground of the picture to the ice-free ground farthest away is about 5 km. The locality is about 75 km north of Kangerlussuaq Airport, West Greenland.

THE ICE AGE

to the reduction in the overlying mass and retreat of the thinning ice (see p. 27).

The uplifted marine terraces are found at heights from a few metres to about 140 m over present-day sea level. By mapping these terraces and determining their ages, it is possible to calculate how much the land has rebounded at different times after the ice melted. The pattern is much the same everywhere in Greenland. The greatest accumulated uplift of around 140 m is found where the ice-free coastal areas are broadest, while the uplift is normally not more than 40–50 m where the strip of ice-free land is narrower. Where the margin of the Inland Ice is close to the coast, the uplift is less than 20 m.

What the Inland Ice can tell us about the past

The precipitation that falls onto the Inland Ice is normally snow. Because of the low temperatures it does not melt and simply stays where it falls. New snow falling onto the older snow buries it, forming annual layers that become more and more compressed the deeper below the surface they are buried; gradually the snow is transformed into ice. At a depth of about 80 m, the snow is compressed completely to ice that contains air bubbles and grains of dust. Since temperatures in the inner part of the Inland Ice have

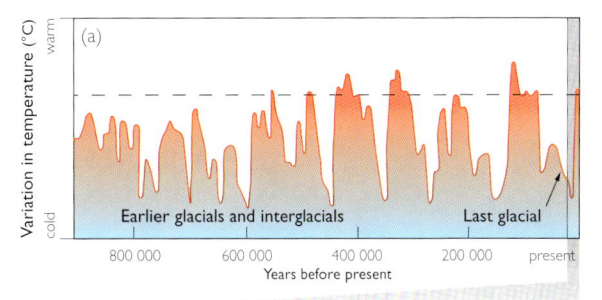

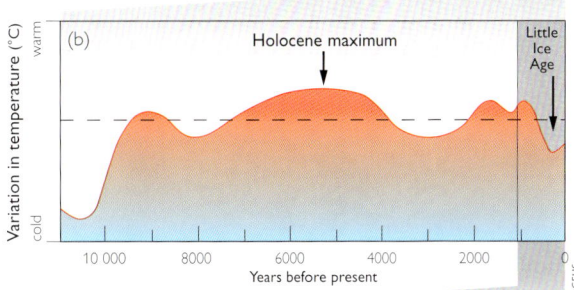

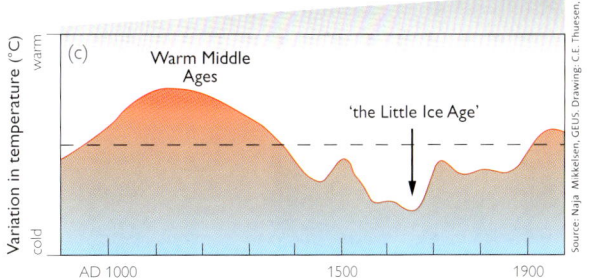

Curves showing the temperature variations during the last 900 000 years. During the last (Weichselian) glacial, the average annual temperature was about 5–25°C colder than today and ice covered Greenland the whole time, even though the margins of the ice migrated back and forth as the temperature varied. The final retreat of the ice began about 11 700 years ago when there was a substantial rise in temperatures. The difference between highest and lowest temperatures shown on the upper curve (a), is about 20–25°C. The two lower curves (b and c) show details of the changes in temperature during the Holocene. The temperature during the Holocene maximum was about 2°C above today's average and in the 'Little Ice Age' the average temperature was about 1°C lower than today's.

Stepped terrain produced by flat-topped marine terraces ('raised beaches') of Holocene age deposited in the coastal zone in front of a steep slope eroded in Palaeozoic sediments. Each terrace contains sediments that were deposited near a former coastline and uplift of the land has now raised them to their present heights, up to nearly 80 m above present-day sea level. The oldest (highest) terrace consists of sandy and silty sediments 9000 years old. Northern Washington Land, North Greenland.

Larch and spruce tree trunks, from about 2.4 Ma ago, from Washington Land in North Greenland. The longest trunk is 176 cm long, indicating that the trees were relatively small. The trunks have been smoothed by transport in a river.

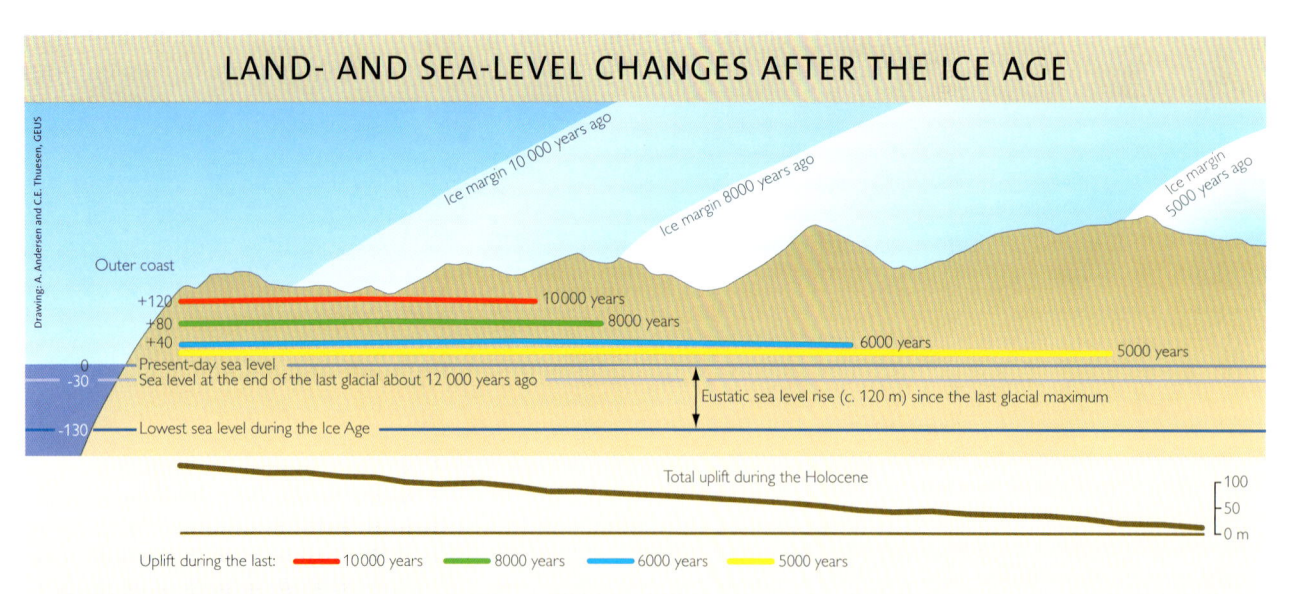

THE ICE AGE

LAND- AND SEA-LEVEL CHANGES AFTER THE ICE AGE

The diagram shows a hypothetical cross-section through part of West Greenland where the interplay between the melting of the ice cap, sea-level change and the consequent uplift of the land is shown schematically. Ten thousand years ago, the edge of the ice lay near the present day outer coast and sea level was about 30 metres lower than today. By 8000 years ago, some of the ice had melted and the ice margin had moved east. By 5000 years ago, the edge of the ice was even farther east. The oldest marine deposits, about 10 000 years old, are found at a height of about 120 metres near the outer coast. The upper limit of marine sediments descends eastwards and lies at 20 metres farthest to the east where the sediments are about 5000 years old. The diagram thus shows how much uplift there has been relative to present-day sea level, from 120 m near the coast to 20 m inland.

The huge volumes of water that were bound up in the ice sheets during the Quaternary glacial maximum caused global sea level to be about 130 m lower than it is today, so many of today's shallow seas (e.g. the North Sea) were dry land. If the ice that still forms the Greenland and Antarctic ice caps today were to melt, global sea level would rise about 60 m and extensive lowland areas of the world would be flooded. These changes in sea level, called eustatic changes, are absolute and can be measured globally in areas of stable continental crust.

Glaciation of large areas of land during the Ice Age caused huge amounts of ice to accumulate to a thickness of 3–4 km. The weight of this ice-mass burdened the crust beneath and caused it to sink until it was in isostatic equilibrium (see p. 27). The weight of the Scandinavian ice cap, for example, caused the crust under its centre to sink about 300 m into the mantle and it has been calculated that the centre of Greenland under the Inland Ice has been pushed down about 800 m.

Melting of the ice during interglacials and during the Holocene happened rapidly, causing global sea level to rise in the space of a few thousand years, while it has taken much longer for the land to rebound after the weight from the ice was removed. The rise in sea level was therefore not synchronous with the uplift of the land and the effects of these processes were felt over different time spans. The combined effects of uplift and sea-level variations mean that it is possible to measure only relative sea-level changes in formerly glaciated areas. The only record left is where the sea encroached upon the land at a given time, and this was a balance between eustatic sea-level rise and post-glacial crustal rebound.

Geologists have used old raised beaches in Greenland to measure how the relative sea level changed during the last 10 000 to 14 000 years. Fossil shells found in these beach deposits give an accurate date for when sea level was in this position. Since then, isostatic rebound has caused the land to rise so that the old beach deposits are now found at various heights above sea level, from a few metres up to 140 m. There are, however, large local differences in the height of the raised beaches. The highest raised beaches are found where large amounts of ice have melted and thus the unloading of the land was greatest.

The gradual retreat of the ice combined with occasional advances and the formation of raised beaches in front of the ice have given geologists a fantastic dataset with which to evaluate the detailed relationships between melting of the ice and uplift of the land. These data, combined with data from lake deposits and their record of plant and animal remains, show how the landscape and climate have changed since the end of the last glacial 11 700 years ago.

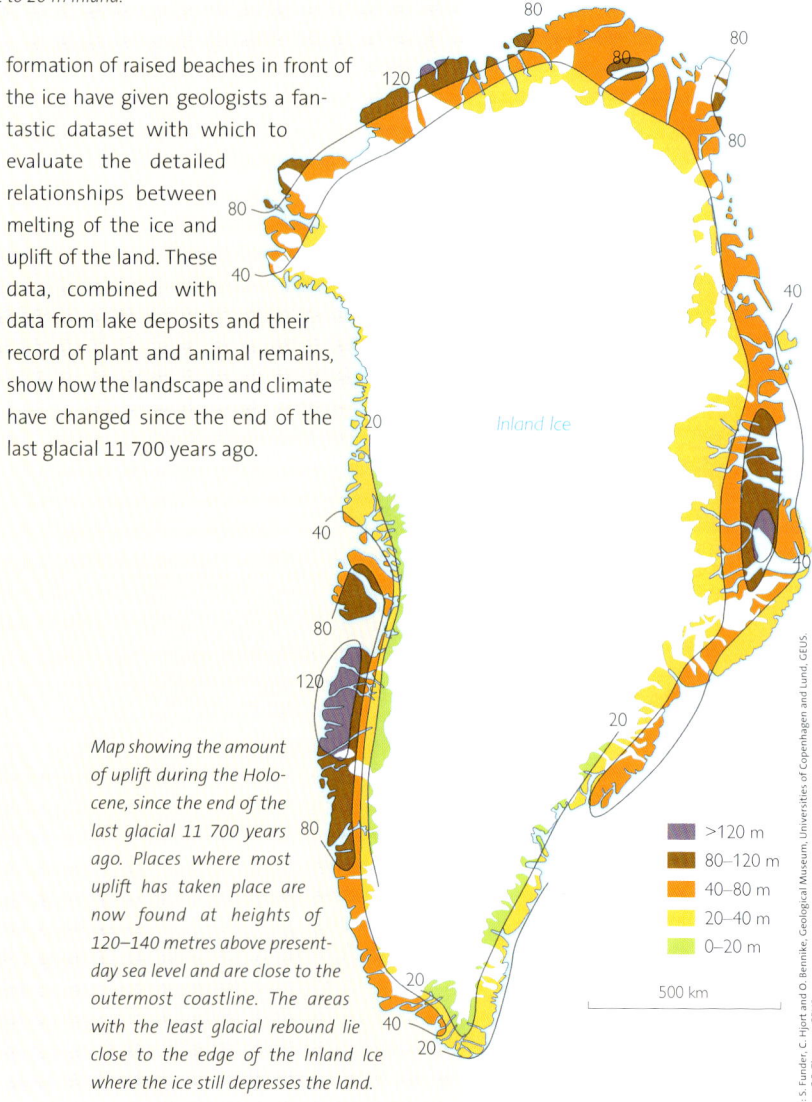

Map showing the amount of uplift during the Holocene, since the end of the last glacial 11 700 years ago. Places where most uplift has taken place are now found at heights of 120–140 metres above present-day sea level and are close to the outermost coastline. The areas with the least glacial rebound lie close to the edge of the Inland Ice where the ice still depresses the land.

been constantly below freezing for several hundred thousand years, the annual snowfall has been preserved and has accumulated to a thickness of about 3 km. The oldest layers at its base are more than about 250 000 years old.

Oxygen in water molecules (H_2O) consists of two different isotopes. The most common is 'normal' oxygen with an atomic weight of 16 (^{16}O), but a small proportion consists of a heavier oxygen isotope with an atomic weight of 18 (^{18}O). The relative amount of the two isotopes in the falling snow depends on its temperature. Snow falling at higher temperatures contains more ^{18}O atoms than snow falling at lower temperatures. The proportion of ^{16}O to ^{18}O atoms can therefore be used as a temperature indicator and past temperatures can be estimated by measuring their ratio in samples of ice. By measuring this ratio in every annual layer, it is possible to track changes in temperature from year to year and thus to evaluate climate change in the past.

The grains of dust that fall onto the snow on the surface of the Inland Ice become frozen into the individual annual layers of ice, so they too can be analysed. The dust grains were transported onto the ice by wind, both from local and from global wind systems. Most of the dust was derived from erosion of the land area surrounding the Inland Ice, but some layers in the ice cores include dust from large volcanic eruptions far away. Such volcanic activity is also recorded in the ice layers by their content of sulphuric acid, carried as aerosols in the global wind systems and precipitated with the snow.

Five deep holes have been drilled into the Inland Ice. The first was in 1966 at Camp Century east of Thule in North-West Greenland where the American military had established a research station. The hole reached the base of the ice at a depth of 1390 m. The next deep hole was drilled at the Dye 3 radar station in South Greenland in 1981 by an American–Danish–Swiss consortium, the Greenland Ice Sheet Program (GISP). It reached basement rock below the ice at a depth of 2038 m. In 1990–1992 a European consortium, the Greenland Ice Core Project (GRIP) used drilling equipment developed in Denmark to drill close to where the Inland Ice is thickest (over 3000 m). It reached the base of the ice at 3029 m. At the same time, an American group (GISP 2) drilled another hole 30 kilometres away, finishing in 1993. The most recent drilling of the Inland Ice finished in 2003 after 7 years of field work when the European group GRIP drilled at the NGRIP (North GRIP) locality reaching the base of the ice at 3085 m.

Cores were cut during all these deep drillings and the material analysed. These data have provided an outstanding picture of the changes in Greenland's climate over the last 250 000 years, covering the last two glacials and two interglacials (see p. 183).

The deeper parts of the ice cores from central Greenland are, however, disturbed by folding and the temperature records are only fully reliable down to an age of a little more than 100 000 years. New data from NGRIP can extend this precise record down to about 125 000 years ago.

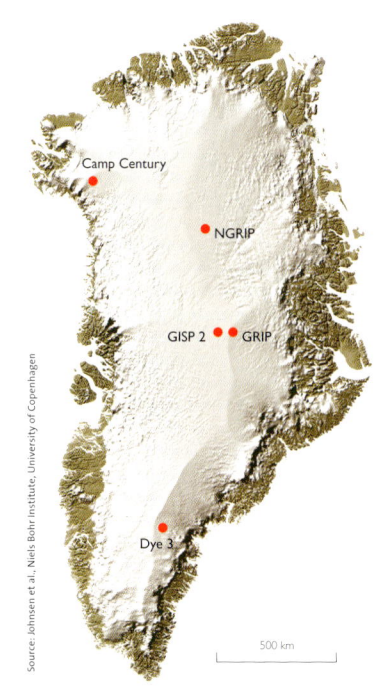

Map showing the locations of the five deep holes drilled into the Inland Ice prior to 2005. There are plans to drill again during the next few years to the north of NGRIP.

Ice cores from one of the deep holes drilled into the Inland Ice. The cores are placed in special trays and kept in a deep freeze.

Ice core with a diameter of 98 mm being extruded from the lower part of a drill pipe. The work has to be carried out below freezing so the scientists are dressed appropriately.

A number of shallower holes has been drilled on local ice caps and near the margins of the Inland Ice to supplement the deep holes. These holes can be drilled using smaller mobile equipment that can reach to depths of several hundred metres during one drilling season. A glaciologist is shown here removing the core from the drill.

Entrance to one of the mine adits in the Nalunaq gold mine, Kirkespirdalen in South Greenland.

Photo: Bureau of Minerals and Petroleum, Nuuk

MINERAL RESOURCES

Subsurface mineral occurrences are difficult to find – but they exist

12

FINDING THE NEEDLE IN THE HAYSTACK

MINERAL RESOURCES

Searching for mineral resources in Greenland takes place as a co-operative venture between the Geological Survey of Denmark and Greenland (GEUS) and private mining companies. GEUS undertakes the primary regional geological investigations on behalf of the Danish state, while private companies carry out detailed studies within allotted areas, based on time-limited licences with work commitments and taxes to the State.

The majority of the mineral resources required by modern society come from geological occurrences found in the upper parts of the crust. Examples of such resources are iron, copper, zinc, chromium and nickel, as well as the precious metals gold, silver and platinum. Other examples of mineral resources are non-metallic materials such as coal, graphite, gypsum and rock salt, as well as gemstones like ruby and diamond. In addition there are construction materials including granite and marble, as well as aggregates such as sand, gravel, limestone and clay.

Field work in connection with exploration for diamonds in the basement rocks of West Greenland south of Kangerlussuaq (Søndre Strømfjord). The geologists live in tents in the field and undertake systematic collection of large samples from localities that have possibilities for containing diamonds. The samples are collected in bags (see main photograph) and sent to a laboratory for analysis. When the results are available the prospects can be reassessed.

In order to find and exploit mineral resources, a detailed knowledge of their mode of occurrence and extent is required, together with an understanding of how to extract and then separate the minerals from the enclosing rock. This is usually achieved by a combination of various geological investigations such as mapping, geophysical and geochemical techniques as well as the engineering aspects of the mining and processing of the raw materials.

Mineral resources usually occur in limited quantities with considerable variation depending upon the type of material. In general ores comprise only a (very) small proportion of the rocks in which they are found. Exploration for minerals must, therefore, be organised as a thorough piece of detective work. Traces of minerals in the host rocks may lead to areas where the potential can be investigated more carefully. At an early stage surface investigations are undertaken by geological mapping and collection of samples for analysis. If these provide positive indications the next stage is drilling to obtain core samples from the hidden part of the ore body. From analysis of the mineral content of these samples an estimate can be made of how much of a particular element is present and the size of the occurrence. With this background, calculations of the economic potential can be made in order to estimate the possibilities of exploitation.

Exploration for mineral resources in Greenland is often a collaboration between GEUS and private mining and exploration companies where, in general, the private companies undertake the detailed surveys in concession areas with geophysical investigations

Diamonds from Greenland. Whilst those found so far are very small, their mere presence is an encouraging sign for further investigation.

NATURAL FORM OF MINERAL RESOURCES

Exploitable mineral resources, known as ores, can contain metallic elements as single minerals (e.g. gold ~ Au), or may comprise combinations of elements (e.g. magnetite ~ Fe_3O_4). To be able to extract the element sought it must be separated from the unwanted components that are of no immediate commercial value. This separation is mostly carried out by industrial processing of the raw ore, which usually takes place close to the site where the ore is mined.

Below are some examples of raw materials as they occur naturally. In the upper row three minerals are shown; gold and graphite are pure elements, while ruby is a combination of two elements. In the bottom row three rocks are shown: the two metallic ores can be worked for their iron and zinc contents, whereas the sandstone is a natural building stone that can be cut for use in the building industry.

GOLD

A quartz vein with a rich concentration of native gold (Au). This sample contains much more gold than an average gold-bearing quartz vein. The gold occurs as a pure element, unlike most ore minerals that are chemical combinations with other elements. The sample is from the gold mine Nalunaq, Kirkespirdalen, South Greenland, and is about 3 cm across.

GRAPHITE

Graphite is a non-metallic ore composed only of the element carbon (C). It occurs in different places in Greenland as bands in metamorphic schists within the basement gneisses. The sample shown is about 5 cm high and comes from Eqalussuit in West Greenland.

RUBY

Ruby is a red-coloured gemstone variety of the mineral corundum (Al_2O_3). The sample comes from the Archaean basement near Qeqertarsuatsiaat/Fiskenæsset, south of Nuuk in West Greenland, where it occurs with other rare minerals in specific horizons of the layered anorthosite complex (see p. 38). Corundum is very hard and is therefore used as an industrial abrasive and polishing material. A blue variety with gem qualities is called sapphire. The ruby shown is about 5 cm in diameter.

BANDED IRON ORE

Quartz-banded iron ore from a large occurrence in the innermost part of Godthåbsfjord in the Nuuk region. The ore comprises alternating thin bands of magnetite (Fe_3O_4) and quartz (SiO_2). It occurs in the basement at Isukasia as a part of the approximately 3800 Ma old Isua sedimentary and volcanic rocks that make up one of the world's oldest supracrustal rock sequences (see p. 37). Magnetite forms the dark layers while the brownish-red layers are quartz. The sample is about 21 cm high.

SPHALERITE

Coarse-grained, dolomitic sedimentary rock (white) with large aggregates of the ore mineral sphalerite (dark). Sphalerite (ZnS) is the main mineral from which zinc is obtained. The sample comes from a Palaeozoic sequence in Washington Land, North Greenland, and is approximately 10 cm across.

IGALIKO SANDSTONE

Red, quartz-rich sandstone – Igaliko sandstone – is used as a building and decorative stone. The rock is about 1350 Ma old and comes from the Gardar Province in South Greenland (see p. 56) where the sandstones were deposited in a continental basin. The red coloration is due to weathering of the iron oxide hematite. The sample is approximately 25 cm high.

MINERAL RESOURCES

Quarry in Ilulissat where gneissic and granitic rocks are extracted for construction purposes, e.g. roads and buildings in towns. Part of the material is shipped out for use in other parts of the district.

and drilling and possibly even construction of mine shafts for test mining. It is popularly said that the State survey can be compared with finding the haystacks in the field, while that of the private companies is finding the needle in the haystack.

Mode of occurrence of raw materials

The mode of occurrence of raw materials in nature is closely related to the way they are formed geologically. The simplest types are materials used for building and construction purposes, which often are derived from occurrences found at the surface and extracted from open pits and quarries. These materials can be located by simple methods and the reserves can be estimated by the contractor undertaking the work. In contrast to this the majority of ores typically occur in very small quantities disseminated in a large volume of rock that may lie hidden underground.

The most important raw materials are: minerals of metals, minerals of non-metals and a group of various rock types (see box p. 197).

Geological environments of mineralisation

Mineral raw materials occur in a series of different geological environments that include sedimentary deposits, metamorphic crystalline rocks and igneous volcanic and plutonic rocks (see p. 34). The process by which an element is concentrated within a limited volume in the crust is called mineralisation. Such natural processes take place either in connection with the primary formation of the rocks or through later secondary enrichment. With many of the sedimentary occurrences the enrichment of the mineral(s) and element(s) takes place through external processes at the same time as the sediment is deposited, and in these cases the enrichment is connected to certain layers in the succession. Examples of this type of mineralisation are concentrations of detrital, heavy mineral sands (gold, tin, zircon and rutile) and mineralisation generated by volcanic fluids and gasses expelled on the sea floor (copper, lead and zinc).

Another category of mineralisation is related to internal processes in crystalline rocks where the ore

Example of the use of Greenlandic marble from Maarmorilik for the façade of Lyngby town hall (north of Copenhagen, designed by architects H.E. Langkilde and I.M. Jensen and built 1939 to 1941). The hall was faced by thin slabs polished to give a smooth and elegant appearance that has proved to be resistant to weathering, keeping its surface in good condition.

Marble quarrying on the island Agpat near Uummannaq in 1934. Excavating methods were primitive at that time, and most of the marble was exported by ship and sold in Copenhagen.

MINERAL RESOURCES

CATEGORIES OF MINERAL RESOURCES

Metals
The majority of metals only occur in minerals where they form chemical compounds with other elements. The pure metal is extracted from the mineral ore by smelting or is separated by other methods. Examples of metals and the ore minerals from which they can be extracted are:

Iron (Fe): from magnetite / lode stone (Fe_3O_4) and hematite (Fe_2O_3)
Copper (Cu): from chalcopyrite / copper pyrites ($CuFeS_2$), cuprite (Cu_2O) and bornite (Cu_5FeS_4)
Lead (Pb): from galena / lead glance (PbS)
Zinc (Zn): from sphalerite / zinc blende (ZnS)
Aluminium (Al): from bauxite (Al_2O_3 + water)
Chromium (Cr): from chromite ($FeCr_2O_4$)
Nickel (Ni): from pentlandite ($(Fe,Ni)_9S_8$), and niccolite / arsenical nickel (NiAs)
Silver (Ag): from argentite / silver glance (AgS) and native silver (Ag)

Some metals occur as pure elements, e.g.
Gold (Au): from native gold (Au)
Platinum (Pt): from native platinum (Pt)

In many instances several minerals occur at the same locality, e.g. galena and sphalerite frequently occur together, while noble metals often occur with other elements; for instance silver may be found as a minor component in the galena of lead ore.

Non-metals
These may occur both as isolated elements and in chemical combination with other elements. Examples are:
Coal: a rock rich in organic material, largely amorphous carbon (C)
Graphite: a low-pressure crystalline form of carbon (C)
Diamond: a high-pressure crystalline form of carbon (C)
Corundum: a mineral composed of aluminium and oxygen (Al_2O_3)
Quartz: a mineral composed of silicon and oxygen (SiO_2)
Cryolite: a mineral composed of sodium, aluminium and flourine (Na_3AlF_6)
Olivine: a mineral composed of magnesium, iron, silicon and oxygen ($(Mg,Fe)_2SiO_4$)

Rocks – aggregates of minerals
These are used mainly as industrial raw materials, mainly in the building and construction industry, e.g.:
Granite: igneous rock dominated by alkali feldspar, plagioclase and quartz
Limestone: rock dominantly made of the mineral calcite ($CaCO_3$)
Marble: a crystalline, metamorphic limestone, dominated by calcite ($CaCO_3$)
Sandstone: rock dominated by quartz (SiO_2)

occurs as fillings in fine-scale, cross-cutting fissure systems, or as more diffuse accumulations within a mineralised zone. Other types of ore concentrations can arise in plutonic rocks through magmatic differentiation when a magma solidifies (see p. 152). These internal processes take place at different levels within the crust and thus under different temperature and pressure conditions. Some mineralisations takes place deep in the crust at high temperatures, other types occur higher in the crust at lower temperatures. Examples of dyke-like concentrations are quartz- and feldspar-rich veins containing metals (e.g. lead, zinc, copper, tin, molybdenum, gold) or rare-earth elements (e.g. lanthanum, niobium, yttrium). More diffuse, high-temperature mineralisation may occur in metamorphic rocks involving both metalliferous ores (copper, iron, gold) and non-metalliferous materials (graphite, asbestos). Some of the precious metals and a series of rare elements (copper, nickel, platinum, gold, zirconium, beryllium, uranium, palladium, tantalum and rare-earth elements) are in general found within magmatic bodies.

Mineral provinces
Nature's mineral resources are closely related to their geological setting. Diverse geological environments give possibilities for several different mineral assemblages and thus different raw materials. Some types of mineralisation are particularly characteristic of

Galena (PbS), the main mineral from which metallic lead (Pb) is obtained. The hand specimen shows the three sets of cleavage planes at right angles to each other.

Sphalerite or zinc blende (ZnS) from which the metal zinc (Zn) is obtained. Sphalerite is the dark brown mineral in the hand specimen.

A large lens-shaped body of olivine-rich ultramafic rock (dunite) occurs in the 3000 Ma old Archaean gneisses of the Akia terrane north of Nuuk. This lens, mainly composed of the mineral olivine, has been quarried since 2005 by the Swedish mining company Minelco AB. The deposit, named the Seqi Olivine Mine, contains at least 100 million tonnes of high quality olivine. Olivine is mainly used in iron-ore smelting processes as an agent for forming slag. Locality: Niaqunngunaq/Fiskefjord (see p. 36), West Greenland.

MINERAL RESOURCES

GREENLANDIC GRANITE FOR A ROYAL MONUMENT

The grave of King Frederik IX and Queen Ingrid at Roskilde Cathedral is covered by a large granite monument quarried in Greenland. The main motif is a large anchor on a wavy background.

By drilling and careful blasting, a large block of granite was detached on the island Paarliit in South Greenland.

The Danish King Frederik IX (1899–1972) and his Queen Ingrid (1910–2000) are buried in an elegant chapel built in the grounds of Roskilde Cathedral. The grave is covered by a large granite gravestone obtained from Greenland, on which is carved an anchor motif set on a wavy background and an inscription that simply records their names.

King Frederik and Queen Ingrid had for many years a close connection with Greenland and, after the King's death, the Queen expressed a wish that her husband's gravestone should come from Greenland. Geologists were asked to help to find a suitable stone that could be transported to Denmark, and there decorated with a motif designed by a Danish artist.

A search was made for a site with a homogeneous rock that could be used by stonemasons and from which a sufficiently large block without fractures or hidden defects could be extracted. Additionally, it had to be possible to transport a block weighing 10–20 tonnes from the locality. A suitable locality was found at the coast on the island Paarllit, about 10 km west of Qaqortoq/Julianehåb. At this locality a pale granitic rock of monzonitic composition is exposed. This rock is part of the 1850 to 1800 million year old Julianehåb batholith, which forms the central zone in the Ketilidian fold belt (see p. 50). The task of cutting the block loose was given to an experienced Norwegian stonemason, and transport was arranged by a contractor from Qaqortoq. The block, which weighed 22 tonnes, was sailed to Qaqortoq in 1982 on a special barge, and from there sailed to Aalborg in Denmark on an ocean-going freighter.

In Denmark the block was cut into two slices of equal size, so that there was a reserve slab in case an accident happened during carving. The sculpture was made by the Danish artist Erik Heide.

The use of Greenlandic granite for this grave monument is a good example of how granite can be used for decorative purposes.

Preparations for collecting the block from the shore of the island.

Sketch map of the Qaqortoq region of South Greenland. The red dot indicates the locality of the island Paarliit where the gravestone was quarried.

The 3 m long block was moved on rollers down to the water's edge and suspended from the bottom of a barge to be sailed to Qaqortoq.

MINERAL RESOURCES

sedimentary deposits whilst others are restricted to areas of metamorphic or igneous rocks. Within the different rock categories there can be other variations depending upon the geological architecture and history of an area. There is, therefore, a clear link between the origin of a particular commodity and the possibilities of finding an exploitable occurrence.

Through basic geological, geochemical and geophysical investigations in Greenland it has been possible to obtain a general overview of how the traces of mineral showings are distributed throughout the country. When these indications are compared with the knowledge of the overall geology, different mineral provinces can be delineated that include areas where the right conditions exist for a certain type of mineralisation. There are thus regions where possibilities for finding raw materials of a certain type are better than in other regions. Conversely, this means that some mineralisation types can be excluded in certain areas during the prospecting phase, thus saving resources.

The geological background of mineral provinces is primarily the distribution of the different elements within the crust. This distribution is mainly revealed through geochemical studies that involve systematic

A sandstone succession in Jameson Land, East Greenland, showing the result of lead mineralisation in part of the sequence. The lead-bearing mineral galena makes up the grey-black portions of the layer at the lower part of the hammer shaft (40 cm long). In this unusual occurrence, galena was deposited at the same time as the sand layers, and is thought to have come from hot, lead-rich fluids emerging from vents on the sea floor.

Example of a mineralised layer in gneisses of the Ketilidian fold belt in South Greenland. The mineralisation occured in old metamorphosed sediments, now a layer of schists, within grey gneisses. At the surface the mineralised locality is seen as a rust zone, formed by weathering of the iron sulphide (pyrite and pyrrhotite) that it contains. This rusty weathering, called a gossan, is a good indicator of mineralisation. One of the rust zones contains trace amounts of gold. The locality is from the southernmost part of South-East Greenland.

MINERAL RESOURCES

THE TONNAGE AND GRADE OF AN OCCURRENCE

An ore sample from the Maarmorilik lead-zinc mine that shows the ore is an assemblage of different minerals. The main ore minerals are sphalerite, galena and iron pyrites, but there is also a little chalcopyrite and pyrrhotite together with quartz, calcite and schistose host rock. The main elements extracted from the ore are zinc and lead contained in the sphalerite and galena. Most of the remaining minerals are without commercial value and are separated during processing of the ore.

The economic value of a mineralised occurrence is estimated from a detailed study of the extent of the occurrence in a specific area and an evaluation of the value of the raw materials balanced against the costs of processing and marketing a saleable product. Two of the controlling factors in these calculations are first, the grade of the ore in terms of the concentration of the valuable elements and second, the evaluation of the total tonnage present with a certain minimum grade. To be able to exploit an occurrence in an economic and profitable manner, low-value materials must occur in large tonnages of high grade, whilst high-value materials can be exploited from a low-grade ore, provided that a sufficient tonnage is present.

Minimum grades of ores for exploiting an occurrence

Element	Minimum grade	
Iron (Fe)	30–40%	
Copper (Cu)	0.5–1%	
Lead (Pb)	4–6%	
Zinc (Zn)	4–6%	
Nickel (Ni)	0.5–1%	(frequently extracted with copper)
Silver (Ag)	50–100 ppm	(frequently extracted as a profitable by-product of lead-zinc ore)
Gold (Au)	6–10 ppm	(1 ppm equates to 1g/tonne)
Platinum (Pt)	2 ppm	
Coal (C)	60–80%	
Graphite (C)	20–30%	
Diamond (C)	0.1–1 carat/tonne	(strongly dependent upon quality and size)
Ruby (Al$_2$O$_3$)	1 carat/tonne	(strongly dependent upon quality and size)

ppm (parts per million) = 1:1 000 000 (one part of a million) ~ 0.0001%
1 karat = 0.2 g

The balance between the amount of the valuable element and the rock in which it is found is called the grade of the ore. For economic exploitation the occurrence must have a minimum grade that is strongly dependent upon the value of the material. The table shows ranges of concentrations for different materials for an occurrence to be worth exploiting. These grades span a range from about 80% (coal) down to a few parts per million (gold and platinum). For an individual occurrence, exploitation depends on the necessary minimum grade and total tonnage, together with any valuable by-products that can be extracted, as well as the geological structure and geographical location.

sampling and analysis of stream sediments (see p. 58). Elevated levels of rare elements, e.g. gold in the South Greenland basement, indicate that such areas have a potential for that particular element.

Other mineral provinces may include areas in which a special rock type occurs, e.g. a series of intrusions of a given age, or a mineral province may comprise several areas in which particular sequences of either sedimentary or volcanic origin occur.

The map shows some of the main mineral provinces in Greenland.

Grade: the balance between raw material and host rock

When minerals and elements are mined from an occurrence it is usually unavoidable to excavate some of the surrounding host rock as well. After mining it is therefore necessary to separate and remove the ore from the host rock. The ore is normally dominated by waste material, and the target mineral comprises only a small part of the total extracted bulk. In order to retrieve the target mineral(s) the ore is processed which involves removal of the waste rock, thus concentrating the ore minerals to a point at

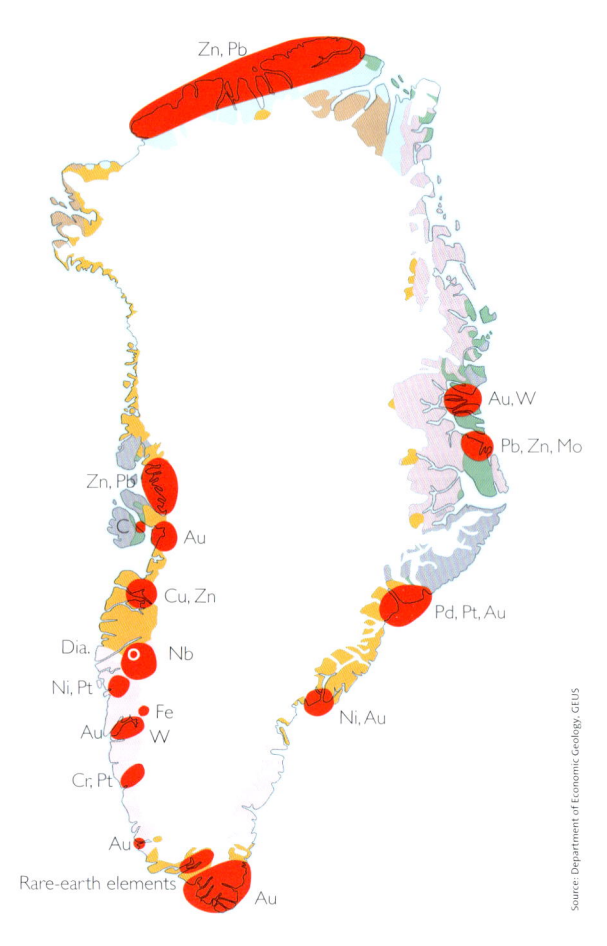

Mineral provinces in Greenland, with designation of the areas where these elements may be found: Au, gold; C, carbon (coal); Cr, chromium; Dia, diamond; Fe, iron; Mo, molybdenum; Nb, niobium; Ni, nickel; Pb, lead; Pd, palladium; Pt, platinum; W, tungsten; Zn, zinc.

which they may be directly used or are worth shipping out for further specialist processing.

This means that much more waste material is extracted than the final quantity of enriched ore produced. The waste rock is therefore separated from the ore and deposited in a spoil tip. The separation of the waste and concentration of the ore usually occur at the mine in a processing plant that first crushes the extracted material and then separates the valuable ore from the waste. The concentrated ore is sent for further processing and smelting in a furnace or separation by a chemical or mechanical process.

Whether an occurrence is exploitable or not is directly related to the value of the mined product which, amongst other things, depends on the quantity of the mineral set against how much material must be discarded as waste. This relationship, known as the grade of the ore, indicates the specific content of the valuable component in the mined material given as a percentage. The possibility of economic production from a mine is further dependent upon the value of the components together with the size of the occurrence, its position, construction costs for the mine, and the cost of extraction, transport and processing. In Greenland these local costs are relatively high due to the arctic climate and the country's scattered infrastructure. Most mines therefore must be based on their own new establish facilities. To all these costs must be added the financial conditions and the estimated price of the final product on the market over the life of the mine. Many of these variable conditions are difficult to predict. However, with this as a background it is possible to indicate a set of minimum levels for particular ore types which must be present before they are considered for extraction. Examples of such levels are: iron ore 30–40%, lead ore 4–6%, zinc ore 4–6%, copper ore 0.5–1%, gold ore 0.001%.

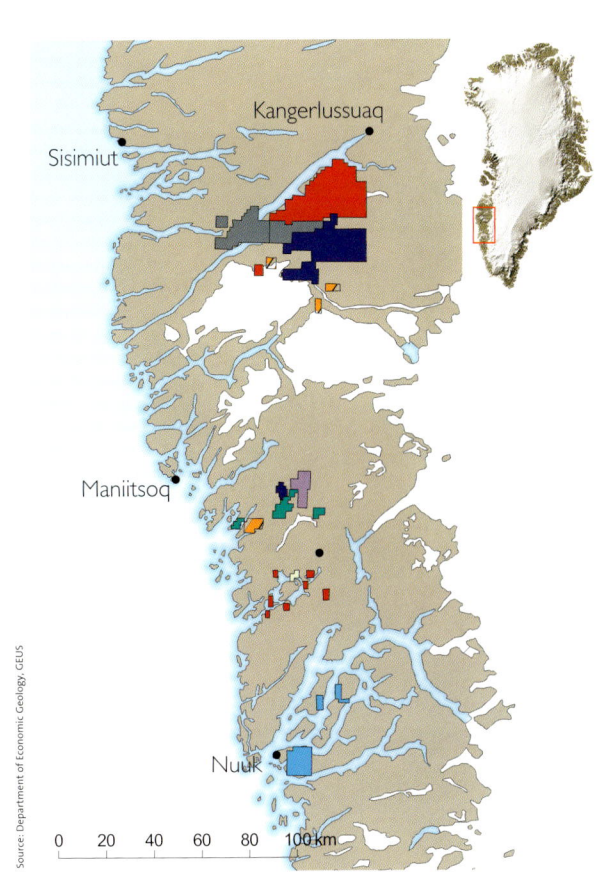

Map of part of West Greenland that shows the distribution of concession areas in 2004. The different colours depict a specific company's concession area(s). The areas of interest are those where the distribution pattern clearly shows that there are mineral showings of particular types. For example, the area south of Kangerlussuaq (Søndre Strømfjord) where it is possible to find diamonds and rare elements such as tungsten.

REGULATIONS FOR PROSPECTING AND MINING IN GREENLAND

Prospecting and mining activities by private industry take place on the basis of legislation that gives companies the right to carry out different activities in a defined prospecting licence area. The concession specifies a series of terms, together with working regulations, taxes and the length of time that a licence will run.

The basis of a company's licence is a division of activities into three phases. First there is a general, preliminary investigation stage that does not exclude other companies from prospecting for other minerals in the same area at the same time. In this phase the company aims at identifying and making a preliminary evaluation of the potential for exploitation of a given raw material. The next phase includes sole exploration rights in the form of an exploration licence that also involves work and investment obligations. At this point the company can work and invest without competition from other companies, and the target of the company is to proceed to a final estimation of the exploitation potential of a given occurrence. The third and final phase is the exploitation licence for a given occurrence of a particular material, including royalties to the State. The general provision of the licencing of the three phases is as follows.

Prospecting licence
This includes all types of minerals except for hydrocarbons and radioactive elements. The licensed area is extensive, often several thousand square kilometres. The licence duration is five years and is not exclusive. There are no fixed work obligations other than reports to the State on the activities carried out and the results obtained.

Exploration licence
This includes all types of minerals, except for hydrocarbons and radioactive elements. The size of the licence area is limited, often from a few to several hundred square kilometres; licence duration in the first instance is five years with the possibility for staged extension to 16 years. The licence is exclusive to one company, and there is an obligation for work to a minimum value of DKK 100 000 per licence (~ 1000 DKK/km^2), gradually increasing the longer the concession is held. In North and East Greenland the licence fee levels are lower. Other obligations to the State are reports on the activities and the results obtained.

Production licence
A company that has located a viable occurrence during the second phase may apply for sole rights to exploit a specific mineral occurrence (excluding hydrocarbons and radioactive elements). The licence is normally given for a small area in which the occurrence is situated, and the licence is normally granted for 30 years and can be extended to 50 years. The mining must take place through a Danish limited company with the necessary expertise and financial background; payment of taxes and fees is according to the rules established by the Home Rule Government for Greenland, but there are no special production taxes (royalties). Special requirements with respect to employment of workers, inspection and control visits etc. are subject to negotiation.

MINERALS AND THEIR USES

An overview map of potentially exploitable mineral occurrences in Greenland (see p. 211) shows that 30 different types of raw materials have been located. A summary of the use of some of these resources follows below.

Antimony: Element (Sb) – a silver-white metal mainly found in the mineral stibnite (Sb_2S_3). Antimony is mainly used in alloys to strengthen other metals.

Barytes: A mineral ($BaSO_4$) used as a filler in paints and paper and in the textile industry. In the oil industry it is used to add density to drilling muds; also a contrast medium in X-ray investigations.

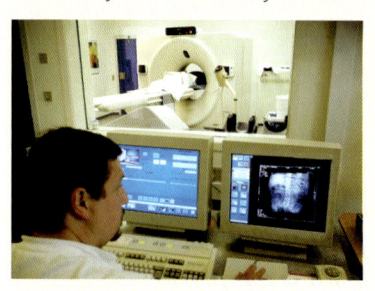

Basalt: A black volcanic rock that can be used as a decorative stone and for building and construction purposes. Used as a component of rockwool.

Beryllium: Element (Be) – a very hard metal that for instance occurs in various minerals as silicates and oxides. Used for very hard metal alloys, for example in the space industry.

Chromium: Element (Cr) – a metal that is nearly exclusively obtained from chromite ($FeCr_2O_4$). Used in many alloys and to harden steels as a protection against corrosion.

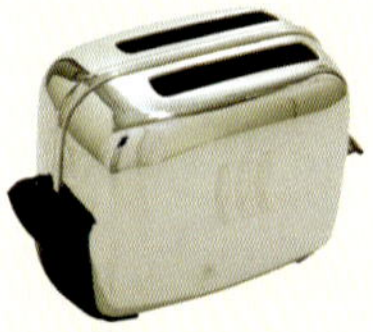

Coal: A rock dominantly comprised of organic carbon (C). Used as a fuel and for many industrial applications.

Copper: Element (Cu) – occurs typically in sulphide minerals such as chalcocite/copper sulphide (Cu_2S), chalcopyrite/copper pyrites ($CuFeS_2$) and bornite (Cu_5FeS_5). Copper is mainly used for electrical wiring and cables, but large amounts are used in alloys such as brass and bronze.

Cryolite: A mineral that is a fluoride-rich salt (Na_3AlF_6). Formerly used as a flux in the electrolytic smelting of bauxite (Al_2O_3) to produce aluminium.

Diamond: Mineral consisting of pure carbon (C). This is the hardest of all minerals and is mainly used for technical applications, in polishing media and particularly in jewellery because of its brilliance. It is rare and has a high value.

Gold: Element (Au) – a precious metal that often occurs in pure form. It is extremely resistant to tarnishing and corrosion and is used for coatings, as an ingredient in alloys and also as bullion and in jewellery.

Granite and gneiss: Pale, hard rocks that are mostly used for building and construction, but also as a decorative facing material.

Graphite: Mineral consisting of pure carbon (C). It is very soft and has a platy crystal structure that slips easily and so is used to enhance lubricants. It is also used in pencils and for crucibles.

Iron: Element (Fe) – occurs mainly in the minerals magnetite (Fe_3O_4) and hematite (Fe_2O_3). Used industrially for iron and steel in many different alloys with other metals.

Iron pyrites: A mineral (FeS_2) that is widespread but can only be mined economically when it occurs in very large bodies or with other metals that may increase its value. Used to some extent in the manufacture of sulphuric acid and as a flux in some smelting.

Lead: Element (Pb) – a grey-black, soft, heavy metal that mainly occurs in the mineral galena (PbS), which often also contains small amounts of silver. Lead is used in batteries, cables, in plumbing installations etc.

Marble: A rock that mainly consists of the mineral calcite ($CaCO_3$). Commonly used as a building stone and for decorative purposes. When powdered it can be used as a filler for paper and in cement manufacturing.

Molybdenum: Element (Mo) – a very soft metal that mostly occurs in the mineral molybdenite (MoS_2). Used in steel alloys for strength and in lubricants to enhance the greasing effect.

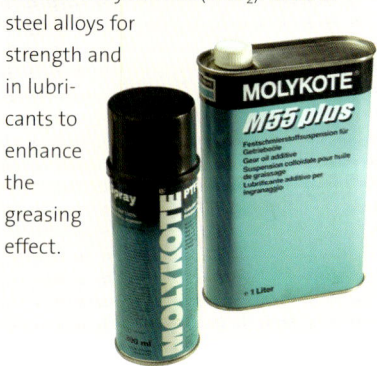

MINERALS AND THEIR USES (continued)

Nickel: Element (Ni) – a metal that occurs in the minerals pentlandite ((Fe,Ni)$_9$S$_8$), niccolite (NiAs) and in other sulphide ore minerals. Used in steel alloys, in exhaust catalysts and to produce special types of batteries.

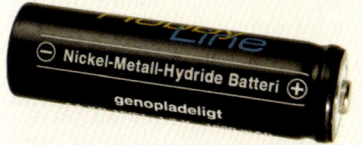

Niobium: Element (Nb) – a heavy metal that, together with tantalum (Ta), occurs in the mineral pyrochlore in carbonatite and in different minerals in alkaline rocks. Used in special steel alloys in nuclear reactors and in electronics (e.g. for mobile telephones).

Olivine: A mineral composed of a silicate of magnesium and iron ((Mg,Fe)$_2$SiO$_4$). Used for its high melting point in the ceramics industry and as a moulding medium.

Palladium: Element (Pd) – precious metal that is chemically resistant. It occurs together with platinum (Pt) as a pure metal and often in connection with other metals in some sulphides. It is used in corrosion-resistant steel alloys and as a component in exhaust catalysts.

Phosphate: Minerals that comprise a salt made from phosphoric acid (H$_3$PO$_4$) and a metal, often calcium (Ca) in phosphorite and apatite. Used for fertilisers and in the chemical industry.

Platinum: Element (Pt) – a chemically resistant precious metal with a very high melting point. Occurs often with chromite and related minerals in basic and ultrabasic plutonic rocks. Used as a catalyst, to make crucibles, in the electronics industry and for jewellery.

Rare-earth elements: Comprise the 15 so-called lanthanides, together with scandium (Sc) and yttrium (Y), which occur in rare silicate minerals such as monazite, xenotime, eudialyte and steenstrupine. Used in catalysts and in special materials for the electronics industry.

Ruby: A red variety of the mineral corundum (Al$_2$O$_3$), which is very hard. Used as a gemstone, but the value is very dependent upon quality and colour. Corundum is mainly used as an abrasive.

Silver: Element (Ag) – a heavy, precious metal that can occur as native silver and as a component in silver minerals, e.g. argentite (silver sulphide) (Ag$_2$S). It often occurs within other minerals, e.g. galena. Used for photographic applications, in the electronics industry, for its resistance to corrosion and as jewellery and silver ware.

Titanium: Element (Ti) – a metal that occurs naturally primarily in the minerals ilmenite (FeTiO$_3$) and rutile (TiO$_2$). The metal is light and very resistant to corrosion and is often used in the aircraft and space industries partly to replace aliminium. An important use for titanium is as a white pigment in paints.

Tungsten: Element (W) – a very heavy metal with a melting point of 3410°C. Occurs in the minerals wolframite ((Mn,Fe)WO$_4$) and scheelite (CaWO4). Because of its high melting point it is used as elements in light bulbs, in alloys to make them resistant to corrosion and also in materials subject to very high temperatures, e.g. furnaces.

Uranium: Element (U) – radioactive heavy metal that occurs in a series of uranium-rich minerals like uraninite (UO$_2$), various carbonates, sulphates, phosphates and silicates. Appears as a trace element in the minerals allanite, monazite and zircon. Used as a fuel in nuclear reactors and in very special dense alloys used in the armament industry.

Vanadium: Element (V) – heavy metal that occurs in a series of different oxide, sulphide, silicate and vanadate minerals. Vanadium atoms can enter into some minerals substituting for iron, chromium and titanium. Used in special steel alloys for example, to make them chemically resistant.

Zinc: Element (Zn) – a metal that occurs in sulphide and carbonate minerals such as sphalerite (ZnS) and zinc spar / smithsonite (ZnCO$_3$). Used mainly to galvanise steel to prevent rusting, but also in alloys with aluminium and copper to make industrial components.

Zirconium: Element (Zr) – a metal found in the mineral zircon (ZrSiO$_4$) and eudyalite (a silicate with Ca, Na, Zr and F); can also occur in small amounts in other minerals. It is very resistant to corrosion and has a high melting point. Used in materials for the nuclear industry, in foundry sands and in special steel alloys.

MINERAL RESOURCES

Mining in Greenland began as early as 1780.

Early mining in Greenland

Industrial scale mining has taken place in Greenland since the middle of the 1800s, when the first commercial mine, the cryolite mine at Ivittuut in South Greenland, was opened. This mine was worked for more than 130 years, until 1987, when it became commercially unviable and was closed. Throughout this long period the Ivittuut mine contributed considerably to the Greenland economy. Other mining has taken place over short periods in both East and West Greenland. A small lead-zinc mine at Mestersvig was open between 1956 and 1962, and another much larger zinc-lead deposit at Maarmorilik in West Greenland was worked from 1973 to 1990. Coal mining for local consumption took place near Qullissat on Disko during the period 1924–1972, and limited amounts of Greenlandic marble were mined at Maarmorilik on the north side of Nuussuaq, West Greenland from 1936 to 1940 and again from 1967 to 1972.

Ships in 1898 waiting to load cryolite at the Ivittuut (Ivigtut) mine before sailing to Copenhagen. In Copenhagen the cryolite was processed and separated from the waste. The ships at this time were mainly steam schooners that used a combination of sail and motor, and could only load a few hundred tonnes of ore. The Cryolite Company owned only a few ships itself, but most shipping was arranged in combination with supply sailings to Greenland, utilising the free capacity on the return trip for cryolite.

There has been a long tradition of mining in Greenland, and in addition to the occurrences named above many other mining possibilities have been investigated and a few have been mined at a small scale. These include occurrences of iron, copper, graphite, chromium, molybdenum, gold, nickel, platinum, uranium and thorium, zirconium, titanium, niobium, tantalum and beryllium. The majority of these occurrences are located in the crystalline basement rocks of West Greenland, but there are mineralised occurrences in the younger fold belts, the younger sedimentary basins and in the Palaeogene volcanic province. However, only a small number of these occurrences have been exploited.

Estimates of the cost of mining, processing, transport, extraction of the final product and the financial investment must be made before a mining operation can begin. This also includes a careful analysis of the size and average grade of the ore, together with a cost calculation for concentrating the ore. These costs must be balanced against the expected sales prices in a changing market, where fluctuations in prices over the anticipated life of the mine must be considered. Such a profitability analysis, including many possible variations, is called a feasibility study. Only when such a

Buildings in the mining town of Qullissat on Disko, with a depot for the storage and shipping of coal. The entrance to the mine was up on the hillside to the left of the photograph, which was taken in 1972, a year before the mine was closed.

study ends with a positive result will a project progress to planning and laying out a mine, and then carry on to extraction. Preparations for a working mine can, therefore, take many years from the first exploration phases to the opening of a mine. All of the mines that were started in the 1900s were based on exploration activities and feasibility studies. This means that most mines have given an economic return and a few have no more than balanced, so that when mining stopped the production costs had just about been covered. However, there are examples of mines that closed due to poor yield.

Four completely different, former mines in Greenland, including their geological background, are described in the following. The cryolite occurrence at Ivittuut was formed by a magmatic intrusion about 1170 Ma ago in the Gardar Province of South Greenland. The lead-zinc ore at Mestersvig was deposited about 55 Ma ago in a fissure system related to the Palaeogene volcanism in East Greenland. The zinc-lead deposit at Maarmorilik in West Greenland was originally a sedimentary ore that became folded and metamorphosed during mountain belt formation about 1900 Ma ago. The coal deposits on Disko in West Greenland originated as organic sediments deposited in a delta about 80 Ma ago. These examples illustrate how broad and varied the possibilities for economic deposits are in Greenland, indicating their extent and potential.

The Cryolite Mine in South Greenland

Cryolite is a non-metallic mineral that chemically comprises sodium, aluminium and fluorine (Na_3AlF_6); it is whitish and derives its name from the Greek kryos (ice) and lithos (stone). Cryolite was mined at Ivittuut between 1854 and 1987. At the start the mineral was used in Denmark in the manufacture of soda and enamel, while from 1887 it was used internationally in the production of aluminium. In this process cryolite was used as a flux in the electrolytic separation of aluminium from its ore material bauxite. During the Second World War cryolite was an important strategic mineral, especially in the aircraft industry.

Cryolite is a rare mineral, and the occurrence at Ivittuut is the only large deposit that has been mined anywhere in the World. During the 130 year long existence of the mine, a total of 3.7 million tonnes of ore with an average content of 57% pure cryolite was mined; this is an extremely pure and high-grade ore. The mining economy relied almost exclusively upon the value of the cryolite itself, even though many secondary minerals also occurred.

The cryolite occurrence at Ivittuut was located on the southern shore of Arsuk Fjord in South Greenland. Geologically the occurrence was formed in the

The opencast cryolite mine at Ivittuut on the south side of the fjord Arsuk in South-West Greenland. The mining over more than 130 years resulted in excavation of a large pit with a diameter of about 200 m and a depth of 60 m. The mine was cut off from the fjord by a dam and the pit was kept dry through constant pumping of the seeping water. When mining and pumping stopped, water gradually seeped in and the hole is now completely full of water.

Cryolite ore. The cryolite is white and the brown mineral is siderite. The sample is about 20 cm across.

Sketch map and profile of the mode of occurrence and position of the cryolite body in the upper part of the 1170 Ma old granite intrusion at Ivittuut. The cryolite ore body had an area of 100 x 200 m and a thickness of 50–60 m. The occurrence was located on the side of a deep-water fjord with possibilities for easy shipping. The Ivittuut cryolite mine was a unique occurrence in the World, the only location where cryolite occurred in sufficient quantities to justify mining.

MINERAL RESOURCES

The zinc-lead deposit at Maarmorilik is located at a height of 600–700 m on the north side of a small fjord opposite the former Maarmorilik marble quarry in the Ummanaq district. When the deposit was exploited a mining village was established at the old marble quarry. Access to the mine is by cableway from the mining village up to the mine entrance, seen as two small black rectangles on the photograph below the 'wing' of the black angle shape.

Lead ore at Mestersvig in East Greenland was excavated from the underground mine and taken to the surface in small tipping wagons. The ore was transported to the coast via road and stored at the coast until the yearly shipment. Access by ship is restricted to about six weeks in the summer due to ice. The underground mining was possible throughout the year, even though the land surface was completely snow-covered.

Gardar period about 1170 Ma ago, as a part of a granitic intrusion that cuts through the surrounding gneissic basement rocks. The cryolite ore occurs as a coherent body in the upper part of a columnar granite intrusion. The granite has a diameter of about 300 m, while the cryolite body in the centre occupied an oval shaped lens with a cross sectional area of 100 × 200 m and a thickness of 50–60 m. The granite is surrounded by a crush zone (breccia) that indicates a forceful emplacement, breaking through the basement rocks. Model calculations suggest that the granite crystallised at temperatures of about 700°C at a depth of about 3 km. Cryolite formation occurred at a late stage, after solidification of the top of the granite. A series of volatile elements separated from the still liquid deeper part of the magma and became concentrated in a residual melt. This enriched residue migrated upwards and invaded the top of the columnar granite body, where the volatile elements reacted with the already solidified part of the intrusion. The resulting fluorine-rich magma phases that formed crystallised as a number of cryolite bodies

with slightly different compositions. The very unusual chemical composition of the cryolite occurrence has meant that the cryolite deposit also contains many rare minerals. A total of 114 different minerals have been recorded of which many were new and described for the first time with Ivittuut as their type locality.

The cryolite was mined near the coast of the fjord from a large cylindrical open pit, that had a final diameter of about 200 m and a depth of 60–70 m. From the bottom of the mine a long inclined tunnel ran up to the surface, through which the ore was transported up to fjord level using special wagons with a 10 tonne capacity.

The high value of the cryolite led to intensive exploration for new deposits from the mid 1950s. This activity was a co-operative venture between the Geological Survey of Greenland (GGU) and Kryolitselskabet Øresund; unfortunately no further cryolite bodies were found, which confirmed the unique character of the occurrence. Today cryolite is manufactured artificially, and there is no longer the same need to discover a natural occurrence.

The Mestersvig Lead Mine, East Greenland

Until now, the only active mine in East Greenland has been the lead-zinc mine at Mestersvig, which was worked from 1956 to 1962. The galena-bearing quartz veins were discovered in 1948 during regional geological mapping. This occurrence was a concentration of lead and zinc sulphides with the minerals galena (PbS) and sphalerite (ZnS) dominant, but both minerals also contained traces of silver. The ore was located in steeply inclined dyke-like bodies with a variable thickness from a few metres to more than 50 m. The dyke material comprised mainly quartz in which an assemblage of the galena and sphalerite occurred. The mining exploited a large underground body containing ore with a concentration of about 10% Pb and 10% Zn.

Geologically the dyke-like bodies developed in association with intrusions in the Palaeogene volcanic province in East Greenland. The dykes fill a fissure system in the surrounding Carboniferous sandstones, the metallic elements having been leached out from the crust by water-rich fluids from the underlying magma chambers and transported up into the fissure systems in the overlying sandstones.

In the short period that the mining took place, a total of 58 000 tonnes of galena and approximately 75 000 tonnes of sphalerite were extracted. Throughout the mining operation the price on the World market fell, and on closure opinion was that there had been a rough balance between income and expenditure.

Black Angel – the zinc-lead occurrence at Maarmorilik in West Greenland

The mining operations at Black Angel have been the largest in Greenland in recent times. The mine was worked from 1973 to 1990 and closed when no more ore could be profitably mined. Throughout the greater part of the mining period about 350 people were employed by the mining companies. When mining ended, the total economic result was not only a substantial contribution to the local community in the form of employment and taxes, but also an operational profit to the mining companies.

The name 'Black Angel' derives from the outline of a dark marking formed by the shape of a folded schist band about 700 m up on the near vertical cliff where the ore is located. The mineralisation occurred at different levels within a folded marble layer, in which rich deposits of the ore minerals sphalerite and galena were found. The marble layers, geologically part of the Mârmorilik Formation of the Karat Group (see p. 47), were deposited in the sea as carbonates on top of the Archaean basement about 2000 Ma ago. Later, the basement and the overlying sedimentary rocks were interfolded during mountain belt formation about 1900 Ma ago, and the mineralisation occurred by concentration of the metallic elements through regional metamorphism under elevated temperatures and pressures. The layers in which the mineralisation occurred were subjected to three phases of deformation, which caused the marble horizons in special places to thick-

A drive in the Maarmorilik deposit, showing a section of the rich zinc-lead ore that occurs as a several metres thick, layered body. The miner is measuring the thickness of the body using a 2 m long rod.

Front-loading mechanical shovel working in a drive in Maarmorilik.

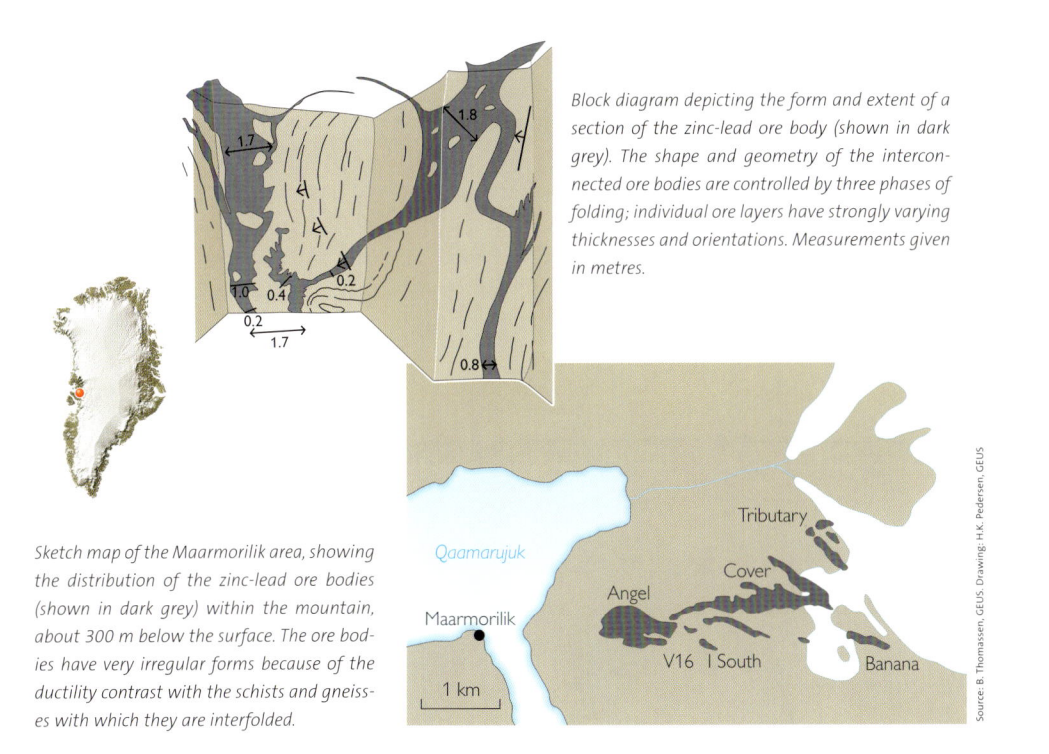

Block diagram depicting the form and extent of a section of the zinc-lead ore body (shown in dark grey). The shape and geometry of the interconnected ore bodies are controlled by three phases of folding; individual ore layers have strongly varying thicknesses and orientations. Measurements given in metres.

Sketch map of the Maarmorilik area, showing the distribution of the zinc-lead ore bodies (shown in dark grey) within the mountain, about 300 m below the surface. The ore bodies have very irregular forms because of the ductility contrast with the schists and gneisses with which they are interfolded.

MINERAL RESOURCES

THE MAARMORILIK MINE

Looking down over the mining works and village of Maarmorilik from the upper part of the cableway (1990). In the foreground the steel cables lead down to the processing plant. The large buildings house the froth-flotation separation system (inset right), other buildings include housing for the personnel, heliport and harbour facilities at the fjord.

After a detailed study phase, including diamond drilling from the surface of the mountain in 1971–1972 and demonstration of a sufficiently large ore body, underground mining was commenced in 1973. The mine was gradually enlarged over the years and there was a constant search for more ore. Fortunately the results of this exploration revealed at least three times as much ore as was originally proven as the basis for opening the mine.

The mine was constructed inside the mountain on several levels. A large main drive tunnel was established as the main thoroughfare, through the mine, a distance of about 3 km. From the drive numerous vertical shafts led to the working faces. The ore bodies were very irregularly distributed and the working faces had to follow the ore through a system of stopes and drives that were gradually constructed in order to extract the ore. A series of pillars of ore were left behind to support the roof and prevent it from collapsing onto the drives and roadways.

Mining was undertaken using modern machinery and blasting techniques. The ore was then transported from the working face, out through the stopes and drives, as far as the main transport tunnel. Small tipping wagons moved the ore to a crushing plant in the outer part of the mine; the cableway transported the crushed ore down to the mine village that lay near sea level on the opposite side of the fjord.

The aerial cableway from the mine entrance over the fjord to the processing plant had a span of 1500 m. The ore was transported in hoppers that had a capacity of 10 tonnes. In the processing plant the ore was concentrated using the froth-flotation method. This took place in a tank in which the pulverised ore was placed in a chemically controlled, frothy liquid and agitated. By pumping air into the liquid from the bottom of the tank, and stirring the mixture, the metalliferous particles from the ore adhere to the surface of the bubbles. The bubbles with the metalliferous grains floated to the surface where they collected as a froth, which was skimmed off into collecting tanks as enriched ore. Using this method the ore could be purified by separating it from impurities and country rock. The concentrate was then washed to recover the flotation liquid and taken to a storage area from which it could be shipped in bulk in the six months of the year (June–November) when the fjord was not closed by ice. The ore was exported from Maarmorilik to smelters in Europe, and in an average year 135 000 tonnes of zinc concentrate (with 58% zinc) and 35 000 tonnes of lead concentrate (with 70% lead and a silver content of 0.5‰) were dispatched.

The mine was manned in the first 12–15 years by about 335 people, but in later years manning levels were reduced to around 250. The Danish state received in total DKK 800 million (~ US $ 120 million) in royalties, and around DKK 250 million (~ US $ 40 million) in taxes from the employees.

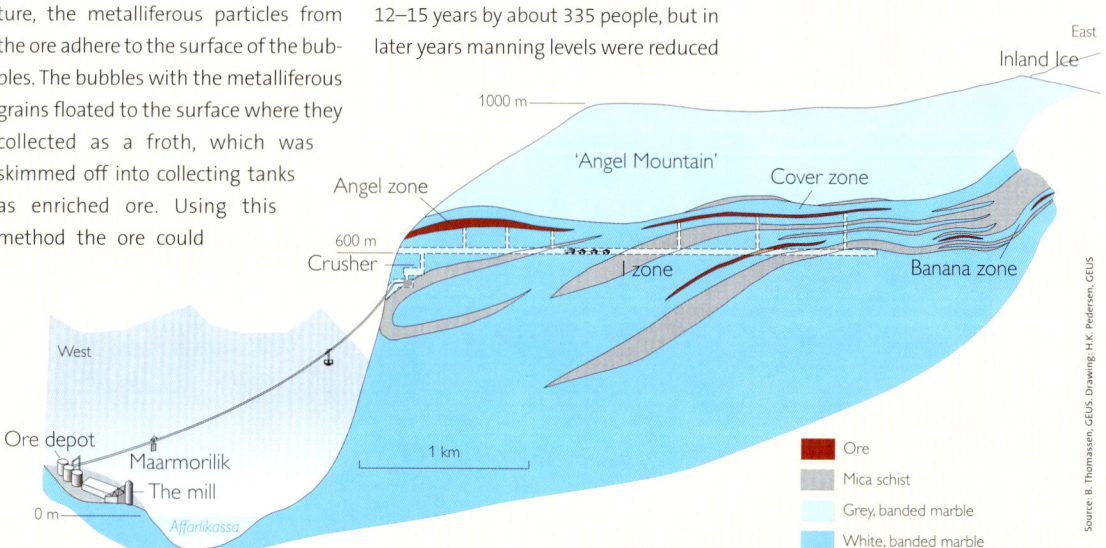

Froth-flotation tanks where pulverised zinc-lead ore is concentrated through separation of the waste rock from the metalliferous component of the ore. Separation takes place through metallic particles adhering to bubbles (the dark froth layer) that are skimmed off the top of the tanks.

Simplified cross section through the Black Angel zinc-lead occurrence at Maarmorilik. It shows how the ore occurs and how the mine has been constructed with the main driveway and shafts. Also depicted is the cableway from the mine to the mining town, where the ore was concentrated before shipment. The sketch is drawn with a variable scale in order to include all the features.

en tectonically up to 1000 m. As a result of the deformation the ore bodies had a very complicated form and extent within the mountain, and mining specialists from Canada and Sweden were employed to carry out some of the work.

The Black Angel ore deposits were mainly located at levels 600–700 m above the fjord, and about 300 m below the plateau surface. The extent of the main ore bodies was approximately 3.5 km long and 0.5–1 km across and with a maximum thickness of the lens-shaped bodies of 30 m. Because of the folding the orientation of the layers varied from near horizontal to vertical. Mining took place underground from mining shafts and adits, which led out to an aerial cableway that transported the ore down to a processing plant established on the opposite side of the fjord in the old Maarmorilik marble quarry.

During its 17 years of operation the mine produced 11.2 million tonnes of ore with an average content of 12.3% zinc and 4% lead, and additionally a small amount of silver contained in the lead. In total the mine produced 1.4 million tonnes of pure zinc, 0.4 million tonnes of metallic lead and 250 tonnes of silver.

Coal mining on Disko

Upper Cretaceous deltaic deposits in the Nuussuaq Basin of West Greenland (see p. 128) contain many thin coal seams in a predominantly sandy sequence. Widespread coal occurrences are found along the eastern side of Disko and along part of the northern and southern coasts of Nuussuaq. The individual coal seams vary from a few decimetres to 1–2 m thick, and the coal is of low quality with a high ash content. Never the less this was a resource that was exploited for local use from 1780, and over the next 200 years. Initially it was used in the town of Uummannaq and later also at the settlement Ritenbenk on the eastern side of Disko Bugt. Extraction methods in these local mines varied from small, open cast pits that were worked with hand tools to small adits, and from 1905 to the beginning of the 1950s mining from small adits was carried out on the north side of Nuussuaq.

In 1924 the Danish state established a mine at Qullissat on north-eastern Disko near the coast. Here an approximately 1 m thick coal seam, known from earlier times, was exploited and, from 1929 'modern' mining methods were employed. In the first 15 years the mine produced about 4000 tonnes/year. The coal was transported by barge from the coast out to ships that distributed it for wider use in Greenland.

After 1939 the mine in Qullissat was moved approximately 1 km towards the south-east, to a better area, with seams up to 1.5 m thick, where a new underground mine was established that was worked up until 1972. Until the beginning of the 1950s this mine produced 6000–8000 tonnes/year, and after an expansion of the new mine, production was increased to 20 000–30 000 tonnes/year. About 60% of the coal was used for household purposes in Greenland, while the rest was so finely crushed that it could only be used for automatic boilers and briquettes. Attempts were made to use these finer fractions for industrial purposes in Denmark.

Mining activites were at their height in 1960, when there were around 1300 people living in Qullissat, which was specifically established to make use of local labour. Both mining and transport have always been very difficult, particularly because there was no proper harbour. As a result, the mine operated for a long period at a deficit. The introduction of oil burners for heating led consumers to choose oil instead of coal, and the decreasing demand for coal led to closure of the mine in 1972. The inhabitants at Qullissat were re-housed in other towns, mainly in the Disko Bugt area.

Over the years of activity more than 600 000 tonnes of coal were taken from the Qullissat mine; the mine provided significant local employment and contributed particularly to the local economy. Although it was not possible to operate the mine profitably, it is reassuring to know that a coal reserve of over 100 million tonnes is present, mainly on the south side of Nuussuaq.

Coal mining in Qullissat in 1929. The mine adits were only 1–2 m high and all work was restricted by the confined conditions. Even though the mine was thoroughly modernised in 1929 and equipped with new machinery, it was by modern standards very primitive in terms of the equipment and tools that were used.

MINERAL RESOURCES

Existence of many mineralised localities in Greenland indicates a good potential for mineral deposits

Mineralisation in Greenland

Mineral prospecting over many years has revealed that there is a great potential for a wide range of different mineral deposits in all parts of Greenland. The most intensive exploration has been undertaken in the basement terrain of West Greenland, and it is therefore no surprise that this is where most mineral occurrences have been found. It is probable that mineral occurrences are equally widespread in the comparable basement areas of South-East Greenland, but prospecting has been less intensive because the area is difficult of access. Mineralisation has also been found in the Palaeozoic fold belts in both North and North-East Greenland, and various possibilities have been demonstrated in strata within the flat-lying sedimentary basins. One of the most promising prospects for future mineral exploration is the Palaeogene volcanic province of East Greenland, where several of the intrusions contain attractive concentrations of rare elements. Mineralisation has thus been found all over Greenland in various geological settings, and exploration by both private companies and public institutions is continuing. One of the most recent developments is the opening of a small but rich gold mine near Nalunaq in South Greenland.

Sample of zinc-lead ore from Maarmorilik in the West Greenland basement (see p. 200). The sample is a coarse-grained type of ore, but in many places in the mine the ore is rather fine grained. The principal components of the ore of the sample are sphalerite, galena (which also contains silver) and iron pyrites together with quartz. The sample is about 1 m across and is displayed in the Geological Museum in Copenhagen.

Mineralisation is connected to the geological setting of the rocks, which in turn relates to the occurrence of the various units. The most important mineralisation showings can be divided into five main types:

1) Mineralisation in the crystalline basement terrain
2) Magmatic mineralisation
   a: Mineralisation in the Gardar Province
   b: Diamonds in the West Greenland basement
3) Mineralisation in Palaeozoic fold belts
4) Mineralisation in younger sedimentary basins
5) Mineralisation related to Palaeogene volcanism

1) Mineralisation in the crystalline basement terrain

The Archaean core of the Greenlandic basement shield, with its more than 2500 Ma old gneisses and different metamorphic schists and intrusive rocks, contains a variety of different mineralisation types. The oldest of these is the banded iron formation found at Isukasia in the inner part of Godthåbsfjord. The formation is part of an old succession of metamorphosed sedimentary and volcanic rocks (see p. 36) and comprises alternating millimetre to centimetre thick bands of magnetite (Fe_3O_4) and quartz that were originally deposited as chemical sediments under marine conditions. The occurrence has been explored in detail, and it contains a minimum resource of two billion tonnes of ore; however, the iron content is not high enough to justify mining under present economic conditions.

Another large Archaean mineral occurrence is found near Qeqertarsuatsiaat (Fiskenæsset), about 150 km south of Nuuk in West Greenland. Here a highly deformed and folded 2850 Ma old layered intrusion contains the chrome-rich mineral chromite. The chromium-rich layers form bands up to a few tens of metres in thickness and can be traced at intervals for many kilometres. Despite intense folding the chromite-bearing units can be followed for more than 200 km. Calculations indicate that there is about 100 million tonnes of low-grade chrome ore, where the ratio between chromium and iron is 0.9/1.0. However, despite the very large size of the deposit the grade is too low to be mined profitably. The intrusion was emplaced into amphibolite and ultrabasic rocks, and in places reactions have led to the formation of ruby (Al_2O_3). Cutting, faceting and polishing tests have been positive, and prospecting is continuing in the hope that this will be an economic resource.

The Archaean basement in the Godthåbsfjord area contains a number of indications of gold mineralisation. These occurrences are hosted in quartz veins in different, altered volcanic and sedimentary rocks that form bands within the surrounding gneisses. The gold potential in this region is promising, and drilling and exploration are continuing.

Overview map of the mineral occurrences in Greenland discovered over 150 years of exploration. Occurrences are shown with a geological map as background, so that the connection between the different types of mineralisation and their geological surroundings is seen.

MINERAL RESOURCES

Finely banded, layered chromitite occurring as dark bands of the chromium-rich spinel chromite, separated by thin, white layers of feldspar rock (anorthosite). The chromite was originally precipitated from a basic magma during magmatic differentiation in a magma chamber (see p. 152), forming a layered rock. Later the entire intrusive complex was folded and metamorphosed during fold-belt formation in the basement (see p. 39). Near Qeqertarsuatsiaat (Fiskenæsset) in southern West Greenland. The hammer is 40 cm long.

A completely different type of mineral deposit, a large olivine-bearing occurrence (a dunite) has been located in the Archaean gneiss terrain some 90 km north of Nuuk. The homogeneous deposit contains at least 100 million tonnes of high-quality olivine. Opencast mining of this industrial mineral began in 2005 (see photograph p. 197).

In the north-eastern part of the Disko Bugt region a number of different mineralised localities occur in Archaean rocks that lie as structural enclaves within the Palaeoproterozoic Rinkian area (see p. 46). The mineralised localities are associated with a 2800 Ma old metamorphosed succession of sedimentary and volcanic rocks that were affected by both Archaean and Palaeoproterozoic events. The occurrences comprise different sulphides of lead, zinc and copper, but most interesting are three smaller areas with gold mineralisation. Both GEUS and private companies have explored the area in detail, but up to the present no exploitable resources have been located.

In the younger part of the basement, mainly formed in the Palaeoproterozoic between 2000 and 1600 Ma ago, there are also mineralised localities. The largest of these is the zinc-lead deposit at Maarmorilik that was mined for 17 years (see previous chapter). The mineralisation found in the Ketilidian fold belt in South Greenland is of similar age and includes several gold showings; the whole of South Greenland is characterised as a gold province. The known gold mineralisation occurs in the boundary zone with the Archaean basement to the north, and in a southern area south of the Julianehåb batholith (see p. 51). This southern area contains the gold occurrence at Nalunaq (Kirkespirdalen) that has been mined since 2003. Gold is found here as native gold in thin quartz veins that form an irregular network cutting through the surrounding metamorphosed sedimentary and basic volcanic rocks. The mineralisation was formed at low temperatures from metal-rich fluids that were associated with the last phase of solidification of the Julia-

Inset shows grains of native gold in a quartz vein at Nalunaq.

The outer part of Kirkespirdalen at the Nalunaq gold mine; the photograph shows the road system from the valley up to the mine entrances. The three mine entrances lie between 350 and 450 m altitude. In the valley a small mining town has been constructed for the workers. The main building (red house) and other installations are seen at the bottom of the photograph. From the mining town a road runs down the valley to the shipment jetty at the harbour on the side of the fjord Søndre Sermilik; the jetty can be used by ocean-going ships.

Mineralisation in Greenland

nehâb batholith about 1800–1770 Ma ago. The main mineral veins have been traced for 1700 m and vary in width from 0.1 to 2 m. The grade of the ore runs at 25 g per tonne of mined rock. Proven reserves are estimated to contain 10 tonnes of pure gold, and there is the possibility of another 5 tonnes from slightly lower grade ore. In 2006 the production rate was planned to reach 500 tonnes of ore per day.

2a) Mineralisation in the Gardar Province

In the Mesoproterozoic the South Greenland basement shield was disrupted by the development of a major rift zone. A series of volcanic rocks were extruded within the rift, and a number of intrusions with unusual chemistry were emplaced into the upper part of the crust around the rift zone (see p. 56). These magmatic rocks include the granite hosting the cryolite occurrence at Ivittuut (see p. 205), together with a number of other intrusions characterised by a high content of alkali elements, such as sodium and calcium, and frequently with low silica contents. These unusual rock types often have a high content of elements that are important raw materials for some modern industrial processes. The rare metals include tantalum, niobium, zirconium, yttrium, lithium, rare-earth elements, beryllium and uranium. Although concentrations are usually only a few hundred grammes per tonne ore, locally they reach up to 1400 g/tonne ore. The possibilities for mining some of the valuable elements have been evaluated, but exploitation has so far not been found profitable.

The rare elements are mainly present in Gardar intrusions, emplaced about 1300 to 1120 Ma ago. Exploration of much of the region has been carried out both by GEUS and private companies, and several of the mineral occurrences have been evaluated. In the 1960s there was a project to exploit the uranium occurrences at Kvanefjeld near Narssaq, and test mining of 20 000 tonnes of ore with about 325 g/tonne uranium was undertaken and the ore sent to Denmark for processing. However, the uncertain political climate regarding the use of atomic power in Denmark ended with a decision in 1985 not to build nuclear power plants, and the mine was never developed. The established potential for the whole region shows that there are large reserves of tantalum and zirconium and possibilities for exploitation of other elements.

2b) Diamonds in the West Greenland basement

In the northern part of the Archaean basement block in West Greenland, south of Kangerlussuaq (Søndre Strømfjord), several thin dykes and sheet-like intrusions (about 570 Ma old) occur; these are composed of the ultrabasic rock known as kimberlite. Kimberlite has a composition similar to peridotite (see p. 141),

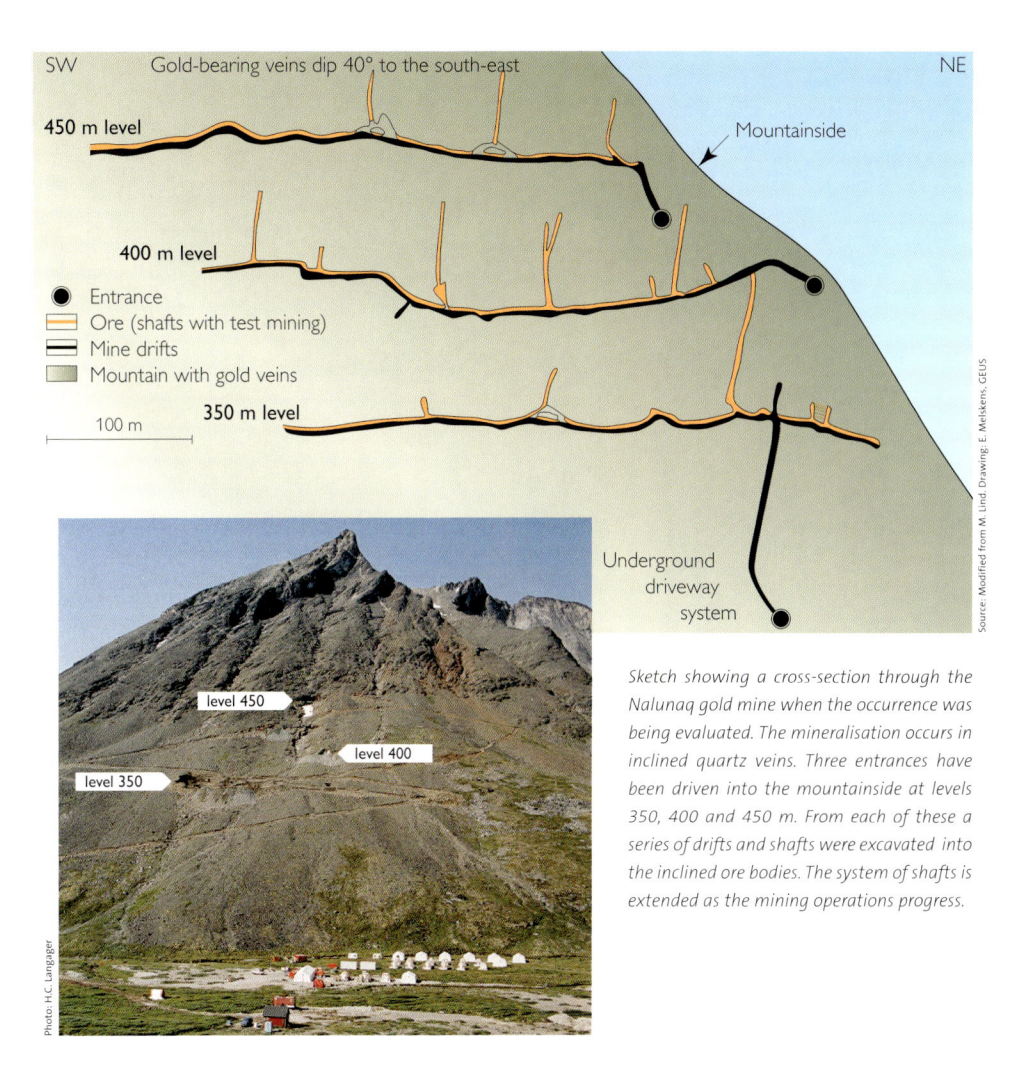

Sketch showing a cross-section through the Nalunaq gold mine when the occurrence was being evaluated. The mineralisation occurs in inclined quartz veins. Three entrances have been driven into the mountainside at levels 350, 400 and 450 m. From each of these a series of drifts and shafts were excavated into the inclined ore bodies. The system of shafts is extended as the mining operations progress.

In the Ilímaussaq intrusion (see p. 57), at Kvanefjeld near Narsaq, a rock called lujavrite contains uranium and thorium. These radioactive elements are contained in the mineral steenstrupine, and a section through the mineralised zone yields about 300 ppm uranium on average. The locality was drilled in the 1960s, followed by test mining and experimental processing undertaken at the atomic research centre at Risø in Denmark. However, the decision not to operate nuclear power stations in Denmark has prevented exploitation. The photograph shows a drilling rig on Kvanefjeld in 1969. The mountains in the background are about 1400 m high, and make up the highest parts of the Narsaq peninsula.

MINERAL RESOURCES

DIAMONDS IN GREENLAND

Loose block of kimberlite, derived from one of the many thin, sheet-like kimberlite dyke intrusions that cut the crystalline basement south of Kangerlussuaq (Søndre Strømfjord). The rocks have a fine-grained groundmass of ultrabasic composition (peridotite-like), with many rounded fragments torn out of the rocks through which the kimberlite magma passed on its way up to the surface. These fragments – known as xenoliths – often constitute more than half of the rock, as in the photograph. Kimberlites may contain diamonds.

Diamond consists of pure carbon (the element C) that crystallised deep in the earth under very high temperature and pressure conditions. The mineral is rare and has special properties which make it particularly valuable. Amongst other things it is used as a gemstone because of its light-reflecting properties and, once facetted and polished, has a remarkable brilliance. It is the hardest of all known naturally occurring materials and is extensively used in cutting and polishing tools.

Diamonds occur as small grains and inclusions in certain special types of volcanic and igneous rocks that are brought up to the surface of the Earth from depths of 150–300 km. Because of their hardness and resistance to erosion, diamonds are also found as a detrital component in sediments that are derived from the weathering and erosion of diamondiferous volcanic rocks. Diamonds are classified by weight given in carats, one carat being equivalent to 0.2 g. The biggest diamond found in the world weighed about 600 g, but the great majority of diamonds are less than one carat in size. Very small diamonds – microdiamonds – also occur, and can only be used for industrial applications. The value of diamonds varies considerably and depends upon size, clarity and colour. To this must be added the cost of cutting and polishing it into a gemstone. For gem quality the price of unpolished diamonds is more than $ 8500 per gramme.

In a geological context the diamond-bearing volcanic rocks of the world occur nearly exclusively in very old, stable basement areas of Archaean age. Such an area is found in West Greenland (see p. 36), and this has been and still is the object of diamond exploration. The volcanic host rock for diamonds, called kimberlite, derives from magmas that formed in the upper part of the Earth's asthenosphere at depths greater than 150 km. The melt is of ultrabasic composition and comprises liquid containing olivine, pyroxene and garnet; the initial magma contains a large amount of gas dissolved under great pressure. When conditions allow, magma – driven by the pressure of the expanding gas – erupts at great speed through to the surface as a relatively narrow, pipe-like volcanic tube. On its way up it rips off and incorporates fragments of the rocks it passes through, and among these may be scattered diamonds. When the intrusive magma reaches the upper part of the crust the material in the pipe comprises a mixture of the original magma and rounded fragments of the rocks bounding the pipe (xenoliths); this mixture forms the rock called kimberlite. Close to the surface the expanding gasses cause the eruption to be explosive and at the surface there is often an explosion crater. Where the pipes cut the surface they may often be only 100–500 m in diameter. In the craters the rocks are often fragmental, soft and crumbly, and easily weathered, so they are usually difficult to find. A different type of intrusion occurs where the kimberlites are emplaced as thin dykes that cut through the upper crustal rocks, which are often part of an old basement. Such types are found in the Archaean basement of West Greenland, where several kimberlite dyke swarms occur south of Kangerlussuaq; no remains of craters have yet been found in this region.

The age of the volcanism giving rise to the kimberlites is often fairly young, although the diamonds themselves are very old. The diamonds can develop through crystallisation of inorganic carbon that dates from the time of the formation of the Earth itself, or from organic carbon derived from the surface that was subducted down to depths of more than 120 km by plate tectonic processes more than 2500 Ma ago. In West Greenland two differently aged suites of kimberlites are known, one from about 570 Ma and the other from about 170 Ma ago.

Sketch cross-section through the outer part of the Earth to show how diamonds form at a depth of 150–300 km, and can be brought to the surface in an explosive volcanic event – a kimberlite intrusion. At the surface a pipe-like body, a sub-circular crater, may be formed and filled with a mixture of volcanic material and fragmentary wall rocks. Such craters above kimberlite pipes are the host of most diamonds found in the world.

Mineralisation in Greenland

with scattered large crystals of olivine and chrome-rich garnet. Kimberlite originates at a great depth in the mantle, and in many places in the world this rock type contains diamonds. Following the discovery of kimberlite dykes in the region between Sisimiut (Holsteinsborg) and Maniitsoq (Sukkertoppen), diamond exploration has been undertaken widely and has revealed that some of the kimberlites have a small content of micro- and macrodiamonds. The results have encouraged continued exploration, and there is considered to be a reasonable potential for diamond occurrences. The special conditions for the formation and mode of occurrence of diamonds are explained in the box opposite.

3) Mineralisation in Palaeozoic fold belts

The Caledonian fold belt in North-East Greenland (see p. 92) comprises interfolded, layered thrust complexes of gneissic basement overlain by thick, sedimentary successions of older basin deposits. During the Caledonian orogeny (see p. 102) these units were deformed and cross-cut by a series of granite intrusions. Numerous smaller mineral showings have been demonstrated in the fold belt, but none of them are large enough to be of immediate interest for more detailed investigation. Mineralisation is primarily connected to the sedimentary successions and to the younger granites, and includes: (1) copper indications in the sedimentary rocks; (2) tin, tungsten and gold mineralisation in fault zones; (3) lead-zinc mineralisation in carbonate veins connected to the granites.

The Ellesmerian fold belt in North Greenland (see p. 104) consists of a series of folded Lower Palaeozoic sediments that were deposited in a shallow-water basin nearest the continental margin and in a deep-water basin further offshore. In this fold belt there are several occurrences of lead, zinc and barium that are related to the original sedimentary deposits, although some of the material was later mobilised and is now concentrated in fault zones.

The largest mineral occurrence currently discovered in the Ellesmerian fold belt in Greenland is the lead-zinc deposit at Citronen Fjord in northern Peary Land. This occurrence is found in the boundary zone between the deep-water and shallow-water parts of the basin. The mineralisation initially developed during sedimentation by deposition of ore-bearing precipitates from fluids and gasses emanating from the seabed. The occurrence has been thoroughly explored and a large drilling project has indicated reserves of more than 350 million tonnes of ore. Of this more than 20 million tonnes of ore contains more than 7% zinc and 1% lead. The main part of the mineralisation consists of iron pyrites (FeS_2) of little value. The main ore deposit has a thickness of up to 50 m

Surface outcrops of the lead-zinc mineralisation at Citronen Fjord in Peary Land, North Greenland. Mineralisation of this type is revealed by the strong yellowish, greyish-white or dark grey coloured weathering zone, a gossan, that forms over a disintegrating ore body due to the breakdown of iron sulphides. Weathering leaches the metals out of the sulphur-rich minerals galena (PbS), sphalerite (ZnS) and iron pyrites (FeS_2), and concentrates the sulphur at the surface. Behind the gossan area in the foreground is a small drilling rig, used to explore the extent of the deposit in 1995.

Copper mineralisation in a vein that cuts Upper Permian calcareous sediments on Wegener Halvø (north-eastern part of the Jameson Land area in East Greenland). In places these Permian sediments are dissected by pale veins of calcite and barytes, some of which are mineralised with copper minerals and smaller amounts of galena and sphalerite. The copper minerals at the surface weather to form the green mineral malachite, which can be compared to verdigris. The mineralisation has been emplaced from circulating watery fluids at a temperature of about 125–150°C, which has leached out the metals from black shales and redeposited them in fissure systems in the sedimentary rocks.

Copper occurrences in the field are often revealed by the highly coloured green-weathering products. The sample here, about 15 cm across, comprises blue azurite and green malachite.

MINERAL RESOURCES

Dissected by glaciers and surrounded by ice-filled fjords the Skaergaard intrusion in South-East Greenland is located in an area that is difficult of access. The intrusion contains very thin layers enriched with the rare elements palladium and platinum, and also gold. The occurrence has been very thoroughly investigated both in connection with the mineral potential, and because it is a celebrated example of a layered intrusion. The peaks in the background are about 1000 m high.

Drill cores from 1011 m depth in the 54.5 Ma old Skaergaard intrusion. The gabbroic rocks include a mineralised zone that contains 2–3 g palladium per tonne. The drill cores have a diameter of about 3 cm.

Extensive diamond core drilling in the Skaergaard intrusion has yielded continuous 3 cm drill cores down to 1100 m depth. The approximately 30 drill holes into the body have a total core length of about 12 km. When operating, the drill string is withdrawn every 3 m in order to extract the drill core. The recovered core is preliminarily described in the field and selected samples are taken for immediate analysis. The remaining core is boxed for transport back to a laboratory for later analysis and storage. The photograph shows drill cores undergoing preliminary investigation and subsequent packing in special boxes for transport.

and extends for almost 3 km. The occurrence is conformable with the sedimentary bedding and is mostly flat lying; it has been traced from the surface down to a depth of 300 m. This relatively large ore deposit is located in a part of North Greenland that is ice-bound most of the year, and as mining costs consequently are very high the deposit has not yet been exploited, but is presently under renewed evaluation.

4) Mineralisation in younger sedimentary basins

In the younger sedimentary basins of central West Greenland and East Greenland, only the coal occurrences in the Disko Bugt area have so far been mined (see p. 209).

In East Greenland the younger basin deposits form flat-lying successions of sediments (see p. 124) many kilometres thick of Carboniferous to Cretaceous age, an age range of 250 Ma. The sediments comprise alternating sands and muds with a few carbonate horizons. The basin deposits are dissected by a series of faults that divide the region into different fault blocks, some of which have influenced the patterns of mineralisation.

In North-East Greenland a number of mineralised localities have been found; because they formed together with the deposition of the sediments they are known as strata-bound. They occur as thin layers that follow the bedding, but none of them have been of a sufficient size to be exploited. The demonstrated mineralisation includes: (1) a sequence of barium-rich limestones south of Mestersvig with about 300 000 tonnes of ore with approximately 72% barium; (2) different lead-zinc mineralisation showings in limestones; (3) a widespread copper mineralisation in the northern part of Jameson Land in Permian black shales; (4) different heavy mineral assemblages (placer deposits) in Jurassic sandstones in the eastern part of Milne Land in the Scoresby Sund area. One of these occurrences contains about 5 million tonnes of ore with 1–4% zirconium and 3–13% titanium.

5) Mineralisation related to Palaeogene volcanism

The most important mineralisation in this group occurs in some of the many plutonic intrusions that are found along a 1000 km stretch of East Greenland in association with the Palaeogene volcanic province (see p. 149). The lead-zinc occurrence at Mestersvig is related to the mineralisation in this zone and is the only occurrence in this group that has so far been mined.

In a few of the Palaeogene intrusions some very significant mineralisation has been demonstrated, and many minor indications have been ascertained in other intrusions. There is good potential for locating other mineral occurrences in this group of intrusions. Of the known mineral occurrences focus has been on high-value metals such as gold, platinum, palladium and molybdenum. These metals occur only in small amounts in specific parts of the intrusions, but because of their high value they have been the target of detailed exploration. Two intrusions have been intensively investigated by diamond-drilling, and the size and distribution of the mineralised zones have been evaluated. The two occurrences are in the Skaergaard intrusion at Kangerlussuaq, and at Malmbjerg south of Mestersvig.

The layered gabbros in the Skaergaard intrusion (see p. 149) contain, at the transition between the Middle and Upper zones, some thin palladium- and platinum-bearing layers that separated from the magma as a type of reef formation. For the most part these layers are present about 1500 m below the roof of the intrusion. A total of more than 200 million tonnes of ore running at 1.68 g/tonnes gold and 1.91 g/tonnes palladium has been proven. In addition there is exploitable titanium and vanadium, which adds significantly to the value of the deposit.

The molybdenum occurrence at Malmbjerg is associated with a 21–26 Ma old granite intrusion emplaced into Carboniferous sandstones. The ore was deposited from hot (approximately 500°C), percolating, metal-bearing fluids and gasses that led to molybdenum enrichment in the upper part of the granite and the immediately overlying sandstone. An ore body of about 217 million tonnes running at a grade of 0.20% molybdenite (MoS_2) has been proven. Most of the mineralisation is at a shallow depth, but the location of the occurrence at the junction between two glaciers makes access and exploitation of the deposit difficult.

View of Malmbjerg from the west across Schuchert Gletscher, showing the pale, dome-shaped, mineralised part of the molybdenum occurrence. The ore deposit was discovered in 1954 during systematic geological mapping of the area. Between 1954 and 1979 the occurrence was investigated in detail, and its extent established by mining companies that put down 147 drill holes with a recovered length of core of 22.8 km.

The molybdenum occurrence at Malmbjerg, about 35 km south of Mestersvig in East Greenland. The occurrence was formed in the upper part of a 26–21 Ma old granite intrusion, emplaced into Carboniferous sandstones. The occurrence lies in an isolated area between two large glaciers in an area known as Werner Bjerge. The photograph shows a dome-shaped, pale-coloured granite, surrounded by a brownish, mineralised zone within the surrounding sandstones. The mineralisation occurs both in the granite and in the overlying sandstones. The junction of the glaciers is at about 700 m altitude, and the peak is 1759 m above sea level.

In 1996, the oil company grønArctic Energy Inc. drilled a well (GRO#3) to a depth of 2996 m on the southern coast of western Nuussuaq. The picture shows the lower part of the drill tower.

Photo: K. Zinck-Jørgensen, GEUS

OIL AND GAS

Carbon and hydrogen from ancient animals and plants form fossil fuels

13

FOUND ONLY IN SEDIMENTS

13 OIL AND GAS

Oil and natural gas, together with coal, are known as fossil fuels. They are formed from organic material buried many millions of years ago within sediments. Because oil and gas are fluids, in contrast to coal, they are mobile and are rarely found in the same place in which they formed; indeed they may be found more than 100 km from the source rock in which they originated. Many sedimentary rocks contain interconnected pores and cracks that allow fluids, including water, oil and gas, to migrate through them. Oil and gas are less dense than the water that normally permeates the pores so, when present, the oil and gas tend to rise upwards, but can also move sideways towards lower pressures. They will continue to migrate through the sediments until they either escape at the surface or become trapped within a porous rock (a so-called reservoir rock) under a non-porous one (a seal), so forming an oil- or gasfield.

Oil and natural gas consist of complex organic molecules containing hydrogen (H) and carbon (C) – hence the name hydrocarbons. They originate from organic material laid down together with other deposits in a sedimentary basin and subsequently buried as the basin continues to subside and more sediment accumulates above it. Marine organic material consists of the remains of planktonic organisms that have settled in large amounts on the bottom of the sea after their death. They can be preserved there if the bottom water is oxygen-free (anoxic). Under these conditions no organisms can live on the sea floor and eat the organic remains, which decompose with little or no oxygen in the reactions. Once buried deeply, a rock rich in organic remains can develop into a source of oil and gas and the rock is referred to as a source rock. There can be as little as 1–2% organic material in a source rock or as much as 5–10% in a good one. Oil and gas accumulations are formed exclusively within sedimentary basins and oil source rocks are found mostly in sediments of Mesozoic or Cenozoic age, but gas accumulations from Palaeozoic source rocks are common.

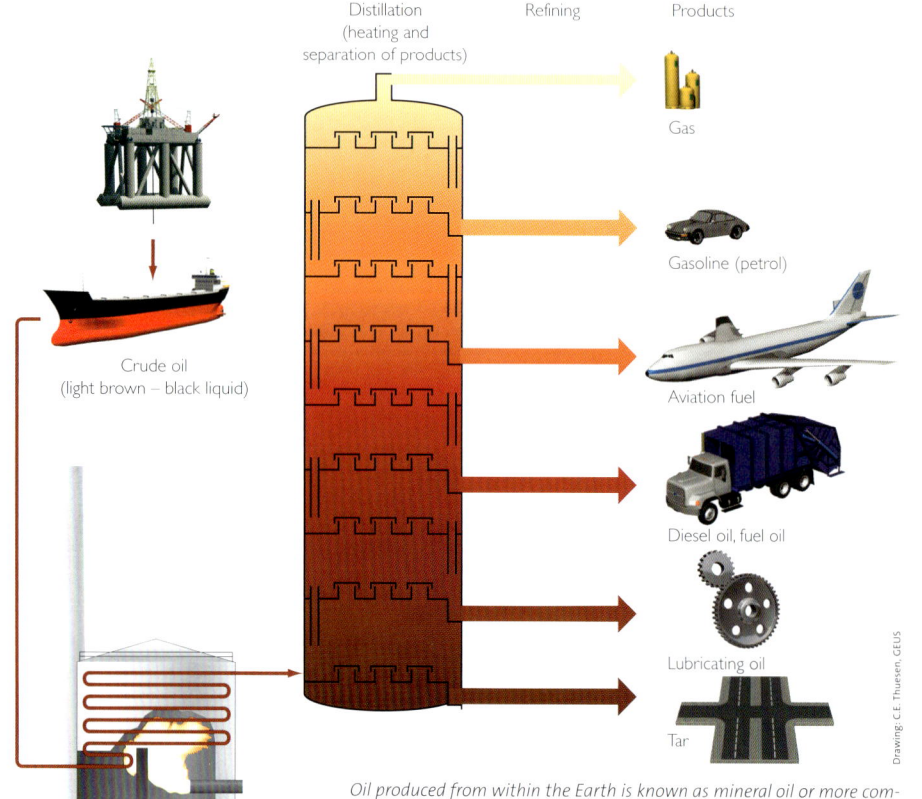

Components
83–87% carbon (C)
10–14% hydrogen (H)
1/10–6% sulphur (S)
1/10–2% nitrogen (N)
1/10–1.5% oxygen (O)
+ trace elements

Oil produced from within the Earth is known as mineral oil or more commonly as crude oil. It consists mostly of large complex molecules of which carbon (C) and hydrogen (H) are the main components. Crude oil is not used in the form in which it is found but is treated in refineries where, by heating it to their respective boiling points, its various components are separated into a series of different products – a process known as distillation. The boiling points range from about 20°C for gas at the low end to about 380°C for heavy bunker fuel at the high end.

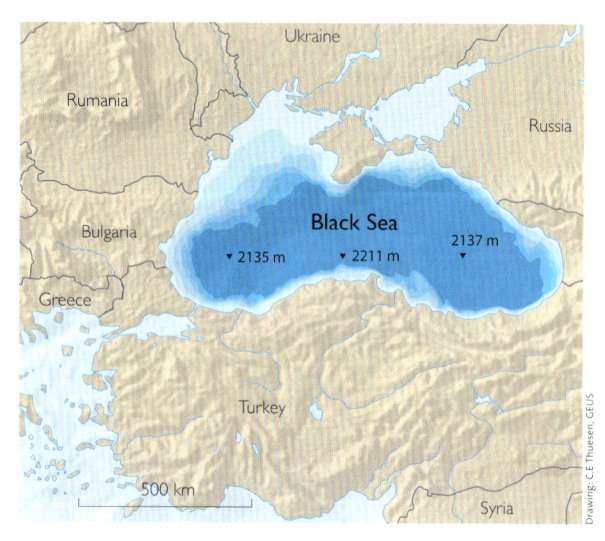

The Black Sea is an example of a present-day, oxygen-poor environment, where large amounts of mud, rich in organic remains, known as sapropelic sediments, are accumulating on the bottom. Once buried, such accumulations may in time become black mudstones that can be future source rocks for oil and gas. The Black Sea has an area of 450 000 km² and is almost isolated from the rest of the world's oceans since it is connected to the Mediterranean only through a very narrow strait. The central parts of the Black Sea are over 2 km deep and fresh water is added only to the uppermost about 150 m where most organisms live. The deeper parts of the Black Sea are, therefore, nearly free of oxygen – conditions known as anoxic. In the absence of oxygen the organic remains falling to the bottom are thus not broken down and are preserved in the sediment.

Oil and gas do not survive exposure to high temperatures (generally not over 160–250°C) and oil- and gasfields are therefore not normally found in metamorphic rocks in fold belts nor in crystalline basement areas. The places to search for oil and gas in Greenland are therefore in the 'Younger Basins', both onshore (see p. 110) and offshore (see p. 156).

Generation of oil and gas

The organic material that is the starting point for the formation of oil and gas can originate from either animals or plants. Whether oil and gas or just gas is produced from a source rock depends on its composition. The source rocks richest in carbon are composed of the remains of land plants and produce mainly methane (CH_4 – a molecule containing one carbon and 4 hydrogen atoms). The plant matter is first transformed into peat that, with burial, is converted into coal consisting almost entirely of carbon with very little hydrogen. Marine source rocks formed from the remains of algae and plankton are much richer in hydrogen. They give rise to longer chain molecules such as C_5H_{12} and C_6H_{12} through a complex series of temperature-dependant chemical changes.

The remains of marine organisms become converted to a material called kerogen when they are buried to a depth of about 1 km and their temperature has risen to about 50°C. With increased burial, the tempe-

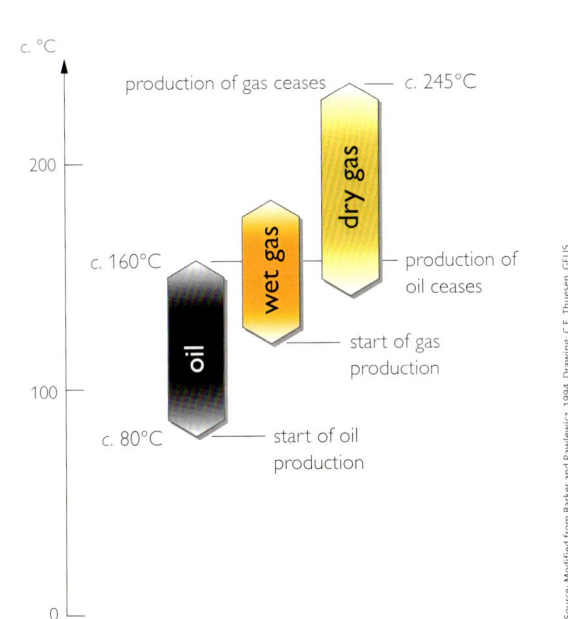

Oil and gas are generated only within a relatively narrow temperature interval, the so-called 'oil- or gas-window', that lies between about 80° and 245°C. The diagram shows the temperature ranges for the formation of oil and for those of wet and dry gas. Measurements of an organic component in the sediment called vitrinite can determine whether the rock is immature (it has experienced only too low temperatures), mature for oil (within the oil-window), mature for gas (within the gas-window) or overmature (it has experienced too high temperatures for gas). Such measurements can determine whether or not hydrocarbons have been generated in a particular area.

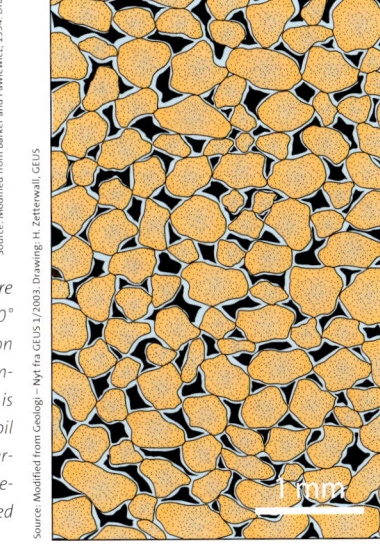

Seeps containing small amounts of oil have been found at more than 100 localities in the Nuussuaq area. These discoveries have led to commercial oil exploration by a small Canadian company who had a concession on western Nuussuaq. Various wells were drilled, the deepest of which reached about 3 km below the surface. The picture shows the drill tower and the other facilities around the GRO#3 well in 1996 (see p. 218). The island of Disko can be seen in the background on the other side of the strait of Vaigat.

Reservoir rocks with oil. The rock is a sandstone with pores between its grains (porosity) into which liquids can penetrate. If there are connections between the pores (permeability), fluids can move through the rock. The rounded sand grains (yellow) are covered by a thin film of water (blue), forming the so-called formation water. Oil (black) lies in the spaces between the film of water. In a reservoir rock containing gas, only small amounts of oil and water are present.

OIL AND GAS

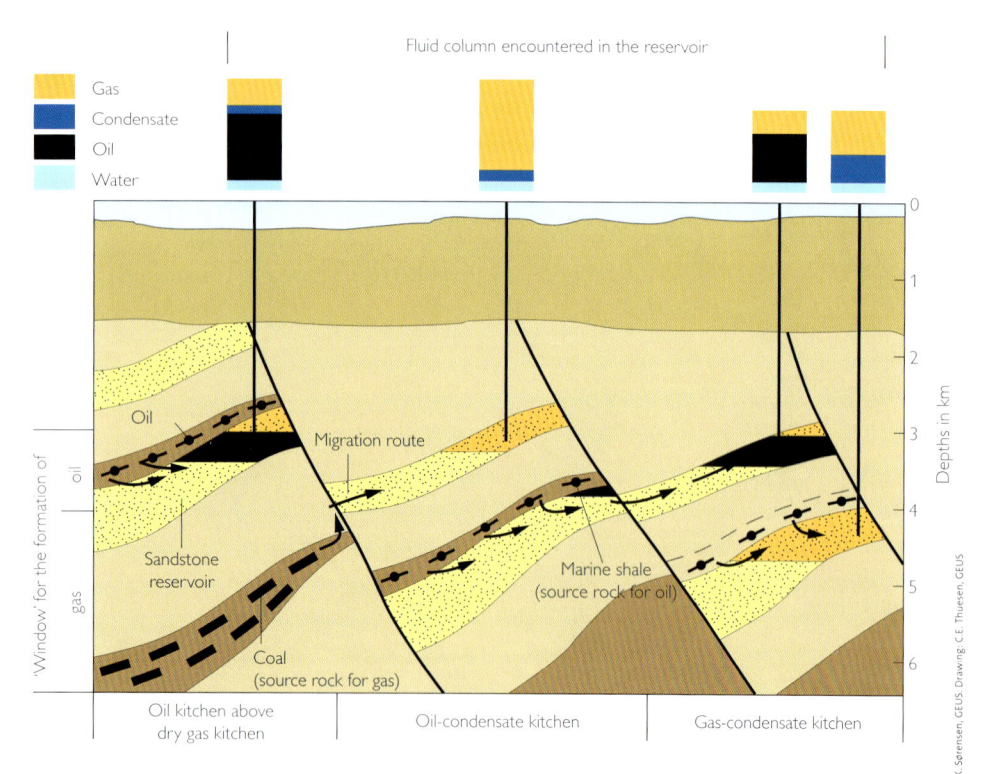

Various scenarios for the generation and trapping of oil and gas. The source rock may be coal or marine mudstones containing organic material. Formation of oil takes place between depths of 2 and 4 km while gas is generated in deeper layers between 3 and 6 km. What happens to these fluids as they reach the surface is dependent both on the depth of the discovery and the nature of the hydrocarbons that migrated from the source rock to the reservoir. Some hydrocarbons that are very light liquids in the reservoir turn into gases under the lower pressure at the surface.

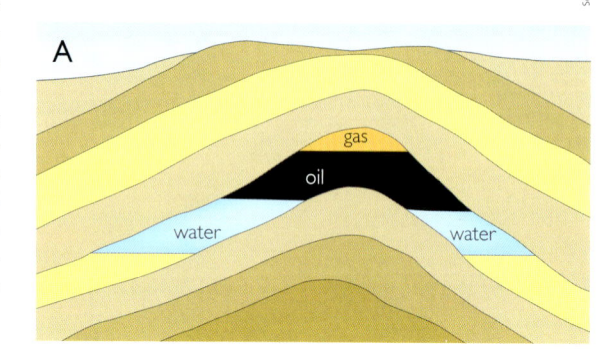

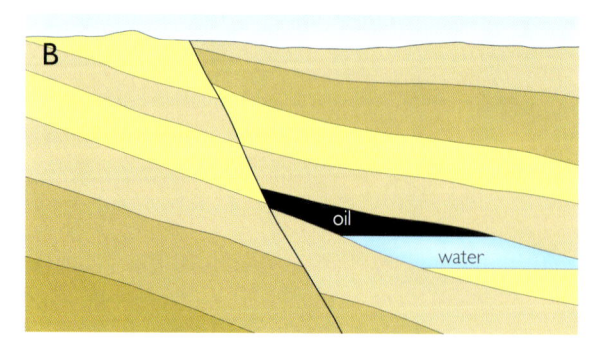

Common traps for hydrocarbons.
A) Trap at the crest of a convex-upwards fold (anticline), where the reservoir is sandstone sealed by an overlying mudstone layer.
B) Trap formed by an impermeable fault; a mudstone seal lies above the sandstone reservoir.
C) Stratigraphic trap in sandstone lenses within a mudstone.
In all cases the boundaries between gas, oil and water are more-or-less horizontal and are not parallel with the sedimentary layers.

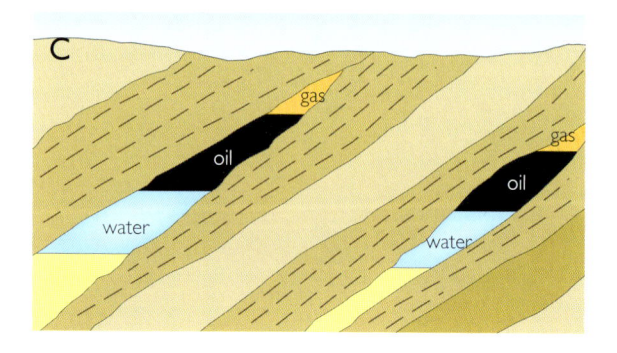

rature rises further (see p. 223). Where the temperature is between approx. 80° and 160°C, at depths of about 3 to 5 km, the large organic molecules of the kerogen break up to form smaller, lighter molecules that constitute oil and small amounts of gas. As temperatures rise even further, up to about 250°C, these molecules, too, break up to form first wet then dry gas. Oil is thus formed when the source rocks attain moderate temperatures, whereas gas is produced at higher temperatures. The temperature that a source rock has experienced is thus of fundamental importance to its ability to generate oil or gas. The changes in the source rock induced by temperature are referred to as its maturity. At low temperatures a source rock is said to be immature. Within a relatively narrow range of higher temperatures, the so-called 'oil-window', oil can be generated and the rock is mature. With still higher temperatures, the rock will become overmature for oil, but gas can still be generated. With even higher temperatures, neither oil nor gas can be produced and the source rock becomes overmature.

Areas where the organic material is converted into hydrocarbons are known as 'oil and gas kitchens'. The hydrocarbon fluids produced there can migrate into rocks containing spaces (pores) between their grains – so-called reservoir rocks. Such rocks are described as porous and their 'porosity' is measured as the percentage of the total rock volume occupied by the pores. For fluids to be able to move through a rock, the pores between its grains must be connected – this is called being permeable. When the hydrocarbon fluids migrate into a reservoir rock, its pores are normally filled with water, most of which will be displaced by the invading hydrocarbons. If both oil and gas move into a reservoir under a sealing cap rock, the hydrocarbons will be trapped and will separate into three zones: gas being lightest in the highest zone, below which is oil and lowest is water. Many fields contain both oil and gas although some fields contain very little gas above the oil and others consist predominantly of gas.

Coal, formed from fossil landplant material, gives rise to natural gas that consists of predominantly the lightest hydrocarbon, methane, with only minor proportions of heavier gaseous hydrocarbons such as ethane, propane and butane. Such gases are often described as 'dry'. The gases associated with oil originating mainly from marine source rocks, are known as 'wet', since they contain higher proportions of heavier gaseous hydrocarbons as well as some liquid hydrocarbons. When wet gases are brought to the surface, the drop in temperature and pressure leads to the formation of two separate phases: a gas and a liquid known as 'condensate'. Oil at depth underground can contain large amounts of dissolved gas. When the oil is raised to the surface, the fall in pressure results in much of the gas coming out of solution, giving considerable engineering problems.

Formation of an oil- or gasfield

The three prerequisites for the formation of an oil- or gasfield are: 1) a mature source rock, 2) a reservoir rock that can contain the oil or gas, and 3) a seal above the reservoir so that the hydrocarbons are trapped. The successful combination of these three elements is known as a hydrocarbon 'play'. Much of the world's oil comes from mudstone sediments of Mesozoic or Cenozoic age that contain large amounts of organic material. Reservoirs consist most commonly of sandstone, conglomerate or limestone and seals are most commonly clay or mudstones through which the fluids cannot flow – they are impermeable. Coal source rocks are most commonly of Palaeozoic or Mesozoic age and give rise to dry gas.

A source rock becomes mature and produces hydrocarbons when it is buried to within a narrow temperature 'window', typically 80°C to about 160°C for oil and up to about 250°C for gas. This happens when the source rock has been buried to a depth of about 3 to 5 km, or at shallower depths if the regional temperature gradient is higher than normal. Oilfields therefore form many millions of years after deposition of the organic material and, since many younger basins are still subsiding, source rocks within them may still be producing hydrocarbons. For example, the principal source rock for oil in the North Sea is of Late Jurassic age (about 150 million years old) while many of the oilfields first formed during the last 10 million years, were caused principally by subsidence of the basin during that time.

Hydrocarbon fields, within which the oil or gas is trapped in a reservoir, form in a number of different

SUBSURFACE TEMPERATURES

To form oil- and gasfields in sediments, it is not enough that there are source rocks, reservoirs and traps. Hydrocarbons have to be formed from the organic material in the sediments and the hydrocarbons have to migrate to the traps. Heat energy is necessary to transform the immature organic material into hydrocarbons. Oil is generated at temperatures between about 80°C and 165°C, and gas is generated at temperatures up to about 250°C. Sediments are heated as they gradually sink underground because the temperature normally increases about 25–30°C for every kilometre below the surface. The temperature increase per kilometre is known as the geothermal gradient and it varies depending on the surrounding geological conditions. To reach the lower limit for generating oil, the sediments need to be buried to about 2.5 to 3 km; where they are buried at more than about 4 km, the temperature is so high that the oil is converted into gas. At depths of more than 5–6 km the gas is eventually burned off, a condition known as postmature.

Heat energy rising from below the surface is a natural process. The flow of heat can be measured at the surface and is normally given in milliwatts per square metre (mW/m²). The average heat flow in continental areas is 57 mW/m², but can vary substantially depending on the geological conditions. One expression of heat emanating from the depths of the Earth is volcanic activity. Some of the heat comes from deep within the Earth that was originally much warmer than today but has gradually lost heat to space. In stable continental areas and in sedimentary basins, however, only a small proportion of the energy is generated from deep within the Earth. In these areas, most of the heat energy is produced by radioactivity; the natural radioactive decay of elements such as uranium, thorium and potassium into their fission products is accompanied by the release of heat. The radioactive elements are principally found in continental rocks and in sedimentary basins in a variety of minerals. Uranium and thorium are the most radioactive but exist only in quite small amounts, while potassium is not very radioactive but make up on average about 3% of rocks.

The flow of heat in a given area is dependent on various conditions, including the ability of the rock to transmit heat (conductivity), local variations in the consistency of the rocks and particular geochemical processes. The structural development of sedimentary basins is another important factor, because the depth to the top of the asthenosphere (a major underground heat source) may vary considerably during different parts of the development process. There can, therefore, be significant variations in the geothermal gradient and consequent variations in how deep a source rock needs to be buried to become mature and produce oil and gas. In all parts of the Earth, the temperature increases with depth below the surface and the heating necessary to produce oil and gas happens entirely naturally.

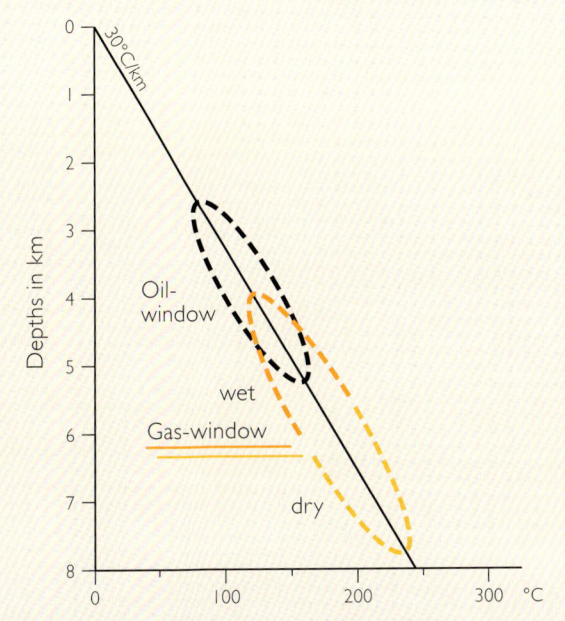

The natural increase of temperature with depth, shown here with an average geothermal gradient of 30°C/km. Oil and gas are generated within particular temperature intervals known as 'windows' that relate to specific burial depths.

OIL AND GAS

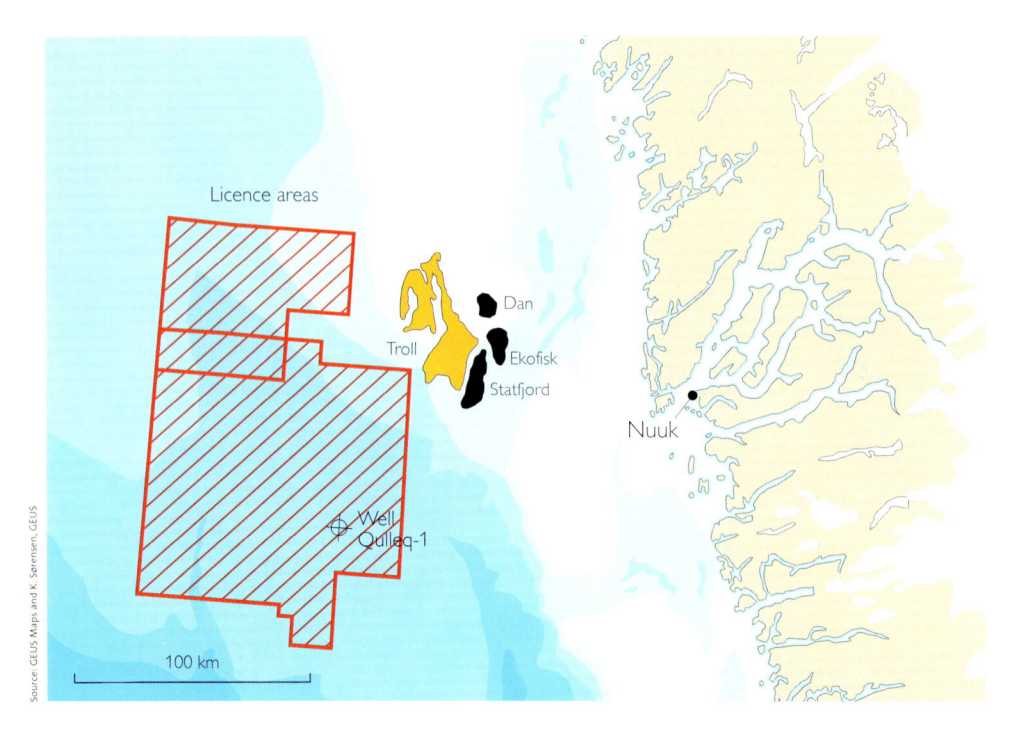

The size of some of the biggest fields discovered in the North Sea shown in relation to the size of an exploration concession west of Nuuk in West Greenland. The Dan Field is Denmark's largest oilfield. The other three fields are all Norwegian. Statfjord is the largest North Sea oilfield and Troll is the largest gasfield in the northern North Sea. Black: oilfield; Yellow: gasfield.

Sizes of fields

Example of field	km²	Productivity %	Produceable oil Millions of cubic metres
Fulmar	10	50	80
Dan	60	20	100
Statfjord	140	60	600
Greenland 'Mammoth field'	100–400	20–50	500

Table showing the sizes of a number of oilfields in the North Sea (the Danish Dan Field, the British Fulmar Field and the Norwegian Statfjord Field) compared with an estimate of the size of a commercially viable field ('Mammoth field') off the west coast of Greenland.

ways. By far the most common type is a 'structural trap', where hydrocarbons are trapped in sediments that have been deformed or displaced after their deposition. If a reservoir rock with an overlying seal has been folded into an arch, a so-called 'anticline', the hydrocarbons may migrate through the reservoir into the anticline and become trapped there like a bubble of air in a tumbler held upside down in a bowl of water. Another common type of structural trap occurs where a reservoir is tilted and faulted so that it terminates against an impermeable seal across a fault. Hydrocarbons moving through the reservoir become trapped against the impermeable seal and cannot cross the fault.

A pure 'stratigraphic trap' occurs where the hydrocarbons are trapped in a reservoir that is sealed both above and below without being moved after deposition. This type of trap is rather unusual and difficult to find, but can occur if a reservoir 'pinches out' i.e. thins away to nothing within sealing impermeable rocks. Rather more common are combined structural and stratigraphic traps, which may be produced when a reservoir between sealing units is tilted then partially eroded away at an unconformity. If an impermeable seal is deposited directly onto the unconformity, hydrocarbons can be trapped under it.

A number of different oil and gas plays have been identified in various areas of Greenland with different source rocks, different reservoirs and different traps. They are described below under the relevant region.

The size of oil- and gasfields

If oil or gas is to be produced economically from any discoveries in Greenland, the fields must be of a very large size, because the lack of infrastructure and difficult operating conditions in some areas will make exploration and production relatively expensive. Production onshore would be less expensive than offshore, but exploration onshore is logistically more difficult and is likely to be more expensive than offshore. While oil prices and costs are difficult to forecast, it is thought that only rather large discoveries will be economic, at least until the infrastructure is built up.

Experience from the North Sea and other parts of the world has shown that a critical factor in generating large accumulations of oil and gas is the existence of an extensive source rock. The actual layer containing the organic material does not have to be very thick, perhaps only a few metres, as long as it covers a large enough area. Another important factor is the presence of thick reservoirs. In the North Sea, the largest fields have reservoirs 200–300 metres thick, which means that there is sufficient room for both a gas and an oil layer above the water in the reservoir. The areal extent of the field is of vital importance. The largest gasfields in the North Sea have areas around 700 km². The largest oilfields, such as the Norwegian Statfjord Field, are of smaller areal extent, around 150 km², but are still deemed giants.

The porosity of a good sandstone lies typically in the range 20–25%, and up to about 80% of the pore volume can be filled with oil. To calculate the total amount of recoverable oil, the reservoir volume must be multiplied by this 'porosity factor' and by yet another, the 'recovery factor', as it is technically impossible to recover all of the oil in a reservoir. The recovery factor can vary greatly. A few decades ago, it was reckoned that only about one third of the oil in a reservoir could be produced, but technical advances mean that this has increased greatly in recent years and now recovery factors of well over 50% are common.

The volumes quoted on the previous page have been given in cubic metres, but the oil industry continues to use a traditional measure called the 'barrel', approximately six of which make up one cubic metre. The largest fields in the North Sea held more than a billion (10^9) cubic metres before start of production, of which somewhat more than half can be produced using modern techniques. For fields offshore Greenland to be brought into production, they would have to contain many hundreds of millions of cubic metres of oil.

Where gas is also present, the amounts are normally quoted in cubic metres of gas measured at atmospheric pressure and temperature.

EVALUATION OF A DISCOVERY

It is not sufficient just to find oil and gas underground in an exploration well, it is also vital to know how much has been discovered, to analyse what kind of oil or gas it is and to find out how rapidly oil can be withdrawn from a reservoir. For this a production test is made in which the oil and gas flow rate is measured from the reservoir to the surface where the oil is stored and the gas is burned off.

If the results of the first test are promising, a number of supplementary 'step-out' wells will normally be drilled to investigate the extent of the field and to calibrate seismic data, which are then used to map the extent of the hydrocarbons.

It is technically impossible to bring all the oil found in a reservoir to the surface. A proportion remains 'trapped' in the reservoir in the narrow throats of the pores in the reservoir rocks. How much can be produced is very dependent on the geological conditions in the reservoir, but is typically 15–50% for oil and as much as 70% for gas. Technological developments during the last decade have increased production factors substantially.

Before deciding to develop a discovery, the total amount of oil or gas that can be recovered over a number of years has to be calculated (known as the recoverable reserves). This depends on how much has been found and how much of that can be produced. The recoverable amount of oil is not fixed, but varies with the geological conditions, the technology utilised and, to a great extent, the economic conditions. A high oil price means that higher production costs can be contemplated and consequently the field reserves can be increased. So the reserves of a field are not constant, but vary almost from day to day as the price of oil varies and that changes with the global political situation. These changing conditions mean that oil companies undertake a risk analysis taking account of a range of geological, technical, economic and political factors before they make a final decision about whether or not to start development of a field. It is, of course, very difficult to forecast which economic and political conditions will apply in 10 or 20 years' time, so it is very important for the decision-making process that the political system is stable.

If oil is produced in Greenland, it is likely to come first from a field offshore central and southern West Greenland where the logistical and practical conditions are different from those known from the large oil and gas fields in the North Sea. In Greenland, there is at present a lack of industrial infrastructure and even though southern West Greenland is an 'open water region', difficulties for sea-bottom constructions can be expected due to drifting icebergs, which may carve trenches up to several tens of metres deep into the sea floor. Substantially greater production costs than are known from other offshore production areas must therefore be taken into account. For this reason, any production from Greenland will have to be from very large and high quality reserves for a field to be economic. This is why the phrase 'mammoth hunting' has been used for exploration offshore Greenland.

An oil rig in the North Sea during a production test of oil from a reservoir. The produced amount of oil and gas is measured – usually over a few hours – and the gas is being burned off. The test was made on the ENSCO70 rig as part of DONG's exploration programme in the North Sea.

OIL AND GAS

> The shelf areas off North-East Greenland are geologically similar to the oil- and gas-producing basins offshore mid-Norway and on the Barents Sea shelf.

Hydrocarbon potential of North and East Greenland

Lower Palaeozoic and younger sedimentary basins are found in both North and East Greenland, whose sediments are exposed onshore and also occur under the continental shelf. Seeps of oil have been found in small amounts in the non-folded Lower Palaeozoic sediments in North Greenland and in more substantial amounts in the Mesozoic sediments of East Greenland. Preliminary petroleum geological investigations financed both by the public sector and by private oil companies have been carried out over the whole of North and East Greenland. One consortium of oil companies carried out a substantial exploration programme in Jameson Land in central East Greenland. The exploration work included seismic surveys but the concession was relinquished without exploration drilling.

This mountainous area on Wegener Halvø, south of Kong Oscar Fjord, reveals a sedimentary series with brownish-coloured sandstones of Devonian age below lighter-coloured limestones and black shales of Permian age. The black shales are one of the source rocks in the Jameson Land Basin. See also the profile on p. 231.

The sedimentary basins exposed in East Greenland continue under the adjacent continental shelf, where geophysical surveys have shown the existence of sediment thicknesses up to 13 km (see p. 229). The exposed sediments can be compared directly with the petroliferous sediments found offshore mid-Norway. Oil and gas accumulations may, therefore, also exist offshore East Greenland, but it would be almost impossible to produce hydrocarbons from under the dense, rapidly moving pack ice off the East Greenland coast with present-day technology.

North Greenland

The Lower Palaeozoic Franklinian Basin extends east–west along the whole length of Greenland's north coast. It consists of a belt of flat-lying, shallow-water carbonate sediments to the south and a northern belt of deep-water folded sediments (see p. 83). Its total area onshore is more than 100 000 km². The basin contains two shaly units, of Cambrian and Silurian ages, rich in organic material that could be source rocks for oil. The region was heated during the Ellesmerian folding in the Late Devonian or Early Carboniferous – enough to mature the source rocks

A lightweight drilling rig that can recover cores from depths down to about 100 m. The large aluminium tank contains water and also acts as a foundation for the drilling tower. An annular diamond drill bit is used to cut cores with a diameter of 31 mm. The drilling equipment can be taken apart and moved using a small helicopter. The picture was taken in North Greenland in 1985.

Localities in North Greenland where small oil seeps (bitumen) in cracks and as impregnations in sediments show that oil has been formed in the area. The probable source rock is over 100 km from the bitumen localities showing that the oil has migrated a long way. The oil is thought to have formed more than 300 million years ago. Farthest south is a belt where the rocks are immature, while they are mature farther north. Farther north again, they have been metamorphosed. The rocks are mature to generate oil only between the two limits shown and could only have generated oil within this area. Black bitumen is seeping out of sediment on the right, showing that source rocks have existed in the area. Southern Wulff Land, north of the Inland Ice. The hammer is 30 cm long.

Wandel Sea Basin

Franklinian Basin
Deep-water sediments
- Silurian
- Cambrian–Ordovician (folded)

Shallow-water sediments
- Ordovician–Silurian
- Cambrian

Proterozoic deposits
- Neoproterozoic sediments
- Palaeo–Neoproterozoic sediments and basalts

Basement and Caledonian rocks
- Predominantly gneiss
- Bitumen
- Thrust
- Fault

OIL AND GAS

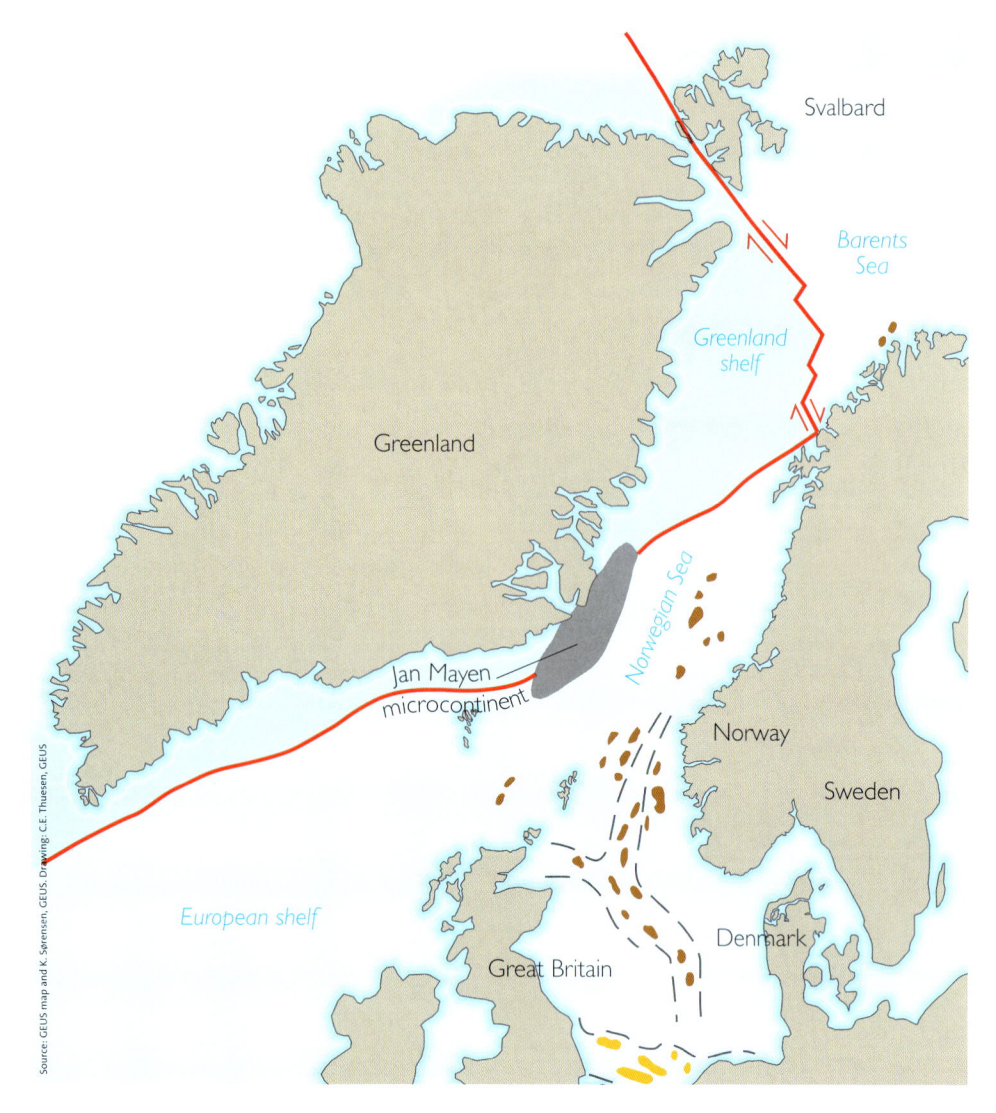

Map showing the relative positions of Greenland and Scandinavia prior to the opening of the North Atlantic. The largest known oil and gas fields are shown. It is clear that the string of fields in the North Sea and west of mid-Norway can probably be extrapolated onto the broad shelf east of North-East Greenland. The fields coloured yellow farthest south are mainly gasfields, whose source is Carboniferous coals. The fields coloured brown in the North Sea and in the Norwegian Sea are oil- and gasfields in reservoirs many of which were formed in association with a Jurassic rift phase and whose source rocks are Upper Jurassic mudstones (Kimmeridge Clay). The red line between Greenland and Scandinavia shows the position of the subsequent mid-Atlantic spreading ridge.

to produce oil. The heating was so intense in the northern area that the oil was destroyed, but oil has been preserved in the southern area and can now be found as asphalt residues in pores and cracks in various rocks. Possible reservoir rocks consisting of sandstones and reef limestones are found in the southern part of the basin.

The total extent of the source rocks could have been enormous and very large quantities of oil could have been generated from them. Traces of oil have been found mainly around the southern margin of the basin close to the Inland Ice, so the oil must have migrated considerable distances from where the source rocks were mature within the basin. Objectively, large hydrocarbon accumulations are unlikely to exist here, since the source rocks and reservoirs are of Early Palaeozoic age and the expulsion of oil from the source rocks took place during the Late Palaeozoic. The geographical location of the basin makes it logistically difficult and expensive to work in, so exploration has ceased for the present.

The younger Wandel Sea Basin in eastern North Greenland (see p. 118) contains several dark shale units, but only one of them, of Jurassic–Cretaceous age, has a limited potential to generate oil. The basin does, however, contain several good potential reservoir rocks in the form of small reefs and carbonates of Early Permian age. The sediments of the Wandel Sea Basin can be compared with those of the Barents Shelf between Svalbard and Norway and with the Sverdrup Basin on Ellesmere Island in northern Canada. Oilfields have been found in both areas.

East Greenland

Any oil and gas found in East Greenland will be in the very thick, non-folded sediments within the younger basins. The sediments onshore are found in a belt along the coast between Scoresby Sund (70°N) and Danmarkshavn (77°N). Equivalent sediments also underlie much of the shelf offshore. They were deposited in rift basins formed during the early stages of break-up of the continental area that once encompassed both Greenland and Norway. This gradual splitting started first in the Late Palaeozoic, and lakes, marine inlets and straits formed in the basin at various times during the Mesozoic. Some of the sediments, deposited in these bodies of water, are sufficiently fine-grained and organic-rich to function as source rocks. Other parts of the sedimentary series comprise coarse-grained deposits of sand and gravel that could function as reservoirs and the extensional movements in some of the basins have also formed large fault-block structures that could function as traps. Overall, the probability of the presence of large hydrocarbon fields appears to be high.

The Upper Palaeozoic and Mesozoic basins offshore mid-Norway and East Greenland were originally contiguous but were split apart by the opening of the North Atlantic during the Cenozoic. Today the Gulf Stream flows over the Norwegian continental shelf and keeps it ice-free all year round. The Greenland shelf, on the other hand, lies under a current of cold water that flows south from the Arctic Ocean, bringing with it large quantities of multi-year sea ice that normally remain over the East Greenland shelf the whole year round. Exploration and development activities during the last few decades have made the large oil and gas accumulations on the Norwegian shelf accessible, but the heavy sea ice on the Greenland side means that even a reconnaissance seismic survey has been difficult to undertake, so the oil and

Hydrocarbon potential of North and East Greenland

gas potential of this side of the North Atlantic is still unproven, although it is highly likely that oil- and gasfields are present.

The most intensive hydrocarbon exploration in East Greenland was carried out onshore in the Jameson Land Basin (70°30´N–72°N), where a consortium of oil companies led by ARCO held a concession from 1984 to 1990. The approximately 10 000 km² basin was surveyed by a regional grid totalling 1800 km of seismic lines that showed that the basin contains Upper Palaeozoic and Mesozoic sediments with a total thickness of over 15 km, of which the Mesozoic sediments are over 4 km thick in places. The basin has not been disturbed tectonically since the Palaeozoic and the sedimentary series is almost complete. Outcrop sampling around the margins of the basin demonstrated the presence of Upper Carboniferous, Upper Permian and lowermost Jurassic source rocks. Reservoir rocks, consisting of both carbonates and sandstones, are known at several stratigraphic levels. Before the seismic data were acquired, the concession holders had expected to find large fault blocks similar to those that contain the giant oilfields in the northern North Sea. The unfaulted nature of the basin came as a surprise and, although the presence of probable reef structures in limestones of Late Permian age could be identified on the seismic data, none of these were large enough to be judged economic at the time, so the concession was relinquished without drilling.

The sedimentary basins continue north of Jameson Land (72–76°N) where they are divided into large fault blocks (see p. 126). Source rocks are found in this area at several levels in the Carboniferous and Permian successions and the presence of an Upper Jurassic mudstone source rock has implications for the offshore region. Several good reservoirs are exposed onshore, mostly of sandstones and conglomerates, and the rotated fault blocks create the potential for structural or combined structural/stratigraphic traps (see p. 222). No seismic surveys have been carried out in this region, so the subsurface can only be inferred from the surface exposures.

The largest potential for discovering hydrocarbons in this region is most probably under the continental shelf offshore North-East Greenland between 75°30´N and 81°N. The shelf here is up to 250 km wide and the sediments are up to 13 km thick (see p. 168). The area has been investigated by an airborne magnetic survey and by reconnaissance seismic surveys using ice-strengthened ships. No drilling has yet been carried out, so all stratigraphic interpretations of these geophysical data are based on analogies with onshore East Greenland and with the shelf areas offshore mid-Norway and in the Barents Sea. The seismic data show that the area is segmented into fault blocks that probably developed during the Late Jurassic or Early Cretaceous. The presence of strata with mature source rocks, similar to the Upper Jurassic sediments that are the source of most North Sea and mid-Norwegian oil, is also likely. Traps containing reservoir rocks probably occur at several levels within the block-faulted structures, so the region as a whole is likely to contain oil- and gasfields. The most promising play (see p. 223) is in the eastern part of the Danmarkshavn Basin, where a large fault-block structure may contain both a thick reservoir and an Upper Jurassic source rock that is located at the right depth to be mature. The play lies to the west of a high structure that contains older sediments.

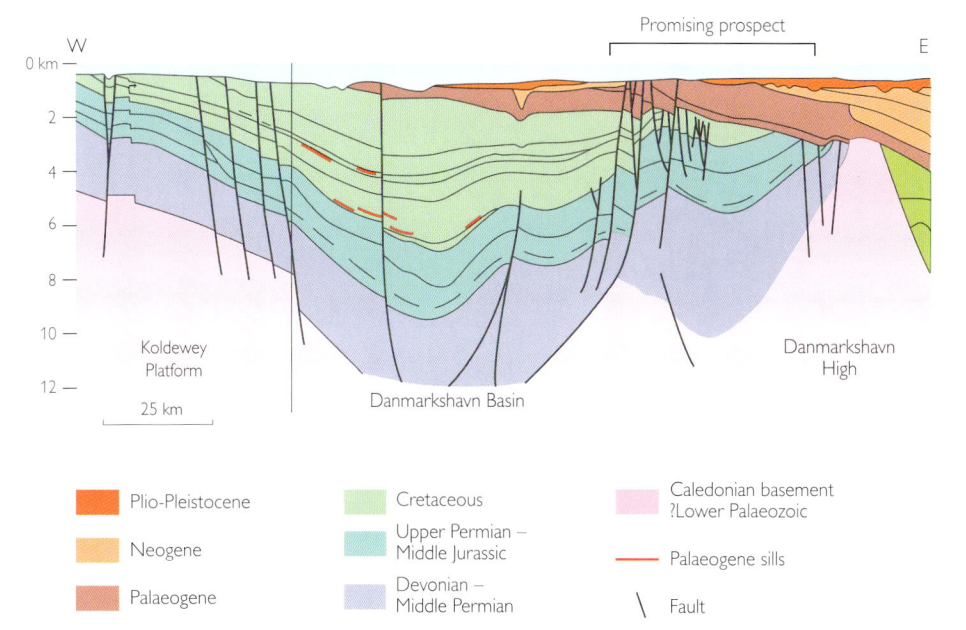

Profile based on seismic interpretation through the Danmarkshavn Basin that contains sediments up to 13 km thick. The greatest chance of finding a large oil accumulation is thought to be in the eastern part of the basin near the Danmarkshavn High where structural traps may exist.

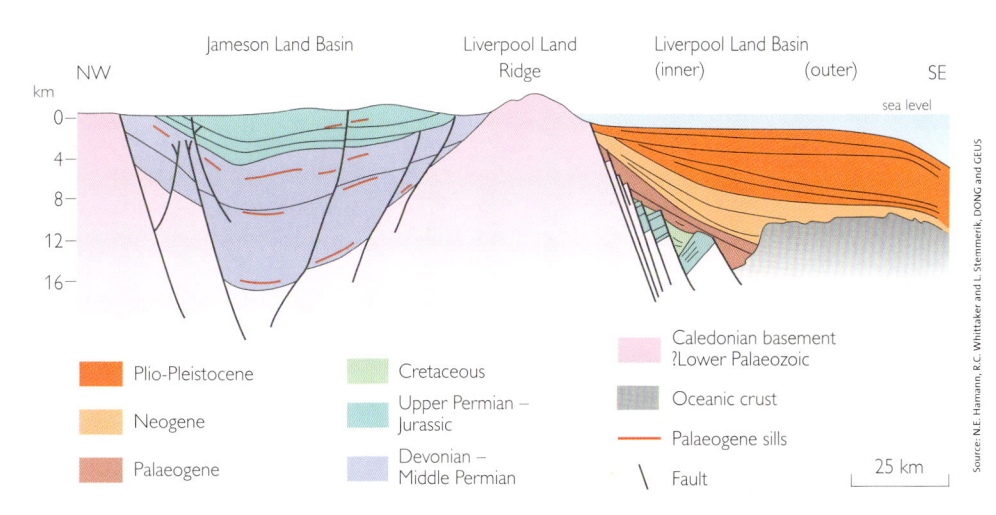

Profile based on seismic interpretations through the Jameson Land Basin and the Liverpool Land Basin just north of the mouth of Scoresby Sund. The Mesozoic sediments in the Liverpool Land Basin have been block-faulted whereas the thick Palaeogene and younger sediments are unaffected by faults. The shelf in this area is rather narrow and the oceanic crust begins close to the coast, but has been covered by thick Neogene and Quaternary sediments (see p. 169). The oil potential is limited and must most likely be sought in the Palaeogene sediments.

OIL AND GAS

Large thicknesses of sediment are also found in two other, more southerly areas; offshore Liverpool Land (70–72°N) and off the Blosseville Kyst (67–70°N) (see p. 229), but the shelf in these areas is much narrower than farther north and most of the sediments are of Cenozoic age, underlain by Palaeogene basalts. Even if source rocks similar to those known from Jameson land were present under the basalts, they would be so deep that they would be overmature, so any hydrocarbon accumulations in these basins would have to form from a Palaeogene source rock and be trapped in a Cenozoic reservoir. Such possibilities are poorly known, because no rocks of equivalent age are exposed onshore in this part of Greenland. It is, however, very unlikely that these southern basins could contain fields comparable in size to those that potentially lie off North-East Greenland.

OIL EXPLORATION IN JAMESON LAND

Seismic data acquisition in the winter was carried out using special, tracked vehicles on which were mounted vibrating plates that sent sound waves down into the ground. The reflected waves were received by geophones (microphones) on 3–5 km long cables and recorded in other tracked vehicles. The picture shows one of the recording trucks operating in Jameson Land at the end of the 1980s.

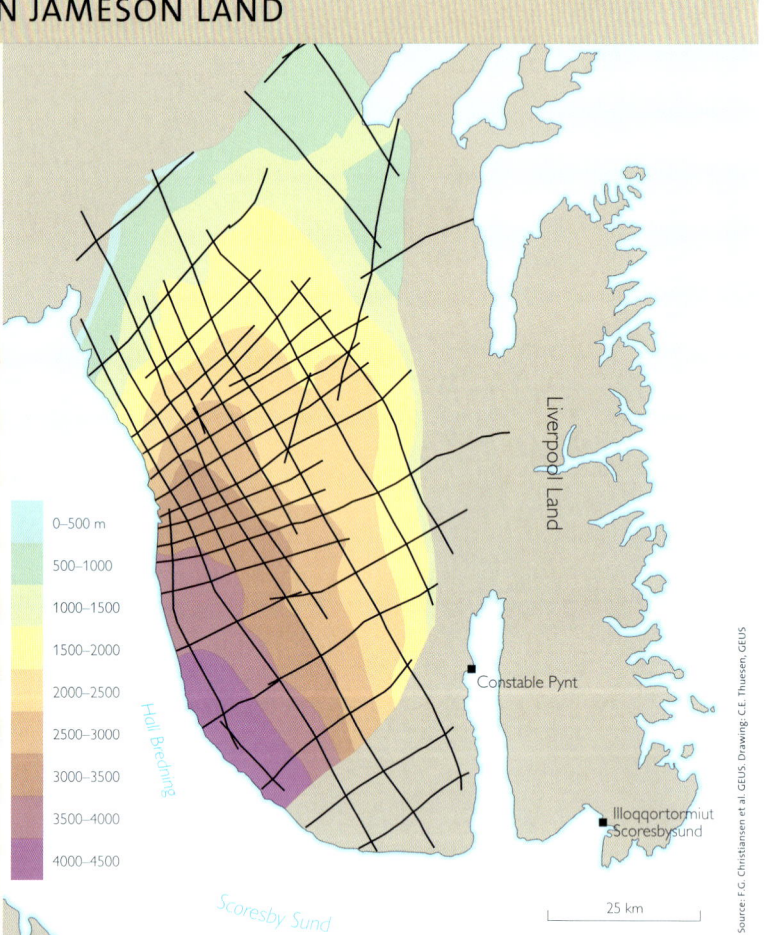

Map constructed from interpreted seismic data showing the thickness of the Mesozoic sediments in the Jameson Land Basin. Depths to the top of the Upper Permian sediments are shown with contour intervals of 500 m that are shaded in different colours. The Mesozoic sediments are 4000–4500 m thick in the deepest part of the basin in the south-west. The seismic lines are shown in black.

Jameson Land in East Greenland was the first onshore area in Greenland where real oil exploration was carried out by a consortium led by Atlantic Richfield Company (ARCO). The Italian company AGIP joined the consortium later. The exploration was carried out between 1985 and 1989, but the consortium relinquished the concession in 1990 after having acquired seismic data but without drilling.

The basis for the exploration was systematic geological investigations and regional geological mapping that had documented a very thick series of Upper Palaeozoic and Mesozoic sediments in the Jameson Land Basin. Several source rocks were identified and there were a number of possibilities for reservoirs.

The area of the Jameson Land concession was around 10 000 km², all of which was uninhabited and there was no infrastructure. Thus the operator had first to establish both an airfield with an operation centre and accommodation and a landing place for ship-borne material at Constable Pynt near the head of the Hurry Inlet fjord. These facilities made it possible to undertake large-scale seismic data acquisition during the next few years. After relinquishment of the concession, the facilities were taken over by the Greenland authorities and today they function as an airfield with a further connection by helicopter to Illoqqortormiut/Scoresbysund.

Seismic data were acquired from a total of 1800 km along a grid of intersecting lines. In summer, the seismic source was small dynamite charges fired in shot-holes drilled by rigs moved by helicopter. During winter, the seismic waves were generated by a special vibrating plate (Vibroseis) mounted on a number of heavy off-road vehicles. Recording cables were laid out in 3–5 km lengths along the seismic profile lines and the sound waves generated by the seismic source were reflected from underground layers and received by geophones (microphones) connected by cables. All of the data were

OIL EXPLORATION IN JAMESON LAND (CONTINUED)

Helicopters were used in summer to lay out the equipment for seismic data acquisition. The recording equipment consisted of geophones (microphones) laid along a cable several kilometres in length and the seismic waves were generated by small dynamite explosions.

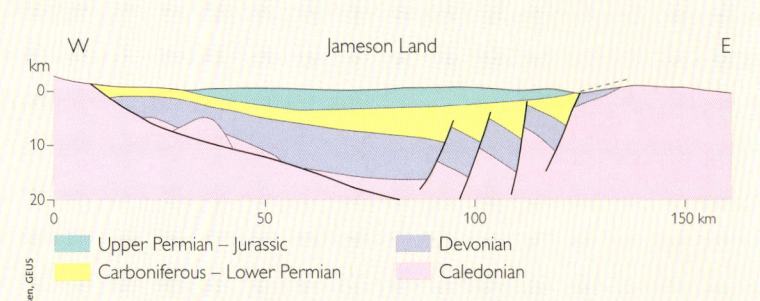

Profile interpreted from the seismic data acquired in the Jameson Land Basin. The basin is underlain by folded Caledonian crystalline rocks. The deeper part of the basin below the Upper Permian layer is heavily intruded by Palaeogene sills (not shown) and affected by faults that cause the thickness of the succession to vary considerably. The deepest part of the basin is under the south-westernmost corner of Jameson Land; compare to the depth map on the opposite page that shows the thickness of the upper, unfaulted sediments.

recorded digitally, processed in a large computer in the USA and then interpreted geologically.

After relinquishment of the concession, the state carried out additional geophysical activities in partnership with a consortium of different companies, including acquisition of offshore seismic data from the fjord and special processing of the combined data. The results showed that the total thickness of sediment in the Jameson Land Basin is around 17 km, the lowest 10–12 km of which consists of a Devonian to Lower Permian sandstone series that was deposited during regional extension after the Caledonian Orogeny (see p. 114). These deep sediments are cut by normal faults and intruded by basic sills which made seismic interpretation difficult. An unfaulted succession of Upper Permian and Mesozoic sediments about 5 km thick lies above the faulted Upper Palaeozoic sediments. Because these younger sediments are unfaulted they lie undisturbed in the fashion of a layer-cake. Analysis of surface samples has shown that very good source rocks are found in this sequence – one of Late Permian age and another of Early Jurassic age. Several good reservoirs have also been recognised in Jurassic and Cretaceous sandstones and in reef-like limestones of Late Permian age.

Before the seismic data were acquired, the concession holders had expected to find large fault blocks similar to those that contain the giant oilfields in the northern North Sea. The unfaulted nature of the Mesozoic sediments came as a surprise and, although the presence of probable reef structures in limestones of Late Permian age could be identified on the seismic data, none of these were large enough to be judged economic at the time, so the concession was relinquished without drilling.

Evaluation of the variation of the temperature of the basin through time and the consequent evaluation of the oil potential have been carried out using analyses of the 'maturity' of the sediments. The results of the analyses and other considerations have shown that the rocks at the surface are overmature (have been subjected to too high temperatures) in the northern part of the basin and are immature in the rest of the area. Only in one sector in the north and a smaller area in the south are the surface sediments mature and therefore able to produce oil. The subsurface sediments are expected to be mature for oil below depths of about 1–2 km and for gas down to depths of about 4–5 km.

The Jameson Land Basin is still one of future potential, and cannot be written off.

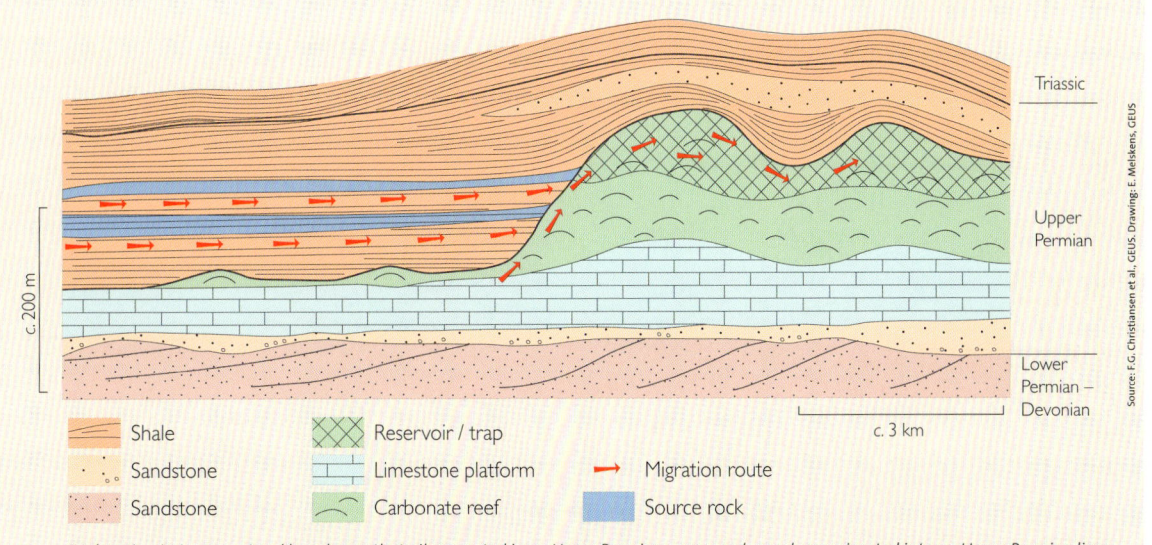

Exploration in Jameson Land has shown that oil generated in an Upper Permian source rock may have migrated into an Upper Permian limestone reef reservoir sealed by overlying mudstones.

OIL AND GAS

Since 1969, about 50 000 line kilometres of seismic data have been acquired and six deep wells drilled. The potential for oil exists, but is still unproven

Hydrocarbon potential of West Greenland

Unlike all the other sea areas around Greenland, the waters offshore southern West Greenland are free of sea ice all year round. There is open navigation south of Disko Bugt and most of Greenland's population and fishing industry are concentrated in this area. A number of large sedimentary basins occur under the seabed off this part of Greenland. The sediments in these basins are related to the deltaic, coal-bearing deposits found onshore in the Disko–Nuussuaq–Svartenhuk Halvø region where oil seeps at the surface were discovered in the early 1990s. The first serious systematic oil exploration activities started in these ice-free open waters in 1969. Since then a large amount of seismic data has been acquired offshore West Greenland, six deep wells have been drilled in the southern Davis Strait between 64°N and 68°N and one onshore on Nuussuaq. No significant discoveries have yet been made, but the substantial exploration means that the subsurface off central and southern West Greenland is now fairly well known and expectations of finding oilfields in the area are high.

Gravity anomalies delineate the sedimentary basins and the highs between them. Faults, interpreted on seismic data run through the whole area, offset the sediments at depth and have caused complex structures. The basins contain sediments mostly of Cretaceous and Palaeogene age. Oil exploration is directed at the sedimentary basins, but accumulations could be discovered bordering some of the highs.

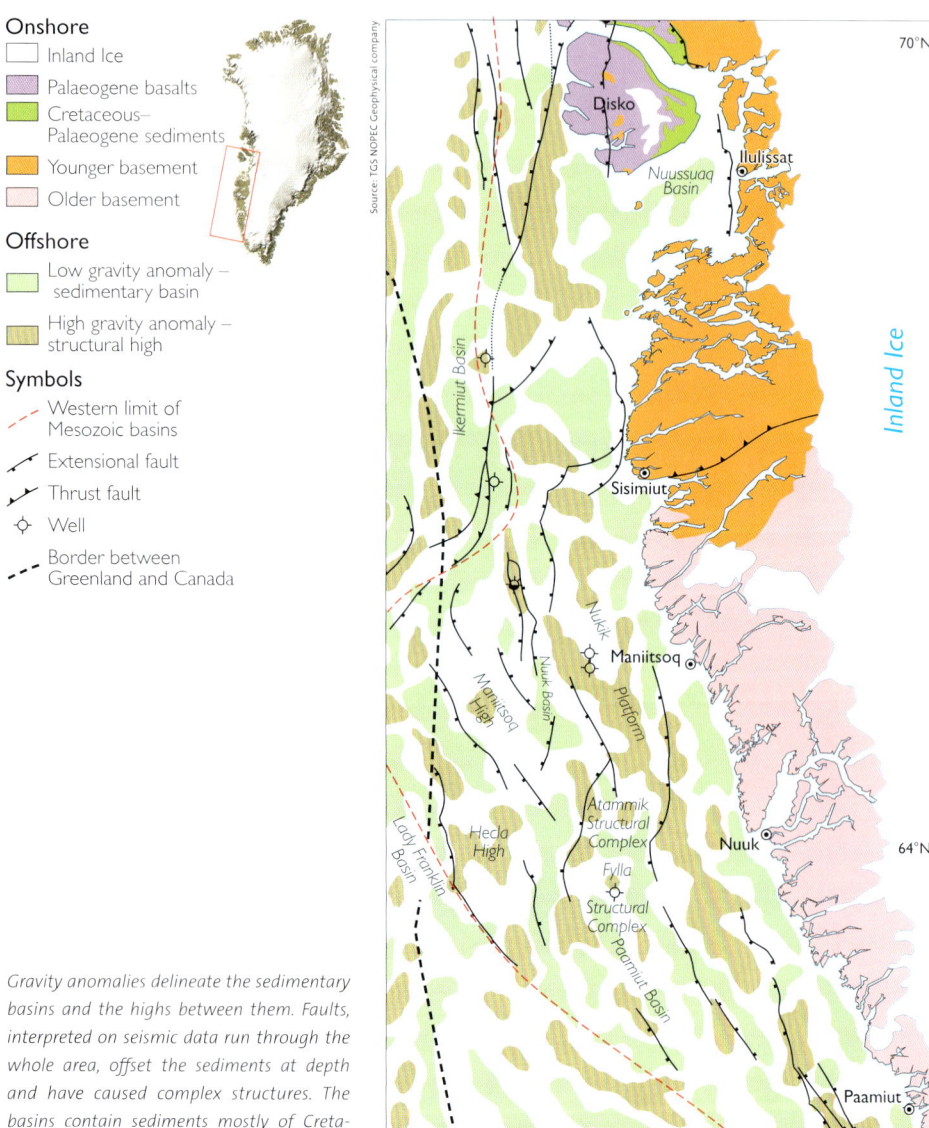

The sea areas offshore northern West Greenland – north of Disko Bugt – are covered in ice for six months of the year and there are many icebergs. Even though thick sedimentary sequences with oil potential have been discovered (see p. 174), they are known only from a sparse grid of ship-borne seismic data. The hydrocarbon potential of the area is therefore poorly understood and the environmental conditions for production of possible discoveries are also poor because of the icebergs.

Exploratory studies for oil and gas have been carried out in recent years in the Disko–Nuussuaq–Svartenhuk Halvø area. A number of shallow wells have been drilled and one deep well was drilled in 1996. Oil has been discovered in seeps and also in the recovered drill cores and cuttings, but no large find has been made. The results are, nonetheless, very important because they demonstrate the existence of oil source rocks in sediments that are expected to extend out under the seabed.

Onshore – the Nuussuaq Basin (69–72°N)

The Nuussuaq Basin contains more than 8 km of Cretaceous–Palaeogene sediments overlain by Palaeogene erupted plateau basalts (see pp. 128 and 142). The sediments were deposited in an ancient delta system that grades into marine-shelf deposits in western Nuussuaq and beyond the present coast. The sand-prone deltaic sediments contain large amounts of organic material in the form of coal. The discovery of oil in pores in some of the overlying volcanic rocks shows that source rocks for oil must exist in the marine sediments in the subsurface, as proved by a deep well drilled onshore on Nuussuaq. These discoveries have led to a significant amount of oil exploration in the area, both by GEUS and by oil companies that held exploration concessions.

Oil seeps have been found within outcrops of volcanic rocks at many places throughout the entire area between Disko and Svartenhuk Halvø. Since oil is not generated from hot volcanic rocks, the seeps are interpreted as an indication of the possible presence of oil accumulations in the sedimentary layers below the basalts. Migration of the oil into the seeps took place relatively recently, many millions of years after the volcanic rocks cooled down. The picture shows black tarry oil deposits in narrow fractures in a light-grey hyaloclastite (see p. 145) on Ubekendt Ejland between Nuussuaq and Svartenhuk Halvø.

The first discovery of oil seeps was made on Nuussuaq in 1992 subsequent to which a range of geological and geophysical studies have been carried out and a number of wells have been drilled. Six of the wells used slim-core drilling to depths up to a kilometre and a conventional well was drilled to a depth of 3 km. The wells found oil-impregnated rocks, several reservoir intervals containing pressurised gas and a source rock of Paleocene age. Geochemical analysis of the oil has shown that it consists of several different types, one of which shows geochemical similarities to an important, well-documented source rock on Ellesmere Island in Canada's north-eastern Arctic. The results from the Nuussuaq Basin suggest that there are possibilities of finding oilfields in the area from Qeqertarsuaq (Disko), over Nuussuaq to Sigguup Nunaa (Svartenhuk Halvø).

Black tarry bitumen that has migrated upwards into a brown basalt and been deposited along fractures in the volcanic rock. The small round white spots are natural pores (vesicles) in the basalt filled with white minerals.

Localities in the area between Disko and Svartenhuk Halvø where oil seeps have been found in both discrete fractures and as bitumen impregnations of the rocks. The discovery of these oil seeps in an area mostly of volcanic rocks was a great surprise and has been one of the most important factors in the renewed interest in oil exploration in West Greenland since the beginning of the 1990s. The oil seeps are interpreted as an indication of the presence of rocks containing oil under the basaltic volcanics and are also a strong indication that oil may be found under the seabed, offshore West Greenland.

OIL AND GAS

GEUS carries out drilling activities using lightweight equipment to drill through sedimentary rocks to depths of several hundred metres and recover sediment cores. The upper picture shows the equipment in operation at a locality on Svartenhuk Halvø. The small picture on the right shows how cores are laid on trays to be measured and later packed to be sent back to the laboratory for analysis and more detailed studies.

The small drilling rig is mounted on a large aluminium water tank that supplies cooling water and at the same time functions as a foundation for the rig. The drill is driven by a small diesel engine from which power is transferred hydraulically to rotate the drill pipe and move it up and down. The drill has a diamond bit that cuts cores with a diameter of 46 mm.

Offshore southern West Greenland (60–68°N)

This is by far the most intensely studied sedimentary basin in Greenland's offshore region. Various seismic companies began exploration in 1969 and exploration licences were awarded in 1975 after which five deep wells were drilled in 1976 and 1977. At the time, all of these wells were declared to be dry – i.e. had no signs of oil – and all licences were relinquished by 1979.

A second phase of activity began in the late 1980s when the Geological Survey (GGU) used reprocessed data to show that Mesozoic basins extend much farther from the coast than was thought during the 1970s. Reinterpretation of the industry seismic data also showed that the basins had been inadequately explored and revealed the presence of many previously unrecognised rotated fault blocks. These developments led to acquisition of new seismic data during the early 1990s by GGU, a commercial seismic company and by the state-owned oil company Nunaoil.

Some of these new seismic data showed so-called 'flat-spots' that were interpreted as possible indicators of large gas accumulations in the 'Fylla structural complex'. One of the structures was subsequently explored in 2000 by a well drilled to a depth of 2973 m. Unfortunately no trace of hydrocarbons was found and the 'flat-spots' were discovered to relate to a transition between mineral phases in the sediments. Despite this disappointment, the well penetrated the oldest sediments known in the area and provided valuable information to guide later exploration.

Commercial seismic companies have acquired more than 50 000 km of regional seismic data since the mid-1990s. These data have shown the presence of very large structures that could contain amounts of oil in excess of a billion cubic metres. Their presence has persuaded the oil industry to undertake additional exploration and, at the time of writing, two exploration licenses are held by a Canadian oil company in the Nuuk and Lady Franklin Basins (see pp. 237 and 241) and seven new licences have been

Hydrocarbon potential of West Greenland

awarded in 2007 to four consortia in the region west of Disko.

The thickest sediments within the basins offshore southern West Greenland are around 9 km thick and consist of Mesozoic sediments within rift basins overlain by several kilometres of Cenozoic sediments (see sections west of Nuuk p. 174). The seismic data have shown the presence of unknown sediments below the Mesozoic sediments, the so-called 'deep sequence'. These older sediments were known even in the 1970s in a limited area, but the more recent seismic data have allowed their presence to be traced over a much larger area. Their identity is unknown, but they could be of Jurassic or Palaeozoic (probably Ordovician) age.

The most likely source rock in the southern West Greenland basins is thought to be of mid-Cretaceous age. Organic-rich rocks of this age are known from the Sverdrup Basin in the Canadian Arctic and some of the oil seeping on Disko and Nuussuaq (see p. 232–234) is thought to originate from such a source rock. Other possible source rocks are lower Palaeogene mudstones that also give rise to the oil seeps, and strata of possible Jurassic or Palaeozoic age in the 'deep sequence'. There are good candidates for reservoir rocks in sandstones of both Cretaceous and Palaeogene ages. The most attractive combination of reservoir, seal and trap is probably a Lower or mid-Cretaceous sandstone within a fault block, sealed by upper Cretaceous mudstones.

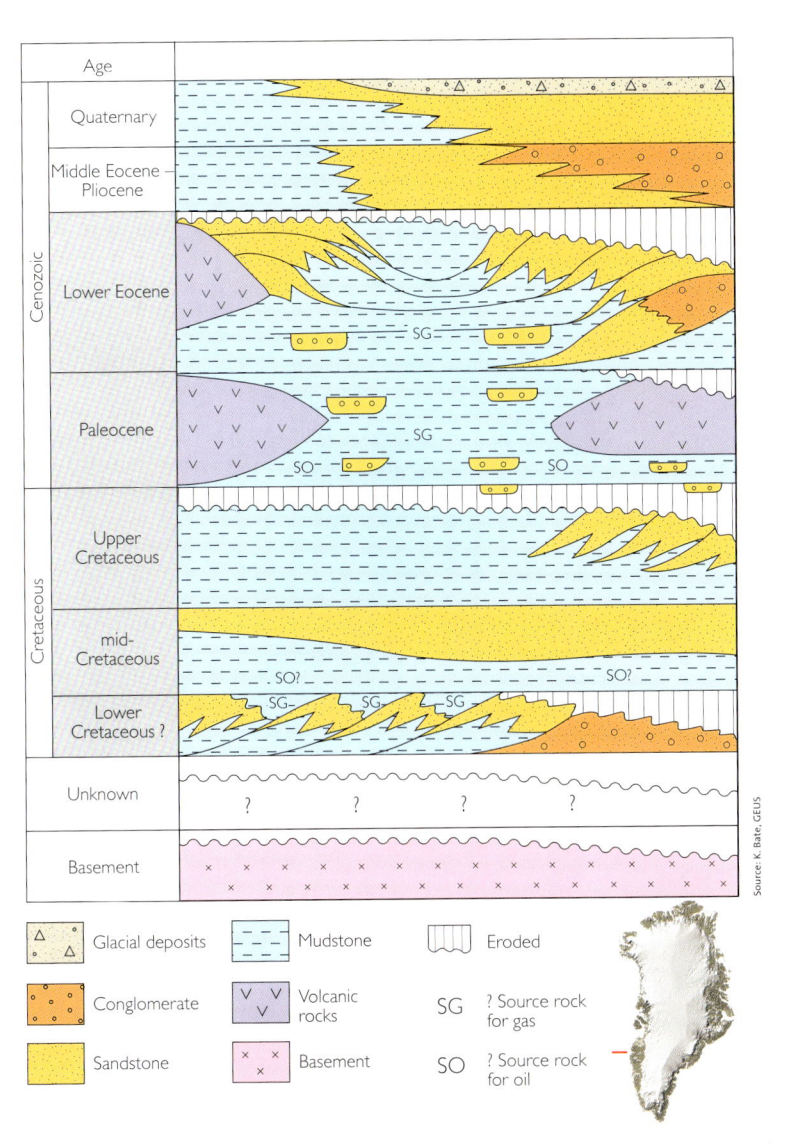

An example of a seismic profile (upper right) acquired offshore West Greenland. After processing, the data are shown as a cross-section from which a geophysicist has interpreted the disposition of the geological layers underground (coloured lines). The interpretation was then calibrated stratigraphically using information from a deep well drilled through the upper sedimentary layers to produce the interpreted geological cross-section (lower right). Integration of the regional geophysical data with geological data from all the wells resulted in the regional stratigraphic column (above) that show schematically the development of the basin.

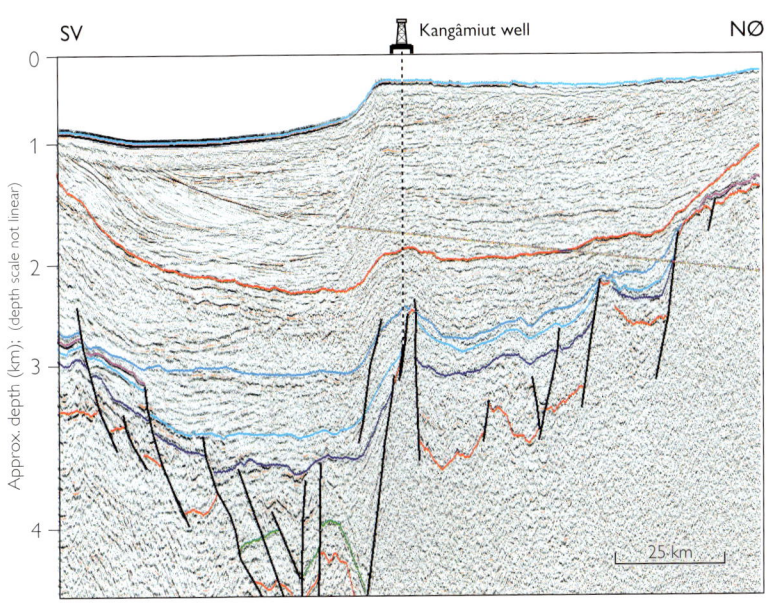

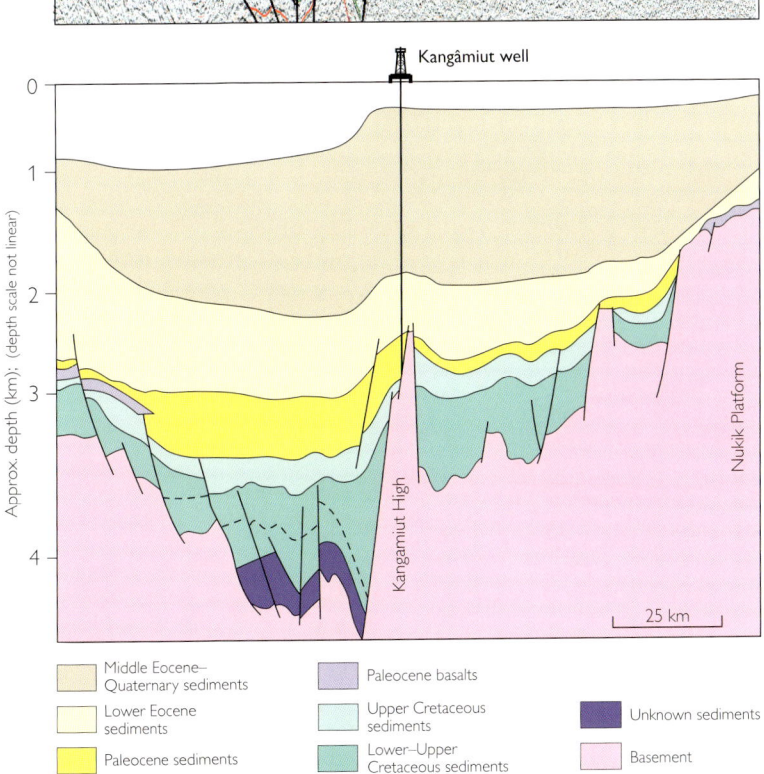

OIL AND GAS

Landslide at Pujoortoq (north side of Nuussuaq) that set fire to the hillside in 1933.

Landslipped hillside with red-coloured burnt shales that were rich in organic material. The landslip allowed introduction of atmospheric oxygen into newly opened cracks, which led to self-ignition and combustion of the rocks. The burnt hillside shows clearly that the sediments contain large amounts of organic material in the form of coal. The locality is on the north side of Nuussuaq.

Offshore northern West Greenland (73–78°N)

Very large sedimentary basins have been shown to exist in this region with sediment thicknesses up to 12 km in the southern part, thinning to about 3 km in the north (see p. 174). The sediments are interpreted to be of Cretaceous to Recent age, with the Mesozoic sediments preserved in a series of large fault-blocks that have been uplifted and folded in the northern area.

No exploration activities other than the acquisition of seismic and other geophysical data have been carried out in this area. Its evaluation is still at a reconnaissance stage, but the area is, nonetheless, thought to be a promising hydrocarbon basin because it lies between oil seeps in the Nuussuaq Basin and the exposed mid-Cretaceous source rock in the Sverdrup Basin on Ellesmere Island. The seismic data have shown the presence of very large rotated fault-blocks that could function as traps.

The area is covered by sea ice every winter and a large number of icebergs are carried into it from the fjords to the south. It is unlikely that commercial exploration will take place in this region within the forseeable future.

Offshore central West Greenland (68–73°N)

The thick sediments below the plateau basalts exposed on Disko, Nuussuaq and Svartenhuk Halvø probably continue offshore into the Davis Strait and southern Baffin Bay. The basalts are exposed at the seabed close to the land, but farther west they dip westwards under a cover of Eocene to Recent sedi-

Hydrocarbon potential of West Greenland

ments (see p. 178). The extent of the basalts can be mapped readily using seismic sections, and the thickness and depositional history of the overlying sediments can also be studied. Determining what lies under the basalts is much more difficult, because seismic waves are scattered by the basalts themselves. Modelling of gravity data has shown that there are probably sediments under the basalts and recently acquired seismic data give a general picture of the major structures found there. These interpretations show that the area is divided into a number of N–S-trending horsts and grabens that contain several kilometres of sediments that were deposited before the basalts were erupted. If this interpretation is correct, it is reasonable to assume the presence of Mesozoic and/or Paleocene source rocks below the basalts.

These considerations have led to oil companies being awarded exploration licences in the area west of Disko (see figure p. 241) in 2007.

Terms and conditions for oil exploration in Greenland

Oil industry activities in Greenland are regulated, as in most other parts of the world, by a set of government imposed terms and conditions. This is done to control the development activities and to protect public interests economically as well as environmentally. At the same time, it is also the wish that such activities be promoted, so the terms and conditions must be sufficiently attractive to encourage the oil industry to invest resources in exploration in Greenland.

The first step in attracting and maintaining the oil industry's interest in the region is to facilitate and support their evaluation of the petroleum potential, i.e. to provide solid documentation of the geological development of the area including access to results from earlier exploration. The results of such studies are held by the Geological Survey (GEUS), who regularly publishes the results of its own investigations and supplies data and information that can be used in the investigations leading to a 'licensing round'.

Licences are agreements between the government and the relevant companies that allow the licence-holders to carry out certain activities in specific areas. *Prospecting licences* are granted on a non-exclusive basis over large areas to enable regional evaluation work to be carried out. They do not permit deep drilling and are primarily used to acquire regional seismic data and sampling of the seabed. *Exploration licences* are granted over smaller areas to a single company or a consortium of companies who believe they have found one or more hydrocarbon prospects using the seismic data acquired under a prospecting licence. Within the area of the concession, they have

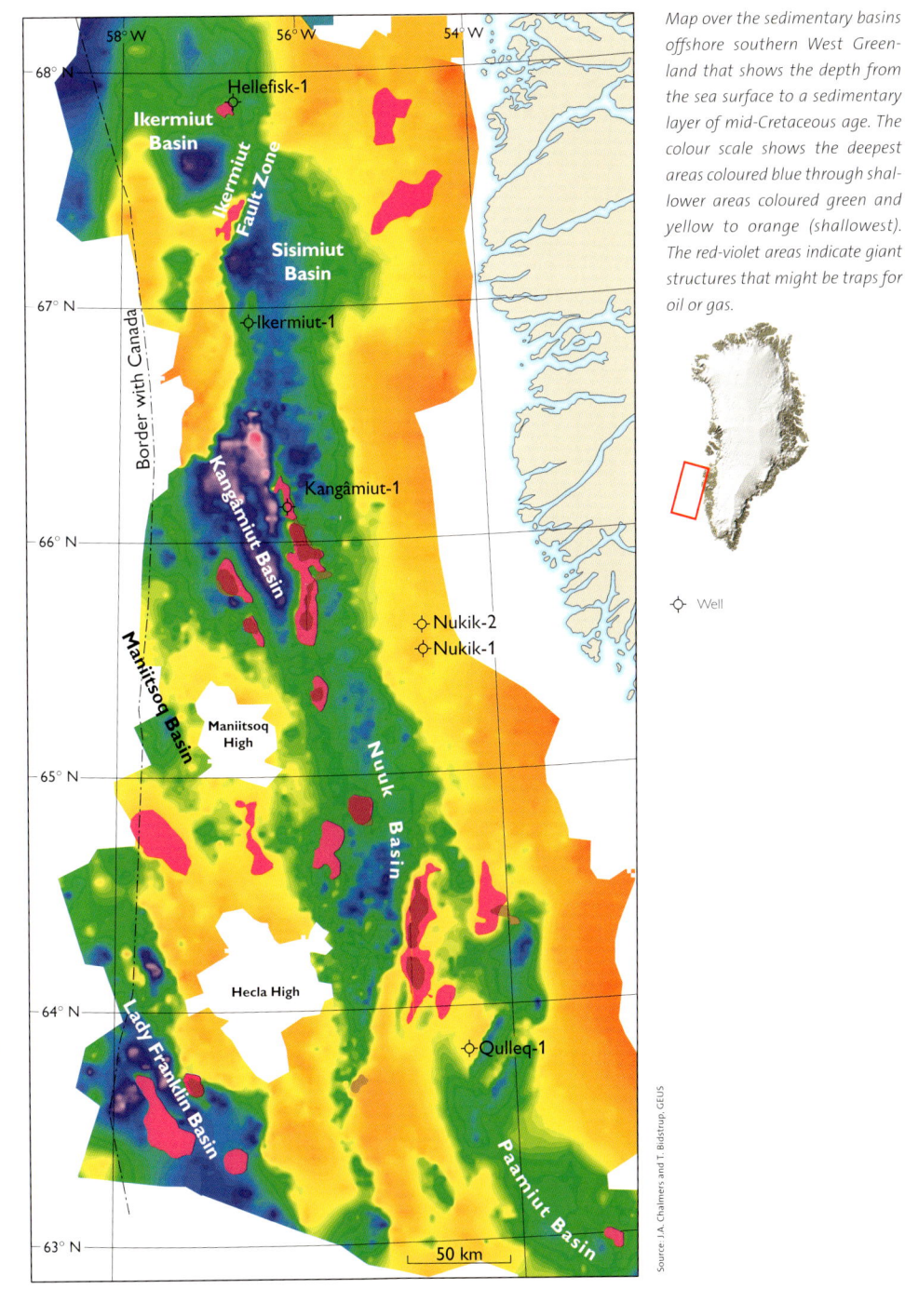

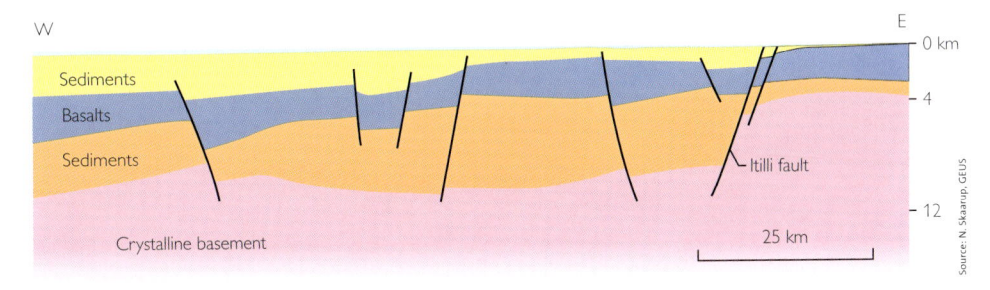

Map over the sedimentary basins offshore southern West Greenland that shows the depth from the sea surface to a sedimentary layer of mid-Cretaceous age. The colour scale shows the deepest areas coloured blue through shallower areas coloured green and yellow to orange (shallowest). The red-violet areas indicate giant structures that might be traps for oil or gas.

Cross-section offshore west of Disko based on a combination of interpretation of seismic data for the shallowest part above the basalts and modelling of gravity data for the deeper part where the seismic signals cannot penetrate because they are scattered by the basalts in this area. The modelling shows the probable presence of sediments 1–5 km thick below the Palaeogene plateau basalts. These deeper layers probably consist of Mesozoic sediments that are of potential petroleum interest. The eastern end of the profile is just west of the west coast of Disko.

an exclusive right to carry out exploration work, including the drilling of deep wells. The evaluation of an exploration licence normally starts with the acquisition of a more detailed seismic survey that is used to decide on one or more drilling locations. One or two deep wells may then be drilled to test if a prospect contains hydrocarbons. Planning, acquiring, processing and interpreting the seismic data followed by drilling will normally take several years.

The exploration licences are granted for a fixed period of time. Licence fees and a minimum work commitment are part of the terms and the national Oil Company of Greenland, Nunaoil, must normally partake as a non-paying partner (12.5–15% in 2007). The authorities have to be informed regularly about the work carried out and the results obtained. As long as the exploration licence is in force, the results may be reported to the general public only by agreement between the licence-holders and the authorities, but the holder's reports can be made public after a licence is relinquished.

Terms and conditions for exploration and production licences
General conditions in 2004

1) **Duration:** An exploration licence lasts 10 years and can be extended to 30 years if production starts.
2) **Public participation:** The publicly owned oil company Nunaoil A/S takes part in an exploration licence as a non-paying partner.
3) **Tax:** Normal company tax is payable in Greenland (30% in 2004), but no royalty is payable. In the event of large profits, a special, negotiated royalty will be payable.
4) **Payment of public expenses:** The licence-holder will pay the cost of administration of the licence incurred by the authorities.
5) **Establishment fee:** A fee of 100 000 Danish kroner is payable to establish the concession.
6) **Annual fees:** An annual fee of one million Danish kroner is payable for an exploration licence.
7) **Training:** The licence-holder will pay a fee of 250 000 Danish kroner per year for training of employees of the Greenland authorities.
8) **Relinquishment of parts of the licence area:** Twenty five per cent of the area of the licence will be relinquished at the end of an agreed period. The authorities can thereafter deal with the relinquished area as they wish.
9) **Reporting:** The licence-holder will report regularly to the authorities about their activities in accordance with the agreed regulations. The reports will be made public either after 5 years or when the licence is relinquished.
10) **Operating conditions:** The terms of the concession include particular rules for how the work is to be carried out and for protection of the environment.

Principal points in the agreement between the public authorities and a consortium of oil companies about the conditions pertaining to an exploration licence offshore West Greenland and production of any discovery made within its area.

Should a commercial discovery be made and the licence-holders submit a satisfactory development plan, they will be granted a permit to develop the discovery. The *development licence* is for a fixed time period and various rules and conditions apply. Nunaoil's participation in this licence requires it to pay its share of the costs.

Exploration areas

The commercial hydrocarbon exploration offshore Greenland that started in the 1970s took place shortly after a dramatic rise in the price of oil and shortly after major discoveries had been made offshore North-West Europe. The interpretation that the wells drilled in 1976 and 1977 were dry (without oil or gas) caused the companies to lose interest and leave the area. The re-evaluation of southern and central West Greenland initiated by GGU led to the acquisition of large amounts of new seismic data and has entirely revised the understanding of the area's potential. The rise in the price of oil since 2005 has also renewed interest.

This re-evaluation has also been extended to other offshore areas of Greenland, resulting in a general understanding of the development of all of them, with the exception of that off North Greenland. Work is now starting in earnest off North Greenland in connection with the extension of the exclusive economic zone to 350 nautical miles and the subsequent negotiation of the political divisions of the Arctic Ocean. The geological understanding has led to the development of a political strategy concerning hydrocarbon exploration. Its primary goal is to attract and maintain the interest of the oil industry in the possibilities of Greenland and to direct exploration and hopefully eventual development into those areas where it can be carried out with available technology. For these reasons, the onshore areas of the Nuussuaq Basin (in West Greenland) and the Jameson Land Basin (in East Greenland) as well as offshore southern and central West Greenland have been chosen as being of current interest.

For administrative purposes, the sea areas off West Greenland have been divided into blocks defined by latitude and longitude. The actual licence areas can encompass several or even parts of such blocks and their work commitment will in part depend on the size of the concession. Examples of the areas of concessions are those held by the Statoil/Phillips group in the late 1990s and early 2000s that were 9487 km^2, 4744 km^2 and 3985 km^2. Normally, part of the initial concession must be relinquished after a number of years.

Licensing rounds

The authorities initiate exploration activity by offering the industry access to exploration licences in a specific area and under conditions announced in advance. The offers are known as licensing rounds. There have been several in recent years, each of them based on the most recent information available from new seismic surveys and other geological data. In order to attract the interest of the industry, the terms and conditions have been different in each licensing round so that they are consistent with the petroleum market at the time.

The initial licensing round offshore southern West Greenland was held in 1975, when 13 exploration licences were awarded to six groups of companies. Five wells were drilled and all licences were relinquished by 1979. A single licence was awarded in 1996 (resulting in one well) and a total of four licences were awarded between 1998 and 2005, two of which have been relinquished. Most recently seven new licences were awarded in 2007 in the region west of Disko. Licences have also been awarded in onshore areas, first in Jameson Land in 1984 (relinquished in 1990) and subsequently on Nuussuaq in 1995 (relinquished in 1998) that resulted in one exploration well (see p. 241).

In addition to the areas that are offered under licensing rounds, there are other, less well-known areas offshore West Greenland, where companies can apply for exploration licences at any time under a so-called 'open-door' policy.

Licence-holders

Exploration licences can be sought for and held by a single company, but often a group of companies agree to form a consortium to hold an exploration licence and work in concert. One of the companies in the consortium is chosen as the operating company, i.e. the one that ensures that the various activities are carried out. Formation of consortia happens for two reasons: 1) to reduce the economic risk faced by one company by spreading it between several parties and 2) to enable smaller companies that would not have the economic potential to find and develop a discovery alone, to participate in exploration and production. All companies in the consortium have access to all the information gathered by the operator.

Examples of consortia that have recently held exploration licences in Greenland are those that held the so-called 'Fylla' and 'Sisimiut-West' licences. The 'Fylla' concession, in which a well was drilled in 2000, was held by a consortium consisting of the four companies Statoil, Phillips Petroleum, Dansk Olie og Naturgas (DONG) and Nunaoil, the national Oil Company of Greenland. Statoil was operator and Nunaoil's exploration expenditure was 'carried' by the other licencees. Phillips Petroleum was operator of the 'Sisimiut West' licence with partners Statoil and DONG, and again Nunaoil participated as a non-paying partner. Not all exploration licence-holders consist of consortia. The exploration licences awarded in 2002 and 2005 were to a single paying company, EnCana, with Nunaoil again participating as a non-paying partner.

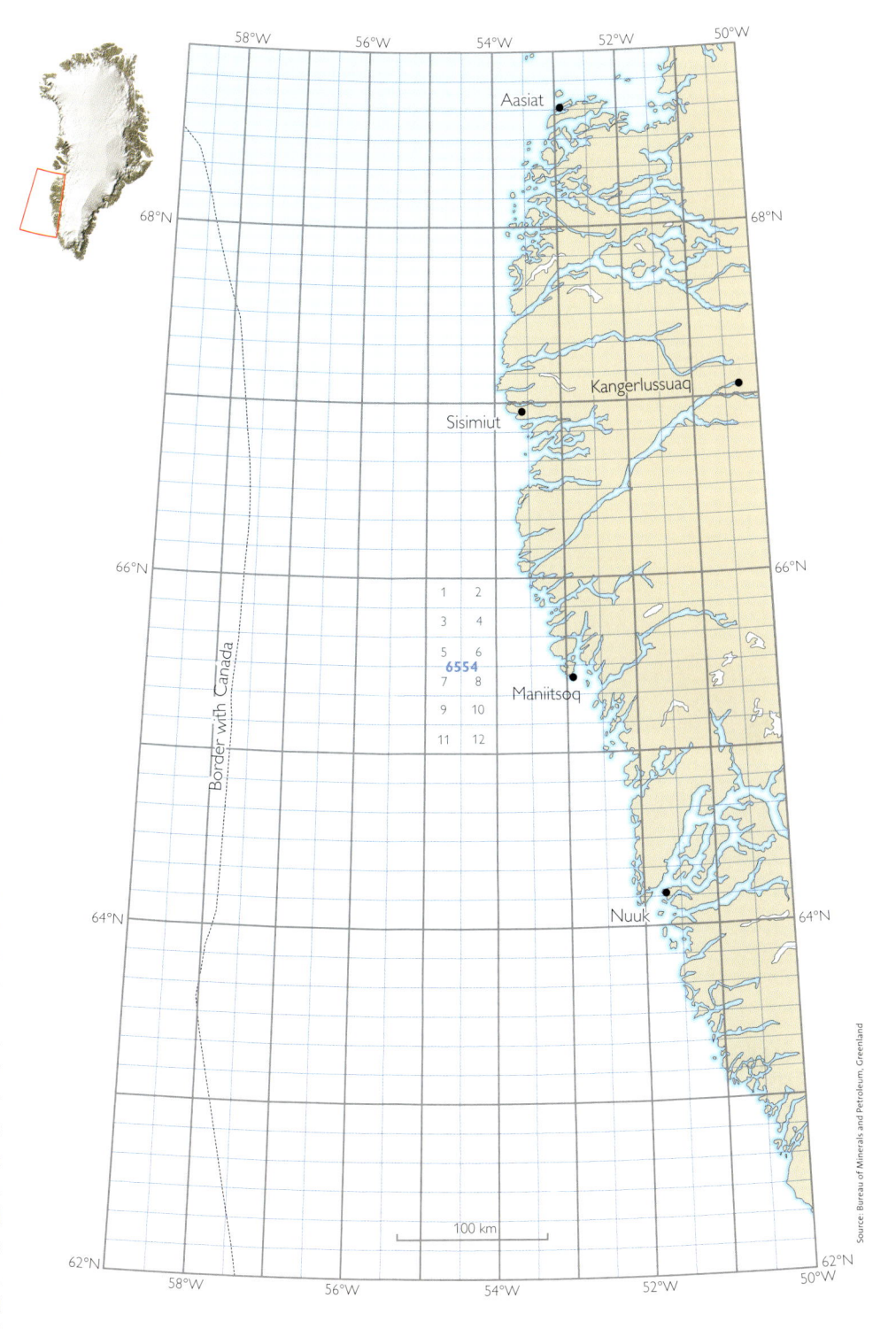

Until now only the areas offshore southern and central West Greenland have been offered for licensing. On the other hand, in order to make sensible long-term strategic planning of hydrocarbon exploration, an evaluation of the other areas around Greenland has been necessary. To this end, a project, known as the 'KANUMAS' project (Kalaallit Nunaat

Block system used for concessions for exploration and eventual production of oil and gas accumulations offshore southern West Greenland. The quadrants are numbered after the latitude and longitude of their south-east corner and divided into blocks with an index number. After partial relinquishment of a concession, its areas may consist of smaller units and the outline of the concession areas can become very complex.

OIL AND GAS

Consortia of oil companies that have drilled in Greenland

Consortium group	Well; year	Participants (% holding, operator in bold)
TGA-Grepco	Kangâmiut-1; 1976	**Total** 29%, Gulf 29%, Aquitaine 29%, Grepco 13%
Chevron	Ikermiut-1; 1977	**Chevron** 30%, BP 30%, Niocden 30%, Saga 10%
Arco	Hellefisk-1; 1977	**Arco** 20%, Cities Serv. 20%, Hispanoil 20%, Hudbay 20%, Fina 20%
Mobil	Nukik-1 og -2; 1977	**Mobil** 25%, Amoco 25%, Deminex 25%, PanCanadian 25%
GrønArctic	GRO#3* plus others; 1996	**grønArctic** 68%, Platinova 17%, Nunaoil 15%
Statoil	Qulleq-1; 2000	**Statoil** 38%, Phillips 38%, DONG 9%, Nunaoil 15%

* Onshore well.

The consortia that have drilled exploration wells in Greenland.

Marine Seismic) ['Kalaallit Nunaat' is Greenlandic for 'Greenland'] was set up to acquire regional seismic data in the areas off North-West, North-East and central East Greenland. Access to these regions is difficult because they are partially or wholly covered in ice for part or all of the year. The KANUMAS project was initiated by GGU but was given to Nunaoil as operator and was carried out in the early to mid-1990s with the participation of a number of oil companies. The project led to the first regional understanding of the development of the areas surveyed and has shown that there are good possibilities for large accumulations of hydrocarbons off both North-East and North-West Greenland, with limited possibilities off central East Greenland. It is, however, difficult or impossible to carry out normal oil exploration activities in the regions with present-day technology and no licensing round is expected in any of the areas in the near future. If and when any such licensing is offered, the KANUMAS participants have a preferential right to obtain exploration licences in the areas from which data were acquired.

Operators and entrepreneurs

The operators of an exploration licence do not themselves normally carry out all the practical work undertaken in a concession. They organise the activities of a large number of specialist entrepreneurial companies that undertake such work as seismic data acquisition and processing, drilling, well-logging and geological consultancy, as well as hiring non-technical service and transport companies.

In addition to the work carried out under the exploration licences, a large amount of preliminary work is carried out under prospecting licences, particularly by seismic companies who acquire and process seismic data to sell to interested purchasers. Much of the recent improvement in understanding of the geological development and prospectivity of southern and central West Greenland has been due to so-called 'speculative data' acquired in this way, particularly by the companies TGS-Nopec and Fugro-Geoteam. Such data are sold to other companies interested in the prospects.

Map showing an interpretation of the structure of the top of the Fylla sand reservoir around the 2749 m deep Qulleq-1 well. The Fylla area is west of Nuuk. The two-dimensional map uses colours and contours to illustrate the depths between 3 and 5 km. Any source rock present just below the reservoir would be mature at the depths shown in green. The yellow and pink areas indicate rocks through which the oil can migrate and the orange arrows show the migration routes. The dark red colour shows where oil might be trapped. Depths in metres; contour interval is 100 m.

During the first phase of oil exploration offshore southern West Greenland from 1969 to 1977, five deep wells were drilled in 1975 and 1976. The picture shows the drillship 'Pelican' in 1975 drilling the well Kangâmiut-1, west of Sisimiut.

Hydrocarbon potential of West Greenland

THE KANUMAS PROJECT 1990–1996

For a number of years during the KANUMAS project, the Danish Navy's fishery-inspection ship F 357 'Thetis' was used to acquire seismic data in the waters around Greenland. To carry out this function, the stern of 'Thetis' was rebuilt to provide working space for seismic equipment that included a streamer with a hydrophone (microphone) cable up to 5 km long, and airguns that used high-pressure air to send sound waves into the sea (see p. 166). The picture shows 'Thetis' during a seismic cruise in the icy waters off northern West Greenland. The red buoys show where the airguns were being towed behind the ship. The ship is breaking a lead between ice floes so that the equipment can be towed through this icefree channel.

The sea areas offshore north-east and north-west Greenland are covered by sea ice for much of the year, including great masses of multi-year ice that drift south from the Arctic Ocean on the East Greenland Current. The continental shelves in these parts of Greenland are therefore nearly inaccessible and can be navigated only during a few summer months with specially ice-strengthened ships or ice-breakers. It has, therefore, been very difficult to study the geology of these areas and to form an impression of their hydrocarbon potential.

To obtain information about the conditions in these partially ice-covered areas, a special project was initiated by GGU with the participation of a number of large oil companies (BP, Exxon, Japan National Oil Company, Shell, Statoil and Texaco). Nunaoil A/S, the national Oil Company of Greenland, coordinated the work and functioned as operator and the Danish Navy (Søværnet) made the ice-strengthened fishery-inspection ship 'Thetis' available to the operation. The ship was partially rebuilt to make room in its stern area for the geophysical equipment and place was found for the electronic equipment within the ship. The ship's own helicopter was used for local ice reconnaissance and regional ice reconnaissance was carried out using the Danish Air Force's Gulfstream aircraft equipped with Side-Looking Airborne Radar (SLAR) so the ice conditions could be ascertained even during conditions of total cloud cover.

The KANUMAS project lasted several years during which a total of 4071 line km of seismic data were acquired offshore northern West Greenland, 5637 km offshore northern East Greenland and 1323 km offshore central East Greenland. These data have been used to establish regional-scale models for the geological structure of the areas and large sedimentary basins have been shown to exist offshore both northern East and northern West Greenland that could have a substantial petroleum potential (pages 168, 229 and 236).

Summary map of areas in Greenland showing the state of concession agreements at the end of 2007 (awarded licence areas shown in red). Bid rounds were held in 2002, 2004, 2006 and 2007. Licences in two smaller areas west of Nuuk were awarded in 2003 and 2005 and a large region west of Disko was awarded in seven block areas to four groups of oil companies in 2007. The green areas are covered by a so-called 'open-door' policy where oil companies can apply for concessions at any time. The yellow areas show where the KANUMAS companies have preferential positions, but where exploration is not expected in the near future.

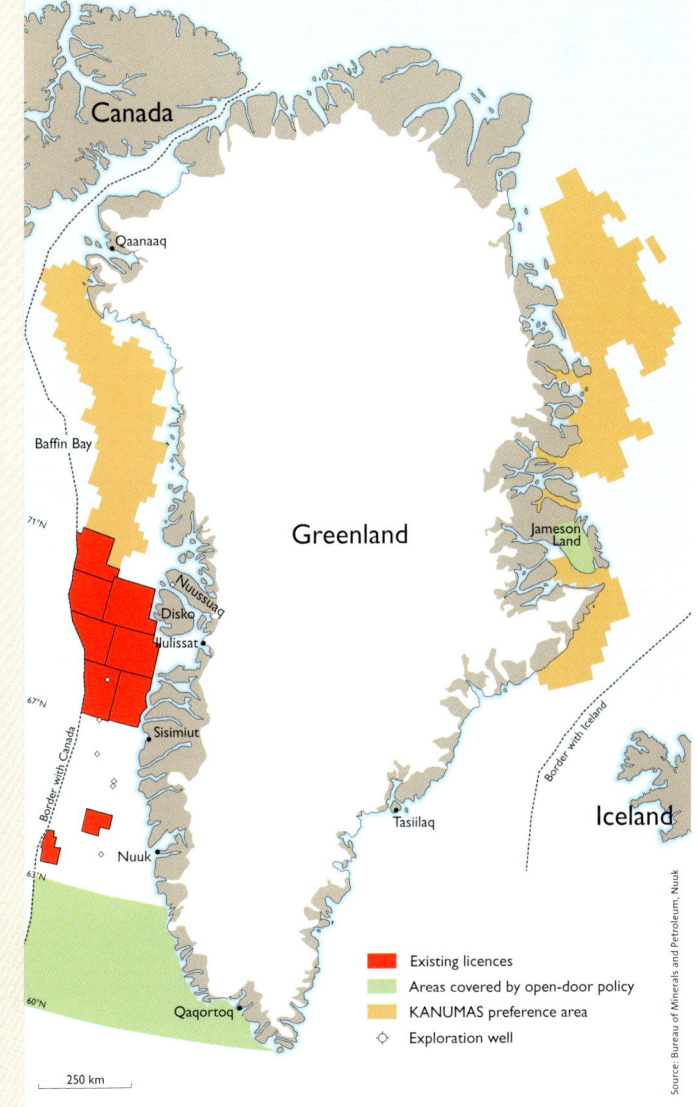

241

BACKGROUND AND ACKNOWLEDGEMENTS

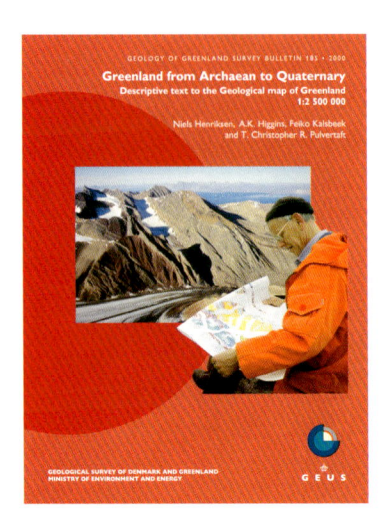

The geology of Greenland has been presented in summary form in the Survey Bulletin 185 (2000). It is a descriptive text to the most recent geological map at a scale of 1: 2 500 000 intended for professional geologists.
Above: the cover of Bulletin 185.
Below: the 1: 2 500 000 map included in the bulletin. Actual map size: 96 x 120 cm.

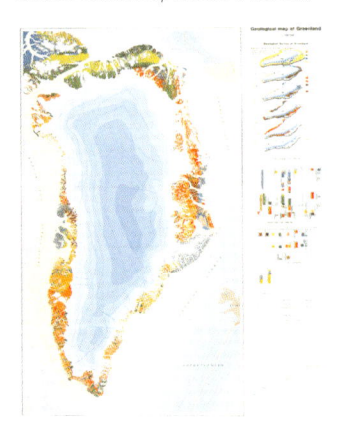

This book is a translation into English of the Danish volume "Grønlands geologiske udvikling – fra urtid til nutid" [the geological development of Greenland – from the beginning of time to the present], first published in October 2005 and reprinted in May 2006. The original publication in Danish was well received by the public, and both the author and publisher have had many requests for an English edition. The marked increase in international tourism, both to the populated parts of West Greenland and the almost unpopulated regions of East Greenland, has certainly raised the level of general interest for information on how the land developed geologically and how geology has influenced its present landforms. Furthermore, the renewed focus in recent years on prospecting for natural resources in Greenland has contributed to a demand for a general description in English of the geology and the background for mineral occurrences and the oil and gas potential. The professional interest in this resource potential is reflected by the present level of prospecting companies' activity and related geoscientific research. By the end of 2006, a total of 40 exploration licences, 14 prospecting licences and two exploitation licences for minerals had been applied for and granted. This is the highest level of exploration activities since 1997. Most of this exploration has been directed towards occurrences of gold, molybdenum, zinc, nickel, rubies, zirconium and diamonds. Petroleum exploration is expected to be a major focus of activity during the coming decade in the region offshore central and southern West Greenland, where some of the world's largest oil companies have been granted exploration licences. It is expected that investment in this region will reach a level of several hundred million dollars in the coming years. These trends suggested that an English edition of the book would be justified. This updated English edition is aimed at readers both with and without a specific knowledge of geology but who have a general interest in Greenland, its nature and the potential for economic exploitation of its natural resources.

This book is a product of more than 50 years of geological research in Greenland. Most of the results have been presented in international scientific journals or in publications and maps produced by the Geological Survey of Denmark and Greenland (GEUS). The background data forming the contents of this book include research results by many hundreds of geoscientists and cover almost the entire spectrum of geoscientific subjects and its many different specialities. In research papers it is normal to acknowledge the specific contributions of other workers at appropriate places. As this book is not an ordinary scientific work, detailed references are omitted in order to make it easier for the non-specialist reader to follow the text. Selected references are presented in the chapter "Further reading" on pages 244–245; a more extensive list is included on the GEUS website: http://www.geus.dk/isbn211.html

The list of the most important literature follows the chapter headings in the book. For references to the primary results, readers are directed to the scientific summary bulletin on the geology of Greenland published by the Survey: Greenland from Archaean to Quaternary. Descriptive text to the Geological Map of Greenland 1:2 500 000, by N. Henriksen, A.K. Higgins, F. Kalsbeek and T.C.R. Pulvertaft. Geology of Greenland Survey Bulletin 185 (2000), 93 pages. This bulletin also contains a folded copy of the most recent, coloured geological map of Greenland at a scale of 1: 2 500 000.

The present English edition follows the layout of the original Danish volume, for which the attractive and informative layout was created by Carsten Egestal Thuesen.

Throughout the preparation of the Danish and English editions of this book I have been greatly helped by a large number of colleagues who have generously provided relevant material and given me access to their own results. It is not possible to give detailed acknowledgement of the specific contributions of all the individuals who have assisted in the production of the volume; I can only express my thanks in summary form below. The individual chapters in Danish were all read through and commented on by colleagues who have specialist knowledge of the specific subject areas. I thank them all for their many constructive comments and discussions that I have had in connection with preparing the material for the book. They include the colleagues named below with respect to the chapters to which they have contributed:

Landscapes: J.M. Bonow and P. Japsen.
The Basement: F. Kalsbeek with contributions from A.A. Garde and J.A.M. van Gool.
The Gardar Province: L. Melchior Larsen with contributions from K. Secher.
Basin Deposition: G. Karup Pedersen.
Older Basins: P.R. Dawes and M. Sønderholm.
The Fold Belts in North and North-East Greenland: A.K. Higgins.
Younger Sedimentary Basins: G. Karup Pedersen, S. Piasecki, S.A.S. Pedersen and M. Larsen, with a special contribution on tetrapods from S.E. Bendix-Almgreen.
Palaeogene Volcanism: L. Melchior Larsen.
Geology Offshore: T.C.R. Pulvertaft and J.A. Chalmers with contributions from P. Japsen, B. Larsen, L. Melchior Larsen and C. Marcussen.
The Ice Age: H.H. Thomsen, with contributions from O. Bennike, S. Funder and N. Mikkelsen.
Mineral Resources: K. Secher, with contributions from B. Thomassen and T.M. Rasmussen.
Oil and Gas: F.G. Christiansen and J.A. Chalmers, with contributions from J. Bojesen-Koefoed, F. Dalhoff and Aa.B. Sørensen.

In addition to those named above, I have received help, information and material of a more specific nature from a large number of other colleagues. First of all, I must thank E. Schou Jensen who has made available material from the Geological Museum in Copenhagen, and supplied the basic outlines for a number of the illustrations. Furthermore, I have consulted and received much useful information from the following (in alphabetical order): Peter W.U. Appel, S. Bernstein, M. Bjerreskov,

BACKGROUND AND ACKNOWLEDGEMENTS

J. Boserup, R.G. Bromley, L.B. Clemmensen, K.S. Frederiksen, J.R. Ineson, T. Nielsen, T.F.D. Nielsen, A.P. Nutman, O.B. Olesen, J. Audun Rasmussen, T.V. Rasmussen, M. Rosing, N.J. Soper, A. Steenfelt, L. Stemmerik, H. Stendal, S. Stouge, H. Sørensen, K. Sørensen, T. Tukainen, W.S. Watt, W.L. Weng, A. Weidick and K. Zinck-Jørgensen.

The following Survey draughtsmen have redrawn the figures of the Danish edition of the book, or provided the layouts for them: A. Andersen, M. Christoffersen, L. Duegaard, J. Halskov, E. Melskens, H.K. Pedersen, S. Sølberg, C.E. Thuesen and H. Zetterwall.

The photographic work for the illustrations used in the book was primarily undertaken by Jakob Lautrup, who also took many of the original photographs in Greenland. I am extremely grateful for his collaboration and also wish to thank Jakob for the many years of close co-operation we have had during fieldwork in Greenland. His superb, illustrative photographs have helped significantly to bring the geology to life and made the geological content easier for the non-specialist to understand.

The original Danish book was developed and produced in close co-operation with members of the Survey's Photo and Graphic Unit, and I acknowledge their significant input towards the production. Henrik Højmark Thomsen, in collaboration with the author, was deeply involved in the realisation of the Danish edition. Throughout the production stage of that edition, he read numerous manuscript drafts and in particular suggested improvements that helped shape the final appearance and content of the volume.

The content of the English version closely follows that of the Danish version, apart from necessary updating of a few sections. The translation into English has been carried out by two of my British colleagues, J.A. Chalmers and C.R.L. Friend, both professional geologists with a long-standing involvement in Greenland geology. The translation was carried out in close consultation with the author and in many places the geoscientific content has been improved thanks to the individual specialist knowledge of the two translators. C.R.L Friend worked on the chapters dealing with crystalline rocks, fold belts, older sediments and mineral resources, whereas J.A. Chalmers translated the chapters on landscapes, younger sedimentary rocks, offshore geology, Ice Ages, the chapter on oil and gas and the glossary.

Following the initial translation, the entire manuscript went through a 'quality control' process carried out by three of my Survey colleagues, A.K. Higgins, J.R. Ineson and W.S. Watt, all British geologists with a very long experience of working with Greenland geology.

I am very grateful for the long and rewarding co-operation of all the individuals involved, with respect to both the original edition in Danish and the translation of the book into English. My thanks for their patience and full support throughout the production phase.

Finally, I would also like to express my thanks to The Geological Survey of Denmark and Greenland (GEUS) for providing the practical and financial support that ensured production of the book. While the Danish edition was also supported by the Bureau of Minerals and Petroleum (The Greenland Home Rule Government), the work with the English edition has been entirely supported by the Survey (GEUS).

Niels Henriksen, GEUS May, 2008.

Rusty schists with bands of black amphibolite cut by pegmatite veins. From the 2830 million years old gold-bearing greenstone belt on Storø in the Nuuk region. The mountain side is about 950 m high.

FURTHER READING

Overviews of Greenland geology

Escher, A. & Watt, W.S. (eds) 1976: Geology of Greenland, 603 pp. Copenhagen: Geological Survey of Greenland.

Henriksen, N., Higgins, A.K., Kalsbeek, F. & Pulvertaft, T.C.R. 2000: Greenland from Archaean to Quaternary. Descriptive text to the Geological map of Greenland, 1:2 500 000. Geology of Greenland Survey Bulletin **185**, 93 pp.

Articles etc. on subjects described in the chapters of this book

Landscapes

Bonow, J.M., Japsen, P., Green, P.F., Wilson, R.W., Chalmers, J.A., Klint, K.E.S., van Gool, J.A.M., Lidmar-Bergström, K. & Pedersen, A.K. 2007: A multi-disciplinary study of Phanerozoic landscape development in West Greenland. Geological Survey of Denmark and Greenland Bulletin **13**, 33–36.

Japsen, P., Bonow, J.M., Green, P.F., Chalmers, J.A. & Lidmar-Bergström, K. 2006: Elevated, passive continental margins: long-term highs or Neogene uplifts? New evidence from West Greenland. Earth and Planetary Science Letters **248**, 315–324.

Sugden, D.E. 1974: Landscapes of glacial erosion in Greenland and their relationship to ice, topographic and bedrock conditions. In: Brown, E.H. & Waters, R.S. (eds): Progress in geomorphology. Institute of British Geographers Special Publications **7**, 177–195.

The crystalline basement

Furnes, H., de Wit, M., Staudigel, H., Rosing, M. & Muehlenbachs, K. 2007: A vestige of Earth's oldest ophiolite. Science **315**, 1704–1707.

Garde, A.A. & Kalsbeek, F. (eds) 2006: Precambrian crustal evolution and Cretaceous–Palaeogene faulting in West Greenland. Geological Survey of Denmark and Greenland Bulletin **11**, 204 pp.

Garde, A.A., Hamilton, M.A., Chadwick, B., Grocott, J. & McCaffrey, K.J.W. 2002: The Ketilidian orogen of South Greenland: geochronology, tectonics, magmatism, and fore-arc accretion during Palaeoproterozoic oblique convergence. Canadian Journal of Earth Sciences **39**, 756–793.

Kalsbeek, F. (ed.) 1999: Precambrian geology of the Disko Bugt region, West Greenland. Geology of Greenland Survey Bulletin **181**, 179 pp.

Kalsbeek, F., Austrheim, H., Bridgwater, D., Hansen, B.T., Pedersen, S. & Taylor, P.N. 1993: Geochronology of Archaean and Proterozoic events in the Ammassalik area, South-East Greenland, and comparisons with the Lewisian of Scotland and the Nagssugtoqidian of West Greenland. Precambrian Research **62**, 239–270.

Nutman, A.P., Bennett, V.C., Friend, C.R.L. & McGregor, V.R. 2000: The early Archaean Itsaq Gneiss Complex of southern West Greenland: the importance of field observations in interpreting age and isotopic constraints for early terrestrial evolution. Geochimica et Cosmochimica Acta **64**, 3035–3060.

The Gardar province

Sørensen, H. (ed.) 2006: Geological guide South Greenland. The Narsarsuaq–Narsaq–Qaqortoq region, 132 pp. Copenhagen: Geological Survey of Denmark and Greenland.

Upton, B.G.J. & Emeleus, C.H. 1987: Mid-Proterozoic alkaline magmatism in southern Greenland: the Gardar province. In: Fitton, J.G. & Upton, B.G.J. (eds): Alkaline igneous rocks. Geological Society Special Publication (London) **30**, 449–471.

Upton, B.G.J., Emeleus, C.H., Heaman, L.M., Goodenough, K.M. & Finch, A.A. 2003: Magmatism of the mid-Proterozoic Gardar Province, South Greenland: chronology, petrogenesis and geological setting. Lithos **68**, 43–65.

Basin deposition

Peel, J.S. & Sønderholm, M. (eds) 1991: Sedimentary basins of North Greenland. Bulletin Grønlands Geologiske Undersøgelse **160**, 164 pp.

Surlyk, F. 1990: Timing, style and sedimentary evolution of Late Palaeozoic – Mesozoic extensional basins of East Greenland. In: Hardman, R.F.P. & Brooks, J. (eds): Tectonic events responsible for Britain's oil and gas reserves. Geological Society Special Publication (London) **55**, 107–125.

Older sedimentary basins

Dawes, P.R. 1997: The Proterozoic Thule Supergroup, Greenland and Canada: history, lithostratigraphy and development. Geology of Greenland Survey Bulletin **174**, 150 pp.

Higgins, A.K., Smith, M.P., Soper, N.J., Leslie, A.G., Rasmussen, J.A. & Sønderholm, M. 2001: The Neoproterozoic Hekla Sund Basin, eastern North Greenland: a pre-Iapetan extensional sequence thrust across its rift shoulders during the Caledonian orogeny. Journal of the Geological Society (London) **158**, 487–499.

Smith, M.P., Rasmussen, J.A., Robertson, S., Higgins, A.K. & Leslie, A.G. 2004: Lower Palaeozoic stratigraphy of the East Greenland Caledonides. In: Higgins, A.K. & Kalsbeek, F. (eds): East Greenland Caledonides: stratigraphy, structure and geochronology. Geological Survey of Denmark and Greenland Bulletin **6**, 5–28.

Sønderholm, M. & Tirsgaard, H. 1993: Lithostratigraphic framework of the Upper Proterozoic Eleonore Bay Supergroup of East and North-East Greenland. Bulletin Grønlands Geologiske Undersøgelse **167**, 38 pp.

Fold belts in North and North-East Greenland

Haller, J. 1971: Geology of the East Greenland Caledonides, 413 pp. London: Interscience Publishers.

Higgins, A.K. & Kalsbeek, F. (eds) 2004: East Greenland Caledonides: stratigraphy, structure and geochronology. Geological Survey of Denmark and Greenland Bulletin **6**, 93 pp.

Higgins, A.K. et al. 2004: The foreland-propagating thrust architecture of the East Greenland Caledonides 72°–75°N. Journal of the Geological Society (London) **161**, 1009–1026.

Soper, N.J. & Higgins, A.K. 1991: Devonian – Early Carboniferous deformation and metamorphism, North Greenland. In: Trettin, H.P. (ed.): Geology of the Innuitian Orogen and Arctic Platform of Canada and Greenland. Geology of Canada **3**, 281–291. Ottawa: Geological Survey of Canada (also The geology of North America **E**, Geological Society of America).

von Gosen, W. & Piepjohn, K. 2003: Eurekan transpressive deformation in the Wandel Hav Mobile Belt (northeast Greenland). Tectonics **22**(4), 13-1–13-28.

Younger sedimentary basins

Chalmers, J.A., Pulvertaft, T.C.R., Marcussen, C. & Pedersen, A.K. 1999: New insight into the structure of the Nuussuaq Basin, central West Greenland. Marine and Petroleum Geology **16**, 197–224.

Stemmerik, L. 2000: Late Palaeozoic evolution of the North Atlantic margin of Pangea. Palaeogeography, Palaeoclimatology, Palaeoecology **161**, 95–126.

Stemmerik, L., Christiansen, F.G., Piasecki, S., Jordt, B., Marcussen, C. & Nøhr-Hansen, H. 1993: Depositional history and petroleum geology of the Carboniferous to Cretaceous sediments in the northern part of East Greenland. In: Vorren, T.O. et al. (eds): Arctic geology and petroleum potential. Norwegian Petroleum Society Special Publication **2**, 67–87.

Surlyk, F. 2003: The Jurassic of East Greenland: a sedimentary record of thermal subsidence, onset and culmination of rifting. In: Ineson, J.R. & Surlyk, F. (eds): The Jurassic of Denmark and Greenland. Geological Survey of Denmark and Greenland Bulletin **1**, 659–722.

Palaeogene volcanism
Larsen, L.M., Waagstein, R., Pedersen, A.K. & Storey, M. 1999: Trans-Atlantic correlation of the Palaeogene volcanic successions in the Faeroe Islands and East Greenland. Journal of the Geological Society (London) **156**, 1081–1095.

Pedersen, A.K., Larsen, L.M., Riisager, P. & Dueholm, K.S. 2002: Rates of volcanic deposition, facies changes and movements in a dynamic basin: the Nuussuaq Basin, West Greenland, around the C27n–C26r transition. In: Jolley, D.W. & Bell, B.R. (eds): The North Atlantic Igneous Province: stratigraphy, tectonic, volcanic and magmatic processes. Geological Society Special Publication (London) **197**, 157–181.

Storey, M., Duncan, R.A., Pedersen, A.K., Larsen, L.M. & Larsen, H.C. 1998: ^{40}Ar/^{39}Ar geochronology of the West Greenland Tertiary volcanic province. Earth and Planetary Science Letters **160**, 569–586.

Tegner, C., Duncan, R.A., Bernstein, S., Brooks, C.K., Bird, D.K. & Storey, M. 1998: ^{40}Ar-^{39}Ar geochronology of Tertiary mafic intrusions along the East Greenland rifted margin: relation to flood basalts and the Iceland hotspot track. Earth and Planetary Science Letters **156**, 75–88.

Offshore geology
Chalmers, J.A. & Pulvertaft, T.C.R. 2001: Development of the continental margins of the Labrador Sea: a review. In: Wilson, R.C.L. et al. (eds): Non-volcanic rifting of continental margins: a comparison of evidence from land and sea. Geological Society Special Publication (London) **187**, 77–105.

Hamann, N.E., Whittaker, R.C. & Stemmerik, L. 2005: Geological development of the Northeast Greenland shelf. In: Doré, A.G. & Vining, B.A. (eds): Petroleum geology: North-West Europe and global perspectives. Proceedings of the 6th Petroleum Geology Conference, 887–902. London: Geological Society.

Sørensen, A.B. 2006: Stratigraphy, structure and petroleum potential of the Lady Franklin and Maniitsoq Basins, offshore southern West Greenland. Petroleum Geoscience **12**, 221–234.

Whittaker, R.C., Hamann, N.E. & Pulvertaft, T.C.R. 1997: A new frontier province offshore northwest Greenland: structure, basin development, and petroleum potential of the Melville Bay area. AAPG Bulletin **81**, 979–998. Tulsa, Oklahoma: American Association of Petroleum Geologists.

The Ice Age
Funder, S. (co-ordinator) 1989: Quaternary geology of the ice-free areas and adjacent shelves of Greenland. In: Fulton, R.J. (ed.): Quaternary geology of Canada and Greenland. The geology of North America **K–1**, 741–792. Boulder, Colorado: Geological Society of America (also Geology of Canada **1**, Geological Survey of Canada).

Funder, S., Jennings, A. & Kelly, M. 2004: Middle and Late Quaternary glacial limits in Greenland. In: Ehlers, J. & Gibbard, P. (eds): Quaternary glaciations, extent and chronology. Part II (North America), 425–430. Elsevier.

Johnsen, S.J., Dahl-Jensen, D., Gundestrup, N., Steffensen, J.P., Clausen, H.B., Miller, H., Masson-Delmotte, V., Sveinbjörnsdottir, A.E. & White, J. 2001: Oxygen isotope and palaeotemperature records from six Greenland ice-core stations: Camp Century, Dye-3, GRIP, GISP2, Renland and NorthGRIP. Journal of Quaternary Science **16**(4), 299–307.

North Greenland Ice Core Project members 2004: High-resolution record of Northern Hemisphere climate extending into the last interglacial period. Nature **431**, 147–151.

Weidick, A. & Bennike, O. 2007: Quaternary glaciation history and glaciology of Jakobshavn Isbræ and the Disko Bugt region, West Greenland: a review. Geological Survey of Denmark and Greenland Bulletin **14**, 78 pp.

Mineral resources
Geological Survey of Denmark and Greenland: Greenland MINEX News. All issues of this newsletter are available at: http://www.geus.dk/minex/minex-dk.htm

Harpøth, O., Pedersen, J.L., Schønwandt, H.K. & Thomassen, B. 1986: The mineral occurrences of central East Greenland. Meddelelser om Grønland Geoscience **17**, 139 pp.

Nielsen, T.F.D., Andersen, J.C.Ø. & Brooks, C.K. 2005: The Platinova Reef of the Skaergaard Intrusion. In: Mungall, J.E. (ed.): Exploration for platinum-group elements deposits. Mineralogical Association of Canada Short Course Series **35**, 431–455.

Stendal, H. (ed.) 2000: Exploration in Greenland: discoveries of the 1990s. Transactions of the Institution of Mining and Metallurgy, section B, Applied Earth Science **109**, B1–B66.

Oil and gas
Bojesen-Koefoed, J.A., Christiansen, F.G., Nytoft, H.P. & Pedersen, A.K. 1999: Oil seepage onshore West Greenland: evidence of multiple source rocks and oil mixing. In: Fleet, A.J. & Boldy, S.A.R. (eds): Petroleum geology of Northwest Europe: Proceedings of the 5th conference, 305–314. London: Geological Society.

Bureau of Minerals and Petroleum: New data are currently being published on: http://www.bmp.gl/petroleum/petroleum.html

Chalmers, J.A. & Pulvertaft, T.C.R. 1993: The southern West Greenland continental shelf – was petroleum exploration abandoned prematurely? In: Vorren, T.O. et al. (eds): Arctic geology and petroleum potential. Norwegian Petroleum Society Special Publication **2**, 55–66.

Christiansen, F.G. (ed.) 1989: Petroleum geology of North Greenland. Bulletin Grønlands Geologiske Undersøgelse **158**, 92 pp.

Geological Survey of Denmark and Greenland: GHEXIS Newsletter. Issues of this newsletter are available at: http://www.geus.dk/ghexis

Gregersen, U., Bidstrup, T., Bojesen-Koefoed, J.A., Christiansen, F.G., Dalhoff, F. & Sønderholm, M. 2007: Petroleum systems and structures offshore central West Greenland: implications for hydrocarbon prospectivity. Geological Survey of Denmark and Greenland Bulletin **13**, 25–28.

Stemmerik, L., Clausen, O.R., Korstgård, J., Larsen, M., Piasecki, S., Seidler, L., Surlyk, F. & Therkelsen, J. 1997: Petroleum geological investigations in East Greenland: project 'Resources of the sedimentary basins of North and East Greenland'. Geology of Greenland Survey Bulletin **176**, 29–38.

Extended further reading (bibliography)

An extended list of literature on Greenland geology is available at:

http://www.geus.dk/isbn211.html

The list will be updated at intervals when relevant new data have been published.

GLOSSARY

List of technical words and expressions together with a short description of selected geological phenomena mentioned in the book. The list is to help non-geological readers with the specialist geological terms that were unavoidable in describing complex geology and geological processes.

A

Absolute age: The age of a rock or geological unit expressed in years – usually in thousands of years (ka = kilo-annum) or millions of years (Ma = Mega-annum).

Acanthostega: A vertebrate genus found in the Devonian sediments of East Greenland representing a transitional form between fish and reptile – known informally as 'four-legged fish'.

Acidic rocks: Group of igneous rocks that are relatively rich in silica, so they are also sometimes known as silicic rocks. Rocks poor in silica are known as basic rocks.

Acritarchs: An informal collective name for a group of single-celled microfossils, commonly with hard spherical skeletons, 20–150 microns in diameter, that are preserved in sediments. They are used as biostratigraphic indicators of Proterozoic and Lower Palaeozoic deposits.

Aeolian deposits: Sediments deposited on land by the wind.

Aeromagnetic surveying: Surveying the Earth's magnetic field using aircraft.

Age: (a) The lowest unit in the geochronological system, e.g. Oxfordian (see box on p. 69). (b) Age is also used informally to indicate how old rocks, structures or events are, e.g. rocks of Mesozoic age.

Agglomerate: Volcanic rock consisting of fragments of lava of various sizes. Agglomerate containing angular fragments is known as volcanic breccia.

Airguns: Devices (usually towed behind a ship) that use high-pressure air to send sound waves from the water down into the Earth. The refracted or reflected waves returned to the surface are received by hydrophones and recorded.

Albite: A variety of the feldspar mineral plagioclase with the chemical composition $NaAlSi_3O_8$ that forms the sodium-rich end-member of a mixed series in which the Ca-rich mineral anorthosite forms the other end-member.

Algal structures: Fossil impressions of algae that comprise a very wide range of plants from tiny single-celled forms to very large seaweeds many metres long.

Alkali feldspars: Feldspar minerals consisting of various mixtures of sodium and potassium feldspars such as albite, orthoclase and microcline.

Allochthonous: Formed somewhere else and transported to where it occurs now. Contrasts with autochthonous.

Alluvial deposits: Sediments deposited on land by streams and rivers.

Alpine fold belts: Mountain ranges in Eurasia, including the Alps and the Himalayas, formed by continental collision during the Palaeogene and Neogene.

Ammonites: Extinct group of marine, shell-bearing cephalopods related to squids and octopuses. The shells are common as fossils in sediments of Jurassic and Cretaceous age.

Ammonite zone: Stratigraphic interval in which a characteristic assemblage of ammonites is found.

Amphibole minerals: Group of green, blue-green and black, iron- and magnesium-rich silicate minerals usually found as prismatic crystals. They commonly form a large part of the dark minerals in metamorphic rocks. Examples are hornblende and tremolite.

Amphibolite: Dark metamorphic rock consisting mostly of amphibole (hornblende) and feldspar minerals (plagioclase). It is formed principally by the alteration (metamorphism) of basic volcanic rocks.

Amphibolite facies: Collective name for the assemblage of minerals formed by metamorphism at temperatures between 450° and 750°C at a pressure equivalent to depths of 20–40 km in the Earth's crust. The assemblage commonly contains the mineral hornblende from the amphibolite group, from where the name comes.

Andalusite and sillimanite: Two aluminium-rich minerals formed during the metamorphism of mudstone. Andalusite forms at low pressures and fairly low temperatures while sillimanite forms at high temperatures.

Andesite: Fine-grained volcanic rock with a composition similar to plagioclase-rich granite.

Angiosperms: Plants with a root, stem and leaves. The seeds are encapsulated in a fruit or otherwise protected. Comprise 80% of present-day trees, bushes and herbaceous plants. Nearly all contain chlorophyll that uses sunlight to make sugar. The group arose during the Cretaceous about 130 million years ago and is descended from plants with naked seeds (gymnosperms).

Anhydrite: Mineral consisting of calcium sulphate ($CaSO_4$). Commonly found associated with other evaporite minerals such as halite (rock salt).

Anomaly: Deviation from a background value. Commonly used in geophysical and geochemical studies for measurements that are either greater than (positive) or less than (negative) the background (reference) level. An anomaly map is used to show the size and geographical distribution of the measured anomalies.

Anorthite: A variety of the feldspar mineral plagioclase with the chemical composition $CaAl_2Si_2O_8$ that forms the calcium-rich end-member of a mixed series in which sodium-rich albite forms the other end-member.

Anorthosite: An igneous rock dominated by the feldspar mineral anorthite.

Anoxic: Oxygen-free conditions within a sediment or within lakes or the sea.

Anticline: A convex-upwards fold. When exposed and eroded, the rocks at its centre are older than those on its margins.

Antiform: A convex-upwards fold in strata whose relative ages are not known.

Antimony (Sb): Metallic element related to arsenic (As).

Apatite: Green, brown or blue mineral that commonly forms prismatic crystals and has the formula $Ca_5(PO_4)_3(F,Cl)$.

Appalachian fold belt: Palaeozoic fold belt parallel to the east coast of North America and named after the Appalachian Mountains. It is the equivalent of the Caledonian and Variscan fold belts in Greenland and Europe.

Aragonite: Mineral with the same composition as calcite ($CaCO_3$), but with a slightly different crystal structure. It is the stable precipitate of seawater and is commonly found in the shells of marine animals, e.g. molluscs.

Archaean: The eon (geochronological unit) between 3800 and 2500 million years ago (see p. 21).

Archaean time: The period between 3800 and 2500 million years ago when the first continents are recognised.

Arkose: Sandstone with a high content of feldspar (>25% of grains).

Asthenosphere: The softer upper part of the ultrabasic silicate mantle under the rigid lithosphere and above the mesosphere. The plates of the lithosphere are able to move around over the asthenosphere because its material deforms easily. Asthenospheric material can flow to or from areas undergoing uplift or subsidence, allowing isostatic compensation to take place.

Atlantic Province: A faunal province defined by an assemblage of marine animals that developed on the eastern side of the Atlantic. It contrasts with the Pacific Province found on the western side of the Atlantic where marine animals developed in genetic connection with those in the Pacific Ocean.

Augen granite/augen gneiss: Rocks with a characteristic structure of large feldspar crystals (augen = eyes in German) surrounded by a matrix of finer-grained crystals (see p. 102).

Augite: Silicate mineral coloured green, blue-green or black. A member of the pyroxene group of minerals. It commonly forms a substantial part of the dark minerals in igneous rocks

Autochthonous: Found in the place in which it formed. Contrasts with allochthonous.

Azurite: Mineral consisting of copper carbonate. It has a characteristic azure blue colour. Like malachite, it forms at depth around the margins of copper bodies and is a readily-recognised indicator of copper mineralisation.

B

Baltica: The continental area of northeastern Europe that consists of Scandinavia, Finland, the Baltic countries and Russia west of the Urals. This area moved as an independent plate prior to the Caledonian Orogeny.

Banded iron-ore: Rock consisting of alternating thin layers of iron-rich minerals and quartzite or chert. Such rocks formed during the Archaean and Early Proterozoic in deep-sea sediments in which layers of iron-minerals were deposited by bacterial activity. The rock is commonly an important iron ore. All occurrences in Greenland are metamorphosed.

Bar: Measure of pressure; one bar is approximately equal to the pressure exerted by the atmosphere. In geology, pressures in kilobars (kbar) are most commonly quoted, as a measure of how deep a rock has been in the crust.

Barrel (of oil): Old-fashioned unit of volume still in common use in the oil industry to measure the amount of oil produced from an oil field. Equivalent to 158.76 litres.

Baryte: Mineral ($BaSO_4$) with the high density of 4.5 g/cm^3.

Basalt: Black volcanic rock consisting of approximately equal amounts of the minerals plagioclase and pyroxene and often containing olivine. Basalt is the most common constituent of the shallowest part of the oceanic crust and frequently forms lava flows on land.

Basement shield: An extensive area of Precambrian gneissic and granitic rocks. The rocks form the roots of Precambrian fold belts that have been welded together. Those in Greenland are all more than 1600 million years old.

Basic rocks: Volcanic rocks relatively poor in silicon and rich in calcium, magnesium and iron.

Basin: See Depositional basin.

Basin analysis: Integrated analysis of the development through time of a depositional basin. The method includes integration of the sedimentological, stratigraphic and structural development and commonly an evaluation of its palaeoenvironmental and palaeogeographical conditions.

Batholith: A body or complex, commonly of plutonic acidic rocks, whose area is greater than 100 km^2. Batholiths commonly form above areas where oceanic crust is being subducted (descending into the mantle), such as along island arcs or some fold belts.

Bed: The smallest lithostratigraphic unit of sedimentary rocks (see p. 69).

Belemnites: Collective name for squid-like cephalopods whose internal dagger-like shells are commonly found fossilised in rocks of Carboniferous to Cretaceous ages.

Benthonic organisms: Bottom-dwelling organisms living on the surface of sea or lake sediments.

Beryl: Hard silicate mineral containing aluminium (Al) and beryllium (Be).

Biogenic deposits: Deposits produced by organisms; examples are corals or algae that can form reefs, collections of shells that may form limestones and remains of woody plants that can form coal.

Biostratigraphy: Stratigraphic classification using the content of biological material in rocks (fossil animals and plants – see p. 69).

Biotite: An important rock-forming mica mineral, brown to black in colour with perfect cleavage. A common mineral in igneous and metamorphic rocks, e.g. granite, gneiss and mica schist.

Biotope: An environment in which a particular assemblage of animals and plants can live.

Bioturbation: Mixing or disturbance of the uppermost, recently-deposited layers of sediment by the activities of burrowing animals such as worms and bivalves.

Biozone: The basic unit in biostratigraphy (see p. 69).

Bitumen: Solid or viscous, naturally occurring asphalt, containing hydrocarbons with molecules consisting of long carbon chains. It is commonly found on the Earth's surface where hydrocarbons have leaked out from below and lighter hydrocarbons with small molecules have evaporated. Large accumulations of bitumen can be utilised as a source of oil products.

Bivalves: Class of molluscs possessing two shells, commonly equal in size and hinged together, e.g. mussels and clams.

Block faults: Structural pattern where the crust is broken into segments by steep faults (see p. 177). Mostly within depositional basins in continental crust.

Blocky lava: A lava flow where the surface consists of lava fragments that produce a rugged, blocky terrain. Also known as aa (pronounced ah-ah) lava.

Boreal Period: The period of the Holocene between 9000 and 7500 years ago when the climate of Greenland and northern Europe was relatively warm and dry.

Boudinage: Deformation by stretching that causes harder layers in a rock to break up into lozenge-shaped fragments that resemble a row of sausages, known as 'boudins' in French (see pp. 33 and 44).

Boulder clay: See Till.

Boundstone: Limestones formed of the skeletal remains of invertebrates that were bound together in life forming a rigid structure (e.g. a reef; see p. 64).

Brachiopods: Invertebrate bivalve organisms with two symmetric hemispherical shells (resembling mussel shells) that are commonly found as fossils. They existed from the Early Cambrian until today.

Braided river or stream: A stream that branches continuously into channels that then reamalgamate with one another, only to branch again. Characteristic of fast-flowing streams carrying large loads of sediment.

Breccia: Crushed rock or a sedimentary rock consisting of coarse angular clasts set in a finer matrix.

Brittle faulting: Deformation of a rock by rupturing along faults where the bodies of rock on opposite sides of the fracture have moved relative to one another.

Bryozoan limestone: Limestone composed of the remains of bryozoans.

Bryozoans: Small colonial marine organisms, mostly with a calcareous skeleton in which each animal is housed in a small chamber. They commonly contribute to the formation of reefs.

Burrows: Traces of the holes or tunnels formed where animals such as worms or bivalves have burrowed into a soft sediment. The traces of the holes are preserved by being filled with silt or sand, often actively back-filled by the organism itself as it moves through the sediment.

C

Caledonian fold belt: Belt of folded and metamorphosed rocks that formed between about 460 and 400 million years ago by the closure of the Iapetus Ocean. Segments of the fold belt are now found on both sides of the present North Atlantic Ocean (see p. 96). Named from Scotland (called Caledonia in Latin), from where part of this fold belt was first described.

Caledonides: The Caledonian fold belt.

Cambrian: The first period of the Palaeozoic, from 542 to 488 million years ago (see p. 21).

Canyon: A deeply incised valley with steep sides, formed by erosion of its bottom by a river. Comparable valleys eroded into the seabed by under-water, sediment-laden currents are known as submarine canyons.

Carat: Unit of weight equivalent to 0.2 g that is used to measure the weight of diamonds.

Carbonate (sediment): Sedimentary rock composed predominantly of calcium carbonate ($CaCO_3$).

Carbonate platform: A large shelf-like area under shallow water formed by the accumulation of carbonate sediments (limestones/dolomites).

Carbonate rocks: Rocks, such as limestone and dolomite, consisting primarily of carbonate (CO_3) minerals e.g. calcite ($CaCO_3$). Most are of sedimentary origin, but a minority (carbonatites) are produced by volcanoes.

Carbonatite: Magmatic rock containing large proportions of carbonate minerals such as calcite ($CaCO_3$).

Carboniferous: The middle period of the Late Palaeozoic, from 359 to 299 million years ago.

Casing: Tubes used to line a drill hole to prevent its sides from collapsing as drilling proceeds.

Cement: Material that binds together the grains of a sediment to form a sedimentary rock.

Cenozoic: The latest era of Earth's history from 65.5 million years ago until the present (see p. 21).

Cephalopods: Class of marine molluscs that includes many fossil groups such as ammonites and belemnites. Its present-day representatives are octopuses, squids and cuttlefish.

Chalcocite: Copper sulphide (Cu_2S). Often found in association with other copper minerals and forms an important constituent of copper ores.

Chert: Hard microcrystalline silica occurring as bands or nodular layers in sedimentary rocks. Flint is a particular form of chert.

Chitinozoans: Group of extinct marine, organic-walled micro-organisms found as fossils in sediments from the Cambrian to the Devonian.

Chlorite: Greenish coloured, iron-, magnesium- and aluminium-bearing silicate mineral that resembles mica. It is widespread in low metamorphic schists and as an alteration product from other dark minerals.

Chron: The time it takes for the Earth's magnetic field to proceed through one cycle of normal and reversed orientation. The chrons are numbered successively backwards in time. The present phase of normal (N) magnetisation is chron 0. The reversed phases are labelled R. Example: Sea-floor spreading in the Labrador Sea started during the normal phase of chron 27 (chron 27N), 62 million years ago, but in the North Atlantic first during chron 24R, 55 million years ago.

Chronostratigraphy: The classification of rocks according to age (see p. 69).

Clastic sediments: Sediments consisting of grains (fragments of minerals and rocks) eroded from other rocks, transported elsewhere and redeposited.

Clay: Sediment with a grain size of less than 0.002 mm, consisting of hydrous aluminium silicates (clay minerals).

Climatic zone: A zone of the Earth characterised by particular climatic conditions (temperature, precipitation etc.). Examples are tropical, sub-tropical, warm temperate, cold temperate zones.

Clinopyroxene (cpx): Group of rock-forming Ca-, Mg- and Fe-bearing silicate minerals e.g. diopside and augite.

Coal seam: A bed of coal (carbon), usually found within sediments deposited in a delta.

Cobbles: Sedimentary clasts with a size greater than 64 mm (see p. 64).

Coccolithophoridae: Marine planktonic microalgae with an outer skeleton of calcite discs or rings (coccoliths). Common as microfossils since the Jurassic. They form the major component of the rock known as chalk.

Columnar basalts: Structure within a basaltic lava flow, a dyke or a sill, consisting of polygonal, commonly hexagonal, columns. The structure formed during cooling of the rock when the contractional forces exceeded the internal strength of the rock and a network of regular polygonal joints is formed (columnar jointing) (see p. 153).

Complex: Any body or system formed from two or more bodies or structures, such as intrusions, types of metamorphic rock or faults.

Compression: The process by which parts of the Earth's crust are pressed together by forces derived from movements in the lithosphere and asthenosphere. On a large scale, this deforms the crust into fold belts.

Concession/licence: Legal agreement by which the government grants a company or group of companies the right to carry out certain commercial activities, such as exploring for and extracting hydrocarbons or ore minerals.

Concordia diagram: Diagram derived from analysing minerals containing uranium and showing the relative proportions of lead and uranium isotopes. A curve in the diagram (concordia) shows how the proportions of the isotopes vary with time. The diagram is used in the determination of radiometric ages (see box on p. 76).

Condensate: A gasoline-like liquid formed at increased pressure underground by hydrocarbons that are gaseous at surface temperature and pressure.

Conglomerate: Sedimentary rock consisting of rounded fragments of other rocks. The fragments (or clasts) may range from small pebbles to large boulders and may be of different compositions because they may be derived from a variety of different rock types (see p. 63).

Conodonts: Tooth-shaped microfossils (0.1–5 mm in size) commonly found in carbonate sediments from the Cambrian to the Late Triassic. The fossils are the mouth parts of a soft-bodied organism shaped like a fish. They are used in stratigraphic classification of sediments and have been much used to classify the Palaeozoic sediments from Greenland.

Consolidated sediment: See Sedimentary rock.

Constructive plate boundary: Boundary between lithospheric plates that are moving away from one another, forming an oceanic spreading zone where new crust is formed by upwelling, melting and eruption of material from the mantle.

Continent: A contiguous area of the Earth's crust, consisting predominantly of granites, gneisses, metamorphic schists and sediments, whose average density is 2.7–2.8 g/cm^3. This density is substantially less than that of the basic and ultrabasic rocks forming the oceans (3.0 to 3.2 g/cm^3), so continents tend to stand higher than oceanic areas and mostly form land areas and their surrounding continental shelves. Continental crust is typically between 20 and 70 km thick.

Continental basin: Sedimentary basin with little or no contact with the sea because it lies within a continental area. For example, basins containing red beds such as the Devonian Basin in East Greenland (see p. 114).

Continental crust: See 'Continent' and 'Crust'.

Continental deposits: Sediments deposited on land or in fresh water, not in the sea.

Continental drift: Old concept now replaced by plate tectonics (see p. 53).

Continental margin: Marginal area of a continent between the shoreline and the bottom of the deep sea. Comprises continental shelf, continental slope and continental rise (see p. 173).

Continental shelf (often just referred to as 'shelf'): The relatively shallow-water area around a continent, usually defined as that region with water depths less than either 200, 500 or 1000 m. It is bordered on the seaward side by the continental slope. The continental shelves are most commonly underlain by continental crust (see p. 173).

Continental slope: The sloping sea floor between the seaward edge of the continental shelf (the shelf break) and the deep ocean floor (the abyssal plain) (see p. 173).

Convection current: A flow of solid, but plastic rock in the asthenosphere or deeper mantle. The convection current flows upwards towards the base of the lithosphere then turns to flow horizontally to where it descends once more into the asthenosphere. Plate tectonic theory proposes that the downwards movement causes subduction, carrying the lithospheric plates into the asthenosphere (see p. 53). Convection currents also occur at smaller scales in, e.g. magma chambers (see p. 149).

Cordierite: Mineral rich in aluminium formed during the metamorphism of clay at high temperature and low pressure.

Core: Earth's core; the central spherical volume of the Earth from its centre (at 6371 km depth) to about 2900 km below the surface. It consists of an inner solid sphere with a radius of about 1200 km and an outer liquid shell. The core is composed principally of iron and nickel (see p. 20).

Coring: See Diamond Drilling.

Craton: An extensive area of the Earth's continental crust that has been stable for a long period. Greenland's basement shield is a part of a craton.

Cretaceous: The youngest period of the Mesozoic from 145 to 65 million years ago (see p. 21).

Cross-bedding: Lamination within a single bed that is transverse or oblique to the main planes of stratification and is generally truncated at the overlying stratum. May form both small and large scale structures (see p. 134). Cross-bedding is created by the migration of sedimentary bedforms such as ripples and dunes.

Crude oil: Naturally-occurring liquid hydrocarbons formed from organic remains buried underground. It is produced commercially from large accumulations known as oilfields and occurs commonly in many sedimentary basins.

Crust: The outermost thin shell of the Earth that is from about 3 to about 70 km thick. Its lower boundary is the Mohorovičić discontinuity (Moho) at the transition to the mantle. The crust consists of two types: one type, consisting predominantly of granites, metamorphic schists and sediments (continental crust) forms the continents and the other type, consisting of basalts and gabbros (oceanic crust), forms the uppermost crystalline rocks below the oceans (see pp. 20, 27, 169).

Cryolite: Fluorite mineral containing sodium and aluminium (Na_3AlF_6). It was mined commercially for a number of years at Ivittuut in southern West Greenland and was of considerable strategic importance during the Second World War in the manufacture of aluminium (see p. 205).

Crystalline basement: The collective name for the gneissic and granitic rocks that have been highly modified by deformation at great depths in the crust during periods of mountain building. All the crystalline basement in Greenland is older than 1600 million years.

Cumulus: Applied to crystals that have precipitated and settled out from a magma.

Curie temperature: The temperature (about 580°C) below which the magnetic minerals in a rock retain their magnetisation. If a rock is heated to above this temperature, e.g. during mountain building, any former magnetisation it may contain is entirely destroyed.

Current ripples: Undulating, wavy, small-scale bedforms in sand–silt grade sediments. Current ripples are caused by water or wind flowing across the surface of the sediment, so the ripples are typically asymmetric and migrate in the direction of flow (see p. 67).

Cuttings: Crushed material produced as a drill bit cuts into rock.

Cyclic: Description of a layered rock (sedimentary or magmatic) that shows a rhythmic repetition of a certain property, for example a rhythmic repetition of sand and clay in a sediment or the proportion of dark minerals in a layered magmatic rock. Cyclic deposits can be at both large and small scale.

D

Declination, magnetic: The horizontal angle between the direction of geographic north and the local direction of the Earth's magnetic field.

Deep-water sediments: Sediments deposited at significant water depths of up to several kilometres. Such sediments are mostly very fine-grained but may also include sand deposits.

Deep-water trough: An elongated depression with gently dipping flanks in the bed of the deep sea.

Deformation: Alteration of a rock's shape, attitude or volume by compression or extension, resulting in folding, displacement or crushing.

Density current: Current due to differences in density of water from place to place caused by changes in temperature and variations in salinity or amount of material held in suspension.

Depocentre: The place in a depositional basin where the sediments are thickest.

Depositional basin: Also known as a sedimentary basin – a depression in the Earth's crust caused by subsidence into which sediments have been deposited.

Depositional rate: Amount of material accumulated in a given time, e.g. the thickness of sediment deposited per year.

Desiccation cracks: Structure that forms on the surface of mud as it dries out and contracts. Takes the form of interlinked polygonal cracks.

Destructive plate boundary: The type of lithospheric plate boundary where one plate is being forced under another to be destroyed gradually in the asthenosphere.

Detachment: A low-angle displacement or fault between an overlying and an underlying unit. Commonly forms the lower boundary plane of a thrust block and sometimes of an extensional block. Synonyms: Décollement, detachment fault or sole fault.

Detrital material: Collective name for the material broken off and transported away from its parent rocks and minerals to be deposited in a sedimentary basin.

Devonian: The first period of the Late Palaeozoic, from 416 to 359 million years ago.

Dextral: Right-lateral. A dextral fault is a strike-slip fault whose side farther from an observer facing the fault plane has moved to the right.

Diagenesis: The process by which the loose grains of a sediment are cemented together to form a rock. Diagenesis includes cementation and compaction and takes place near the Earth's surface at relatively low temperatures and pressures, so many of the original features of the sediment are unaltered.

Diamond-drilling (also called coring): The process of drilling using a diamond drill bit.

Diamond drill bit: A drill bit consisting of a cylindrical tube with diamonds on its cutting edge that cuts an intact core of rock within the cylinder

Diapir: A cylindrical or mushroom-shaped mass of light rock that is forced upwards through overlying rocks while forcing them apart. Salt and granite are examples of rocks that may form this type of structure.

Diatoms: Single-celled algae whose cell-walls are impregnated with silica. They occur in both marine and freshwater environments. Diatoms are widespread as microfossils and may be used for stratigraphic analysis of sediments from the Cretaceous and Cenozoic.

Dinoflagellates: Single-celled, chiefly marine organisms with both plant and animal affinities. Mostly between 20 and 150 micrometres long. Generally considered to be plants as they have cellulose in their cell walls. The remains of their cysts are resistant and are commonly preserved as fossils (see pp. 164–165), so they are often used for stratigraphical purposes. Dinoflagellates occur mostly from the Trias to the present.

Dinosaurs: An extinct sub-class of reptiles that existed from the Triassic to the Cretaceous. The animals were of very different sizes, from as small as mice to giants more than 20 m long. There were both herbivorous and carnivorous species.

Diopside: Green silicate mineral of the pyroxene group containing both calcium and magnesium ($CaMgSi_2O_6$), often formed by metamorphosis or alteration due to circulation of hot water.

Diorite: Plutonic rock with 0–5% quartz and commonly 5–15% dark minerals, but dominated by the feldspar mineral plagioclase. The rock is classified as intermediate, i.e. with a composition between acid granite and basic gabbro.

Dipping strata: Layered rocks whose layering is not horizontal. The 'dip' is the angle between the horizontal plane and the plane of the layering.

Disconformity: Surface within layered sedimentary or volcanic rocks representing a period of non-deposition (hiatus). The strata above and below the boundary are parallel to one another.

Discordance: See Unconformity.

Dolerite: A rock resembling gabbro or basalt that commonly forms dykes and sill intrusions in the upper crust.

Dolomite: (1) A mineral, calcium magnesium carbonate ($CaMg(CO_3)_2$), commonly formed by alteration (dolomitisation) of limestone ($CaCO_3$) by the addition of magnesium (Mg). (2) A rock that consists principally of this mineral.

Drill bit: The cutting tool mounted at the bottom of a drill string that is used to cut into the ground. The drill bit can consist either of a number of chisel-like projections on a cone that crushes the rock into which it drills (see p. 162) or a cylindrical tube with diamonds on its cutting edge that leaves an intact core within the cylinder (see p. 216).

Drilling: Boring into the subsurface using a drill bit consisting of a rotating cone with a steel chisel mounted on it. The cutting tool is mounted at the bottom of a drill string that is used to cut into the rock. The drill bit crushes the rock into which it drills (see p. 162). When cores are taken, the process is known as coring.

Drilling mud: A fluid that is pumped into a drill hole and then returned to the surface. It is used to maintain pressure so that any high-pressure fluids (e.g. hydrocarbons) encountered cannot escape to the surface. It is also used to lubricate and cool the drill bit and return cuttings to the surface (see p. 161). Drilling mud consists of either water or light oil (diesel or kerosene) to which has been added various clays and heavy minerals to give it the right density.

Dropstone: A stone or block within finer-grained laminated sediments, dropped from floating ice or tree roots.

Dry well: Well drilled to explore for oil or gas that found none.

Ductile deformation: Plastic deformation of a rock so that its shape alters without fracturing.

Dunite: A plutonic rock composed of more than 90% olivine.

Dyke: Steeply-dipping sheet of magmatic rock formed by the intrusion of magma into a fissure in other rocks. When exposed, it often forms a resistant, wall-like feature where the surrounding rocks have been eroded away (see p. 142).

Dyke swarm: Group of usually sub-parallel dykes within a limited region that were intruded during a relatively short period of time.

Dynamic positioning system: A complex of navigation systems and propellers mounted on different axes used to keep a ship or floating drill-rig stationary at a given place.

E

Earth's core: See Core (see p. 20).

Earth's crust: See Crust.

Earth's mantle: See Mantle (see p. 20).

Earth Sciences: The study of the solid Earth; geology, geophysics, geochemistry, geography, glaciology etc.

Eclogite: Strongly altered metamorphic rock with a basic composition formed under very high pressures and temperatures at depths of 40–80 km under the surface. It consists mostly of high-pressure versions of the minerals garnet and pyroxene.

Eclogite facies: Assemblage of minerals formed at temperatures between about 400° and 800°C and pressures corresponding to depths of 40–80 km under the surface. It is characterised by the occurrence in some of the basic rocks of the same minerals that form the rock eclogite.

Ediacaran: The youngest period of the Neoproterozoic (636–542 million years ago).

Eemian: Interglacial 130 000–115 000 years ago between the Saale and Weichselian glacials.

Electron microscope: A device that permits very high magnifications by using beams of electrons instead of light. The electron beam is sent through a thin section of the object being studied and is imaged on a fluorescent screen. The magnification can be up to 1000 times greater than in a normal microscope.

Ellesmerian Orogeny/fold belt: Palaeozoic fold belt trending east–west across Ellesmere Island in northern Canada and northern Greenland. The fold belt formed between the early Devonian and early Carboniferous.

Eoarchaean: The earliest period of the Archaean, from 3800 to 3600 million years ago (see p. 21).

Eocene: The middle epoch of the Palaeogene, 55.8–33.9 million years ago.

Eon: The largest division of geological time, for example the Phanerozoic eon from 542 million years ago until the present.

Eonothem: The rocks formed during an eon, e.g. Phanerozoic rocks (see p. 69). The highest chronostratigraphic unit.

Epicontinental basin: Sedimentary basin that developed within a continental region.

Epoch: Division of geological time corresponding to part of a period, such as the Late Jurassic epoch (see p. 69).

Era: Division of geological time in scale between an eon and a period, for example the Mesozoic era (see p. 69).

Erathem: The rocks formed during an era, e.g. Mesozoic rocks (see p. 69). The next highest chronostratigraphic unit.

Erosion: The process, by which material at the surface of the Earth is loosened, removed and transported away. In this way the terrain is worn down and the eroded material is deposited elsewhere, both temporarily en route and finally in a sedimentary basin.

Eruption: Ejection of volcanic material onto the Earth's surface.

Escarpment: A long steep slope that separates areas at different heights; may occur above or below sea level. A prominent submarine example occurs in North Greenland (see pp. 86–88).

Euramerica: Former continent consisting of the shield areas Laurentia (North America including Greenland), Baltica and Siberia (see p. 103).

Eurasia: The present-day supercontinent comprising Europe and Asia (see p. 129).

Eurekan fold belt: Cenozoic fold belt trending east–west across Ellesmere Island in northern Canada, a small part of North Greenland and Svalbard. It formed during the latest Paleocene and the Eocene, when Greenland moved north relative to both Ellesmere Island and Svalbard.

Eustatic sea-level changes: Changes in sea level that take place simultaneously over the entire world (see p. 66). They take place for two main reasons: 1) changes in the rate of formation of new oceanic crust and 2) changes in the amount of land ice in glaciers and ice sheets. Should all present-day ice sheets, ice shelves and glaciers melt, global sea level would rise by about 80 m.

Evaporites: Deposits that form from the evaporation of salt water, such as salt, gypsum and anhydrite. Evaporites form in shallow water and coastal areas where the climate is hot and dry.

Exclusive economic zone: Sea area surrounding a country to an agreed border with a neighbour or to 200 nautical miles, within which the country has exclusive rights to all economic activities such as fishing and exploitation of minerals and petroleum (see p. 157).

Exhumation: Uplift and erosion that causes deep seated rocks to be buried less deeply and end by being exposed at the Earth's surface.

Exploitation licence: Exclusive license to extract hydrocarbons within a given area for a specific period of time.

Exploration licence: Permit to explore for minerals or hydrocarbons within a given limited area for a fixed period.

Extension: Deformation of part of the Earth's crust by becoming wider and usually thinner, leading to faulting and the formation of depositional basins in the upper crust and plastic flow in the lower crust.

Extensional fault: See normal fault (see p. 94). Fault formed due to extension.

Extrusive rocks: Volcanic rocks erupted at the surface and solidified, e.g. lava.

F

Facies, metamorphic: Mineral assemblage in rocks that have been subjected to alteration in a certain interval of pressure and temperature. Generally caused by orogenic processes (fold belt formation) (see p. 35).

Facies, sedimentological: Characteristic assemblage of sediments deposited under certain environmental conditions, e.g. deep sea facies, red bed facies, evaporite facies.

Fault: A crack or break in the Earth's crust, where the rocks on one side have moved relative to those on the other. Three types of fault are recognised: 1) strike-slip faults with a nearly vertical fault plane where the rocks on either side have moved sideways relative to one another, 2) normal or extensional faults with a dipping fault plane where one side has moved down relative to the other and 3) reverse or thrust faults with a low-angle or even horizontal fault plane where the upper side has moved upwards or across the other (see p. 94).

Faunal assemblage: A natural assemblage of animals. In palaeontology, the same fossil faunal assemblages can be used to correlate between sedimentary units in different parts of the world.

Faunal province: A geographical area within which a distinct group of living or extinct animals (fauna) can be recognised.

Feasibility study: Analysis and evaluation of how practical it is to carry out a planned project and its economic and environmental consequences. Carried out prior to starting any large mining or hydrocarbon project.

Feeder channels: The conduits from a magma chamber through which a magma can pass on its way either to an intrusive body or to eruption at the Earth's surface.

Feldspars: Group of rock-forming silicate minerals with two main types: alkali feldspars and plagioclase. The first are a mixture of sodium and potassium feldspars and the second a mixture of sodium and calcium feldspars. Most ordinary rocks, including granite and gabbro, consist to a great extent of feldspars.

Fission track: Microscopic linear dislocation in the crystal structure of some minerals forming tracks when an atom of uranium decays spontaneously. The tracks are all the same length (15.8 microns) when they form, but shorten at a rate that is dependent on temperature. Statistical analysis of the lengths of the tracks in a crystal can be used to determine its temperature history.

Fissure eruption: A volcanic eruption from an elongated crack in the crust, not from a central vent.

'Fjord zone basins': Shorthand for the group of Neoproterozoic to Ordovician sedimentary basins found in the fjord zone of North-East Greenland (71°–76°N). They consist of the Eleonore Bay Supergroup, the Tillite Group and the Kong Oscar Fjord Group of sediments that have a combined thickness of about 20 km.

Flat spot (in petroleum geology): Reflection on a seismic section indicating the presence of a horizontal surface within dipping sediments. Such a surface may indicate the presence of an accumulation of oil or gas in the sediments.

Flexure: A large-scale linear structure where layers within the crust bend from fairly flat-lying on one side of the flexure, dip within the flexure and lie flat again on the other side. The structure can be thought of as one side of a fold; a monocline.

Flood basalt (also known as plateau basalt): Subaerial basalt lava flow that has spread over a large area (see p. 138).

Flotation (froth-flotation): Method of concentrating ore. It is used industrially after the rock is crushed to separate the required metallic ores from the country rock that is discarded. The process involves suspending the crushed ore in a liquid so that the metallic ores float to the surface as a foam or froth where they can be separated.

Flow structure: A structure within a bed of rock, e.g. a lava, indicating that the material moved as a flow.

Fluviatile: Related to a river or stream, e.g. sediments deposited by a river or stream.

Fluvioglacial sediments: Sediments deposited by meltwater rivers emanating from glaciers or from ice caps.

Flux: In chemistry, a catalyst: material taking part in and facilitating, a chemical reaction without being used up in it.

Flysch: A thick series of sediments derived from an adjacent mountain range as it develops and the deposits are laid down on its foreland or in intermontane positions. The sediments consist commonly of conglomerates and sandstones.

Fold belt: See orogenic belt. Large-scale zone, in which rocks have been compressed, folded and thrust over one another. This process often forms mountain belts.

Fold phase: One distinct phase of folding during a period of mountain building. Later phases of folding may refold structures formed during earlier phases. The phases are often numbered 1, 2, 3, etc., with the oldest first.

Foraminifera: Single-celled animals, most commonly with a calcite shell. Most species live in the sea, where they are a substantial component of both the planktonic and the bottom-dwelling fauna. Foraminifera are important microfossils that can be used to determine palaeoceanographic environmental conditions.

Foraminiferous limestone: Limestone composed principally of foraminiferal shells.

Foreland (to a fold or mountain belt): The non-folded region in front of (outside) a fold or mountain belt.

Foresets: Dipping layers of sediment deposited where a current carrying the sediment crosses a break in slope, such as at the front of a delta. They also form where bedforms such as dunes or ripples are migrating in a current.

Formation: The basic unit of lithostratigraphic classification. A formation is defined by a profile at a type locality (see p. 69).

Fossil: The remains of a plant or animal preserved in rock. See fossilisation.

Fossilisation: Preservation of the structural shape of the remains of an animal or plant. Hard parts such as shells and parts of skeletons are often preserved by impregnation with calcium carbonate, silica or other inorganic minerals. Impressions of the animal or plant can also be preserved within or upon the sediment, and trace fossils are indications of animal activity such as burrows, tracks or footprints.

Four-legged fish: Vertebrate animals with a structure transitional between that of fish and amphibians. Found as fossils in Devonian deposits in Greenland (see p. 116).

Fractional crystallisation: Crystallisation of minerals from a magma where the crystallised minerals are prevented from reacting with the remaining liquid (see p. 152).

Fracture zone: A large zone of discontinuity in oceanic crust that offsets spreading axes (see p. 171).

Franklinian Basin: Large elongate sedimentary basin exposed for a distance of 2000 km from eastern North Greenland to Arctic Canada. The basin contains a sedimentary succession 4–8 km thick, ranging in age from Cambrian to earliest Devonian.

G

Gabbro: Plutonic magmatic rock with the same chemical and mineralogical composition as basalt but with coarser crystals because it cooled and solidified slowly at depth.

Gabbro-anorthosite: Plutonic rock with a composition similar to gabbro but containing more of the feldspar mineral anorthosite.

Gal: Unit of measurement of the acceleration of gravity. One gal is $1\ cm/s^2$ (0.01 m/s^2).

Gardar period: An interval of time between 1350 and 1120 million years ago when there was rifting in South Greenland accompanied by volcanism and the deposition of sediments. Large dykes and central complexes were intruded.

Gardar Province: Area of South Greenland (Ivittuut to Igaliku and east thereof) containing rocks from the Gardar period.

Garnet: Group of silicate minerals containing magnesium, iron, calcium and aluminium. They are common in metamorphic schists.

Gas: See Natural gas.

Gastropods: A class of molluscs that move on a single muscular foot. Some (snails) posses a coiled calcareous shell, others (slugs) do not.

Geochemistry: The study and analysis of the chemical composition of rocks and minerals and the fluids contained within them.

Geochronology: The study of geological time including both relative and absolute ages (see pp. 48 and 49).

Geological map: Map showing the occurrence and extent of various rocks that crop out at the surface of the Earth or the bed of the sea. The various units are commonly shown by a colour code and are displayed against a topographic or bathymetric background map (see p. 92).

Geological profile: Vertical cross-section through part of the Earth's crust (see box p. 107).

Geological timescale: The division of the Earth's 4600 million year history into a hierarchy of units. The units are commonly named after places where rocks of that age were originally studied. Originally it was possible to construct only a relative timescale, but now it is possible to attribute absolute ages to most units in millions of years ago (often shortened to Ma). As an example, the Cambrian period is named after Wales (Latin: Cambria) and lasted from 542 until 488 Ma.

Geophone: Device similar to a microphone that is placed on or pushed into the ground to register the seismic waves produced artificially during a seismic survey on land. See pp. 230–231.

Geophysics: The word has two meanings. 1) The study of the physical properties of the Earth, for example its magnetism, gravity, seismic activity and radioactivity. 2) Use of the methods of physics to study the rocks of the crust and upper mantle.

Geothermal gradient: The naturally occurring rise in temperature with depth into the Earth. Its value varies depending on the local geological conditions but is commonly around 25°C/km, although it is much greater in volcanic areas.

Glacial deposits: Sediments deposited by or in association with ice caps, glaciers, icebergs or sea ice. Commonly used as a term for material that has been transported by ice to be redeposited on land or in the sea.

Glacial rebound: Isostatic uplift of the land after the melt or retreat of an ice cover (see p. 190).

Glacials: The colder periods during the Ice Age of the last 2.4 million years when large parts of the northern continents were covered by ice. The periods between glacials, such as today, are called interglacials (see p. 183).

Glaciation: The formation and covering of large regions of the Earth by ice sheets or glaciers.

Glacier: An accumulation of snow and ice over many years to form a plastic mass that can move towards lower ground. The snouts of some glaciers are afloat in the sea or in lakes and can break up to form icebergs.

Glaciology: The study of naturally occurring snow and ice in all its forms.

Glauconite: Green, iron-bearing mineral of the mica group, formed as small grains on the seabed. It serves as an indication for marine sediments that have been deposited slowly. The sediment 'greensand' contains large amounts of this mineral.

Gneiss: Light-coloured or banded, crystalline rock formed during mountain building by the metamorphism of former near-surface rocks (supracrustal rocks) or plutonic rocks (infracrustal rocks). Gneisses commonly have a granitic or granite-like composition.

Gondwana or Gondwanaland: A former supercontinent in the southern hemisphere consisting of the present-day continents of Africa, South America, Australia, India and Antarctica. Gondwana formed the southern part of the supercontinent Pangaea whose northern part was Laurasia (North America, Europe and part of Asia). Gondwana and Laurasia separated during the Triassic and Gondwana disintegrated during the Jurassic (see pp. 123, 127).

Gossan: Strongly-coloured weathering product of sulphide mineralisation where the sulphide minerals are oxidised at the Earth's surface causing the leaching of sulphur and metals. Commonly used during mineral prospecting as a guide to the presence of a sulphide deposit.

Graben: An elongate segment of the Earth's crust that has sunk relative to its surroundings along two opposing normal faults (see p.177). Formed by extension of the crust at right angles to the length of the graben causing the crust to rift, so they are also known as rift valleys. Grabens can be small, but may also be many tens of kilometres wide and hundreds of kilometres long.

Grade: See Ore grade.

Graded bedding: Sorting of grain sizes within a sedimentary bed and commonly seen in sediments deposited by turbidity currents, reflecting varying current strength. The coarsest material lies at the base of the bed with successively finer material upwards, so that the finest grains are found at the top.

Grainstone: Carbonate rock composed wholly of carbonate grains (shell fragments), typically of sand size, and lacking mud (see p. 64).

Granite: Plutonic magmatic rock dominated by alkali feldspars and quartz together with biotite and other dark minerals. Together with gneiss, one of the most common types of rock in the Earth's continental crust.

Granodiorite: A plutonic rock related to granite but containing some plagioclase instead of the alkali feldspar in granite.

Granules: Rock fragments with sizes 2–4 mm across (between sand and pebbles; see p. 64).

Granulite facies: An assemblage of metamorphic minerals formed at temperatures between about 750° and 950°C and pressures corresponding to depths of 20–40 km in the Earth's crust. The assemblage consists entirely of minerals that do not contain water, so this assemblage is called dry.

Graptolites: Extinct group of colony-forming invertebrate organisms that lived from the Middle Cambrian to the Carboniferous (see p. 89).

Graptolite zone: Stratigraphic interval characterised by the presence of one or more particular graptolites or assemblage of graptolites (see p. 89).

Gravimeter: Instrument for measuring the variation in strength of the Earth's gravity field.

Gravimetry: The study of gravity.

Gravity surveying: Measurement of the variation from place to place of the gravitational attraction of the Earth. The variations can be used to elucidate the structure underground (see p. 232)

Greenschist facies: An assembly of metamorphic minerals formed at temperatures of 350°–500°C and pressures corresponding to depths of 10–20 km in the Earth's crust. The assembly commonly contains the green mineral chlorite, the main constituent of greenschists, hence the name.

Greenstone: A compact dark green rock consisting of dark minerals such as amphibole and chlorite together with feldspar and quartz. It forms by the metamorphism of a basic magmatic rock. The rock occasionally forms large contiguous units known as greenstone belts that are conspicuous in some basement shields.

Grenvillian orogeny: Mountain-building event during the later part of the Mesoproterozoic and the early Neoproterozoic. It can be traced through parts of eastern North America, Scotland, Baltica and locally in East Greenland, all parts of the former supercontinent of Rodinia (see p. 99).

Ground moraine: Usually known as till. Unsorted sediment deposited at the base of a glacier.

Group: Lithostratigraphic unit consisting of two or more formations (see p. 69).

Gymnosperms: Plants with naked seeds in cones and not in ovaries. The group arose during the Late Devonian about 375 million years ago. Examples are ginkgo, cycad, pine, fir and spruce.

Gypsum: Mineral consisting of hydrated calcium sulphate ($CaSO_4 \cdot 2H_2O$). Formed by precipitation from evaporating sea water and commonly found in association with rock salt.

H

Hadean: The oldest era of Earth's history from about 4600 to about 3800 million years ago. Its boundary with the succeeding Archaean era is only informally established. In the early Hadean, the Earth was glowing red hot and later it was subjected to major bombardment by asteroids (see p.12).

Half-graben: A part of the Earth's crust that has sunk relative to its surroundings along a single fault, as opposed to a graben that has sunk along two opposing faults. The down-faulted area always dips relative to the upstanding area on the other side of the fault (see p. 177).

Half-life: The time taken for half of the atoms of a radioactive isotope to decay. There are enormous differences between the half-lives of different isotopes, from many millions of years to a tiny fraction of a second.

Halite: Mineral (NaCl) also known as common salt. When occurring in large quantities in massive form it is called rock salt.

Harzburgite: An ultrabasic plutonic rock. A peridotite containing mainly olivine and orthopyroxene (see p. 141).

Heat flow (more properly geothermal flow): The amount of heat that flows from the interior of the Earth, measured in joules/cm^2/sec. The primary source of heat is the decay of radioactive elements such as uranium, thorium and potassium.

Heavy minerals: Minerals with a density of more than 2.9–3.0 g/cm^3. Usually minerals with a large content of metallic elements. Commonly concentrated in placer deposits.

Hematite: Red to black mineral, iron oxide (Fe_2O_3). An economic ore of iron.

Hercynian fold belt: Another name for the Variscan fold belt.

Hiatus: A gap in the temporal succession of layered rocks, for whatever reason. The hiatus can be caused either by erosion of formerly deposited rocks (an unconformity) or because there was no deposition during the period.

High-grade rocks: Rocks that have been transformed to amphibolite or higher facies by regional metamorphism at high temperatures and pressures.

Highstand (of sea level): The period during which sea level is at its highest during a cycle of sea-level change (a term used particularly in sequence stratigraphy; see p. 130). The coastline is at its highest and most landward position and marine sediments may be deposited across the entire continental shelf.

Hinge line (of a fold): Line of maximum curvature of a fold.

Holocene: The youngest epoch of the Quaternary, since the end of the last glacial about 11 700 years ago.

Hornblende: Silicate mineral containing iron and magnesium, common in e.g. amphibolite and gabbro.

Horst: An elongated part of the Earth's crust that is elevated structurally relative to its surroundings along two opposing faults (see p. 177).

Hot spot: An area of basaltic volcanism, usually 100–200 km in diameter, that is thought to occur above a broadly cylindrical column of hot rock rising up through the mantle (a mantle plume or mantle diapir; see p. 139). The continued movement of a lithospheric plate across such a hot spot causes a succession of volcanic centres to form along a line, lying like pearls on a string.

Humid (climate): Climate with high rainfall.

Hyaloclastite: Volcanic rock consisting of fragments of volcanic glass. Formed by the submarine eruption of basaltic lavas in shallow water, or lavas flowing into water (see p. 145).

Hydrocarbons: Crude oil and natural gas; material consisting of hydrogen and carbon.

Hydrophone: Device, similar to a microphone, that is towed behind a ship to register the seismic waves produced artificially during a seismic survey at sea (see p. 160).

Hydrothermal alteration: Alteration of a rock by hot water flowing through it. Some minerals are dissolved, others precipitated.

Hydrous fluids: Fluids containing a high proportion of water.

I

Iapetus Ocean: The Proto-Atlantic. The ocean that, prior to the Caledonian orogeny, divided North America, Greenland and the northern British Isles on one side from the rest of northern Europe on the other (see pp. 81 and 83). The Iapetus Ocean lasted from late in the Neoproterozoic until late in the Silurian (about 600–415 million years ago).

Ice Ages: A popular term for periods when substantial amounts of ice exist on the Earth. Apart from the present (Cenozoic) ice age (that started in Antarctica about 34 million years ago), ice ages occurred during the late Neoproterozoic (610–570 Ma), in the late Palaeozoic (450–400 Ma) and in the Carboniferous–Permian (350–250 Ma). Traces of earlier Precambrian ice ages are also known.

Ice cap: An ice sheet less than 50 000 km^2 in area.

Ice sheet: A continuous thick mass of ice covering a land area of more than 50 000 km^2. It moves outwards under its own weight and its surface is dome-shaped.

Ice shelf: A continuous sheet of ice, up to several hundred metres thick floating in the sea originating from glaciers or an ice sheet on an adjacent land area. An ice shelf has a flat, even upper surface.

Ice-rafted debris: Gravel, stones and boulders that have been transported out to sea by sea ice or icebergs then dumped on the seabed as the ice melted. The presence of such material in an area is often used to conclude that floating ice had previously existed in the area.

Ichthyostega: Genus of 'four-legged fish' whose fossils are found in Devonian sediments in East Greenland (see p. 116).

Igneous intrusion: A body of magma that has intruded already existing rocks. Intrusions have a wide range of sizes and shapes, from large bodies (batholiths, stocks, sheets) to smaller, sheet-like forms such as steeply dipping dykes and flat-lying sills.

Ignimbrite: Volcanic rock formed from a flow of hot ash that cools to form a welded tuff.

Illinoian (glacial): The American designation for the penultimate glacial (about 400 000–130 000 years ago) that is known as the Saalian glacial in Europe.

Immature sediment: Clastic sediment consisting of a heterogeneous mixture of clasts whose shape, size and composition are widely varied, interpreted to have been deposited only a short distance from their source (see p. 76).

Impermeable: Material through which fluids (liquids and gases) cannot flow. A term commonly used in petroleum geology to describe rocks that can form seals to oil and gas accumulations (see p. 223).

Inclination, magnetic: The vertical angle between the local geomagnetic field and the horizontal. This angle increases from 0° at the magnetic equator to 90° at the Earth's magnetic poles.

Inglefield Land fold belt: A fold belt of Palaeoproterozoic age in northernmost West Greenland (see p. 48).

Inland Ice: The large contiguous ice sheet that covers most of the interior of Greenland, but not the many surrounding, isolated, smaller ice caps. Its volume is about 2.9 million km^3 and its greatest thickness about 3.4 km. Ice streams and glaciers drain its margins.

Inoceramus: Genus of mussels that includes the largest mussel ever found anywhere; on Nuussuaq, West Greenland (see p. 131).

Intercumulus liquid: The liquid remaining between crystals after their separation from the original magma.

Interglacials: Intervals of warmer conditions separating glacials during an ice age. Interglacials typically last 10 000 to 20 000 years whereas glacials are typically 100 000 years long (see pp. 183–184).

Intra-continental basin: See Continental basin.

Intramontane basin: Sedimentary, usually continental, basin that develops within a mountainous area so that input to the basin is usually of very coarse material. Forms commonly during the late, extensional phases of a mountain-building episode (e.g. the Devonian basin in East Greenland, see p. 114).

Intrusion: See Igneous intrusion.

Intrusive rocks: See Plutonic rock.

Invertebrates: Animals without backbones.

Iron pyrites: Mineral (FeS_2) also known as pyrite.

Island-arc: An arcuate line of volcanoes formed on oceanic crust over a subduction zone. The rocks in these volcanoes are formed from the melting of oceanic crust mixed with sedimentary material derived from the continents and large amounts of water.

Isomorphous series: Minerals that have analogous chemical compositions and the same crystal structures. For example, olivine can contain varying numbers of atoms of magnesium (Mg) and iron (Fe) with the general formula $(Mg,Fe)_2SiO_4$ between the end-members Mg_2SiO_4 (forsterite) and Fe_2SiO_4 (fayalite).

Isopach map: Map showing the variations in the thickness of a sedimentary or volcanic unit, usually by means of contours.

Isostasy: The condition in which the outer, lighter parts of the Earth's lithosphere 'float' in gravitational equilibrium on the denser asthenosphere below (see p. 27).

Isotope: Variants of the same element that have different atomic weights. For example, uranium-235 and uranium-238 are isotopes of uranium with different numbers of neutrons in their cores, but the same number of protons. All isotopes of an element react the same way chemically, but may behave differently physically.

J

Jameson Land Basin: Mesozoic sedimentary basin in East Greenland (see pp. 124–127).

Joint: A minor fracture in a rock whose sides are not displaced relative to one another. Joints commonly occur in sets that reflect the stress regimes that produced them.

Jurassic: The middle period of the Mesozoic, from 199.6 to 145.5 million years ago.

Juvenile crustal material (granitic): A granite-like material that separates from the upper mantle as a result of primary partial melting.

K

Kakortokite: A syenitic rock built up of alternating black, red and white layers (respectively dominated by the minerals black arfvedsonite, red eudialyte and white feldspar). It is found in the southern part of the Ilímaussaq complex (see p. 57).

Kangerlussuaq Basin: Sedimentary basin in South-East Greenland, of Cretaceous to Palaeogene age (see p. 132).

Karst: Type of landscape formed by the dissolution of limestone by acidic water running through it, often in the form of underground rivers. The water dissolves the rock to form caves and sinkholes and redeposits it in caverns as stalagmites and stalactites.

Kerogen: Fossilised organic material derived from the remains of animals and marine algae, typically in fine-grained sediments. When buried and heated it changes into crude oil and natural gas.

Ketilidian fold belt: Fold belt that formed in the Palaeoproterozoic between 1850 and 1725 million years ago. It forms a large part of the crystalline basement in South Greenland. Named from the Viking name Ketils Fjord, known as Tasermiut Kangerdluat in modern Greenlandic (see pp. 50–52).

Kimberlite: Fine-grained volcanic rock with an ultrabasic composition similar to peridotite (see p. 141). Kimberlites are porphyritic, containing large grains of olivine, and can also contain xenoliths (rock fragments from the walls of the intrusion). Kimberlites are formed at great depths in the mantle, and some contain diamonds (see p. 214).

Kitchen (oil and gas kitchen): An expression used in petroleum geology for regions of the subsurface where conditions are suitable for the conversion of kerogen into oil or gas (see pp. 220–225).

Klippe: Term used in structural geology to describe a remnant of a thrust sheet left as an isolated unit by erosion of the adjacent parts of the thrust sheet. It usually consists of older rocks that lie on a younger rock unit.

Kronprins Christian Land Orogeny: An informal name for the local, late Cretaceous deformation event in eastern North Greenland (see p. 121).

L

Lacustrine deposits: Sediments deposited in a lake.

Lamination: Small-scale (1–2 mm) layering within a bed, commonly consisting of slight variations in grain size of the sediment.

Late glacial maximum: The time at which the ice was at its greatest extent during the last glacial of the Quaternary, in Greenland about 21 000–22 000 years ago.

Lateral moraine: Also known as side moraine. A moraine deposited along the side of a valley glacier.

Laurasia: Former supercontinent consisting of the continents Laurentia, Baltica and Siberia welded together (see p. 123). It formed the northern part of the supercontinent Pangaea until the end of the Triassic, when it separated from Gondwana.

Laurentia: Former continent consisting of North America, Greenland, Scotland and part of Ireland (see p. 83). It existed from the Late Proterozoic until the Silurian.

Lava: Volcanic rock that flows onto the Earth's surface or seabed while molten. The thickness of individual lava flows is from a few metres to more than 50 m (see p. 140).

Lava flow: A contiguous bed of lava that has flowed from a single eruption.

Layered intrusion: A plutonic body with layers of varying composition, formed by repeated injections of magma and magmatic differentiation (see pp. 149–152).

Licensing round: Offer by the licensing authority, in Greenland the government, to the oil and gas industry to apply for permits to explore for hydrocarbons within a designated area and for a limited time. The interested oil companies apply with details of their financial and technical capabilities of carrying out such an undertaking and the government awards permits to those applicants judged best.

Lichen: A plant composed of a symbiotic coalescence of alga and fungus. Grows on the surface of rocks, often covering the exposure, making it difficult to study the rock itself (see p. 48).

Limestone: Sediments consisting predominantly of calcium carbonate ($CaCO_3$).

Limnic sediments: Sediments deposited in lakes.

Lineament: A linear structural element on the Earth's surface such as a fracture or fault or a line of volcanoes.

Lithosphere: The outer relatively rigid part of the Earth, consisting of the crust and the upper part of the mantle, lying above the more plastic asthenosphere. The thickness of the lithosphere varies considerably, from only a few kilometres at an oceanic spreading ridge to about 120 km below an old continent. The lithospheric plates move relative to one another in the outer levels of the Earth's shell (see p. 53).

Lithospheric plate: A discrete body of relatively rigid lithosphere (crust and upper mantle) that is moved across the Earth's surface by plate tectonic processes as an integral unit (see p. 53). The present-day Earth has seven large and a series of smaller such plates.

Lithostratigraphy: The system of stratigraphic division based on the composition of rocks (see p. 69).

Little Ice Age: The period between A.D. 1500 and 1900 when the climate was colder than both today and the preceding Mediaeval warm period between A.D. 1000 and 1400 (see p. 189).

Llandovery: The oldest epoch in the Silurian, from 444 to 428 million years ago. Named after a town in Wales.

Load structures: Structures within a sediment that have been deformed by loading by overlying sediments. For example, where sands overlie muds in interbedded successions, the sand may sink into the underlying mud, forming isolated balls and lenses.

Lowstand (of sea-level): The period during which sea level is at its lowest during a cycle of sea level changes (a term used particularly in sequence stratigraphy, see p. 130). During the lowstand the coastline is at its most seaward position and much of the continental shelf may be exposed and subject to erosion.

Lujavrite: A dark green syenitic rock found in the southern part of the Ilímaussaq complex (see p. 57).

M

Ma: Mega-annum = million years.

Macrofossils: Fossils large enough to be seen with the naked eye.

Magma: Any type of molten rock in the upper mantle, crust or erupted onto the Earth's surface. Basic silicate magmas are hot and contain large amounts of iron and magnesium. Acidic (or granitic) magmas melt at a lower temperature and are rich in silica, sodium and potassium.

Magma chamber: Body within the crust or upper mantle containing liquid magma.

Magmatic differentiation: Process causing a magma to separate into components with different chemical compositions (see p. 152).

Magmatic foliation: Structure within a former magma consisting of parallel sheets of minerals. The structure formed while the magma was entirely or partly liquid.

Magmatic layering: Layering within a solidified magma chamber caused by different minerals crystallising and

Magmatic rocks

falling to the base of the magma chamber as the magma cooled (see pp. 149, 152).

Magmatic rocks: Rocks produced from a magma.

Magnetic anomalies (on oceanic crust): Pattern of linear bands of high and low magnetic anomalies developed on oceanic crust caused by alternating bands of normally- and reversely-magnetised basaltic volcanic rocks. The bands develop at the central spreading ridge in response to periodic changes in the polarity of the Earth's magnetic field. These magnetic stripes are one of the strong arguments for the theory of plate tectonics (see p. 53).

Magnetic anomaly number: The number assigned to an ocean magnetic anomaly formed by volcanic rocks that was erupted during a chron with the same number. A chron is a time span in which the Earth's magnetic field changes from normal through reverse to normal magnetic direction.

Magnetic field strength (total magnetic intensity): The strength of the Earth's magnetic field measured at a point on or above the Earth's surface. Measured in a unit called Tesla, which is very large, so nanoTeslas ($1nT = 10^{-9}$ Teslas) are more commonly used in practise. The unit called gauss was formerly used.

Magnetic polarity: The overall orientation of the Earth's magnetic field at a given time. When the Earth's north and south magnetic poles have the same polarity as today, its orientation is called normal. When they have the opposite polarity from today, the orientation is called reversed (see p. 113).

Magnetite: Black magnetic mineral with the composition Fe_3O_4.

Magnetometer: Instrument used to measure the Earth's magnetic field.

Malachite: Copper carbonate mineral with a characteristic green colour. It forms near the surface of copper ores and is a good indicator of their presence in the subsurface (see p. 215).

Mantle: Earth's mantle; that part of the Earth between the crust and the core (see p. 20). It occurs at depths from 3 to 90 km at its shallowest to about 2900 km at its inner boundary with the core. There is a major internal boundary within the mantle at a depth of about 660 km where the ultrabasic material of the mantle changes from one crystalline state to another. The asthenosphere lies above this boundary and the mesosphere below it.

Mantle plume: A mass of hot rock, roughly cylindrical in cross-section, that has risen up from deep in the mantle (see p. 139). The material spreads out to form a broad head when it encounters the base of the lithosphere.

Marble: Metamorphic rock consisting of recrystallised aggregates of calcium carbonate ($CaCO_3$) in sedimentary limestones. Pure marble is white but impurities commonly give it various colours and help to define its characteristic structures, such as bands and flow structures (schlieren).

Marine geology: The study of unconsolidated sediments at and just beneath the seabed (see pp. 156–157).

Marine sediments: Sediments deposited in the sea.

Marine terrace (raised beach): Platform (terrace) of loose marine sediments deposited near a shoreline and exposed by uplift or lowering of sea level (see p. 189).

Matrix: The finer-grained material in a sediment surrounding coarser grains, e.g. in a limestone where a mud-grade calcite matrix forms the volume between the coarser grains in the rock (see p. 64).

Mature sediment: A clastic sediment consisting of well-rounded grains caused by the transport of the sediment from a distant source (see box p. 76).

Maturity (in petroleum geology): A concept used to describe how readily a source rock may produce oil or gas (see pp. 221–222).

Mauritanides: Fold belt in Mauritania in north-western Africa. Its Early Palaeozoic age corresponds with that of the Caledonides in Europe and Greenland (see p. 96).

Mean sea level: The mean level of the sea once tidal changes are allowed for. Sea level has changed substantially during geological time; these changes are commonly plotted relative to modern mean sea level (see p. 66).

Meandering river: A river with many sinuous bends, curves, turns and loops. Typical fluvial pattern in low-gradient, mud-rich river systems.

Median line: Political boundary between the exclusive economic zones of two countries, drawn half-way between their coasts. Greenland has median lines with Canada in the west and Iceland and Jan Mayen (Norway) in the east.

Meltwater sediments: Sediments deposited from meltwater within or derived from glaciers, large ice caps and ice sheets.

Meltwater terrace: Flat terrace composed of unconsolidated sediments deposited from meltwater streams (see p. 185).

Member: The unit in the lithostratigraphic classification between formation and bed (see p. 69).

Meso: Prefix meaning middle.

Mesoarchaean: The middle era of Archaean time, from 3200 to 2800 million years ago (see p. 21).

Mesoproterozoic: The middle era of Proterozoic time, from 1600 to 1000 million years ago (see p. 21).

Mesosphere: The lower mantle below the asthenosphere at depths between about 660 and 2900 km. The mesosphere is comparatively strong and is not affected by the convection processes taking place in the asthenosphere (see p. 20). The mesosphere probably consists of olivine-like minerals with the general composition $(Mg,Fe)SiO_3$.

Mesozoic: The middle era of Phanerozoic time from 251 to 65 million years ago.

Metamorphic facies: Mineral assemblage in rocks that have been subjected to alteration in a certain range of pressure and temperature. Generally caused by orogenic processes (fold belt formation) (see p. 35).

Metamorphic schist: See Schist.

Metamorphism: The alteration of rocks by recrystallisation at high temperature and often high pressure. New suites of minerals commonly form during metamorphism and these can be used to deduce the temperatures and pressures that the rock has experienced. Metamorphism may grade into partial melting.

Metasediments: Former sediments altered by metamorphism.

Mica: Various types of rock-forming silicate minerals with a pronounced cleavage so that they split easily into thin plates. Most common types are the dark-coloured biotite and the light-coloured muscovite.

Mica schist: Metamorphic rock formed by the recrystallisation of clay-rich sediment. The rock consists of mica, quartz and feldspar.

Microcline: Potassium feldspar ($KAlSi_3O_8$). Very common mineral in metamorphic rocks.

Microcontinent: An isolated small area of the continental crust surrounded by oceanic crust. Microcontinents occur commonly as submarine plateaus (see p. 171).

Microfossils: Fossils too small for the naked eye to see, such as many single-celled organisms (e.g. foraminifers and radiolaria), which have to be studied with a microscope.

Micron: 1 μ (pronounced myu) = one micron = 0.001 mm.

Micropalaeontology: The study of microfossils.

Mid-ocean spreading axis: Boundary between tectonic plates that are moving away from one another, where new crust is formed by upwelling, melting and eruption of material from the mantle. The spreading zone is commonly a ridge 1–3 km high with a central graben. Ocean spreading axes form the largest single structure on Earth with a contiguous length of more than 80 000 km (see p. 53).

Migmatisation: The process of forming a migmatitic rock.

Migmatite: Metamorphic rock with two components – the remains of a former rock (the palaeosome) that has been thoroughly intruded by a granite-like component (the neosome). The rock commonly forms where the original rock starts to melt forming a mobile phase that penetrates the rock and solidifies as veins and diffuse zones (see p. 34).

Migration: The movement whereby hydrocarbons migrate from a source rock to a reservoir.

Milankovitch cycles: Cycles of climate caused by regular changes of the Earth's orbit and attitude in relation to the sun. The three main parameters are the obliquity of the Earth's orbit, the angle the Earth's axis makes with its orbital plane and the time of year when the Earth has its closest approach to the sun. Glacials and interglacials are one of the consequences of these changes.

Milligal: A unit commonly used to measure gravity anomalies. One milligal is one thousandth part of a gal (1mGal = 0.001 Gal) and about 1 millionth part of the gravity attraction of the Earth at its surface.

Mineral: Inorganic substance formed naturally by geological processes. Most minerals are crystalline with a distinct atomic structure, have a characteristic chemical composition and have specific physical properties.

Mineral province: An area with a characteristic geological identity that contains economically valuable concentrations of specific elements. Examples are gold provinces, lead-zinc provinces and rare-earth provinces (see p. 199).

Mineralisation: Commonly used term for the natural process of concentrating economically valuable minerals in a rock so that they become large enough to be exploited. The process commonly involves precipitation of the valuable mineral in veins or as impregnations and replacements of the host rock (see p. 199).

Mineralogy: The study of the occurrence, composition and structure of minerals.

Miocene: The first epoch of the Neogene, 23.0–5.3 million years ago.

Moho: Commonly-used but shortened form of Mohorovičić discontinuity. The sharp boundary between the Earth's crust and its mantle, normally recognised by an increase of seismic p-wave velocities from around 6.5–7.2 km/sec in the crust to around 8.0–8.2 km/sec in the mantle.

Molasse: A thick series of sediments deposited on the foreland or in an intermontane position during a late phase of fold-belt formation; commonly consisting of conglomerates with a smaller content of finer-grained clastic sediments (see pp. 114–117).

Molybdenum: Metal element (Mo) that occurs in the mineral molybdenite (MoS_2).

Monocline: Open fold consisting of one steep flank between gently-dipping limbs (see flexure).

Monzonite: Plutonic rock with approximately equal amounts of plagioclase and alkali feldspar and with little free quartz. Augite is usually the dominant dark mineral.

Moraine: Unsorted material that has been deposited directly by ice or formed by a glacier. A moraine ridge formed at the snout of a glacier, by pushing together material at and under its snout, is called an end moraine and those along the flanks of a glacier are called side or lateral moraines.

Mudflow: Flow consisting of a water-saturated mass of sand, mud and rock debris.

Mudstone: A sedimentary rock consisting primarily of clay and silt.

N

Nagssugtoqidian fold belt: A fold belt in West Greenland that formed 1900–1800 million years ago (see pp. 42–45).

NanoTesla (nT): A unit used to measure the strength of a magnetic field. One Tesla is 10^9 nT. The Earth's total field strength is around 30 000 nT and magnetic anomalies are from a few tens to a few thousands of nT.

Native iron: Pure iron occurring naturally, often as the result of the reduction of iron oxides by organic acids from coal or oil.

Natural gas: The term used for all naturally occurring hydrocarbons that are gaseous under normal atmospheric pressure. The most common constituent is methane (CH_4) (see pp. 221–222).

Naujaite: A very coarse-grained syenite found in the southern part of the Ilimaussaq complex in South Greenland (see p. 57).

Nautilus: Present-day free-swimming mollusc with a coiled shell that resembles those of ammonites (see p. 125).

Neo (prefix): New or late. Used for the youngest part of an interval of time e.g. Neoproterozoic, Neogene.

Neoarchaean: The latest era of Archaean time, from 2800 to 2500 million years ago (see p. 21).

Neogene: The latest period of Cenozoic time, from 23 million years ago until the present.

Neoproterozoic: The latest era of the Proterozoic, from 1000 to 542 million years ago (see p. 21).

Neosome: A granite-like component in a migmatite (see p. 34).

Nepheline: Alkali feldspar-like mineral with less silica (SiO_2) than normal feldspar.

Nepheline syenite: Plutonic rock dominated by alkali-feldspars and nepheline.

Nonconformity: A boundary similar to an unconformity, but between layered sedimentary or volcanic rocks above and eroded basement of magmatic or metamorphic rocks below (see p. 146).

Normal fault: A fault produced by fracture and movement of rocks in response to extensional forces. The fault plane usually dips at angles of about 45°–60° and the block above the fault plane has moved downwards relative to the block below it (see p. 94).

Normal magnetisation: Natural remanent magnetism with the same polarity as the present-day Earth's magnetic field.

Nunatak: Land completely surrounded by glacier ice so that it appears as an island in the ice. Occurs commonly in the marginal zone of the Inland Ice (see p. 138).

Nuussuaq Basin: Sedimentary basin of Cretaceous to Palaeogene age in West Greenland that also contains thick Palaeogene volcanic rocks (see pp. 128–131).

O

Ocean-floor formation: Process by which new oceanic crust is formed (see p.169).

Ocean-floor spreading (also called seafloor spreading): The widening of the oceans by formation of new ocean floor at the mid-oceanic spreading axis. The new crust moves away from both sides of the axis (spreads) at a rate of less than 1 to 10 cm per year (see p. 53).

Oceanic basalts: Basaltic volcanic rocks formed at the mid-oceanic spreading zones. These rocks have a characteristic composition called tholeiite that can be described most simply as basalts with a small content of quartz. It is often called MORB (Mid-Ocean Ridge Basalt).

Oceanic crust: The crust under the oceans consisting of oceanic basalts with a density of around 2.9 g/cm^3 and a thickness from 5 to 25 km (see p. 169).

Oceanic sediments: Sediments deposited under deep water in the oceans. Most of them are very fine-grained oozes.

Offshore: The area under the sea.

Offshore basin: Sedimentary basin found today in an offshore area.

'Old Red Sandstone' (ORS): Thick continental sediments of Devonian to Early Carboniferous age found in many places on both sides of the North Atlantic (see p. 117). Forms extensive basins in North-East Greenland.

Oligocene: The youngest epoch of the Palaeogene, from 33.9 to 23.0 million years ago.

Olivine: Green to black-brown silicate mineral that is very common in volcanic magmas and is a major component of the Earth's mantle. Its composition is $(Mg,Fe)_2SiO_4$.

Olivine basalt: Basalt with higher content of olivine and a lower content of silica (SiO_2) than normal basalt.

Olivine gabbro: Gabbro with higher content of olivine and a lower content of silica SiO_2) than normal gabbro.

Ooids: Small spherical or ovoid carbonate grains formed by concentric growth around a seed particle (nucleus) in the centre of the grain. Ooids are 0.25–2 mm in diameter and form the dominant grain type in certain limestones (oolitic limestones or oolites). The grains are thought to be primarily inorganic in origin and most commonly form in shallow marine environments affected by waves.

Open door policy: Licensing policy in Greenland for oil and gas under which companies can apply for concessions in certain areas at times other than during licensing rounds (see p. 241).

'Open water area': Sea area off southern West Greenland (about 61–68°N) that is unaffected either by new winter sea ice or by pack ice drifting from the Polar Basin via East Greenland.

Ordovician: The second oldest period of the Palaeozoic, from 488 to 444 million years ago (see p. 21).

Ore: Naturally occurring rocks containing elements or minerals of potential commercial value. The term is also used for the minerals present in such rocks. The term is commonly used to specify the particular content of economic worth, as in lead ore, iron ore, copper ore etc.

Ore grade: Measure of the percentage of economically valuable element or mineral that can be extracted from the rock in which it is found (see p. 200).

Orogenic belts: Areas, usually substantially longer than they are wide, in which orogenesis has taken place.

Orogeny, orogenesis: Mountain-building involving the formation of fold belts by compression accompanied by deformation of the crust by folding and overthrusting and by metamorphism and magmatic activity.

Orthoceratite: A nautiloid from the extinct genus Orthoceras; a non-coiled, shell-bearing cephalopod related to present-day squids and octopuses (see p. 81).

Orthoclase: Potassium feldspar ($KAlSi_3O_8$) found in various plutonic rocks such as granite.

Orthogneiss: Gneiss formed by the deformation and metmorphism of a plutonic rock, often with a granitoid composition. See also paragneiss.

Orthopyroxene (opx): Group of rock-forming Mg- and Fe-bearing silicate minerals belonging to a mineral group called pyroxenes. They occur commonly in basic plutonic rocks. Examples are hypersthene and enstatite.

Ostracods: Sometimes known as 'seed shrimps'. Tiny crustaceans that are entirely enclosed in a bean-shaped shell. There are both marine and freshwater forms. These (micro)fossils are found in rocks from the Lower Palaeozoic to the present (see p. 163).

Overthrust: A low-angle thrust fault with a displacement on a scale of kilometres.

P

Pacific Province: The faunal province of marine animals found on the western side of the Atlantic that has been in genetic connection with the Pacific Ocean. The faunal province on the eastern side of the Atlantic is called the Atlantic Province.

Packstone: Limestone composed of granular carbonate grains in contact with one another in which the spaces between the grains are filled with lime mud (see p. 64).

Pahoehoe lava: See Ropy lava.

Palaeo (prefix): Older or oldest. Used for the oldest part of an interval of time e.g. Palaeozoic, Palaeogene and to designate studies of past things and conditions, e.g. palaeontology, palaeogeography.

Palaeoarchaean: The oldest era of the Archaean, from 3600 to 3200 million years ago (see p. 21).

Palaeogene: The older of the two periods of the Cenozoic, from 65.5 to 23.8 million years ago (see p. 21).

Palaeogeography: The study of the changes to continents, oceans and climate belts through geological time and reconstruction of the relative positions of land and sea at various times (see p. 112).

Palaeomagnetism: The study of the Earth's former magnetic field (see p. 113).

Palaeontology: The study of past animal and plant life using their remains preserved as fossils.

Palaeoproterozoic: The oldest era of the Proterozoic, from 2500 to 1600 million years ago (see p. 21).

Palaeo-Tethys (ocean): An ocean east of the supercontinent Pangaea during Permo-Triassic time (see p. 123).

Palaeozoic: The first era of Phanerozoic time, from 542 to 251 million years ago (see p. 21).

Paleocene: The oldest epoch of the Palaeogene, from 65.5 to 55.8 million years ago (see p. 169). The word is derived from Palaeo-Eocene, hence Pal-eocene, not Palaeo-cene.

Paleosome: Remains in a migmatite of the former rock from which it formed (see migmatite and p. 34).

Palladium: Metallic element (Pd) related to platinum.

Pangaea: A supercontinent that existed for about 100 million years during the Permian and Triassic. Pangaea began to break up about 250 million years ago and its remains form the present-day continents (see p. 123).

Pannotia: A late Proterozoic continent in the southern hemisphere that resulted from the break-up of the supercontinent Rodinia.

Paragneiss: Gneiss formed by the deformation and metamorphism of sedimentary rocks. In contrast to orthogneiss formed from plutonic rocks.

Passive continental margin: A margin between continental and oceanic crust that formed by extension and rifting. Such a margin is no longer active once the spreading axis migrates away from it, hence passive.

Pebble: A rounded stone in a sediment in the size fraction between 4 and 64 mm in diameter (see p. 64).

Pegmatite: A very coarse-grained magmatic rock found as sheets, dykes and lenses in plutonic rocks and their surroundings.

Pelagic sediments: Marine sediments consisting of the remains of marine plankton and deposited at great depths of water. Usually very fine-grained oozes.

Peneplain: A flat landscape near sea-level formed by long-term erosion by rivers. Remnants of former peneplains can be seen at heights well above sea-level in areas that have been uplifted after their formation (see pp. 27 and 179).

Peridotite: Ultrabasic rock consisting of about 70% olivine and 30% pyroxene (see p. 141).

Period: The formal fundamental unit in the geological time scale (see p. 69). A period lasts typically several tens of millions of years; e.g. the Jurassic, that lasted from 199.6 to 145.5 million years ago.

Permeability: A measure of the ease with which a fluid, such as water, oil or gas, can move through a rock. Pores and fractures within a rock must be connected into a network for this to happen.

Permian: The youngest period of the Palaeozoic, from 299 to 251 million years ago (see p. 21).

Petrology: The study of the formation, composition, internal structure and development of plutonic and metamorphic rocks.

Phanerozoic: The youngest eon of Earth's history, from 542 million years ago until the present (see p. 21).

Phenocryst: A relatively large single crystal in a more finely grained magmatic rock. The presence of phenocrysts means the rock is a porphyry.

Pillow breccia: Volcanic rock formed from fragments of pillow lava.

Pillow lava: Basaltic volcanic rock with a structure resembling a heap of pillows. The pillows form when lava extrudes into water, e.g. onto the sea floor, and its surface is cooled very quickly by the cold water. Each pillow has a dense, commonly glass-like, chilled outer margin and a somewhat coarser interior. Individual pillows are up to about a metre across, but can accumulate to form thick layers (see pp. 46 and 169).

Pisoids: Spherical or ellipsoidal carbonate grains larger than 2 mm in diameter formed by concentric growth around a nucleus. Pisoids form in quiet conditions, in contrast to ooids. A rock consisting of pisoids is called a pisolitic limestone.

Placer deposit: Concentration of grains of metallic or other economic minerals in a stream sediment due to their relatively high density.

Plagioclase: Feldspar mineral with a composition intermediate between the two end members sodium feldspar (albite – $NaAlSi_3O_8$) and calcium feldspar (anorthite – $CaAl_2Si_2O_8$).

Plankton: Small animals and plants that drift passively in water. Most plant plankton consists of algae, while animal plankton consists of both small single-celled organisms such as foraminifers and the larvae of sessile animals as bivalves and sea anemones.

Plant fossils: Fossilised remains of plants, occurring as stems, leaves etc. or their impressions in soft sediment.

Plate tectonics: The theory that the Earth's crust and upper mantle are formed of a number of internally rigid plates that can move relative to one another (see p. 53). Boundaries between the plates are of various types; volcanic sea-floor forms where the plates are moving away from one another and fold belts form where plates move towards one another.

Plateau lavas or Plateau basalts: A large area of flat-lying lava flows. The lavas are usually of basaltic composition and flow relatively easily. Their eruptions are usually from elongated fissures and each lava flow can extend more than 10 km from its source (see p. 72).

Platform: An area bordering a rift basin where crystalline basement is shallow and covered only by no or relatively thin, flat-lying sediments.

Play (in petroleum geology): A set of conditions that have to be fulfilled for oil or gas to be present in a reservoir. The necessary conditions are: 1) that there be a mature source rock, 2) that there be a reservoir of sufficient volume and with enough porosity to contain economically valuable amounts of hydrocarbon and 3) that it has enough permeability for the hydrocarbons to flow into it and out again during production, 4) that the reservoir be formed into a shape where hydrocarbons can accumulate, 5) that there be an impermeable seal over the reservoir and 6) that there be a route by which the hydrocarbons produced by the source rock can enter the sealed reservoir.

Pleistocene: The older of the two epochs of the Quaternary, from 1.8 million years ago until 11 700 years ago (see p. 169).

Pliocene: The younger epoch of the Neogene, from 5.3 to 1.8 million years ago (see p. 169).

Plume: See Mantle plume.

Plutonic rock: Coarse-grained crystalline rock formed by the slow solidification of a magma at depth (see p. 34).

Polarised light: Light whose vibrations take place in one plane (plane polarised). Used to investigate thin sections of rock using a polarising microscope.

Polarity time scale: The division of geological time using the reversals in polarity of the Earth's magnetic field (see p. 113).

Pore volume: The total volume of the small spaces between the grains of a sedimentary or other rock.

Porphyritic: Mineral structure in a magmatic rock containing some coarse crystals in a finer-grained matrix.

Porosity: Measurement of how much open space there is between the grains of a rock. Quoted as a percent of the total volume of the rock.

Post (prefix): After. For example, 1) post-depositional: happened after the end of sedimentation; 2) post-rift: after the end of rifting.

Postglacial time: 'After the end of the Ice Age'. Synonym for the Holocene epoch from 11 700 years ago to the present (see p. 169).

Ppm: Parts per million. Used to show how much of a rare material is found in a more common one, such as the amount of gold within its host rock. 1 ppm gold is equivalent to 1 gram of gold per tonne of host rock.

Pre (prefix): Before. For example, 1) pre-rift: before the start of rifting; 2) Precambrian: before the start of the Cambrian period.

Precambrian: The time 'before the Cambrian', i.e. before 542 million years ago. The Precambrian comprises the Hadean, the Archaean and the Proterozoic (see p. 21).

Processing plant: Industrial plant for the processing of metal ores to concentrate the ore that can then be sent to a smelter where the actual metal is extracted from the ore.

Production licence: Exclusive licence to produce hydrocarbons, minerals or other materials (e.g. sand, quarry stone) within a given area for a period of time.

Profile, reconstructed: The result of using the principles of structural geology to restore a vertical section through a deformed, layered succession of rock to show how it was prior to the deformation.

Profiles: See Geological profile.

Progradation: The process by which a locus of sedimentation migrates laterally. For example, a river delta building out into the sea is said to prograde seaward.

Prospect (in economic or petroleum geology): A site where a potential economic mineral deposit or accumulation of hydrocarbons may be found.

Prospecting licence: Permission to carry out geological or geophysical surveys, but with no right to exploit any discovery of minerals or hydrocarbons.

Proterozoic: The era from 2500 to 542 million years ago. The Proterozoic is divided into three epochs, the Palaeo-, Meso- and Neoproterozoic (see p. 21).

Provenance studies: Studies of the origins of the constituent grains of clastic sediments and the determination of the areas from which they were originally eroded (see p. 76).

Pseudo-escarpment: The appearance on a reflection seismic section of the landward termination of a basalt flow within the sediments. The abrupt termination of the strong reflections from the basalts creates an impression resembling an escarpment. This is a characteristic feature of the boundary zone between ocean basalts and shelf sediments offshore East Greenland (see p. 170).

Pteridophytes: Primitive green land plants that reproduce by means of spores instead of seeds. They were the earliest dominant group of land plants during the Silurian to Triassic periods when they commonly grew to the height of large trees. Modern examples include club mosses, horsetails and ferns.

Pterosaurs: An order of extinct flying reptiles. Their remains occur as fossils in Mesozoic sediments.

Pull-apart basin: Sedimentary basin formed where one strike-slip fault dies out and displacement along it is taken up on a parallel but offset fault. The resulting basins are commonly narrow, elongated and deep and can contain more than 10 km of primarily coarse sediments, breccias, conglomerates and sandstones (see p. 120).

Pyrochlore: Economically important sodium and potassium oxide mineral containing niobium and tantalum.

Pyroxene: Group of green, brown and black silicate minerals with large contents of magnesium and/or iron and most also contain calcium or sodium. Examples are augite and aegirine. An important constituent of many plutonic rocks.

Pyroxenite: Ultrabasic rock consisting of about 20% olivine and 80% pyroxene (see p. 141).

Q

Quartz: One of the most common minerals; silicon dioxide (SiO_2). A major component of sandstone, quartzite and granite.

Quartz diorite: Plutonic rock with a composition dominated by plagioclase and with between 5 and 20% quartz. Dark minerals occur in appreciable amounts.

Quartz gabbro: Basic plutonic rock with a gabbroic composition but with small amounts of quartz present.

Quartzite: Compact, hard rock formed by the recrystallisation of sandstone and consisting predominantly of quartz.

Quaternary: The youngest period of Earth's history from 1.8 million years ago until today. It is divided into the Pleistocene (the Ice Age), from 1.8 million to about 11 700 years ago, and the Holocene (post-glacial) from about 11 700 years ago until today.

R

Radioactive elements: Elements that are naturally unstable and convert (decay) to another isotope of the same element or to some other isotope of other elements by the emission of particles and radiation. Natural radiation gives off energy into the surroundings. About 15 radioactive elements occur naturally, but those that emit most energy in the Earth's crust are potassium-40, thorium-232 and uranium-238 and -235. Uranium and thorium are the most radioactive of these elements.

Radiolarians: Marine planktonic organisms with a spherical shell, normally with a siliceous composition. They are common microfossils from the Cambrian until today.

Radiometric age dating: Determination of the absolute age of a mineral or rock using the time that a proportion of the atoms in naturally-occurring radioactive elements take to decay (their half-live) (see p. 48).

Raised beach: Also known as a marine terrace. A linear accumulation of water-worn sand, silt and gravel with a flat top that is at a constant or smoothly-changing height above present-day sea-level. It represents a former beach that has been raised to its present height by uplift of the land (see p. 190).

Rapakivi granite: Coarse-grained granite characterised by large rounded feldspar crystals with cores of greyish or reddish potassium feldspar surrounded by a thin rim of white plagioclase, recording a shift in magmatic chemistry.

Rare-earth minerals: Minerals containing elements from the so-called rare-earth series (lanthanides) and a few others. The lanthanides have atomic numbers from 57 to 71. Examples of rare-earth elements are: neodymium (Nd), samarium (Sm), erbium (Er), ytterbium (Yb), scandium (Sc) and yttrium (Y). The latter two are rare-earth elements, but not lanthanides.

Recrystallisation: Process by which the elements in rocks are rearranged to form new minerals. Takes place mainly during metamorphism at elevated temperatures and high pressures.

Recumbent fold: A fold with an almost flat-lying axial plane (see p. 94).

Reflected seismic wave: Sound wave travelling steeply upwards through the subsurface after having been reflected from the boundary surface between two units of rock with different physical properties (see p. 160).

Reflector: A surface between two bodies of rock in the subsurface with different physical properties, often two layers of sediment, that gives rise to a reflection of sound energy that is visible on a reflection seismic section (see p. 158).

Refracted seismic wave: Sound wave in the subsurface that has been refracted to travel along the boundary between two units of rock with different physical properties. The passage of the wave disturbs the boundary causing it to emit new waves that travel upwards to the surface of the Earth, where they can be detected.

Regional metamorphism: Metamorphism that takes place over a large volume of rock, normally at 5–50 km depth during formation of a fold belt.

Regression: Seaward migration of a coastline. May be a result of a fall of global sea level or because of an uplift of the land. Shallow-marine sediments deposited at earlier times become exposed above sea-level and may be eroded. Deltas build out (prograde) into the sea and shallow-water sediments may be deposited on top of deeper water sediments. The opposite of transgression (see p. 130).

Relative chronology: Placing of events in a time sequence of the relation within rocks, structures and fossils, but without assigning absolute ages to them (see p. 49).

Relinquished licence: Exploration licence that has been given up by the licence holder who no longer wishes to explore for minerals or hydrocarbons within the area of the licence (see p. 238).

Remanent magnetism: Traces of the Earth's former magnetic field naturally preserved in rocks. Studies of these preserved magnetic fields have shown that the Earth's magnetic field has changed in strength and polarity at different times. The remanent magnetic fields preserved in oceanic crust have been the strongest evidence that the continents have drifted relative to one another (see p. 113).

Reserves (in petroleum geology): The amount of oil or gas that can be produced from an oil or gas reservoir (see p. 225).

Reservoir (in petroleum geology): A rock that is both sufficiently porous and permeable for hydrocarbons to accumulate in it and be produced from it in sufficient quantities to be economically viable (see p. 221).

Reverse fault: A fault with a moderately- or steeply-dipping fault-plane along which the overlying block has moved upwards relative to the underlying block (opposite of normal fault; see p. 94).

Reverse magnetisation: Natural remanent magnetism with opposite polarity to that of the present-day magnetic field of the Earth (see p. 113).

Rhyolite: Fine-grained volcanic rock with a granitic composition.

Rift, rift valley: Elongated zone of the Earth's crust where faulting has caused blocks to move downwards, forming grabens and half-grabens. Rifting is often accompanied by volcanism. Rifts are caused by extension of the crust and may occur within a continent or be the early stages of continental splitting leading to sea-floor spreading.

Rift basin: A sedimentary basin formed by rifting. May develop initially as a continental basin, but if rifting develops towards continental splitting the sea enters the region between the separated continental blocks and marine sediments are deposited in the basin.

Rinkian area: Area of northern and central West Greenland where the basement consists of rocks metamorphosed during Palaeoproterozoic times (see p. 46). The Rinkian area is now considered to be a northern continuation of the Nagssugtoqidian fold belt that is found south of Disko Bugt. Named after Rink Isbræ (glacier).

Riphean: An informal name for the early part of the Proterozoic from 1600 to 800 million years ago.

Ripple lamination: Wavy, small-scale, cross-stratification in sediments with alternating sand and silt layers.

Rock: An aggregate of one or more minerals that have consolidated into a contiguous solid. There are hundreds of named kinds, but the commonly known examples are sandstone, shale, basalt and granite, each with its own characteristic assemblage of minerals and internal structures (see p. 34).

Rodinia: Former supercontinent about 1200–1000 million years ago consisting of most of today's continents (see p. 99).

Ropy lava: Lava with a runny consistency whose surface forms a ropy, wrinkled structure on solidifying (see p. 148). Also known as pahoehoe lava (a Hawaiian word). Compare with blocky lava.

Ruby: Red precious stone consisting of the mineral corundum (Al_2O_3) with impurities of chromium that give it its red colour (see p. 195).

S

Saalian (glacial): The penultimate glacial, from 400 000 to 130 000 years ago (see p. 183). Corresponds to the Illinoian in North America

Salt diapir: See Diapir and p. 170.

Salt lake: Lake in hot dry climate where the small amounts of salt entering via rivers are concentrated in the water due to evaporation and precipitated as layers of salt (e.g. gypsum, halite).

Sand: Clastic sediment in which the grains are in the size range 0.06–2.0 mm. Consists most commonly of quartz and feldspar grains.

Sandstone: Clastic sedimentary rock formed of silicate sand grains (mainly quartz and feldspar) cemented together by a matrix of finer grained material.

Sapropel: Sediment containing organic material deposited on the bottom of a lake or an enclosed sea whose waters are free of oxygen so the organic material is preserved intact in the sediment and not broken down. Bacteria convert this material into kerogen that can become a source for oil and gas (see pp. 220–221).

Scale: The ratio between the sizes of objects depicted on a map and their real size. A map at a scale of 1:100 000 shows one kilometre in the real world as one centimetre on the map.

Scanning Electron Microscope (SEM): Instrument that uses electrons (instead of a light beam in an ordinary microscope) reflected from a surface to produce images of the surface at very high magnifications, up to 100 000 times (see p. 20).

Schist: Foliated metamorphic rock that can be cleaved easily in one direction (schistosity). Schists are most commonly formed by the metamorphism of previously sedimentary or other supracrustal rocks.

Scree: An accumulation of angular, mostly coarse rock fragments that have fallen from a steep cliff or slope and it typically forms a wedge at its base (see p. 143).

Sea-floor spreading: The theory that oceanic crust spreads apart along a central spreading zone accompanied by the upwelling of basaltic material in the rift zone. The forces causing the movement are generated by convection currents in the asthenosphere (see p. 53).

Sea-level change: See Eustatic sea-level change.

Seal (in petroleum geology): An impermeable sediment through which liquids are unable to flow. In combination with a suitable reservoir rock and structure, a seal can form a trap for oil or gas (see p. 222).

Seaward-Dipping Reflector Sequence (SDRS): Offshore sequence of layered basalts that were produced subaerially (on land), but which now occur below the sea floor due to subsidence and downwards deflection. The SDRSs produce a characteristic seaward-dipping reflection pattern on reflection seismic data, hence the name.

Section log: Diagram showing the results of measuring a one-dimensional section through a layered rock sequence. It shows the thickness of the layers, the rock types, the structures, the occurrence of various features such as fossils, minerals and sedimentological features etc. (see p. 78). Most commonly used to record sediments but can also be used to record other layered rock sequences such as volcanic successions.

Sediment: Unconsolidated material such as sand, silt, mud and ooze deposited by gravity from water or by wind.

Sedimentary basin: Also known as a depositional basin. A depression in the Earth's crust caused by subsidence in which sediments have been deposited.

Sedimentary rocks: Consolidated sediment (see p. 34). Most sedimentary rocks form thick sequences within a sedimentary basin.

Sedimentation rate: The speed at which sediment accretes vertically in a sedimentary basin. Typical magnitudes are from well under one metre per million years to several hundreds of metres per million years.

Sedimentology: The study of sediments and sedimentary rocks, their deposition and consolidation.

Seepage (of oil): Outflow of oil at the surface. The volatile components normally evaporate into the atmosphere leaving bitumin residues, often in cracks and pores, but in large amounts the bitumin can flow onto and across the surface. Seeps are a clear indication that oil has been generated in the neighbourhood (see p. 233).

Seismic section/profile: Cross-sectional diagram showing the time taken for seismic waves to descend into the subsurface, be reflected and return to the surface again. The diagram resembles a cross-section showing the structure of the upper part or the whole of the crust and upper mantle (see p. 158).

Seismic surveys: Surveys using low-amplitude, artificially-produced sound waves (seismic waves) transmitted into the Earth to study the structure and rocks of the crust and upper mantle (see p. 158–161).

Seismology: The study of the large-scale structure of the Earth's crust, mantle and core using sound waves (seismic waves) generated naturally by earthquakes or artificially by large explosions.

Sequence: A body of sediment deposited during a full cycle of sea-level change (low – high – low sea-level; see p. 130).

Sequence stratigraphy: A discipline of stratigraphy in which the sedimentary rock record is interpreted in relation to 'sequences' of sedimentary deposits which reflect cycles of changes in sea-level (see p. 130).

Series: (a) A stratigraphic unit equivalent to the chronostratigraphic unit the epoch, e.g. Upper Jurassic (see p. 69). (b) Informally, any consistent body of sediments.

Serpentinisation: Metamorphism of ultrabasic rocks reacting with hydrous fluids or water to form serpentinite, a hydrated magnesium silicate rock.

Shale: A very fine-grained (mud-grade grain size), consolidated or low-grade metamorphosed sediment that splits easily into thin sheets.

Shear zone: Zone across which rocks have moved relative to one another, but within which the movement has been ductile with little or no fracturing of the rocks (see pp. 90–91). A shear zone at depth may be a fault at shallower depths where movements take place by fracturing, sometimes causing earthquakes.

Shearing: Internal planar movements in a rock causing re-arrangement of its minerals by rotation and associated recrystallisation, so that their orientation becomes conformable to the planes of movement.

Shelf: See Continental shelf.

Shelf break: The change in angle between the sub-horizontal continental shelf and the continental slope that dips towards the ocean (see p. 173).

SHRIMP: Acronym for Sensitive High-Resolution Ion MicroProbe. An instrument that uses mass spectroscopy to determine precisely the relative proportions of elements in a small sample. Commonly used to determine accurately the ages of minerals such as zircon by measuring the proportions of radioactive elements and their products (see p. 48).

Silicates: Minerals in which silicon oxide tetrahedra (SiO_4) form the basic building blocks. They are the most common minerals in the lithosphere (examples are quartz, feldspar, pyroxene, amphibole and mica).

Siliciclastic sediments: Clastic sediments composed primarily of silicate mineral grains with little or no carbonate component, e.g. sandstones and shales.

Sill: A sub-horizontal body of igneous rock whose thickness is much smaller than its lateral extent. Sills occur most commonly as sheets intruded conformably to stratification in sedimentary successions (see p. 71).

Sillimanite: See Andalusite.

Silt: Clastic sediment with a grain size between 0.002 and 0.06 mm (see p. 64).

Siltstone: Sedimentary rock formed of consolidated clastic silt-sized sediment.

Silurian: The latest period of the Early Palaeozoic, from 444–416 million years ago (see p. 21).

Sinistral: Left-lateral. A sinistral strike-slip fault is a fault whose side farther from an observer facing the fault plane has moved leftwards (see p. 94).

Skolithos: A trace fossil found in Lower Palaeozoic, shallow-marine sandstones. It consists of vertical tubes, a few millimetres in diameter, that are probably the vertical burrows of marine worms.

Slate: A low-grade metamorphic rock of fine-grained material showing well-developed schistosity with a similar composition to shale.

Slump structures: Internal deformation (folding/contortion of layering) due to slumping and sliding that has formed underwater in unconsolidated sediment; from a few centimetres to many tens of metres in scale.

Smelter: Device used in an industrial process by which metal is extracted from its ore, normally at high temperatures so the metal is produced as a melt.

Sodalite: Silicate mineral containing sodium and chlorine. Rather rare but occurs in some of the nepheline syenites in the Gardar Province in South Greenland.

Soft-bodied animals: Animals with no hard parts. They normally leave no fossilised trace after death, but under special circumstances impressions may be left in fine-grained sediment (see p. 86).

Source rock: Organic-rich sedimentary rock from which oil and/or gas may be generated (see pp. 220–222).

Speculative seismic data: Reflection seismic data acquired and processed by a commercial seismic company without a specific customer for the data. Most oil industry seismic data are acquired for a specific customer, but when a seismic company has a ship not otherwise occupied, it may use it to acquire data to sell non-exclusively to several customers. Many of the data off southern West Greenland have been acquired in this way.

Spreading axis: The zone where oceanic crust forms and from which newly-formed oceanic crust spreads apart (see pp. 53, 171).

Spreading ridge: See Spreading axis. Oceanic crust is normally hot when it is newly formed, and stands up as a ridge around the spreading axis. For example, the spreading axis in the mid-Atlantic Ocean forms a two kilometre high submarine ridge.

Stage: The shortest division of geochronology applied to a rock sequence e.g. Oxfordian (see p. 69).

Staurolite: Metamorphic silicate mineral containing aluminium and some iron. It is often found in mica schists and, together with kyanite and sillimanite, characterises various amphibolite-facies rocks.

Stratigraphy: The study of the succession of rocks through time (see p. 69). Classical stratigraphy uses the variations in the populations of fossils in sedimentary rocks to classify and correlate them, but other types of stratigraphy are now also used, such as magnetostratigraphy, chronostratigraphy and sequence stratigraphy.

Stream sediments: Sediments found within a present-day water course. They are used to indicate what types of rock may be found upstream of the point at which they were collected (see box p. 58).

Strike-slip fault: Fault with a near-vertical plane whose opposite sides have moved nearly horizontally with respect to one another (see p. 94). The displacement of the side of the fault opposite to an observer facing the fault plane can be either to the left (sinistral) or to the right (dextral).

Strike-slip mobile belt: A zone at regional scale with deformations that include both faulting and local folding. For example, the Wandel Sea Strike-Slip Mobile Belt (see p. 120).

Stromatolites: Branched, columnar and dome-shaped, laminated sedimentary structures formed in shallow water by micro-organisms such as bacteria and single-celled algae that live as a sheet on the surface of the sediment. Fine grains adhere to this sheet and the structure grows both vertically and laterally (see p. 68). The cross-sections of stromatolites are very varied. They were common in Precambrian times, but are now rare.

Structural geology: Study of the three-dimensional structures of rocks such as folds, faults and internal mineral orientations and their development by deformation.

Sturtian: An informal name for the middle part of the Neoproterozoic from about 800 to 610 million years ago.

Sub (prefix): Meaning 'under'.

Subaerial: Taking place at the land surface under the open air (as opposed to subaqueous – under water).

Subduction: Plate tectonic process by which oceanic crust slides under another plate and descends into the asthenosphere where it is consumed (see p. 53).

Subduction zone: The zone of contact between two lithospheric plates where one of them is being subducted under the other.

Submarine canyons: Underwater canyons with steep sides caused by submarine erosion.

Sub-tropical zone: Climatic region between the tropical and temperate climate zones. Its limit to the tropics is defined where the mean temperature of the coldest month is 15–17°C; its limit towards the temperate zone is irregular because of transport of heat by currents in the sea. The mean temperature of the warmest month in the sub-tropical zone is over 20°C and of the coldest month is between 3° and 10°C.

Sulphides: Metal salts containing sulphur such as copper sulphide (Cu_2S), zinc blende/sphalerite (ZnS), galena (PbS) and pyrite (FeS_2) (see p. 197).

Supercontinent: A continent of much larger size than those of today, consisting of two or more continents of the size of today's continents. For example, Pangaea included most continents on Earth in the Triassic (see p. 123).

Supergroup: The highest lithostratigraphic rank. The successions in supergroups are usually many kilometres thick (see p. 69).

Supracrustal rocks: Rocks such as sediments and extrusive volcanic rocks formed at the Earth's surface.

Suture line: Line on the inner surface of the shell of an ammonite showing where the walls of the chambers met the external shell (see p. 124).

Sverdrup Basin: Sedimentary basin in Arctic Canada containing sediments from Carboniferous to Palaeogene age (see p. 118).

Sveconorwegian orogeny: Fold belt in southern Norway and Sweden that formed during the late Mesoproterozoic and the early Neoproterozoic. The orogeny took place at the same time as the Grenville Orogeny in the supercontinent Rodinia (see p. 99).

Syenite: Plutonic magmatic rock dominated by alkali feldspars and lacking free quartz.

Syn (prefix): Meaning 'together with'. For example, syndepositional means 'happening while deposition took place'.

Syncline: A concave-upwards fold. When exposed and eroded, the rocks at its centre are younger than those on its margins (see p. 94).

Synform: A concave upwards fold in strata whose relative ages are not known.

System: The stratigraphic unit equivalent to the chronostratigraphic unit Period, e.g. Jurassic (see p. 69).

T

Taxon (plural: taxa): In biology, a group of organisms at any level in a formal (taxonomic) system of classification. Examples are phylum, family, genus or species.

Tectonic window: Structure within a fold belt where part of an overlying thrust unit or thrust sheet has been eroded away so that the underlying rocks are exposed and visible, as through a window.

Tectonics: The branch of geology dealing with the large-scale structure and development of the outer part of the Earth. Structural geology deals with similar, smaller-scale, phenomena.

Telluric iron: Iron whose origin is within the Earth (Greek: Tellus = Earth) as opposed to iron from meteorites.

Tension: Force that pulls on a material and causes it to stretch.

Terminal moraine: Moraine, often in the form of a ridge, at the snout of a glacier (see p. 182).

Terrane: Concept used in plate tectonics to distinguish geological areas now united in a single continent. They were previously microcontinents or island arcs that moved independently of one another, each with their own geological history. Plate tectonic processes amalgamated them by continental collision (see p. 39).

Terrestrial deposits: Material deposited on land or in freshwater as opposed to in the sea. The depositing agents can be wind, glaciers, rivers or lake water.

Tesla: See NanoTesla.

Tethys Ocean: A former ocean that partly separated the supercontinents of Gondwana and Laurasia in Mesozoic time (see p. 127). Its closing caused formation of the Alpine fold belts that stretch from the European Alps through the Himalayas and beyond. Today's Mediterranean is a remnant of the Tethys Ocean.

Tetrapods: Informal name for vertebrate animals with four legs, such as amphibians, reptiles and mammals, as opposed to fish and birds and many dinosaurs. The 'four-legged fish' found in North-East Greenland (see p. 116) were the first tetrapods.

Thermal subsidence: Subsidence caused by the cooling and consequent contraction of the crust, often following a period of rifting. The subsiding crust leads to the formation of a sedimentary basin above and beyond the limits of a rift basin.

Thick-skinned tectonics: Style of deformation in the deeper parts of a fold-belt where the folding pattern is characteristically intense and plastic (see p. 95).

Thin-skinned tectonics: Style of deformation in a fold-belt involving only the upper crust so that the deformation within each thrust unit or thrust sheet is independent of those above and below it. The amount of deformation is usually moderate (see p. 95).

Tholeiite: Basalt with a small excess of free silica (SiO_2) in the form of quartz.

Thrust fault: A reverse fault with a gentle dip of 45° or less, where the block above has moved upwards along the slope of the fault.

Thrust plane: The low-angled or flat dislocation surface over which a thrust unit moves relative to the underlying rocks.

Thrust sheet/Thrust unit: Slab-shaped body of rocks pushed over other rocks along a low-angled or flat thrust plane or fault. Thrusting is a common process during the formation of mountain belts (orogenies). Thrust units can travel hundreds of kilometres across the rocks below (see p. 100).

Till: Boulder clay. Sediment consisting of an unsorted mixture of silt, sand, gravel and boulders deposited at the base of a glacier or ice sheet.

Tillite: A rock formed of consolidated till. Interpreted as evidence for pre-Quaternary ice ages. An example is the approximately 600 million-year-old tillite found in northern East Greenland (see p. 79–80).

Tonalite: Plutonic rock rich in quartz and in which plagioclase (calcium feldspar) comprises more than 90% of the total amount of feldspar.

Trace fossils: Traces left by organisms as they move or live within a sediment. Examples are footprints, tracks, burrows and excrements.

Trachyte: Volcanic rock with the same composition and mineralogy as syenite, but more fine-grained than syenite.

Transform fault: A type of transverse fault found in oceanic crust that displaces the spreading axis (see p. 177).

Transgression: Landward migration of a shoreline, due either to a rise in global sea-level or to subsidence of the land. The zone in which coastal and shallow-marine sediments are deposited moves landwards and the sediments lap onto underlying deposits, commonly with an unconformity between the two units. The opposite of regression.

Transition zone: Zone between continental and oceanic crust where continental rocks have been stretched, thinned and ruptured so the mantle has risen diapirically between the residual continental blocks. The mantle rock has reacted with the seawater and has been hydrated to serpentine.

Trap (for oil or gas): A combination of structure, reservoir and seal in the subsurface that can cause oil or gas to be trapped.

Triassic: The oldest period of the Mesozoic, from 251 to 200 million years ago (see p. 21).

Trilobites: Extinct class of invertebrates (arthropods) that lived throughout the Palaeozoic from the Early Cambrian until the Permian. Their body was segmented into three lobes (hence the name), a head, an abdomen with articulated legs, and a tail, all covered by segmented armour made of calcium carbonate and chitin (see p. 81). Most trilobites lived on the sea bed, but pelagic forms also existed.

Tropical belt: Climate zone around the equator where the mean temperature in its coldest month is not under 15–17°C, i.e. there is no winter.

Trough: Elongated depression in the sea bed, of kilometre-scale or up to hundreds of kilometres long, usually with moderately-inclined margins (see p. 88).

Tuff: Consolidated volcanic ash.

Tugtupite: Red semi-precious stone (see p. 59) found in the Gardar Province. It is Greenland's national gemstone.

Tungsten: Metal element (W), in German known as wolfram, with chemical properties similar to chromium and molybdenum. Occurs in the mineral wolframite $(Mn,Fe)WO_4$.

Turbidite: Clastic sediment deposited by a waning turbidity current so that the grains in the lower part of the bed are much coarser than those higher up and the finest grains are at the top; creating graded bedding (see pp. 47, 87).

Turbidity current: Chaotic flow within the sea or a lake consisting of a mixture of sediment and water that moves so rapidly downslope that it lifts off the bottom and becomes a turbulent cloud. As the speed of the flow slows, the coarser particles settle out first followed by finer and finer grains, forming a turbidite. The flow originates from the slump or slide of unconsolidated deposits on a sloping sea or lake bed.

U

Ultrabasic rocks: Magmatic rocks, very poor in silica and consisting of dark minerals such as olivine, pyroxene and magnetite. Also known as ultramafic rocks.

Ultramafic rocks: See Ultrabasic rocks.

Unconformity: Surface within a stratigraphic succession representing a time gap during which formerly existing strata have been removed by erosion (see p. 84). Where the strata below and above the surface are oriented differently, the boundary is called an angular unconformity.

Uplift: Movement of a segment of crust vertically upwards relative to sea level. An example is the isostatic uplift after formation of a mountain belt (see p. 27). Sea level is not constant, however, but varies as the total amount of water in the sea changes, for example by variations in the size of ice caps. These eustatic changes have to be taken into account when calculating the amount of uplift and subsidence.

V

Varangian*: An informal name for an interval of the Vendian during Neoproterozoic time from about 610 to 570 million years ago, during which much of the world, including Scandinavia and Greenland was glaciated (The Snowball Earth).

Note in proof: In recent stratigraphic revisions the Varangian is now broadly equivalent to the Marinoan glacial event with a revised age of about 650–630 Ma.

Variscan: Period of mountain building and folding during the Carboniferous and Permian in what is today central and north-western Europe and northwest Africa, and in the Appalachians of eastern North America where it is known as the Acadian. The mountain-building was caused by the collision of Gondwana with Laurasia to form Pangaea. It is also known as the Hercynian in Europe.

Vein: One of various forms of thin sheet of rock, of short lateral extent, implaced into a foreign rock. The rock in the vein is different from that in the host rock, e.g. granitic veins in a gneiss (see p. 31) or mineralised fillings of cracks.

Vendian†: Period during the latest Neoproterozic between 610 and 542 million years ago. The Vendian is divided into the Varangian, when large parts of the Earth, including Greenland were glaciated (about 610–570 million years ago) and the Ediacaran (about 570–542 million years ago).

†Note in proof: In recent stratigraphic revisions the Vendian is now about 650–542 Ma, broadly equivalent to the Marinoan glacial event (650–630 Ma) plus the subsequent Ediacaran period (630–542 Ma).

Vergence: The horizontal direction in which a fold is inclined or overturned. This corresponds to the direction towards which the material has been transported during thrusting (see p. 105).

Vertebrate: An animal with a backbone.

Vibroseis: Method of producing seismic waves on land in which a large plate on the ground is vibrated using a sequence of pre-determined frequencies. The resulting waves travel into the ground and the waves refracted or reflected there return to the surface to be received by geophones and recorded (see p. 230).

Viscosity: The resistance of a fluid to deformation. In geology, low viscosity magmas flow easily, while high viscosity magmas are stiff. Acid magmas are more viscous than basic magmas. Viscosity falls as temperature rises.

Vitrinite: Material found in hydrocarbon source rocks and in coals whose bright vitreous lustre varies with the temperature to which the vitrinite has been subjected. Vitrinite reflectance is used in petroleum geology to determine the maturity of an oil source rock (see pp. 221–223).

Volatile components: Water as steam, carbon dioxide, hydrogen sulphide and other gases in a magma.

Volcanic province: A region of volcanic rocks that were erupted during a distinct period of time e.g. the Palaeogene volcanic province in East Greenland (see p. 146).

Volcanic rock: Rock formed by the cooling and solidification of liquid magma at or near the surface. Because the cooling takes place rapidly, the rock is fine-grained.

W

Wackestone: A carbonate sediment consisting of at least 10% carbonate grains (shells, ooids etc.) dispersed within a lime mud matrix (see p. 64).

Wandel Sea Basin: Carboniferous to Eocene sedimentary basin with Upper Cretaceous volcanics rocks in eastern North Greenland (see pp. 118–121).

Wandel Sea Strike-Slip Mobile Belt: Zone of deformation, including both faulting and folding, that formed in eastern North Greenland (see p. 119) during the Late Jurassic – Early Palaeogene.

Warm water sediments: Sediments deposited in tropical or subtropical conditions, including evaporites such as halite or rock salt.

Wave ripples: Small-scale wavy bedforms often preserved on bedding planes. They form in water-lain sands/silts by the action of waves in shallow water when the sediment is unconsolidated. The ripple pattern is generally symmetric, in contrast to current ripples that are asymmetric.

Weathering: A process occurring at the Earth's surface that breaks up and corrodes the rocks due to the influence of weather. The surface layer of a rock is modified either mechanically or chemically so that its structure, composition, colour, cohesion and shape change. Solid rocks break down to loose grains that can be transported away by wind or water. The weathering zone is a few millimetres to a few metres thick in temperate or arctic regions, but can be many tens of metres thick in the tropics.

Weichselian: The last glacial in Europe lasting from 115 000 to about 11 700 years ago. The equivalent to the Wisconsian in North America.

Welded tuff: Also known as ignimbrite. Volcanic rock formed from a hot stream of ash that cools and becomes welded to form a coherent tuffaceous rock.

Well: Term used within the oil industry for a drillhole or borehole.

Wilson cycle: The plate tectonic cycle that starts with continental break-up, proceeds through the formation of ocean crust by sea-floor spreading followed by subduction, formation of island arcs, continental collision and mountain building. The total cycle takes several hundred million years to complete and can repeat several times in the same region.

Window (oil or gas window in petroleum geology): The temperature interval between about 80°C and 245°C within which oil and gas is generated from its source rock (see p. 221).

Wisconsian: See Weichselian.

X, Y, Z

Xenolith: Fragment of foreign rock within a magmatic rock, often from the wall of the magma chamber or from rocks adjacent to the feeder channels. Xenoliths can be from a few centimetres to several kilometres in size.

Younger Dryas: A brief period of cold climate 12 700–11 700 years ago at the end of the Pleistocene and preceding the Boreal period of the Holocene. It is named from the plant 'Dryas' characteristic of the flora in high arctic regions. Deposits from the Younger Dryas period are poorly represented in Greenland.

Zechstein Sea: Sea area in northern Europe during the Permian, roughly where the North Sea is today. Sediments deposited in the Zechstein Sea are typically evaporites including large amounts of salt (halite).

Zircon: An important constituent of many magmatic rocks with the composition $ZrSiO_4$. Commonly red in colour. Used in determinations of absolute ages (see p. 48) by measuring the proportions of uranium and lead isotopes present in the mineral.

ACRONYMS – for organisations and projects

(The) Danish Expeditions to East Greenland: A series of large geological and other scientific expeditions to northern East Greenland in the period 1926–1958. Led by Lauge Koch.

DGU: Danish acronym for the Geological Survey of Denmark that was responsible for geological surveys of Denmark and the Faroe Islands. It was united in 1995 with the Geological Survey of Greenland (GGU) to form the Geological Survey of Denmark and Greenland (GEUS).

DLC: Acronym for the Danish Lithosphere Centre. A Danish Research Institute from 1995 to 2005, associated with GEUS and the Geological Institute of Copenhagen University.

DONG Energy: The largest Danish energy company, presently engaged with oil and gas exploration and production activities in north-west Europe and Greenland.

DSDP: Acronym for Deep Sea Drilling Project. An international project from 1964 to 1984 that investigated the deep ocean basins by drilling, normally coring. Its drilling ship was 'Glomar Challenger' (see pp. 164–165).

GEUS: Danish acronym for the Geological Survey of Denmark and Greenland.

GGU: Danish acronym for the Geological Survey of Greenland. It was united in 1995 with the Geological Survey of Denmark (DGU) to form the Geological Survey of Denmark and Greenland (GEUS).

GISP: Acronym for the Greenland Ice Sheet Program. A joint American-Danish-Swiss consortium that drilled a 3058 m ice core through Greenland's Inland Ice in 1991–1993 (see p. 191).

GRIP: Acronym for the Greenland Ice Core Project. A European consortium that drilled a 3028 m ice core through the Greenland Inland Ice close to its highest point (Summit) in 1990–1992 (see p. 191).

IODP: Acronym for the Integrated Ocean Drilling Program. An international project that began in 2000. It operates the 'JOIDES Resolution', and from 2007 the 'Chikyu', a drilling ship that uses a riser, as well as ships or other vehicles specific to each mission. The main aim of the programme is to continue and extend the DSDP and ODP investigations of deep ocean areas by recovering cores of sediments and crystalline rocks of the lithospheric plates (see pp. 164–165).

KANUMAS project: Acronym for Kalaallit Nunaat Marine Seismic Project. 'Kalaallit Nunaat' is the Greenlandic name for Greenland. Industry-financed project that acquired reflection seismic data off North-East and North-West Greenland during the early 1990s (see p. 241).

NGRIP: Acronym for North Greenland Ice Core Project. A multinational research programme that drilled a 3080 m long ice core through Greenland's Inland Ice at 75°N. The project began in 1999 and was completed in 2003 (see p. 191).

Nunaoil A/S: Greenland's National Oil Company. Founded in 1985 and owned equally by the government of Greenland and the Danish National Energy Company, DONG. By law, Nunaoil has a share in all hydrocarbon licences in Greenland.

ODP: Acronym for the Ocean Drilling Program. An international project that continued the investigations of the deep ocean begun by DSDP. ODP started in 1984 and continued until early 2000 using the drillship 'JOIDES Resolution'. This program was succeded by IODP (see pp. 164–165).

SUBJECT INDEX

A

Aasiaat **8**, 42, 239
absolute age determination 48
abyss 173
abyssal ooze 130
Acanthostega 116
acid rocks **141**, 152
acritarchs 13, **73**
Aegir Ridge **171**, 172
aegirine 59
aeolian deposits 65
aeromagnetic surveying 45, **159**
African Plate 53
Age (stratigraphic) 69
agglomerate 121
AGIP (oil company) 230
Agpat, marble quarry 196
airguns **160**, 241
Akia terrane 12, **36**, 37, 197
Akulleq terrane **36**, 37
Alert 5
algal structures/constructions 13, **73**, 83
alkali feldspar **141**, **152**
alkali granite 57
allochthonous sediments 64
alluvial deposits 65
Alpha Ridge **118**, 167
Alpine Orogeny 93, **111**
aluminium 197
amazonite 59
Ameralik 36
Ameralik dykes 37
Amîtsoq 211
Amîtsoq gneisses 12, **36**–**38**
Ammassalik **8**, 211
ammonites 15, **125**, 127, 129, 131, 132
Amoco (oil company) 240
amphibole 152
amphibolite **31**, 32, 34
amphibolite facies **35**, 106
Amundsen Bassin 167
Amundsen Land Group 107
Anap Nunaa 41
andalusite 35
andesite 141
Andrée Land 8
Andrée Land Group **77**, 79
angular uncomformity 49
anomaly map 159
anorthosite complex **37**–**39**, 195, 212
anoxic 220
Antarctic Sund 92
Antarctic Plate 53

antiformal folds 93, **94**, 222
antimony **202**, 211
apatite 203
Appalachians **96**, 110
Appat 211
Appat/Kitsisut sediments 175
Aptien 132
Aquitane (oil company) 240
aragonite 125
Archaean 5, 12, 21, **30**, 36, 43
Archaean basement 5, 30, **36**, 39, 56, 98
Archean orogenies 12, **30**, 97, 103
Archaean sediments 40
ARCO (Atlantic Richfield Company, oil company) 229, **230**
argentite **197**, **203**
arkose 64
Arsuk Fjord 205
Arveprinsen Ejland 41
asthenosphere **20**, 27, 44, 53
Ataa, Ataa Sund 41
Atâ granite 40
Atammik Structural Complex 232
Atlantic Richfield Company (ARCO) 229, **230**
Atlantic Province 81
augen gneiss 34
augen granite 102
augite syenite 57
Australian Plate 53
autochthonous sediments 64
azurite 215

B

Baffin Bay **5**, 8, 174, 175, 178, 236
Baffin Bay Group 73
Baffin Bay shelf 174
Baffin Bay crust 174, **178**
Baltica 14, 15, **81**, **96**, 99, 117
Banded iron formation 12
banded iron ore **195**, 210
Barents Sea Platform **118**, 228
Barents Sea shelf 118, **167**, 168, 171, 228
barrel of oil 225
barytes **202**, 211
basalt magma 72, **152**
basaltic rocks 34, 141, **152**, 176, 202
basement rocks 5, 26, **30**
basement shield 5, 97
basement, older 12, 21, **36**
basement, palaeoproterozoic 40
basement, younger 13, 21, **40**
basic dykes 97

basic rocks **141**, 152
basin analysis 78
basin deposition 62
Bastion Formation 78
batholith 50
bauxite **197**, 202
bedding 67
beech 131
belemnites **129**, 132
benthonic fauna (fossils) 84
beryllium **197**, **202**, 204, 211, 213
Bessel Fjord 74
biogenic sediments 62–64
biostratigraphical division 89
biostratigraphy 69, 89, 125
bioturbation 68
biozone 69
bitumen 227, **233**
bivalves 81, 131, **132**, 185
Bivrost Fracture Zone 168
Black Angel zinc-lead occurrence 207, **208**, 209
Black Sea 220
blocky lava 143
Blosseville Kyst **8**, 24, 147, 230
Blosseville Kyst Bassin 170
blow-out 163
blue mussel (Mytilus edulis) 186
blue-green algae 68
body fossils 81
Boreal Period 188
boreholes, geophysical and geological studies 163
Borgtinderne 149
bornite **197**, 202
Bothriolepis 14, **115**
boudinage 33, 38, **44**
boulder clay 183
boundstone 64
Boyd Bastion fault 100
BP (British Petroleum, oil company) 240, 241
brachiopods 78, **81**, 129
braided river system 65
Bredefjord 8
Brønlund Fjord 63
Brønlund Fjord Group 63, **86**
bryozoan limestone 65
bryozoans **81**, 129
Buen Formation **86**, 107
Buksefjorden 36
Burgess Shale fauna 86
burrows 68, 73
butanes 222

C

calcite 197, **202**
Caledonian deformation 75, 77, 90–95, **96**, 97-103
Caledonian fold belt 5, 14, 90–95, **96**, 97-103, 156, 215
Caledonian fold belt, chronology 103
Caledonian foreland 100
Caledonian granites 100–**102**
'Cambrian explosion' 89
Cambrian **14**, 21, 77
Camp Century 191
Canadian Basin **118**, 167
canyons 179
carbon (element) 38, 202, **220**
carbonate debris cones 87
carbonate shelf 86
carbonatite 203
Carboniferous (period) **14**, 21
Carlsberg Fjord 122
casing of a drill hole 163
celestite 211
Celsius Bjerg Group **114**, 115
cement in sediments 64
Cenozoic 15, 21, **169**
Centrumsø 84
cephalopods 125
chalcocite 202
chalcopyrite **197**, 202
channels 68
Charcot Bugt Formation 108
chemical sediments 62, 63
chert 64, 88
Chevron (oil company) 240
chitinozoans 165
chlorite schist 35
chrome ore 210
chromite **197**, **202**, 210, 211, 212
chromium **197**, **202**, 204, 211
chron 113
chronology, in the basement 49
chronostratigraphy 69
Cities Services (oil company) 240
Citronen Fjord 211, **215**
clastic (detrital) sediments 62, **64**, 65, 76
Clavering Ø **74**, 211
clay 64
climate changes during the Ice Age **184**, 189
climate zones 79, 111, **112**
climatic conditions 79, **110**, 184
clinopyroxene 152
coal 126, 197, 200, **202**, **209**, 211, 220

SUBJECT INDEX

coal mine   204, **209**
cobbles   64
coccolithophoridae   165
collision zone   43
columnar jointed basalt   140, **153**
compression   94
compressional faults   174
concessions (minerals)   201
concordia diagram   76
condensate   222
conglomerate   34, 63, **64**
conodonts   165
Constable Pynt   230
constructed profiles   107
constructive plate boundaries   53
continent   27, **53**
continental basins   70
continental collision   44
continental crust   **20**, 53, 156, 157
continental deposits   71
continental drift   **21**, 111, 112
continental margin   151, **173**
continental sediments   **65**, 114, 117
continental shelf   5, 156, 157, 172, **173**
continental slope   85, **173**
convection cell   53, **139**
convection current   **20**, 53, 149–151, 169
convergent plate-tectonic margins   85
convoluted lamination   68
copper   197, 200, **202**, 204, 211, 215
coral   **88**, 129
coral limestones   65, 119, 122
core of the Earth   **20**, 139
corundum   **195**, 197, 203
craton   13
Cretaceous (period)   **15**, 21
crinoid   78
cross-bedding   68, 79, 132, **134**
crude oil   220
crust of the Earth   **20**, 152, 169
crust, plutonic rocks   152
crustaceans   132
cryolite   57, 59, 197, **202**–206, 211
Cryolite Mine   **205**, 206, 211
crystalline basement   30
cumulus   150
cuprite   197
Curie temperature   113
current ripples   **67**, 75, 79
cuttings from drilling   162
cyanophyta   68
cyclic sedimentary sequence   67

D

Dan Field   224
Dana (survey ship)   154
Daneborg   **5**, 8
Danell Fjord   52
Danian stage   131
Danish Expeditions to East Greenland   20
Danish Geological Survey (DGU)   9
Danmark Fjord   8
Danmarkshavn   8, 74
Danmarkshavn Basin   156, **167**–170, 229
Danmarkshavn High   **168**, 169, 229
Dansk Olie og Naturgas (DONG Energy)   **225**, 240
Davis Strait   5, 8, **174**, 175, 178
dawn redwood   131
debris cones   86
declination, magnetic   113
Deep Sea Drilling Project (DSDP)   164
deep-sea drillings   **164**, 165
deep-water boreholes   165
deep wells   174
deep-water sediment   85
deep-water trough   **88**, 105
deltaic deposits   62, **65**, 129, 130, 131
Deminex (oil company)   240
Denmark Strait   **5**, 8, 149, 165, 173
depositional structures   68
desert conditions   **112**, 124
desert sandstone   122
desiccation cracks   **68**, 79
destructive plate margins   53
detritus   76
development licence (oil/gas)   238
Devonian   14, 21
Devonian basin   100, **114**–117
diagenesis   66
diamond   194, 197, 200, 202, 210–**214**
diapir   168
diatoms   **163**, 165
dinoflagelates   **164**, 165
dinosaurs   15, **122**
diorite   152
Disko Bugt area   5, 8, 41, 42, 129, 135, 138–**143**–145, 176, 209, 232, 233
Dohrn Bank   173
dolerite   44
dolomitic limestone   66
DONG (DONG Energy)   **225**, 239, 240
Dove Bugt   8
dredge samples   178
drill bit, - cores, - pipes   162
drill ships   **161**, 164
drilling fluid   163
drilling in ice   21
drilling mud   161, **162**, 163
drilling tower   **161**, 164, 218, 227, 234
drillings, deep   161, **162**, 164, 227
drillings into the Inland Ice   191
Dronning Louise Land   8
dropstone   173
dry gas   **221**, 223

'dry wells'   234
ductile deformation   33
Dundas Group   73
dunite   197
Dye 3   191
dyke complex in oceanic crust   169
dyke swarms   151
dykes   **151**, 152
dynamic conditions in the mantle   20
dynamic positioning system   162, **164**

E

Earth climate   110
Earth core   **20**, 139
Earth crust and mantle   **20**, 139, 152
Earth's average temperature   **110**, 184
Earth's geological evolution   21
Earth's magnetism   113
Earth's oldest rocks   37
East Greenland   8
East Greenland younger basins   110, 111
East Greenland, oil geology   228
eclogite facies   **35**, 101
economic geology   21
economic zone   19, **157**
Ediacaran   79
Eemian   **183**, 186, 187
Ekofisk Field   224
electrical resistance measurements   163
electron microscope photo   163
Eleonore Bay Basin   13, **74**
Eleonore Bay Supergroup   67, 74, **77**, 79, 92, 98, 100–103
Eleonore Sø deposits   **100**, 101
Eleonore Sø window   100
Ella Ø   78, **92**
Ella Ø Formation   78
Ellesmere Island   **5**, 72, 167
Ellesmerian deformation   105
Ellesmerian fold belt   5, 92, 95, **104**, 106, 107, 156, 215
Ellesmerian orogeny   14, 82, **104**
EnCana licence   239
end-moraines (terminal moraines)   **183**, 186, 188
Eoarchaean   **12**, 21
Eocene   169
eolian sediments   115
Eon (stratigraphic)   69
Eonothem (stratigraphic)   69
epi-continental basins   71
Epoch (stratigraphic)   69
Eqalussuit   211
Eqi   211
Era (stratigraphic)   69
Erathem (stratigraphic)   69
erosion   64

erosion products   76
erosion, glacial   183
erosional structures   **68**, 179
escarpment   86, 87
Etah Group   49
ethane   222
eudialyte   203
Euramerica   14, **103**
Eurasian Basin   118
Eurasian Plate   53
Eurekan deformation   **106**, 107, 119
Eurekan Fold Belt   119, 121, 168
eustatic sea-level changes   190
evaporitic deposits   **64**, 122, 124, 170
exclusive economic zone   **157**, 167
exhumation   103
exploration licence (oil/gas)   237, **238**
exploration, minerals   194, **201**, 210
exploration (olie/gas)   **238**, 239
extension   103
extensional fault   174
extrusive rocks   34
Exxon (oil company)   241
eye structures in gneiss   33, **34**

F

Faroes (Faroe Islands)   134, **148**
fault blocks   177
fault scarp (structure)   107
faulting offshore   177
faults   56, **94**, 177
faunal assemblages   125
faunal provinces   125
feeder dykes and channels   72
Fina (oil company)   240
Fiskefjord   36
Fiskefjord fault   36
Fiskenæsset anorthosite complex   12, **38**
Fiskenæsset area   8, **38**, 195, 210–212
fission track   179
fissure eruptions   15
Fjord region fault   100
'Fjord Zone Basin'   **76**, 97, 101
Flammefjeld   211
'flat spots'   234
flexuring of the continental margin   151
flood basalts   72
flotation   208
flutes   68
fluvioglacial (meltwater) sediments   65, 183, **185**, 186
fold belts   92
fold phase   39
folds; compressed, overturned   94
foraminifera   **131**, 163, 165
foraminiferal limestones   65
foreland   100

SUBJECT INDEX

foresets   145
Formation (stratigraphic)   69
fossil fuel   220
fossils   78, **89**, 131
'four-legged fish'   115, **116**
foyaite   57
fractional crystallisation   152
fracture zone   56
Fram Strait   172
Franklinian Basin   14, **82**-88, 92, 103, 104, 226, 227
Franz Joseph thrust sheet   100, **101**
Franz Joseph unit   100, **101**, 102
Frederick E. Hyde Fjord   105
Frederik VII Mine   211
Frederikshåb Isblink   8
Frigg Fjord Mudstone   107
froth-flotation   208
Fugro-Geoteam   240
Fulmar field   224
Fylla area   232, **240**
Fylla fault complex   175
Fylla licence   239
Fylla Structural Complex   **232**, 234, 240
Fyns Sø Formation   **83**, 84

G

gabbro   34, **141**, 152, 169
gabbro anorthosite   38
Gakkel Ridge   **118**, 167, 172
gal   161
galena   **197**, 199, 200, 202, 206, 207
gamma radiation   162
Gardar dykes   56
Gardar intrusions   5, **56**
Gardar Province   18, 30, 54, **56**, 58, 59
Gardar Province, mineralisations   213
Gardar rift zone   13, **56**
Gardiner intrusion   211
garnet   141
gas, natural   220, **221**–223
Gåseland   24
gastropods (snails)   81, **131**
gas-window   223
Gauss Bank   173
geochemical map   52, **58**
geochemical provinces   58
geochemistry   21
geochronology   21
geographic poles   113
Geographical Society Ø   117
geological map of Greenland   5
geological mapping   18, **21**, 92
Geological Survey of Denmark and Greenland (GEUS)   9, **20**
Geological Survey of Greenland (GGU)   9, **20**

geological terrane   39
geophone   161, **230**, 231
geophysical investigations   **21**, 159, 163
geophysical surveys   158, **159**
geophysics   21
geothermal gradient   **35**, 223
GEUS   9, **20**
GGU   9, **20**
'ghost' structures   33
glacial deposit   15, 65, 79, **182**–189
glacial erosion   183
glacials   **182**-185
glaciation   15, 181, **182**
glaciology   21
glauconite   78
global sea level   190
global temperature   110
Glomar Challenger (drill ship)   164
gneisses   **31**, 34, 202
Godhavn (Qeqertarsuaq)   143, 233
Godthåb (Nuuk)   36
Godthåbsfjord   8, 26, 30, **36**
gold   52, 195–197, 200, 202, 204, 210, 211, **212**, 213, 215, 217
goldmine   210-**212**
Gondwana   14, 83, 96, 97, **103**, 110, 111, 117 123, 127
gossan   **199**, 215
graben   177
grade of ore   **200**, 201
grain shapes   76
grains (in sediments)   **64**, 76
grainstone   64
Grandjean Fjord   96
granite   34, 102, 141, 152, 197, 198, **202**
granite gravestone   198
granules   64
granulite facies   35
graphite   **195**, 197, 200, 202, 204, 211, 214
graptolites   85, **89**
gravity acceleration   161
gravity anomalies offshore West Greenland   232
gravity investigations   **161**, 232
Greenland Fracture Zone   168, **171**
Greenland Ice Core Project (GRIP)   191
Greenland Ice Sheet Program (GISP)   191
Greenland Sea   5, **8**, 156
Greenland shield   30
Greenland, size   5, 18, **19**
Greenland, size of shelf area   156
Greenland's climate in geological time   112
Greenland's national gemstone   59
greenschist facies   **35**, 105
greenstone belt   46, 47
Grenville fold belt   **99**, 103

Grenvillian orogenic event   13, **97**
Grepco (oil company)   240
greywacke   64
GRO#3 (well)   218, **221**, 240
Grønnedal   56, **211**
Grønnedal syenite   56
grooves   68
Group (stratigraphic)   69
grønArctic (oil company)   **218**, 240
Gulf (oil company)   240
Gunnbjørn Fjeld   **8**, 24, 147
Gymnosperms   15
gypsum   63, **64**

H

Hadean   **12**, 21
Hagar Bjerg thrust sheet   **100**–102
Hagar Bjerg unit   100, **101**, 102
Hagen Fjord   8
Hagen Fjord Basin   13, **82**
Hagen Fjord Group   83
half-graben   82, 126, 174, **177**
'half-life'   48
Hall Bredning   230
Hall Land   **8**, 87, 88, 185
halysites   88
Hammer Dal   211
Harder Fjord Fault Zone   118, **119**
Hareøen   **129**, 233
harzburgite   169
heat conductivity   223
heat flow   223
heavy minerals   **196**, 216
Hecla High   175, 176, **232**, 237
Hekla Sund   84
Hekla Sund Basin   13, 77, **82**, 83
Hellefisk-1 well   **237**, 240
hematite   57, 113, **197**, 202
hiatus   67, 79, 83, **84**, 114
high stand (in sequence stratigraphy)   130
Himalayan mountain chain   93
hinge lines   94
Hispanoil (oil company)   240
Hochstetter Forland   8
Hold with Hope   **8**, 146, 149
Holm Land   **8**, 119
Holocene   169, **182**–190
Holocene temperature maximum   189
horst   177
hot spot   112, 138, 140, 151, 152, 173
hot-water drill   21
Hovgaard Fracture Zone   168
Hudbay (oil company)   240
Hudson Fault Zone   176
Humboldt Gletscher   8
Hurry Inlet   135

hyaloclastites   128, 133, 142, 144, **145**, 147, 233
hydrocarbons   **220**, 222, 223, 226, 232
hydrogen   220
hydrophone   161
hydrothermal fluids   52
Hyolithus Creek Formationen   78

I

Iapetus Ocean   14, 81, 82, **83**, 96, 99, 103
ice age   15, 24, **182**
Ice Age, the   182
ice caps   **182**, 183
ice core drillings   191
ice cores   191
ice rafted debris   184
ice sheet   **182**, 183
ice shelf   15, **183**
ice-age deposit   79, 84
ice-margin deposits   186
Ichthyostega   116
Igaliko Sandstone   **57**, 195
Igaliku   **8**, 56
ignimbrite   **34**, 121
Ikeg   8
Ikermiut Basin   156, **232**, 237
Ikermiut Fault Zone   237
Ikermiut-1 well   **237**, 240
Ikersuaq   8
Ikka   211
Ikkattoq gneisses   37
Ilimaussaq intrusion   **56**–59, 213
Ilivertalik granite   37
Illimmaasaq   211
Illinoian   186
Illoqqortoormiut (Scoresbysund)   5, **8**, 230
Illorsuit   211
Illukunnguaq   **143**, 211
ilmenite   203
Ilorput   205
Ilulissat (Jakobshavn)   5, **8**, 129, 196, 233
Imilik   149
immature sediments   76
impermeable seal (oil/gas)   224
inclination, magnetic   113
Independence Fjord   **8**, 70, 71
Independence Fjord Basin   13, **70**
Independence Fjord Group   13, 70, **71**, 72, 82, 84, 88, 103
Inglefield Land   5, **8**, 30, 49, 211
Inglefield Land fold belt   5, **30**, 46, 48
Ingolf Fjord   70, **84**
Inland Ice   8, 19, **25**, 182, 186–188, 191
Inoceramus   129, **131**
Integrated Ocean Drilling Program (IODP)   164
intercumulus   150

interglacials **182**–185
intra-continental basins **71**, 85, 117
intra-montane basins 115
intrusion 149
intrusive rocks **141**, 149
invertebrates 89
Irminger Basin 165
iron 197, 200, **202**, 204, 210, 211
iron pyrites 200, **202**, 203, 206, 215
island arc 44, **45**, 50
isoclinal fold 96
isomorphous series 152
isopach map 158
isostasy **27**, 103, 182, 187, 189, 190
isotopes 48
Isua 211
Isua sedimentary and volcanic rocks 12, **37**, 38, 195, 210
Isukasia 8, **36**-38, 211
Itilli Fault 237
Itilliarsuk 211
Itilliup Qeqertaa 211
Ivigtut granite 57, **205**
Ivisaartoq **36**, 211
Ivittuut **8**, 56, 59, 204, 205, 211
Ivigtut cryolite 59, **205**
Ivittuut region 51

J

J.P. Koch Fjord 8
Jakobshavn (Ilulissat) **8**, 233
Jameson Land **8**, 24, 110, 111, 229, 230
Jameson Land Basin 15, 111, 112, 114, **123**-127, 168, 170, 226, 229–231
Jameson Land, oil exploration 229–**231**
Jan Mayen 166, **171**, 172
Jan Mayen Fracture Zone 168, **171**, 172
Jan Mayen microcontinent **171**, 228
Japan National Oil Company 241
Jarners Kulmine (coal mine) 211
'jigsaw method' 112
JOIDES Resolution (drill ship) 164
joints 153
Jøkelbugten 8
Jørgen Brønlund Fjord 63
Josva Minen 211
Julianehåb (Qaqortoq) 8
Julianehåb batholith **50**–52, 56, 198
Jurassic **15**, 21
juvenile crustal material 98

K

kakortokite **57**, 58
Kangâmiut dykes **42**–**45**
Kangâmiut High 235
Kangâmiut well **235**, 237, 240
Kangeq sediments 175
Kangerluluk 52, **211**
Kangerlussuaq (East Greenland) **8**, 138, 146, 148, 149, 173, 217
Kangerlussuaq (Søndre Strømfjord) **8**, 42, 45, 211, 239
Kangerlussuaq Basin (East Greenland) 15, 111, 112, **132**, 133, 135
Kangaamiut Basin **156**, 237
KANUMAS project 239–**241**
Kap Ammen 182
Kap Cannon thrust 121
Kap Edvard Holm 149, 211
Kap Farvel (Nuna Isua) **8**, 52
Kap Graah Group **114**, 117
Kap Gustav Holm **8**, 149
Kap København 135, 183, **185**–**187**
Kap Kolthoff Group **114**, 117
Kap Morris Jesup **8**, 121
Kap Parry 149
Kap Simpson 149
Kap Washington 8, **119**, 121
Kap Washington volcanics 119, **121**
Kara Platform 167
Karrat Group **46**–48
Karrat Isfjord 46
karst **68**, 83
Karstryggen 211
Kejser Franz Joseph Fjord 8
Kempe Fjord 92
kerogen 221
Ketilidian basement 56
Ketilidian fold belt 5, 13, 18, 30, 31, **50**–52
Ketilidian sedimentary and volcanic rocks 50
Kialineq 149
Kilen 8, **119**, 121
kimberlite 213, **214**
Kimmeridge clay 228
King Frederik IX 198
Kirkespirdalen 192, 195, **212**, 213
kitchens (oil/gas) **222**, 240
Kivioq Basin **156**, 175
Kivioq Ridge 175
klippe (geological structure) 94
Kløftelv Formation 78
Knipovich Ridge 171
Kobbefjord fault 36
Kobberminebugt 59
Kolbeinsey Ridge **171**, 172
Koldewey Platform **168**, 229
Kong Oscar Fjord **8**, 23, 74
Kong Oscar Fjord Basin 14, **74**
Kong Oscar Fjord Group 74, **77**, 78, 80, 81, 92, 103
Kronprins Christian Land **8**, 70, 82, 83, 84, 87, 95, 100, 119, 121, 167

Kronprins Christian Land Orogeny 121
Krummedal Basin 13, **74**, 75
Krummedal supracrustal sequence 12, 13, **74**–**76**, 97–103
Kryolitselskabet Øresund 206
Kvanefjeldet 59, **213**
Køge Bugt 8

L

Labrador Sea **5**, 8, 138, 174, 175, 178, 232
lacustrine deposits 65
Lady Franklin Basin **156**, 175, 232, 237
lake sediments 62
Lambert Land 8
lamination 68
land level changes 190
landscapes 24
Langø 211
lanthanides 203
lanthanum 197
last glacial **186**–188
last glacial maximum 187
Late Archaean basement 37
lateral moraines 183
Latra Bank 173
Laurasia 15, 123, *127*
Laurentia 14, 80, 81, **83**, 96, 97, 99, 103, 117
lava 140
lava flows 72, **140**, 141, 147
Layer (stratigraphic) 69
layered gabbro 152
lead 196–**202**–212, 215, 216
lead isotopes 48
lead-zinc mineralisation 215
lherzolite 169
licence areas (oil/gas) 224, 230, 237–239, **241**
licences 201
licences (oil/gas) **238**–240
licensing rounds (oil/gas) 237, **238**
Liloise 149
limestone 63–**66**, 197
limestone reef 86
limpet snail 131
Lincoln Sea **5**, 8, 227
Lincoln Sea Basin **156**, 167
Lindenow Fjord **8**, 52
lithium 213
lithosphere **20**, 27, 44, 53
lithospheric crust and mantle **27**, 44
lithospheric plates 53
lithostratigraphy 69
Little Ice Age 189
Liverpool Land 230
Liverpool Land Basin 156, **168**, 170, 229
Liverpool Land Ridge 229

lizzard 15
Llandovery 89
load structures **68**, 79
lode stone 197
Lomonosov Ridge **118**, 167
low stand (in sequence stratigraphy) 130
lujavrite 57
Lyell Land Group **77**, 79
Lyngby town hall 196

M

Ma (mega-annum) = million years 21
Maarmorilik 8, 46, 206, *207*, 208, 211
Maarmorilik lead zinc deposit 200, 204 – 206, **207**, 208, 210–212
Maarmorilik marble **196**, 208
Maarmorilik Mine 207, **208**, 211
Målebjerg 99
Målebjerg window 100
macrofossils 65, **89**
magma **141**, 152
magma chamber 56, **72**, 149, 152, 169
magmatic (plutonic) rocks 34
magmatic crystallisation 152
magmatic differentiation 140, 149, **152**
magmatic layering 58, 149, **150**
magnetic anomaly **45**, 159
magnetic intensity **45**, 159
magnetic minerals 113
magnetic polarity 113
magnetic poles 113
magnetic storms 160
magnetic stripes 113
magnetic surveying **45**, 159
magnetic polarity pattern (oceanic crust) 113, **171**
magnetite 113, 197, **202**, 210
magnetometer **45**, 113
Majuagaa 211
Makarov Bassin 167
malachite 215
Mallemuk Mountain Group 119
Malmbjerg 211, **217**
mammals 15
'Mammoth Field' (oil/gas) 224
Maniitsoq (Sukkertoppen) 232, **239**
Maniitsoq Bassin 237
Maniitsoq High **232**, 237
mantle of the Earth **20**, 139, 152
mantle parent rock **141**, 152
mantle plume 112, **138**–140, 165, 171, 173
mapping 92
marble 34, 196, **197**, 202, 207, 208, 211
marble quarry 196
marginal moraines **183**, 185
marine sediments 65

SUBJECT INDEX

marine terrasses (raised beaches) 184, **189**, 190
Mârmorilik Formation 46
matrix, in limestone 64
mature sediments 76
maturity (oil/gas) **221**, 223, 231
Mauritanides **96**, 110
median line between coasts 157
melting temperature (rocks) 141
meltwater sediments 183
meltwater terraces 185
Melville Bay Fault 175
Melville Bay Graben **156**, 174, 175
Melville Bay Ridge 175
Melville Bugt 5, **8**, 156, 161, 174, 175, 211
Member (stratigraphic) 69
Mesoarchaean *12*, 21
Mesoproterozoic 13, 21, 40, 56
Mesoproterozoic sedimentation 74
Mesozoic **15**, 21
Mestersvig 8, 77, 205, *206*, 211, 216
Mestersvig lead mine **206**, 207, 211
metals 196
metamorphic facies 35
metamorphic rocks 32, **34**, 35
metamorphism 35
metasediment 110
Metasequoia 131
mica schists **32**, 51
microfossils 65, **89**, 163–165
micropalaeontological studies 164
microscope section (thin section of rock) 35, **141**, 145
Middle Age warm period 189
mid-oceanic spreading axis 53, 139, 148, **171**
Midsommersø Dolerites **71**, 72
migmatite 33, **34**, 99, 102
migmatitic amphibolite 34
migmatitic gneiss 34
migration route (oil/gas) **231**, 240
Milankovich cycles 184
milli-gal 162
Milne Land **8**, 24, 108, 126, 211
mineral exploration 194
mineral occurrences **210**, 211
mineral oil 220
mineral provinces 197, 199, **200**
mineral resources 194
mineralisations **210**, 211, 213, 215
mineralogy 21
minerals and their uses **202**, 203
mines; abandoned, active 211
mining 201, **204**
mining, general regulations 201
Miocene 169
Mobil (oil company) 240
Mohns Ridge **171**, 172

Mohorovičić discontinuity (Moho) 95, **100**, 169
molasse 114
molybdenite **202**, 217
molybdenum 197, 200, **202**, 211, 217
monazite 203
monocline (structure) 107
monsonitic granite 198
moraines 65, **183**
Morænesø Formation 84
Moriusaq 211
Morris Jesup Plateau **167**, 171
Motzfeldt intrusion 54
Motzfeldt Sø 54, **211**
mudstone 64

N

Naajaat Lake 144, **145**
Nagssugtoqidian fold belt 5, 13, 30, 31, **42**, 44, 45
Nagssugtoquidian orogeny 43
Nalunaq gold mine 52, 192, 195, 211, **212**, 213
Nanortalik *8*, 52
nanoTesla (nT) 45, **159**
Nansen Basin 167
Nares Land 106
Nares Strait **5**, 8, 168
Nares Strait Group 73
Narhvalsund 92
Narsaq 56
Narsaq intrusion 56
Narsarsuaq 8, **56**
Narssârssuk Group 73
Nasca Plate 53
Nassuttooq 8
Naternaq 211
Nathorst Land Group **77**, 79
native iron **143**, 144
naujaite 57
Navarana Fjord 87, **211**
Navarana Fjord escarpment **86–88**
Neoarchaean *12*, 21
Neogene **15**, 21, 169, 179
Neogene uplift 176, **178**, 179
Neoproterozoic 13, 21, 40, 77, 79, 80
Neoproterozoic basin **82**, 103
neosome 34
nepheline syenites 57
Niaqornaarsuk 211
niccolite 197, **202**
nickel 197, 200, **202**, 204, 211
Niggli Spids segment 101
Niggli Spids thrust sheet **100**, 101
niobium 58, 197, 200, **203**, 204, 211, 213
Niocden (oil company) 240
Nioghalvfjerdsfjorden 84

non-metals 197
Nordlandet 36
Nordre Strømfjord (Nassuttoq) **8**, 42, 43
normal faults 94
normal magnetisation 113
North American Plate 53
North Atlantic Ocean 81
North Greenland 8
North Greenland fold belt **104**, 106
North Greenland hydrocarbon potential 226
North Greenland offshore geology 167
North Greenland Rift Basin 119
North GRIP (NGRIP) 191
North Pole 167
North Qôroq intrusion 56
Norwegian Sea 228
Nualik 149
Nûkavsaq Formation 46
Nukik platform **232**, 235
Nukik-1 and -2 wells 162, **237**, 240
Nunaoil (oil company) **238**–241
Nunap Isua 8
nunataks **24**, 138
Nuuk 5, **8**, 36, 232, 239
Nuuk Basin **156**, 175, 232, 234, 237
Nuuk region 36, 184, 232
Nuussuaq 8, 129, 138, 142, 176, 209, 211, 233, 236
Nuussuaq Basin 15, 111, 112, **128**, 129, 131, 135, 142, 144, 209, 232, 236
Nyeboe Land 8, 87

O

ocean 20
Ocean Drilling Program (ODP) 162, **164**
ocean floor 53
ocean-floor spreading **53**, 139, 169, 171
oceanic basalts 156
oceanic crust 53, 156, 157 165, **169**, 171, 174, 178
oceanic sediments **65**, 165
offshore basins 15
offshore continental shelves 5
offshore drillings 161
offshore geology **156**, 166, 167, 174
oil **220**–221
oil- and gasfields, North Sea 228
oil- and gasfields, size **224**, 225
oil and gas generation **220**–222
oil and gas licence areas 241
oil and gas window **221**–223
oil exploration, terms and conditions 237, **238**
oil kitchen 222
oil seeps 227, 232, **233**
oil traps **222**, 231

Old Red Sandstone 114, **117**
Older basement **12**, 21
Older sedimentary basins 14, 21, **68**, 70, 74, 82
oldest traces of life 38
Oligocene 169
olivine 141, **152**, 197, 203, 211, 212
olivine basalt 140, **141**, 152
olivine gabbro 141
ooids 64
'open door policy' 241
open folds 33
'open water region' **225**, 232
Ordovician **14**, 21, 77, 81
Ore 195, 197, **200**
Orogenic belts 5
orogeny 32
orthoceratite 81
orthogneisses **31**, 49
orthopyroxene 152
ostracods **163**, 165
overturned folds 94
oxygen isotopes 191

P

Paakitsoq 41
Paakitsup Nunaa 41
Paamiut **8**, 232
Paamiut Basin **156**, 232, 237
Paarliit 198
pachnolite 59
Pacific Plate 53
Pacific Province 81
packstone 64
pahoehoe lava 142
Palaeoarchaean **12**, 21
Palaeogene **15**, 21, 169
Palaeogene basalts 5, **138**–153
Palaeogene deformation 106
Palaeogene intrusions **149**, 216
Palaeogene vulcanism 5, 15, **138**–153
palaeogeographical maps 83, 96, 97, 99, 111, **123**, 127, 129, 139
palaeogeography 112
palaeomagnetism 112, **113**
palaeontology 21, 89
Palaeoproterozoic 5, **13**, 21, 30, 40
Palaeoprotorozoic basement **40**, 43
Palaeoproterozoic fold belts **13**, 45
Palaeoproterozoic orogeny **97**, 98, 103
Palaeoproterozoic sediments 47
palaeosome 34
Palaeo-Tethys Ocean 15, *111*, 123
Palaeozoic **14**, 21
palaeozoic fold belts **92**, 96, 104
Paleocene **169**, 178
palladium 197, 200, **203**, 211, 216, 217

SUBJECT INDEX

PanCanadian (oil company) 240
Pangaea 15, 96, 103, 110, 111, 118, **123**, 124, 127
Pannotia 81
Paradisfjeld Group 107
parent magma 152
passive margin 85
Peary Land 8, 87, 104–106, **119**, 185
pebbles 64
pegmatite 49
pelagic sediments 65
Pelican (drillship) 240
peneplain 27, **179**
pentlandite 197, **202**
peridotite 141
Period (stratigraphic) 69
permeable rocks (oil/gas) **221**, 222
Permian **14**, 21, 119, 123, 124
Petermann Bjerg 100
petroleum geology 21
petrology 21
Phanerozoic **14**, 21
Phillips Petroleum (oil company) 239, **240**
phosphate **203**, 211
pillow lava **46**, 141, 142, 145, 169
pisolitic limestones 79
placer deposits 216
plagioclase **141**, 152
planktonic faunas (fossils) 85, **164**
plant fossils 14, 15, 126, 131, **135**
plants 89
plastic folds 33
plate tectonics 20, 44, **53**, 112
plateau basalts 72, 132, 138, **142**–144, 147, 148, 170, 237
Platinova (prospecting company) 240
platinum 197, 200, **203**, 211, 216, 217
play (oil/gas) **223**, 229
Pleistocene **169**, 182, 183
Pliocene **169**, 183
plutonic rocks 34, **141**, 152
polarity timescale 113
polarity, magnetic 53, **113**
Polarstern, RV (survey ship) **166**, 167
Polkorridoren Group 107
pore space in sediments 64, **224**
porosity in rock 163, **221**
Portfjeld Formation 107
positive magnetic polarity 113
postdepositional structures 68
post-glacial time **182**, 186
posttectonic sedimentation 177
Precambrian **12**, 18, 21
Precambrian basement shield 30
predepositional structures 67
pretectonic sedimentation 177
Prins Christian Sund 8
production factor (oil/gas) 225

production licence (minerals) 201
production licence (oil/gas) 238
production test (oil/gas) 225
Professor Logachev (survey ship) 178
profile construction 107
progradation 65
propane 222
prospecting licence **201**, **237**
prospection 199, **201**, 210
Proterozoic 5, **12**, 21, 40
proterozoic basement 5, **40**, 41
proterozoic mountain belts **40**, 41
provenance studies **76**, 134
pseudo-escarpment 170
pterosaurs 15
Pujoortoq 236
pulaskite 57
pull-apart basins 118, **120**
pyrochlore 203
pyroxene 141
pyroxenite 141

Q

Qaanaaq (Thule) 5, *8*
Qaarsuarsuk 211
Qaqortoq (Julianehåb) 5, **8**, 50, 52
Qaqqaarsuk 211
Qeqertarssuaq Formation 46
Qeqertarsuaq (Godhavn) 129, **143**
Qeqertarsuaq (Disko) 8, 129, **143**, 233
Qeqertarsuatsiaat (Fiskenæsset) 195, **210**–212
Qeqertarsuatsiaq 8
Qorqut granite 37
quadruped fish 14, 115, **116**
Quaternary **182**, **183**, 186
Quarternary deposits 183
quartz 141, 152, **197**, 200, 210, 212
quartz diorite 141
quartz gabbro 141
Queen Ingrid 198
Qulleq-1 (well) 161, 184, 224, **237**, 240
Qullissat 204, **209**, 211

R

radioactive decay 223
radioactivity measurements 16
radiolaria 165
radiometric age determination **48**, 76
raised beaches 189, **190**
rapakivi granite 50, **51**
rare minerals 59
rare-earth elements 197, 200, **203**, 211, 213
raw materials, minerals 195
reaction series (minerals) 140, **152**

recoverable reserves (oil/gas) 225
recovery (oil/gas) 224
recumbent fold 47
reef limestone 88
refinery (oil/gas) 220
reflection seismic techniques 161
refraction seismic techniques 161
regional metamorphism 39
regression 130
regulations for mining 201
relative chronology 49
remanent magnetism 113
Renland 24
reptiles 14, *15*
reservoir (oil/gas) **222**–224
reservoir rock (oil/gas) **220**–224, 229, 231
reverse faults 94
reversed magnetised 113
Reykjanes Ridge **171**, 172
'Rhaetic-Liassic'stage 135
Rheic Ocean 97
rhyolite 121, **141**
rift basins, North-East Greenland 15, 112, **122**
rift zone 56
Rink Isbræ 8
Rinkian area 5, 13, 30, 31, **46**, 47
Riphean 77
riser (drill pipe) 163
river desposits 62
Rivieradal Basin 84
Rivieradal Group **82**–84
rock groups 34
rock magnetism 113
rock salt 64
Rodebay 41
Rodinia 99
root zone **27**, 103
ropy lava 148
royal monument 198
royalties (from concessions) **201**, 208
ruby **195**, 200, 203, 210, 211
rust zone 199
rutile 196, **203**

S

Saalian **183**, 186
Saga Petroleum (oil company) 240
salt deposits **63**, 64, 169
salt diapir 169, **170**
salterella 78
sand 64
sandstone 34, **64**, 197
sapropel 220
Saqqaq **143**, 211
Sârdloq fault zone 52
Sarfartoq 211

scandium 203
Scanning electron microscope 20
scheelite 203
schist 32
scientific voyages of discovery 20
Scoresby Sund (the fjord) **8**, 24, 74, 230
Scoresby Sund (the region) 24, 98, 123, 138, 146, 187
Scoresbysund (the town) (Illoqqortoor-miut) 5, **8**, 230
scours 68
seabed topography 173
sea-floor spreading 15, 53, 165, 167, **171**, 172, 174
sea-floor spreading anomalies 178
sea-level changes 66, 78, 130, 182, 190
seaward-dipping reflector sequence 165
section log 78
sedimentary basins 5, **62**, 85, 110, 111
sedimentary environments 62
sedimentary profiles 78
sedimentary rocks 34
sedimentary structures **67**, 77, 78
sedimentation rates **66**, 71
sedimentology 21
sediments; types, classification 62, **64**
sediments; mature, immature 76
seismic investigations 158, **160**, 179, 230, 231, 235, 241
Senja Fracture Zone 171
Seqi Olivine Mine **197**, 211
sequence stratigraphy 130
Series (stratigraphic) 69
serpentinised ultrabasic rocks 178
shallow-water shelf 15
Shannon 8, **138**, 146
sheared rocks 33
shelf **85**, 88
shelf areas 157, **173**
Shell (oil company) 241
SHRIMP method 48
Sigguup Nunaa 233
siliciclastic sediments 65
sillimanite 35
sills **72**, 151
silt 64
siltstone 34
Silurian **14**, 21
silver 197, 200, **203**, 208, 211
silver glance 197
Sinarsuk 211
Sirius Passet 86
Sisimiut 5, 8, **42**, 239
Sisimiut Bassin 237
'Sisimiut-West' licence 239
Skaergaard Intrusion **149**, 150, 211, 216, 217
Skagen Group 107

SUBJECT INDEX

Skarvefjeld   143
Skjoldungen   8
skolithos   78
Skærgårdshalvø   8
slide structures   68
slumps   68
Smith Sound Group   73
snails, gastropods   81, **131**
'Snowball Earth'   80
Sofia Sund   117
soft-bodied fauna   86
Søndre Strømfjord (Kangerlussuaq)   **8**, 42, 45
sonic velocity   162
sørensenite   59
Sortebakker Formation   119
source rock (oil/gas)   **220**, 222-224, 227–231, 235, 236
South American Plate   53
South Greenland   8
South Qôroq intrusion   56
'speculative data' (oil/gas)   240
sphalerit   195, 197, 200, **203**, 207, 208
Spitzbergen Fracture Zone   121
sponges   129
spreading axis (ridge)   **53**, 139, 156, 169, 171, 177
spreading pattern, ocean floor   53
Stage (stratigraphic)   69
Statfjord Field   224
Station Nord   5, **8**, 156
Statoil (oil company)   **240**, 241
Stauning Alper   25
staurolite   35
steenstrupin   **203**, 213
stibnite   202
Storø   211
stratigraphic traps (oil/gas)   **222**, 224
stratigraphy   21, **69**, 89
stream sediments   52, **58**
Strede Bank   173
strike-slip faults   **94**, 120, 177
stromatolites   **68**, 73, 84
structural geology   21
structural traps (oil/gas)   **222**, 224
structures in the basement   33
Sturtian   77
subduction   50, **53**
subduction zone   39, 51, **53**, 98
submarine avalanches   133
submarine canyons   **179**, 183, 184
submarine eruptions   176
Suess Land   92
Sukkertoppen Iskappe   42
supercontinent   80, 81, **93**
Supergroup (stratigraphic)   69
supracrustal rocks   51
survey ships   **154**, 166, 178

suture lines   **124**, 125
Svalbard   74, 76, 96, **118**, 167, 171, 228, 235, 236
Svartenhuk Halvø   **8**, 129, 142, 233, 234
Sveconorwegian orogeny   **97**, 99, 103
Sverdrup Basin   107, **118**, 228
syenite   34
syenite complex   56
syndepositional structures   67
synformal folds   **93**, 94
syntectonic sedimentation   177
System (stratigraphic)   69

T

Taartoq   211
tantalum   58, **197**, 204, 211, 213
Tasersuatsiaq   211
Tasiilaq (Ammassalik)   5, *8*
Tasiusarssuaq terrane   36, 37
taxa (biological)   69
tectonic window   **94**, 100
telluric iron   144
temperature variations during the Ice Age   **184**, 189
temperatures in the subsurface   223
terranes   12, **36**, 37, 39
Tethys Ocean   15, 122, 123, **127**, 129, 132
tetrapods   **115**, 116
Texaco (oil company)   241
TGA-Grepco (oil company)   240
TGS-Nopec (geophysical entrepreneur)   240
Thetis (ship)   241
Thetis Basin   156, **168**, 169
thick-skinned tectonics   **95**, 100
thin section of a rock   141
thin-skinned tectonics   **95**, 100, 107
tholeiite   141–143, **152**
thrust plane   43, **94**
thrust sheets   74, **94**, 100
thrusts   94
Thule   *8*, 70, 72, 73
Thule Air Base   8
Thule Basin   13, **70**, 73
Thule Supergroup   60, 67, 70, **72**, 73
till deposits   **183**, 186
Tillite Group   13, 74, 77, **79**, 80, 84, 92, 98, 101, 103
Timmiarmiut   8
tin   196, **215**
titanium   **203**, 211, 216
Tobias islands   **167**, 168
tonalite intrusion   37
Torsukattak   41
Total (oil company)   240
trace fossils   13, **68**, 73
Traill Ø   8

transform faults   174, **177**
transgression   **130**, 132
transition zone ocean/continent   **156**, 174
traps for hydrocarbons   **222**–224, 229, 237
Triassic   **15**, 21, 124
trilobites   14, **81**, 89
Troll Field   224
Trolle Land Fault Zone   118, **119**
Trolle Land Group   119
trough (sediment basin)   88
tugtupite   59
Tugtutoq   56
tungsten   200, **203**, 211, 211
Tunulliarfik   56
Tuppiap Qeqertai   **167**, 168
turbidite deposits   47, 62, 65, 86
turbiditic sandstones   86, 105, 127, 130
turbidity currents   **47**, 86, 133, 173

U

Ubekendt Ejland   **129**, 142, 233
Uiffaq   143
ultrabasic rocks   72, **141**, 152
unconformity   130
Upernavik   5, *8*
Upernavik Isfjord   8
uplift (of land)   **27**, 103, 178, 188, 190
uplift of crust   178, **179**
uraninite   203
uranium   57, 197, **203**, 204, 211, 213, 223
uranium isotopes   48
U-shaped glacial valley   **24**, 183, 185
Utoqqaat   211
Uummannaq   47, **129**, 233
Uummannaq Fjord   233

V

Vaigat   **129**, 143, 221, 233
vanadium   **203**, 211
Varangian (period)   79
Variscan   96
Variscan fold belts   **93**, 110
Vendian (period)   77, **79**, 84
verging (structure geology)   104
vertebrate fossils   **115**, 122, 124
vertebrates   89
vesicular basalt   34
vibroseis   230
Victoria Fjord   105
Vilddal Group   **114**, 117
viscosity   141
volcanic ash   141
volcanic island arc   **44**, 50, 85
volcanic rocks   34, 138, **141**, 152
volcanic provinces   **138**, 142, 146

W

wackestone   64
Wandel Sea   5, **8**, 119, 227
Wandel Sea Basin   15, 106, 107, 111, 112, **118**–121, 156, 167, 168, 227, 228
Wandel Sea Strike-Slip Mobile Belt   **119**, 120
Wandel Valley Formation   84
Warming Land   19
Washington Land   **8**, 26, 189, 211
Wegener Halvø   **211**, 215, 226
Weichselian   183, **186**, 189
well data   162
well head   163
wells offshore   **156**, 162, 232, 235, 240
wells onshore   218, **221**, 233
Werner Bjerge   **149**, 217
West Greenland   **8**
West Greenland, oil geology   **232**-240
West Navion (drill ship)   161
Western fault zone   100
wet gas   **221**–223
Wilson cycle   **93**, 111
window (oil/gas)   **221**, 223
windows (structural geology)   94
Wisconsinian   183, **186**
wolframite   203
Wollaston Forland   **8**, 111, 126
Wollaston Forland Basin   112, **126**, 127
Wulff Land   8, 185, *227*

X, Y, Z

xenolith   214
xenotim   203
Yermak Plateau   **167**, 171
Ymer Ø   66, **211**
Ymer Ø Group   **77**, 79
Younger basement   13, 21, **40**
Younger fold belts   **5**, 21
younger sedimentary basins   **5**, 14, 15, 21, 68, 110, 111
yttrium   197, **203**, 213
Zechstein Sea   120
Zig-Zag Dal Basalt Formation   13, 70, **71**, 72, 82, 103
zigzag folds   105
zinc-lead deposit, Maarmorilik   206, **207**, 211
zink   197, 200, **203**, 204 – 212, 215, 216
zinkspat   203
zircon   76, 196, **203**
zircon crystals   48
zirconium   57, 58, 197, **203**, 204 21